AF352768

Molecular Aspects of Plant Salinity Stress and Tolerance

Molecular Aspects of Plant Salinity Stress and Tolerance

Editors

Jen-Tsung Chen
Ricardo Aroca
Daniela Romano

MDPI • Basel • Beijing • Wuhan • Barcelona • Belgrade • Manchester • Tokyo • Cluj • Tianjin

Editors

Jen-Tsung Chen
Department of Life Sciences
National University of
Kaohsiung
Kaohsiung
Taiwan

Ricardo Aroca
Departement of Soil
Microbiology and Symbiotic
Systems
Estación Experimental del
Zaidín (CSIC)
Granada
Spain

Daniela Romano
Department of Agriculture, Food
and Environment (Di3A)
Università degli
Catania
Italy

Editorial Office
MDPI
St. Alban-Anlage 66
4052 Basel, Switzerland

This is a reprint of articles from the Special Issue published online in the open access journal *Int. J. Mol. Sci.* (ISSN 1422-0067) (available at: www.mdpi.com/journal/ijms/special_issues/ SSTolerance).

For citation purposes, cite each article independently as indicated on the article page online and as indicated below:

LastName, A.A.; LastName, B.B.; LastName, C.C. Article Title. *Journal Name* **Year**, *Volume Number*, Page Range.

ISBN 978-3-0365-1380-5 (Hbk)
ISBN 978-3-0365-1379-9 (PDF)

Contents

International Journal of
Molecular Sciences

Editorial

Molecular Aspects of Plant Salinity Stress and Tolerance

Jen-Tsung Chen [1,*], Ricardo Aroca [2] and Daniela Romano [3]

[1] Department of Life Sciences, National University of Kaohsiung, Kaohsiung 81148, Taiwan
[2] Department of Soil Microbiology and Symbiotic Systems, Estación Experimental del Zaidín (CSIC), 18008 Granada, Spain; ricardo.aroca@eez.csic.es
[3] Department of Agriculture, Food and Environment, University of Catania, 95123 Catania, CT, Italy; dromano@unict.it
* Correspondence: jentsung@nuk.edu.tw

check for updates

Citation: Chen, J.-T.; Aroca, R.; Romano, D. Molecular Aspects of Plant Salinity Stress and Tolerance. *Int. J. Mol. Sci.* **2021**, 22, 4918. https://doi.org/10.3390/ijms22094918

Received: 23 April 2021
Accepted: 4 May 2021
Published: 6 May 2021

Publisher's Note: MDPI stays neutral with regard to jurisdictional claims in published maps and institutional affiliations.

Salinity is one of the major abiotic stresses that inhibit the growth, development, and productivity of crops, particularly in hot and dry areas of the world. It is an intensive topic on which many studies have been conducted, with the aim of understanding the physiological and molecular responses involved in plant salinity stress. In recent years, with the rapid progress of molecular technologies, scientists have acquired more advanced tools to reveal in-depth mechanisms and to establish crop breeding programs for enhancing plant salinity tolerance. This Special Issue, entitled "Molecular Aspects of Plant Salinity Stress and Tolerance", collected 13 innovative publications which could enrich our knowledge about the molecular mechanisms of plant salinity stress.

It has been shown that investigations using advanced analyzed tools associated with multi-omics are reliable to gain new insights into the mechanisms associated with responses to salinity stress. Song et al. profiled the transcriptome together with the evaluation of photosynthetic efficiency of photosystem II and the amount of free amino acids in watermelon seedlings when exposed to short-term salinity stress [1]. They revealed that certain genes were found to express differentially in response to salinity, and these genes might code for transcription factors, or were proposed to be related to primary metabolism, endocytosis, hormonal pathways, and transporters. Isayenkov et al. investigated strategies for adaptation to salinity in seaside barley, using nuclear magnetic resonance imaging associated with multi-omics approaches [2]. The authors proposed that seaside barley has developed specific mechanisms involving the capability to regulate concentrations of Na^+ and Cl^- in leaves and morphological and physiological adaptations of roots when subjected to salinity stress.

For monitoring the status of the cell wall, plants have evolved a system, namely, cell wall integrity (CWI) [3]. Liu et al. provide a review on the roles of CWI in salinity tolerance and propose that the genetic engineering of CWI-related genes using genome editing technology might generate salt-tolerant varieties for applications in the future.

Jasmonates (JAs) are lipid-based plant hormones that regulate an array of processes in plants, particularly involved in defense mechanisms and stress tolerance. Delgado et al. refined the role of JAs in plant salinity tolerance, chiefly based on a genome-wide association study, and concluded that MYC2 transcription factor and JASMONATE ZIM-DOMAIN repressors are key components in JA signaling [4]. The authors provide a perspective that the knowledge of JAs against plant salinity stress might be applied as a guide in breeding programs.

Rice is one of the most important staple crops, as well as an intensively studied model for plant salinity stress and tolerance. Ponce et al. contributed a review focusing on the molecular mechanisms of salinity tolerance in rice [5]. The authors stated that the most investigated plant hormone in plant salinity stress is abscisic acid, and the recently identified mechanisms together with the key genes involved are critical to the breeding programs of highly salt-tolerant cultivars in the future.

Theoretically, biotechnological tools could be applied to identify candidate genes and subsequently alter the patterns of gene expression using the classical gene transfer method or targeted genome editing for acquiring salinity tolerance or inducing mutation to obtain salinity-tolerant genotypes. Previously, calmodulin-like proteins (CMLs) were found to be involved in salinity stress tolerance in Arabidopsis. Zhang et al. isolated a CML gene, *MpCML40*, from Pongamia, and found that its heterologous expression could improve salinity tolerance in yeast cells and enhance the rate of seed germination and the length of roots when exposed to salt and osmotic stresses in Arabidopsis [6]. The authors suggested that *MpCML40* contributed to the proline accumulation for reducing the damage caused by reactive oxygen species. A stress-induced aquaporin, *ZxPIP1;3*, was identified by Li et al. from succulent xerophyte, is belonging to the PIP1 subgroup of plasma membrane intrinsic proteins (PIPs) [7]. The overexpression of *ZxPIP1;3* in Arabidopsis showed certain improvements in growth and physiological attributes which lead to more efficient photosynthesis under salinity and drought stresses. Tran et al. tested the ion transport capacity of a PIP gene, *HvPIP2;8*, when expressed in the oocytes of *Xenopus laevis*. It was shown that the expression of *HvPIP2;8* enhanced the activity of water as well as Na^+/K^+ transport and might be involved in the responses when subjected to salinity stress in barley [8]. Kawakami et al. isolated one gene, *SvHKT1;1*, coded for a sodium transporter from a halophytic turf grass, *Sporobolus virginicus* [9]. The authors found that under severe salinity stress, SvHKT1;1 could prevent the excess accumulation of shoot Na^+ in *S. virginicus*. Zhang et al. investigated the role of soybean transcription factor GmbZIP15 in regulating gene expression under abiotic stresses [10]. The resulting data showed hypersensitivity to abiotic stress in soybeans when overexpressing *GmbZIP15*, and thus they proposed that GmbZIP15 might be a negative regulator in plant salinity and drought stresses.

In agriculture, the application of genetic engineering to introduce genes for acquiring salinity tolerance has certain limitations. An alternative method to enhance plant salinity tolerance is using green inoculants of plant growth-promoting rhizobacteria (PGPR), such as salinity-tolerant (halophilic) plant-associated bacteria. Miller and Nielsen summarized the molecular mechanisms involving the interactions between plants and salinity-tolerant bacteria and proposed their potential applications in improving crop production under salinity stress [11]. In the review by Ha-Tran et al., the authors identified that PGPR is a promising agent for seed bio-priming to enhance seed vigor and germination and the uniformity of subsequent seedling growth under salinity stress [12]. The advantages of the PGPR together with the fluctuation of antioxidants and osmolytes in PGPR-treated plants were comprehensively discussed, and the authors suggested that further investigation on more complex interactions between plants and PGPR needs integrative multi-omics and systems biology, and other advanced tools.

Plant breeding programs for developing salinity-tolerant crops are crucial for improving their growth, yield, and product quality in salt-affected fields. In rice breeding, it has been recognized that salinity tolerance is contributed by multiple genes; consequently, it is a complex quantitative trait which is usually difficult to be evaluated. Qin et al. reviewed the advances in rice breeding for salinity tolerance, including the progress of molecular markers, genetic mapping, genetic engineering, and the interventions of biotechnology tools [13]. The authors suggested that the resulting high-salinity-tolerant germplasm is valuable for rice breeding programs in the future.

In conclusion, the advances in plant salinity stress and tolerance presented in this Special Issue include mechanistic insights revealed using powerful molecular tools together with multi-omics, and gene functions studied by genetic engineering and advanced biotechnological methods. Additionally, the use of plant growth-promoting rhizobacteria in the improvement of plant salinity tolerance and the underlying mechanisms and progress in breeding for salinity-tolerant rice are comprehensively discussed. Clearly, the published data have made significant progress in expanding our knowledge in the research field of plant salinity stress, and the full results are valuable in developing salinity-stress-tolerant crops, benefiting their quality and productivity, and eventually, supporting the sustainability of the food supply in the world.

Funding: This research received no external funding.

Conflicts of Interest: The author declares no conflict of interest.

References

1. Song, Q.; Joshi, M.; Joshi, V. Transcriptomic Analysis of Short-Term Salt Stress Response in Watermelon Seedlings. *Int. J. Mol. Sci.* **2020**, *21*, 6036. [CrossRef] [PubMed]
2. Isayenkov, S.; Hilo, A.; Rizzo, P.; Tandron Moya, Y.A.; Rolletschek, H.; Borisjuk, L.; Radchuk, V. Adaptation Strategies of Halophytic Barley *Hordeum marinum* ssp. *marinum* to High Salinity and Osmotic Stress. *Int. J. Mol. Sci.* **2020**, *21*, 9019. [CrossRef] [PubMed]
3. Liu, J.; Zhang, W.; Long, S.; Zhao, C. Maintenance of Cell Wall Integrity under High Salinity. *Int. J. Mol. Sci.* **2021**, *22*, 3260. [CrossRef] [PubMed]
4. Delgado, C.; Mora-Poblete, F.; Ahmar, S.; Chen, J.-T.; Figueroa, C.R. Jasmonates and Plant Salt Stress: Molecular Players, Physiological Effects, and Improving Tolerance by Using Genome-Associated Tools. *Int. J. Mol. Sci.* **2021**, *22*, 3082. [CrossRef] [PubMed]
5. Ponce, K.S.; Guo, L.; Leng, Y.; Meng, L.; Ye, G. Advances in Sensing, Response and Regulation Mechanism of Salt Tolerance in Rice. *Int. J. Mol. Sci.* **2021**, *22*, 2254. [CrossRef] [PubMed]
6. Zhang, Y.; Huang, J.; Hou, Q.; Liu, Y.; Wang, J.; Deng, S. Isolation and Functional Characterization of a Salt-Responsive Calmodulin-Like Gene *MpCML40* from Semi-Mangrove *Millettia pinnata*. *Int. J. Mol. Sci.* **2021**, *22*, 3475. [CrossRef] [PubMed]
7. Li, M.; Li, M.; Li, D.; Wang, S.-M.; Yin, H. Overexpression of the *Zygophyllum xanthoxylum* Aquaporin, ZxPIP1;3, Promotes Plant Growth and Stress Tolerance. *Int. J. Mol. Sci.* **2021**, *22*, 2112. [CrossRef] [PubMed]
8. Tran, S.T.H.; Horie, T.; Imran, S.; Qiu, J.; McGaughey, S.; Byrt, C.S.; Tyerman, S.D.; Katsuhara, M. A Survey of Barley PIP Aquaporin Ionic Conductance Reveals Ca^{2+}-Sensitive *HvPIP2;8* Na^+ and K^+ Conductance. *Int. J. Mol. Sci.* **2020**, *21*, 7135. [CrossRef] [PubMed]
9. Kawakami, Y.; Imran, S.; Katsuhara, M.; Tada, Y. Na^+ Transporter SvHKT1;1 from a Halophytic Turf Grass Is Specifically Upregulated by High Na^+ Concentration and Regulates Shoot Na^+ Concentration. *Int. J. Mol. Sci.* **2020**, *21*, 6100. [CrossRef] [PubMed]
10. Zhang, M.; Liu, Y.; Cai, H.; Guo, M.; Chai, M.; She, Z.; Ye, L.; Cheng, Y.; Wang, B.; Qin, Y. The bZIP Transcription Factor GmbZIP15 Negatively Regulates Salt- and Drought-Stress Responses in Soybean. *Int. J. Mol. Sci.* **2020**, *21*, 7778. [CrossRef] [PubMed]
11. Miller, A.K.; Nielsen, B.L. Analysis of Gene Expression Changes in Plants Grown in Salty Soil in Response to Inoculation with Halophilic Bacteria. *Int. J. Mol. Sci.* **2021**, *22*, 3611. [CrossRef] [PubMed]
12. Ha-Tran, D.M.; Nguyen, T.T.M.; Hung, S.-H.; Huang, E.; Huang, C.-C. Roles of Plant Growth-Promoting Rhizobacteria (PGPR) in Stimulating Salinity Stress Defense in Plants: A Review. *Int. J. Mol. Sci.* **2021**, *22*, 3154. [CrossRef] [PubMed]
13. Qin, H.; Li, Y.; Huang, R. Advances and Challenges in the Breeding of Salt-Tolerant Rice. *Int. J. Mol. Sci.* **2020**, *21*, 8385. [CrossRef] [PubMed]

International Journal of
Molecular Sciences

Review

Analysis of Gene Expression Changes in Plants Grown in Salty Soil in Response to Inoculation with Halophilic Bacteria

Ashley K. Miller and Brent L. Nielsen *

Department of Microbiology & Molecular Biology, Brigham Young University, Provo, UT 84602, USA;
miller.ashley.kay@gmail.com
* Correspondence: brentnielsen@byu.edu; Tel.: +1-801-422-1102

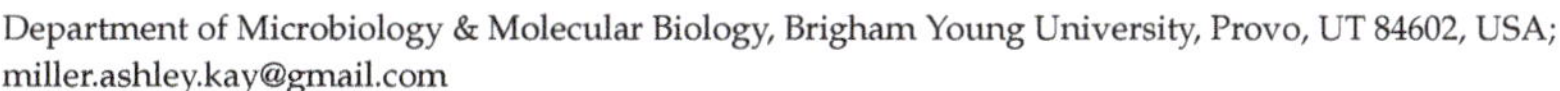

Abstract: Soil salinity is an increasing problem facing agriculture in many parts of the world. Climate change and irrigation practices have led to decreased yields of some farmland due to increased salt levels in the soil. Plants that have tolerance to salt are thus needed to feed the world's population. One approach addressing this problem is genetic engineering to introduce genes encoding salinity, but this approach has limitations. Another fairly new approach is the isolation and development of salt-tolerant (halophilic) plant-associated bacteria. These bacteria are used as inoculants to stimulate plant growth. Several reports are now available, demonstrating how the use of halophilic inoculants enhance plant growth in salty soil. However, the mechanisms for this growth stimulation are as yet not clear. Enhanced growth in response to bacterial inoculation is expected to be associated with changes in plant gene expression. In this review, we discuss the current literature and approaches for analyzing altered plant gene expression in response to inoculation with halophilic bacteria. Additionally, challenges and limitations to current approaches are analyzed. A further understanding of the molecular mechanisms involved in enhanced plant growth when inoculated with salt-tolerant bacteria will significantly improve agriculture in areas affected by saline soils.

Keywords: halophiles; plant growth-promoting rhizobacteria (PGPR); RNA sequence analysis (RNA-seq); quantitative reverse transcriptase PCR (qRT-PCR)

 check for updates

Citation: Miller, A.K.; Nielsen, B.L. Analysis of Gene Expression Changes in Plants Grown in Salty Soil in Response to Inoculation with Halophilic Bacteria. *Int. J. Mol. Sci.* **2021**, *22*, 3611. https://doi.org/10.3390/ijms22073611

Academic Editors: Jen-Tsung Chen, Ricardo Aroca and Daniela Romano

Received: 17 March 2021
Accepted: 27 March 2021
Published: 31 March 2021

Publisher's Note: MDPI stays neutral with regard to jurisdictional claims in published maps and institutional affiliations.

1. Introduction

1.1. Soil and Crop Loss Due to Rapidly Increasing Concentrations of Salinity Buildup in Soil

According to the U.S. Department of Agriculture (USDA), about 10 million hectares of land are lost yearly to soil salinization. The rate of soil salinization is influenced by many factors, including the practice of poor irrigation techniques and local climates. In arid climates such as Utah, the rate of salinization has been documented to occur about three times faster than in temperate climates [1]. Most plants do not have the biochemical or structural mechanisms needed to survive in high salt environments, leading to decreased plant yield.

Salinity causes several negative effects on plant growth and development due to water stress, cytotoxicity from elevated levels of sodium and chloride ions in the cytoplasm, nutritional imbalance, the production of ethylene, reactive oxygen species and other oxidative damage. These disruptions can seriously affect plant productivity [1,2]. Unless these stressors are at least partially alleviated, severe decreases in plant yield can result.

1.2. Limitations of Bioengineering Salt Tolerance

As improving plant salt tolerance is of mounting importance around the world, a number of approaches have been taken to address this problem. Introducing genes to confer salinity tolerance to plants is one approach. It has been reported that transformed tobacco plants expressing the gene for ectoine showed improved growth in salty soils [3]. Others suggest that *Medicago truncatula* production could be improved in salty soils when transformed with the rstB gene [4]. Some native plant genes have also been shown to play a role in plant salt tolerance, such as PPR40 in Arabidopsis [5], ethylene response factors [6], K+ transporter genes [7], and the chimeric ryegrass gene OsDST-SRDX, among others [8]. Still, there is much that is not understood about plant salt tolerance. For example, it is likely that multiple genes may be needed to provide meaningful increases in salinity tolerance in transgenic plants, and different genes may be required for different species. In addition, transcriptional control elements may also differ between species [9]. These and many more questions still need to be addressed in greater depth.

There are other potential limitations outside of genetic engineering. One of these limitations is the approval process. In the United States there are three federal agencies that oversee the approval of bioengineered crops: (1) the Food and Drug Administration (FDA), (2) the Environmental Protection Agency (EPA), and (3) the U.S. Department of Agriculture (USDA). Each of these agencies must be involved in the approval process of new genetically modified organisms (GMOs) [10]. The approval process is often long and difficult. There is one additional and significant limitation to a GMO salt-tolerant crop, and that is the public perception. In an age where the safety of animal feed and the direct human consumption of GMOs is questioned and avoided by many, public perceptions of bioengineered salt-tolerant crops could potentially limit their use and impact [11].

1.3. Survival Skills: How Halophytes Survive in Salty Soil

While most plants lack necessary mechanisms of salt tolerance, there are uniquely adapted plants named halophytes that complete their entire life cycle in soils containing 200–500 mM salt [1]. There are four main mechanisms of halophyte salt tolerance including: (1) the secretion of salt through salt glands, (2) the regulation of cellular ion homeostasis and osmotic pressure, (3) the detoxification of reactive oxygen species, and (4) alterations in membrane composition [6,7,12–15]. Each mechanism listed above is energetically expensive. As a result, most plants will need to increase photosynthesis to maintain homeostasis under saline conditions [1,12–15].

However, plants are not alone in their battle against increasing salinity. The plants' microbiota also play an important role in overall salt tolerance (see Figure 1; effects shown in B–C).

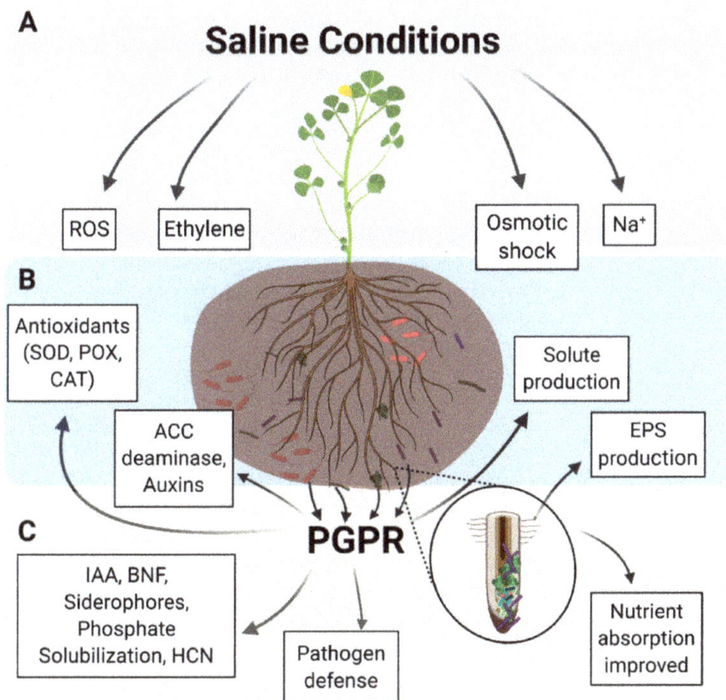

Figure 1. Salt tolerance-related plant growth promoting rhizobacteria (ST-PGPR) and their role in enhanced salt tolerance in halophytes. (**A**) Saline soils cause an abundance of toxic molecules such as ethylene and reactive oxygen species (ROS) to form and impede plant processes, leading to disease and death in plants without sufficient mechanisms of salt tolerance. Na^+ ions disrupt the function of plant ion channels, leading to plant osmotic shock. (**B**) Direct mechanisms by which ST-PGPR can enhance plant salt tolerance. Each ST-PGPR produces different antioxidants or solutes to help fight salt toxicity. (**C**) Indirect methods of ST-PGPR plant salt tolerance enhancement. Not all ST-PGPR rhizobium produce all substances mentioned above. Image produced based on ideas adapted from [15].

2. The Role of Halophilic Plant-Associated Bacteria in Plant Growth under Saline Conditions

Halophyte microbiomes contain plant growth promoting microbes both within the plant (endosymbionts) and outside the plant around their root system (ectosymbionts). These symbionts are often involved in promoting plant growth under stress and are referred to as plant growth promoting rhizobacteria (PGPR). Some strains also assist halophyte survival under salty conditions; such salt tolerance-related PGPR are noted as ST-PGPR [16]. ST-PGPR often fall into one of the salt-loving genera such as *Halomonas* and *Halobacillus*, though microbes from the genus *Kushneria* and other genera have similarly been shown to promote salt tolerance [1].

In recent years, there has been a significant increase in the number of publications that demonstrate the ability of ST-PGPR to improve salt-sensitive (glycophyte) crop yield in saline soil [1,14,16–19]. Many of these glycophytes are members of the grass family including rice, wheat, and barley, though similar success with legumes including soybean have been reported [19–21]. In rice, it was reported that two strains of endemic *Halobacillus* including *Halobacillus dabanensis* strain SB-26 and *Halobacillus* sp. GSP-34 greatly increased salt tolerance and plant yield when inoculated into saline soils [2]. Similarly, in wheat,

ST-PGPR including *Halomonas* sp. demonstrated a 62.2% to 78.1% increase in the length and wet mass of root and shoot tissues when compared to plants grown in the same salty soil but without inoculum [18]. Nakayama et al. [3] showed that *Halomonas elongata* could be used to improve the salt tolerance of tobacco plants. Similarly, ST-PGPR strains from the microbiome of native Utah halophytes, when added to salty soil containing non-salt-tolerant *Medicago sativa* (alfalfa) seedlings, reportedly increased plant yield when compared to controls [1]. Some of these ST-PGPR strains were isolated from the rhizosphere closely associated with roots, while other bacteria were isolated from within plant tissue.

2.1. Potential Mechanisms for Plant Growth Stimulation by ST-PGPR

Mechanisms by which endophytes enhance plant growth are thought to include improved nutrient acquisition and changes in gene expression. For example, ACC (1-aminocyclopropane-1-carboxylase) deaminase is a bacterial enzyme found in many endophytes that stimulates nutrient acquisition and plant growth by reducing the amount of ACC converted to ethylene. Ethylene is a known inhibitor of plant growth that is produced by the plant in response to salt, drought, and other environmental stresses [16,22]. *Burkholderia phytofirmans* is another endophyte that alters plant gene expression to enhance the growth of multiple cultivars of switchgrass [23,24]. Inoculation with this strain was found to induce wide-spread gene expression changes in the plant host, including the altered expression of some transcription factors that are known to regulate the expression of plant stress genes [25]. Other bacterial endophytes (species of *Sphingomonas*, *Pantoea*, *Bacillus* and *Enterobacter*) have been identified as enhancing the salt tolerance of hybrid elephant grass [26], likely because of enhanced nutrient acquisition and/or gene expression changes. Plants inoculated with *Halomonas elongata* (accession number MK873884) and 1% NaCl demonstrated an average increase of 2.4-fold in the biomass of alfalfa plants grown without inoculum [1]. These data suggest that this strain of *Halomonas* influences plant salt tolerance, raising the question of how these bacteria stimulate plant growth under saline conditions. *H. elongata*, a moderate halophile, produces a well-studied osmolyte named ectoine. Ectoine is produced by the bacteria in direct relationship with how much salt is in the environment surrounding the microbe. As levels of salt in the environment increase, so do the ectoine levels in the bacteria. This increase in ectoine helps the bacteria regulate osmotic pressure and fend off salt toxicity [3,16,17,19]. Though it remains unclear whether ectoine is among or is the mechanism of observed salt tolerance conferred from *H. elongata* to plants, there is growing evidence that ectoine plays some role in improving salt tolerance. A study with tobacco plants showed that significant salt tolerance did occur in plants inoculated with *H. elongata* producing ectoine vs. an ectoine knockout mutant; plants inoculated with the knockout mutant showed a reduction in salt tolerance [3].

The development, understanding, and application of well-characterized halophilic bacteria isolated from halophytes could help alleviate many of the challenges of soil salinity, without the use of genetic engineering. Additionally, since these isolates are in their native state, using them to inoculate plants would be considered an organic treatment. PGPR technology has been applied to crop production for decades [27,28]. The focus in this review is on the effects of ST-PGPR inoculation on gene expression in the plants that lead to enhanced plant growth in salty soils.

2.2. Changes in Plant Gene Expression in Salt-Grown Plants Inoculated with ST-PGPR

The stimulation of plant growth in salty soil, in response to inoculation with halophilic ST-PGPR, infers that the expression of at least some plant genes are altered to enhance plant growth. In general, glycophytes exhibit a significant decrease in photosynthesis when exposed to salinity stress, along with an increase in the production of reactive oxygen species (ROS). One mechanism by which ST-PGPR stimulate plant growth is to offset the decrease in photosynthesis by stimulating the expression of proteins involved in this process in the plant [29,30]. One such ST-PGPR stimulated gene is auxin, a plant hormone involved in regulating growth. Increased auxin production was reported for

Salicornia plants inoculated with actinobacteria and irrigated with seawater [29]. Taj and Challabathula [30] observed an increased rate of photosynthesis and a decrease in ROS in both tomato and rice inoculated with *Staphylococcus sciuri*.

Some other plant gene expression pathways likely to be involved are those that regulate redox potential, ion homeostasis, leaf gas exchange, ion transport, osmolyte production and other genes involved in the stress response. Several studies have shown that inoculation with halotolerant bacteria enhances plant growth by inducing increased expression of antioxidant enzymes and proteins. For example, an increase in plant osmolyte production has been shown to occur in tomatoes inoculated with *Streptomyces* [31], *Staphylococcus sciuri* [30], *Limonium* (a coastal halophyte) and *Bacillus flexus* [32]. Several of these genes are components of stress response pathways. Transcription factors that control these pathways are also likely differentially expressed in inoculated plants exposed to salinity stress. In maize and chickpea, some genes involved in these pathways have demonstrated differential expression in response to microbial inoculation [33,34].

Bharti et al. [35] showed that several genes involved in salinity tolerance were differentially expressed in both roots and shoots of wheat in response to inoculation with a strain of *Dietzia natronolimnaea* and growth of the plants in salty soil. Their studies identified the role of an abscisic acid (ABA)-signaling cascade which led to the upregulation of TaABARE and TaOPR1 and the induction of TaMYB and TaWRKY transcription factors, leading to the stimulation of several stress-related genes. In addition, they found that TaNHX1, TaHAK and HaHKT1 ion transporter genes were differentially expressed in these plants, while antioxidant enzymes were upregulated under these same conditions.

A similar study with wheat inoculated with *Arthrobacter nitroguajacolicus* was conducted by Safdarian et al. [36]. This study involved the comparative transcriptomic analysis of un-inoculated and inoculated wheat roots under salt stress. They found that 152 genes were significantly upregulated, and five genes were downregulated under these conditions. Many of the upregulated genes are involved in antioxidant biosynthesis, flavonoid, porphyrin, and chlorophyll metabolism.

Another study showed the increased expression of two ZmPIP (plasma membrane aquaporin protein) isoforms in response to inoculation with ST-PGPR [37]. ST-PGPR reportedly induce osmolyte accumulation and phytohormone signaling, allowing plants to overcome osmotic shock caused by salinity. The inoculation of halophilic *Bacillus amyloliquefaciens* SN13 onto rice (Oryza sativa) plants enhanced salt tolerance when the rice was exposed to salinity (200 mM NaCl). The expression of 14 genes were affected by the following: SOS1, ethylene responsive element binding proteins EREBP, somatic embryogenesis receptor-like kinase SERK1, and NADP-malic enzyme (NADP-Me2). In the presence of these elements, 14 genes experienced upregulation, while two (glucose insensitive growth GIG and (SNF1) serine-threonine protein kinase SAPK4) were downregulated in plants grown hydroponically in response to salinity [37].

Some studies have used the model plant *Arabidopsis thaliana* to examine differential gene expression in response to inoculation with salt-tolerant bacteria. Baek et al. [38] showed that plants inoculated with *Bacillus oryzicola* exhibited increased chlorophyll production and the activation of the SOS1-dependent salt signaling pathway in comparison to uninoculated plants. SOS1 is a plasma membrane-localized Na^+/H^+ antiporter. Liu et al. [39] performed transcriptome profiling in Arabidopsis inoculated with *Bacillus amyloliquefaciens*, resulting in 1024 upregulated and 264 downregulated genes in plants grown in 100 mM NaCl compared to uninoculated plants grown at the same salinity. Upregulated plant genes included those involved in auxin-mediated signaling, SOS scavenging, sodium ion transport, photosynthesis, and the production of osmoprotectants, including trehalose and proline. They also analyzed hormone pathway mutants and determined that ethylene/jasmonic acid signaling but not ABA signaling may be altered in inoculated plants to increase their salt tolerance.

3. Overview of RNA Sequencing and Data Analysis

The regulation of gene expression is not just an on/off switch but can be better compared to the volume dial on a radio, where expression can be dialed up or dialed down. In the cell, the gene expression dial is adjusted according to environmental change and cellular need [40]. Thanks to next generation sequencing (NGS) and bioinformatics, it is easier than ever to analyze the volume of gene expression [41]. Gene expression analysis is a multi-step process with two main parts: 1) sequencing, and 2) analysis of sequence data [42]. In the studies discussed below, we focus on the analysis of differential gene expression in response to the inoculation of plants with halophilic bacteria, compared to uninoculated plants, grown in salty soil. Both sequencing and analysis necessitate multiple sub-steps which will be briefly described.

Sequencing begins with the isolation of plant RNA, which is critical for all subsequent steps. RNA is a notoriously unstable molecule due to its single-stranded nature and 2′ hydroxyl group on the ribose sugar, making the RNA molecule vulnerable to degradation. This means that isolated RNA samples must always be kept on ice when working in the lab and at −80C when stored. Careful isolation also includes avoiding contamination by anything that contains RNase or will inhibit downstream steps [43]. Plant RNA can be purified using a Trizol method or using an RNA isolation kit such as from Qiagen or Invitrogen, coupled with using a shredder spin column to improve RNA recovery from plant tissue. Following isolation, the RNA undergoes a quality assessment by Qubit or High Sensitivity Fragment Analysis to ensure that the RNA is intact and to determine its concentration. Once RNA quality has been assessed and is acceptable, the next step is reverse transcription to produce cDNA, which is significantly more stable than RNA. At this point, two options present themselves: (1) perform library prep for sequencing in lab, or (2) send cDNA to a sequencing center for further processing [44]. Either option will require the cDNA to undergo three additional processing steps: (1) cDNA fragmentation, (2) size selection of cDNA fragments, and (3) NGS on the appropriate instrument (See Figure 2A). The output of NGS is a large file of raw reads which represent base pairs inferred by the sequencer during processing. After processing, reads are then quantified and analyzed using bioinformatic tools.

As with all data analysis, the first step is to check quality. Since adapters are ligated onto the cDNA during library prep, it is important to remove these sequences before further processing [45]. Here, reads are mapped to an annotated transcriptome. If the plant of interest has an annotated transcriptome this step is rather simple and straightforward. If the percent of reads mapped is low, it is wise to re-evaluate the earlier quality control steps. If the percent of reads mapped are high, then generally moving on to further analysis is permissible. Moving on, however, requires the answer to the question, "which genes are differentially expressed among samples?" Uncovering the answer is assisted through building comparative table(s) and plotting data to help visualize fold changes in gene expression. Further, performing gene ontology analysis is another great tool for understanding gene function and potential interaction pathways (See Figure 2B).

Unfortunately, not every model organism has an assembled transcriptome, let alone a well annotated transcriptome. What then? Thanks to bioinformatic packages and assemblers such as trinity and SOAPdenovo-trans, there are many ways to create a de novo assembly [46]. A de novo assembly, however, would not have any gene annotation. For this reason, de novo assemblies are not very informative on their own. If, however, there is a related organism with an annotated transcriptome, the de novo transcriptome can be compared with that of its relative. This comparative analysis helps to determine gene name, function, expression levels, and potential ontologies. This process is generally referred to as "lift over" and will be further described in the next section. This next section outlines some of the approaches used to analyze differential plant gene expression in response to the bacterial inoculation of alfalfa plants grown in the presence or absence of salt.

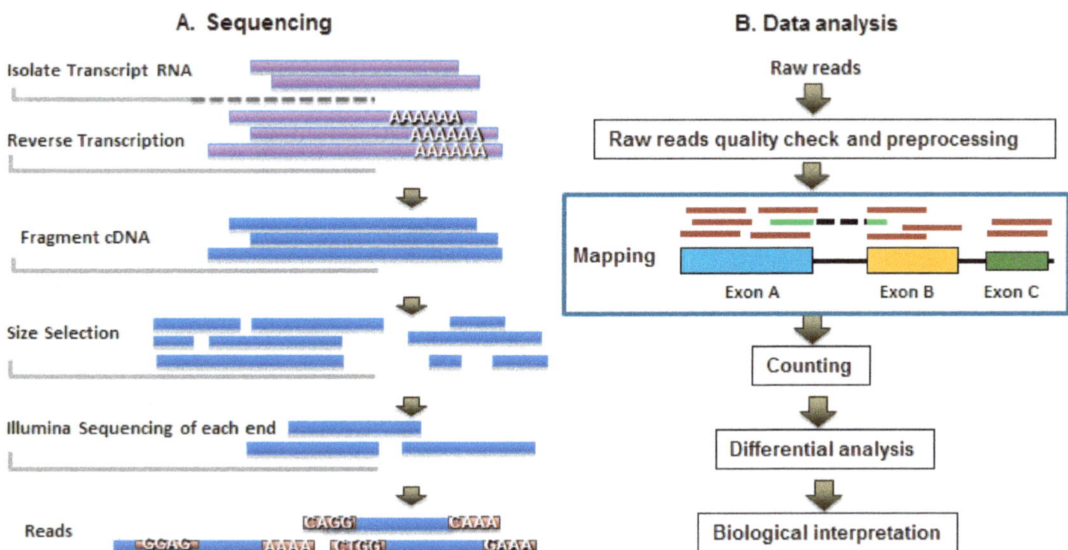

Figure 2. General overview of the steps involved in RNA- sequencing and subsequent analysis. (**A**) The major steps of RNA processing, sequencing, and read output. (**B**) Main steps involved in RNA- sequence analysis. Figure from "Bioinformatics for RNA- Seq Data Analysis" opensource article [42].

3.1. Approaches for Analyzing Changes in Plant Gene Expression in Response to Bacterial Inoculation

All approaches used to determine differentially expressed genes (DEGs) in plants start with a well-designed growth trial. Alfalfa grown with or without the inoculation of *Halomonas elongata* (accession number MK873884) under four soil conditions: (1) no salt or bacterial inoculant, (2) salt (1% NaCl) without bacteria, (3) bacteria without salt, and (4) bacteria and salt assist in the analysis of gene expression changes for each growing condition. After about 8 weeks of growth, plants can be harvested for root and shoot tissues. These tissues can help determine whether differential gene expression occurs, and if so, which of the four growing conditions and/or tissue types the gene expression change is associated with. In order to assess the tissues for DEGs, RNA needs to be extracted. One option is to grind tissues samples individually in liquid nitrogen for total RNA isolation using a Purelink RNA Mini kit with a homogenizer for plant tissue (Invitrogen). After grinding, tissue samples from each growing condition are lysed and homogenized. The lysate is then passed through a spin column to remove genomic DNA and washed to remove contaminants. Total RNA is then eluted from the column and subjected to quality testing, often performed on a Bioanalyzer. Quality RNA is then converted into a library of cDNA, which is either sequenced in house or sent to a sequencing center.

Sequence data files (reads) are subsequently reviewed for quality using a package such as FASTQC. Sequence reads are then aligned to the *M. sativa* transcriptome on a bioinformatics platform such as R. If using R, Rsubread is a robust splice-aware package that will align, quantify, and analyze RNA-seq data. After alignment, genes and their associated quantifications are determined for each condition and tissue type bioinformatically. Due to the incomplete nature of the alfalfa gene ontology (GO) annotation, converting alfalfa genes to their most similar homolog in *Arabidopsis*, in a process generally referred to as "lift over," can improve expression analysis. The results of the GO analysis, when collected into a super-table containing all results from each tissue type and growth condition, allows for easy visualization and comparison among different growth conditions and tissues.

3.2. Mining Sequence Data for Candidate Genes

As previously mentioned, the inoculation of *Halomonas elongata* (accession number MK873884) demonstrated an increase in alfalfa growth under salt-stress when compared to the control. From this observation developed the hypothesis that the increase in plant growth was likely due to changes in plant gene expression induced in the presence of the bacterial inoculum [1]. To test such a hypothesis careful analysis of RNA-seq data representative of each growing condition and plant tissue type must be acquired. Once tables representing all data types are produced, candidate features are easily isolated via the quick-and-dirty method of table sifting in Excel, or more eloquently, through bioinformatics. Candidate features should represent both significantly up- and downregulated genes with low Q-values (i.e., $<1 \times 10^{-5}$). A Q-value is a false discovery rate (FDR) adjusted p-value. This value adjusts the original p-value to increase statistical stringency and further reduce the presence of false-positive results. The lower a given Q-value, the more significant a feature (i.e., up- or downregulated gene) is within the study [47]. Once gene candidates are identified, it is common practice to validate targets by performing quantitative reverse transcriptase PCR (qRT-PCR) [48]. To do so, RNA is isolated (as before described) from root and shoot tissue of alfalfa grown in the presence or absence of the bacterial inoculum and salt. Then, cDNA is prepared from isolated RNA using the Thermo Fisher Superscript IV. With the sample now prepared, primers specific to the gene candidates are designed using the analysis results from RNA-seq data. To determine potential changes in gene products among different alfalfa growth treatments and tissue types, qRT-PCR is used to quantify gene candidate amplification. The ABI StepOnePlus Real-Time PCR System is one of the commonly used instruments to carry out this process. There are two methods of measuring qPCR amplifications: 1) by an intercalating dye, or 2) via a fluorescently labeled probe [49]. Intercalating dyes are generally the more cost-efficient method of qPCR amplification when many gene targets are being verified. Probes, however, are generally considered the cleaner and more simplistic mode of quantification, as they generally require less optimization. One example of a commonly used intercalating dye for qPCR is the standard PowerUp SYBR Green Master Mix. One important note about SYBR is that it binds to any double stranded DNA, including primer–dimers [50]. As a result, melt curves should always be reviewed for each well after qPCR. If more than one melting curve presents, it may be due to primer–dimer amplification. One way to confirm the correct product and absence of primer–dimers is by analyzing qPCR products via gel electrophoresis. Gel visualization is an excellent tool for testing if multiple melting curves are associated with more than one amplicon. If the gel returns only one band then likely the multiple melting curves shown on the qPCR machine are artifacts of intermediate steps in the amplification process and are not due to primer–dimer formation. Alternatively, programs such as uMelt curve prediction software (a free online tool) can accurately predict melting curves. This type of predictive software serves as a comparative tool against the melting curve produced by the qPCR instrument [51]. As always, selecting an appropriate control transcript as reference, generally a house keeping gene such as the 18S rRNA transcript, is critically important. For the amplification curve and general results, the $2^{-\Delta\Delta CT}$ method is selected for comparison. Biological and technical replicates (three for each) are generally worked into the qRT-PCR set-up, improving confidence in and analysis of later results.

3.3. Limitations of RNA Sequencing and Analysis

Before exploring the limitations of RNA sequencing and analysis, it is appropriate to clearly state that these technologies have been revolutionary and their impact critical in many technological and procedural advances [42]. Even so, these tools have limitations and weaknesses that have not yet been overcome.

One such limitation to the success of RNA-seq and analysis is user error during library preparation [52]. For instance, one important factor in good library preparation for RNA-seq is the size selection of cDNA fragments [53]. Selecting the correct size fragments improves the odds that most low molecular weight dimers are not included in sequencing, which improves the efficiency and accuracy of the sequence analysis [54]. Similarly, sequencing depth must be sufficient to cover low-abundance transcripts multiple times. Doing so improves sequencers' base pair inference, which also helps to improve sequencing accuracy [55].

Further, commonly used second-generation sequencers (i.e., Illumina) have some serious flaws. Short reads produced by this generation of instruments can become a problem if performing a de novo assembly is desired [56]. Short read sequencing (SRS) uses fragmented DNA generally 75–400 bp long. These short read lengths can become a problem as the size of the genome being sequenced increases. For instance, larger complex genomes such as that of humans contain many repetitive sequences, and due to the nature of PCR, these sequences are preferentially amplified. As preferential amplification increases, there is a risk of failing to generate enough sequence overlap between fragments for a quality de novo assembly. This lack of sequence overlap makes stitching scaffolds into one contiguous sequence very difficult, as an incredibly high number of reads are produced during short read sequencing [57].

Another common limitation to RNA-seq analysis is understanding bioinformatic packages well enough to select the best fit. For instance, three common comparative genomics aligners are: (1) Bowtie2, (2) TopHat, and (3) Rsubread [58,59]. All three of these tools serve to align sets of reads to a reference genome, but each does so with different strengths and weaknesses [60]. Bowtie2, while especially good at aligning large sets of reads to a reference genome, often fails to capture rare transcripts. TopHat on the other hand, is a tool that builds upon Bowtie2, improving the splice variant problem and reliably capturing rare transcripts [61]. Additionally, Rsubread is a package with robust read mapping of both small and large genomes and performs quantification and variant analysis [59]. Though a great option, Rsubread requires some understanding of the R platform and language to use packages such as Rsubread [62].

4. Conclusions and Future Challenges

The potential for using salt-tolerant plant-associated bacteria as inoculants of glycophyte crop plants to enhance growth and yield is clear from an increasing number of reports in the literature. However, understanding the mechanisms driving plant growth stimulation by these bacterial inoculants is currently limited. Photosynthesis is reduced in plants under salinity stress, while inoculation with ST-PGPR often results in increases in photosynthesis, as discussed in this review. Clearly, the expression of many genes is altered in the inoculated plants, resulting in plants with an increased tolerance to salinity and enhanced growth. While additional potential bacteria and plant interactions need to be studied, it appears that there may be some bacterial species–plant specific interactions, making it difficult to make general conclusions about the mechanisms involved. The properties of the bacterial inoculant may directly provide some help in reducing the effect of salinity by stimulating photosynthesis (see Figure 1), but different plants may exhibit up- or downregulation of different genes in response to the bacteria. From the currently available literature, it appears that gene pathways involving the regulation of redox potential, ion homeostasis, leaf gas exchange, ion transport, osmolyte production, and other genes involved in stress responses such as SOS pathways, as well as transcription factors all have some control over the expression of these genes. The above pathways, or some combination of them, appear to be involved in enhanced plant growth after inoculation with ST-PGPR.

Improvements in sequencing technology and quantitative analysis of differential gene expression have improved our capabilities to analyze changes in plant gene expression in response to various signals and stresses. This includes the ability to analyze changes in the gene expression of plants inoculated with salt-tolerant bacteria compared to uninoculated plants. With time, these expression data will better elucidate the mechanisms by which ST-PGPR enhance plant growth under saline conditions. However, there are still challenges to address, which include the isolation of high-quality RNA from plant tissue and the use of a dependable computational pipeline to accurately measure differences in gene expression. The new knowledge generated as more plants and bacterial inoculants are examined will help farmers who have land affected by salinity increase their crop yield by using salt-tolerant bacterial inoculants. This will also be of great benefit for increasing agricultural productivity to feed the growing world human population, as good agricultural land is quickly decreasing.

Author Contributions: Conceptualization, A.K.M. and B.L.N.; writing-manuscript preparation, review and editing, A.K.M. and B.L.N. All authors have read and agreed to the published version of the manuscript.

Funding: This research received no external funding.

Acknowledgments: We thank the BYU Department of Microbiology and Molecular Biology for support.

Conflicts of Interest: The authors declare no conflict of interest.

References

1. Kearl, J.; McNary, C.; Lowman, S.J.; Mei, C.; Aanderug, Z.; Smith, S.; West, J.; Colton, E.; Hamson, M.; Nielsen, B.L. Salt-tolerant halophyte rhizosphere bacteria stimulate growth of alfalfa in salty soil. *Front. Microbiol.* **2019**, *10*, 1849. [CrossRef] [PubMed]
2. Rima, F.S.; Biswas, S.; Sarker, P.; Islam, M.D.; Seraj, Z. Bacteria endemic to saline coastal belt and their ability to mitigate the effects of salt stress on rice growth and yields. *Ann. Microbiol.* **2018**, *68*, 525–535. [CrossRef]
3. Nakayama, H.; Yoshida, K.; Ono, H.; Murooka, Y.; Shinmyo, A. Ectoine, the compatible solute of Halomonas elongata, confers hyperosmotic tolerance in cultured tobacco cells. *Plant Physiol.* **2020**, *122*, 1239–1247. [CrossRef]
4. Zhang, W.; Wang, T. Enhanced salt tolerance of alfalfa (Medicago sativa) by rstB gene transformation. *Plant Sci.* **2015**, *234*, 110–118. [CrossRef] [PubMed]
5. Zsigmond, L.; Szepi, A.; Tari, I.; Rigo, G.; Kiraly, A.; Szabados, L. Overexpression of the mitochondrial PPR40 gene improves salt tolerance in Arabidopsis. *Plant Sci.* **2012**, *182*, 87–93. [CrossRef]
6. Müller, M.; Munné-Bosch, S. Ethylene Response Factors: A Key Regulatory Hub in Hormone and Stress Signaling. *Plant Physiol.* **2015**, *169*, 32–34. [CrossRef]
7. Li, J.; Gao, Z.; Zhou, L.; Li, L.; Zhang, J.; Liu, Y.; Chen, H. Comparative transcriptome analysis reveals K+ transporter gene contributing to salt tolerance in eggplant. *BMC Plant Biol.* **2019**, *19*, 67. [CrossRef]
8. Cen, H.; Ye, W.; Liu, Y.; Li, D.; Wang, K.; Zhang, W. Overexpression of a Chimeric Gene, OsDST-SRDX, Improved Salt Tolerance of Perennial Ryegrass. *Sci. Rep.* **2016**, *6*, 27320. [CrossRef] [PubMed]
9. Bohnert, H.; Golldack, D.; Ishitani, M.; Kamasani, U.; Rammesmayer, G.; Shen, B.; Sheveleva, E.; Jensen, R. Salt Tolerance Engineering Requires Multiple Gene Transfers. *Ann. N. Y. Acad. Sci.* **2006**, *792*, 115–125. [CrossRef]
10. How GMOs Are Regulated for Food and Plant Safety in the United States. Available online: https://www.fda.gov/food/agricultural-biotechnology/how-gmos-are-regulated-food-and-plant-safety-united-states (accessed on 8 March 2021).
11. Public Opinion about Genetically Modified Foods and Trust in Scientists. Available online: https://www.pewresearch.org/science/2016/12/01/public-opinion-about-genetically-modified-foods-and-trust-in-scientists-connected-with-these-foods/ (accessed on 10 March 2021).
12. Deinlein, U.; Stephan, A.B.; Horie, T.; Luo, W.; Xu, G.; Schroeder, J.I. Plant salt-tolerance mechanisms. *Trends Plant Sci.* **2014**, *19*, 371–379. [CrossRef]
13. Xu, C.; Tang, X.; Shao, H.; Wang, H. Salinity tolerance mechanism of economic halophytes from physiological to molecular hierarchy for improving food quality. *Curr. Genom.* **2016**, *17*, 207–214. [CrossRef] [PubMed]
14. Meng, X.; Zhou, J.; Sui, N. Mechanisms of salt tolerance in halophytes: Current understanding and recent advances. *Open Life Sci.* **2018**, *13*, 149–154. [CrossRef]
15. Sáenz-Mata, J.; Palacio-Rodríguez, R.; Sánchez-Galván, H.; Balagurusamy, N. Plant Growth Promoting Rhizobacteria Associated to Halophytes: Potential Applications in Agriculture. In *Sabkha Ecosystems*; Springer Science and Business Media LLC: Berlin/Heidelberg, Germany, 2016; Volume 48, pp. 411–425. [CrossRef]
16. Egamberdieva, D.; Wirth, S.; Bellingrath-Kimura, S.D.; Mishra, J.; Arora, N.K. Salt-Tolerant plant growth promoting rhizobacteria for enhancing crop productivity of saline soils. *Front. Microbiol.* **2010**, *10*, 2791. [CrossRef] [PubMed]
17. Etesami, H.; Beattie, G.A. Mining halophytes for Plant Growth-Promoting Halotolerant bacteria to enhance the salinity tolerance of non-halophytic crops. *Front. Microbiol.* **2018**, *9*, 148. [CrossRef]
18. Orhan, F. Alleviation of salt stress by halotolerant and halophilic plant growth-promoting bacteria in wheat (Triticum aestivum). *Braz. J. Microbiol.* **2016**, *47*, 621–627. [CrossRef]
19. Kundan, R.; Pant, G.; Jadon, N.; Agrawal, P.K. Plant growth promoting rhizobacteria: Mechanism and current prospective. *J. Fertil. Pestic.* **2015**, *6*, 155. [CrossRef]
20. Malik, K.A.; Rasul, G.; Hassan, U.; Mehnaz, S.; Ashraf, M. Role of nitrogen fixing and growth hormone producing bacteria in improving growth of wheat and rice. Nitrogen Fixation with Non-Legumes. In Proceedings of the Sixth International Symposium

on Nitrogen Fixation with Non-Legumes, Ismalia, Egypt, 6–10 September 1993; Hegazi, N.A., Fayez, M., Monib, M., Eds.; American University Cairo Press: Cairo, Egypt, 1993; pp. 409–422.

21. Cattelan, A.; Hartel, P.G.; Fuhrmann, J.J. Screening for plant growth-promoting rhizobacteria to promote early soybean growth. *Soil Sci. Soc. Am. J.* **1999**, *6*, 1670–1680. [CrossRef]
22. Glick, B.; Penrose, D.; Li, J. A model for the lowering of plant ethylene concentrations by plant growth-promoting bacteria. *J. Theor. Biol.* **1998**, *190*, 63–68. [CrossRef]
23. Kim, S.; Lowman, S.; Hou, G.; Nowak, J.; Flinn, B.; Mei, C. Growth promotion and colonization of switchgrass (*Panicum virgatum*) cv. Alamo by bacterial endophyte *Burkholderia phytofirmans* strain PsJN. *Biotechnol. Biofuels* **2012**, *5*, 37. [CrossRef]
24. Lara-Chavez, A.; Lowman, S.; Kim, S.; Tang, Y.; Zhang, J.; Udvardi, M.; Nowak, J.; Flinn, B.; Mei, C. Global gene expression profiling of two switchgrass cultivars following inoculation with *Burkholderia phytofirmans* strain PsJN. *J. Exp. Botany* **2015**, *66*, 4337–4350. [CrossRef]
25. Li, X.; Geng, X.; Xie, R.; Fu, L.; Jiang, J.; Gao, L.; Sun, J. The endophytic bacteria isolated from elephant grass (*Pennisetum purpureum* Schumach) promote plant growth and enhance salt tolerance of hybrid Pennisetum. *Biotech. Biofuels* **2016**, *9*, 190. [CrossRef]
26. Yan, J.; Smith, M.; Glick, B.; Liang, Y. Effects of ACC deaminase containing rhizobacteria on plant growth and expression of TocGTPases in tomato (*Solanum lycopersicum*) under salt stress. *Botany* **2014**, *92*, 775–781. [CrossRef]
27. Kumari, B.; Mallick, M.; Hora, A. Plant growth-promoting rhizobacteria (PGPR): Their potential for development of sustainable agriculture. In *Bio-Exploitation for Sustainable Agriculture*; Trivedi, P.C., Ed.; Avinskar Publishing: Jaipur, Rajasthan, 2016; pp. 1–19.
28. Ilangumaran, G.; Smith, D. Plant growth promoting rhizobacteria in amelioration of salinity stress: A systems biology perspective. *Front. Plant Sci.* **2017**, *8*, 1768. [CrossRef] [PubMed]
29. Mathew, B.T.; Torky, Y.; Mourad, A.H.; Ayyash, M.M.; El-Keblawy, A.; Hilal-Alnaqbi, A.; AbuQamar, S.F.; El-Tarabily, K. Halotolerant marine rhizosphere-competent actinobacteria promote *Salicornia bigelovii* growth and seed production using seawater irrigation. *Front. Microbiol.* **2020**, *11*, 552. [CrossRef]
30. Taj, Z.; Challabathula, D. Protection of photosynthesis by halotolerant *Staphylococcus Sciuri* ET101 in tomato (*Lycoperiscon esculentum*) and rice (*Oryza sativa*) plants during salinity stress: Possible interplay between carboxylation and oxygenation in stress mitigation. *Front. Microbiol.* **2021**, *11*, 547750. [CrossRef] [PubMed]
31. Gong, Y.; Chen, L.J.; Pan, S.Y.; Li, X.W.; Xu, M.J.; Zhang, C.M.; Xing, K.; Sheng, Q. Antifungal potential evaluation and alleviation of salt stress in tomato seedlings by a halotolerant plant growth-promoting Actinomycete *Streptomyces* sp. KLBMP5084. *Rhizosphere* **2020**, *16*, 100262. [CrossRef]
32. Xiong, Y.W.; Li, X.W.; Wang, T.T.; Gong, Y.; Zhang, C.M.; Xing, K.; Qin, S. Root exudates-driven rhizosphere recruitment of the plant growth-promoting rhizobacterium *Bacillus flexus* KLBMP 4941 and its growth-promoting effect on the coastal halophyte Limonium sinense under salt stress. *Ecotoxicol. Environ. Saf.* **2020**, *194*, 110374. [CrossRef]
33. Marulanda, A.; Azcón, R.; Chaumont, F.; Ruiz-Lozano, J.; Aroca, R. Regulation of plasma membrane aquaporins by inoculation with a *Bacillus megaterium* strain in maize (*Zea mays* L.) plants under unstressed and salt-stressed conditions. *Planta* **2010**, *232*, 533–543. [CrossRef] [PubMed]
34. El-Esawi, M.A.; Al-Ghamdi, A.A.; Ali, H.M.; Alayafi, A.A. *Azospirillum lipoferum* FK1 confers improved salt tolerance in chickpea (*Cicer arietinum* L.) by modulating osmolytes, antioxidant machinery and stress-related genes expression. *Environ. Exp. Bot.* **2019**, *159*, 55–65. [CrossRef]
35. Bharti, N.; Pandey, S.; Barnawal, D.; Patel, V.; Kalra, A. Plant growth promoting rhizobacteria Dietzia natronolimnaea modulates the expression of stress responsive genes providing protection of wheat from salinity stress. *Sci. Rep.* **2016**, *6*, 34768. [CrossRef]
36. Safdarian, M.; Askan, H.; Shariati, V.; Nematzadeh, G. Transcriptional responses of wheat roots inoculated with *Arthrobacter nitroguajacolicus* to salt stress. *Sci. Rep.* **2019**, *9*, 1792. [CrossRef] [PubMed]
37. Nautiyal, C.S.; Srivastava, S.; Chauhan, P.S.; Seem, K.; Mishra, A.; Sopory, S.K. Plant growth-promoting bacteria Bacillus amyloliquefaciens NBRISN13 modulates gene expression profile of leaf and rhizosphere community in rice during salt stress. *Plant Physiol.* **2013**, *66*, 1–9. [CrossRef]
38. Baek, D.; Rokibuzzaman, M.; Khan, A.; Kim, M.C.; Park, H.J.; Yun, D.-J.; Chung, Y.R. Plant-growth promoting *Bacillus oryzicola* YC7007 modulates stress-response gene expression and provides protection from salt stress. *Front. Plant Sci.* **2020**, *10*, 1646. [CrossRef] [PubMed]
39. Liu, J.; Ishitani, M.; Halfter, U.; Kim, C.; Zhu, J. The *Arabidopsis thaliana* SOS$_2$ gene encodes a protein kinase that is required for salt tolerance. *Proc. Natl. Acad. Sci. USA* **2020**, *97*, 3730–3734. [CrossRef]
40. Gene Expression. Available online: https://www.nature.com/scitable/topicpage/gene-expression-14121669/ (accessed on 12 March 2021).
41. Wang, Z.; Gerstein, M.; Snyder, M. RNA-seq: A revolutionary tool for transcriptomics. *Nat. Rev. Genet.* **2009**, *10*, 57–63. [CrossRef] [PubMed]
42. Zhao, S.; Zhang, B.; Zhang, Y.; Gordon, W.; Du, S.; Paradis, T.; Vincent, M.; Von Schack, D. Bioinformatics for RNA-seq data analysis. *InTechOpen* **2016**. [CrossRef]

43. RNA Sample Collection, Protection, and Isolation Support-Troubleshooting. Available online: https://www.thermofisher.com/us/en/home/technical-resources/technical-reference-library/nucleic-acid-purification-analysis-support-center/rna-sample-collection-protection-isolation-support/rna-sample-collection-protection-isolation-support-troubleshooting.html (accessed on 14 March 2021).

44. Methods of RNA Quality Assessment. Available online: https://www.promega.com/resources/pubhub/methods-of-rna-quality-assessment/#bioanalyzer (accessed on 14 March 2021).

45. Conesa, A.; Madrigal, P.; Tarazona, S.; Gomez-Cabrero, D.; Cervera, A.; McPherson, A.; Szczesniak, M.; Gaffney, D.; Elo, L.; Zang, X.; et al. A survey of best practices for RNA-seq data analysis. *Genome Biol.* **2016**, *17*, 13. [CrossRef]

46. Rana, S.; Zadlock, F.; Zhang, Z.; Murphy, W.R.; Bentivegna, C.S. Comparison of de novo transcriptome assemblers and K-mer strategies using the Killifish, fundulus heteroclitus. *PLoS ONE* **2016**, *11*, e0153104. [CrossRef]

47. Storey, J.D.; Tibshirani, R. Statistical significance for genome wide studies. *Proc. Natl. Acad. Sci. USA* **2003**, *100*, 9440–9445. [CrossRef]

48. Adamski, M.; Gumann, P.; Baird, A. A method for quantitative analysis of standard and high-throughput qPCR expression data based on input sample quantity. *PLoS ONE* **2014**, *9*, e103917. [CrossRef]

49. The Luna qPCR Mixes? Available online: https://www.neb.com/faqs/2016/11/14/what-is-the-difference-between-the-probe-and-dye-versions-of-the-luna-qpcr-mixes#:~|}:text=qPCR%20is%20typically%20measured%20in,sequence%20in%20the%20PCR%20amplicon (accessed on 19 March 2021).

50. TaqMan vs. SYBR Chemistry for Real-Time PCR. Available online: https://www.thermofisher.com/us/en/home/life-science/pcr/real-time-pcr/real-time-pcr-learning-center/real-time-pcr-basics/taqman-vs-sybr-chemistry-real-time-pcr.html (accessed on 18 March 2021).

51. Explaining Multiple Peaks in qPCR Melt Curve Analysis. Available online: https://www.idtdna.com/pages/education/decoded/article/interpreting-melt-curves-an-indicator-not-a-diagnosis (accessed on 19 March 2021).

52. Paszkiewicz, K.; Studholme, D.J. *De novo* assembly of short sequence reads. *Brief. Bioinform.* **2010**, *11*, 457–472. [CrossRef] [PubMed]

53. Roberts, A.; Trapnell, C.; Donaghey, J.; Rinn, J.L.; Pachter, L. Improving RNA-seq expression estimates by correcting for fragment bias. *Genome Biol.* **2011**, *12*, R22. [CrossRef] [PubMed]

54. Head, S.; Komori, H.; LaMere, S.; Whisenant, T.; Van Nieuwerburgh, F.; Salomon, D.; Ordoukhanian, P. Library construction for next-generation sequencing: Overviews and challenges. *BioTechniques* **2014**, *56*, 61–68. [CrossRef] [PubMed]

55. Considerations for RNA-seq Read Length and Coverage. Available online: https://support.illumina.com/bulletins/2017/04/considerations-for-rna-seq-read-length-and-coverage-.html (accessed on 13 March 2021).

56. Baker, M. De novo genome assembly: What every biologist should know. *Nat. Methods.* **2012**, *9*, 333–337. [CrossRef]

57. LRS vs. SRS Genomics. Available online: https://genome.vinbigdata.org/documentation/blog/2020/12/17/long-read-sequencing.html (accessed on 13 March 2021).

58. Tuxedo Suite. Available online: https://support.illumina.com/help/BS_App_RNASeq_Alignment_OLH_1000000006112/Content/Source/Informatics/Apps/TuxedoSuite_RNASeqTools.htm (accessed on 13 March 2021).

59. Rsubread. Available online: https://bioconductor.org/packages/release/bioc/html/Rsubread.html (accessed on 14 March 2021).

60. RNA-seq Work Flows. Available online: http://bioconductor.org/help/course-materials/2014/SeattleOct2014/B02.1_RNASeq.html (accessed on 12 March 2021).

61. Trapnell, C.; Pachter, L.; Salzberg, S. TopHat: Discovering splice junctions with RNA-Seq. *Bioinformatics* **2009**, *25*, 1105–1111. [CrossRef]

62. Liao, Y.; Smyth, G.; Shi, W. The R Package Rsubread is easier, faster, cheaper, and better for alignment and quantification of RNA sequencing reads. *Nucl. Acids Res.* **2019**, *47*, e47. [CrossRef]

International Journal of
Molecular Sciences

Article

Isolation and Functional Characterization of a Salt-Responsive Calmodulin-Like Gene *MpCML40* from Semi-Mangrove *Millettia pinnata*

Yi Zhang [1,†], Jianzi Huang [2,†], Qiongzhao Hou [3], Yujuan Liu [3], Jun Wang [4] and Shulin Deng [1,5,6,*]

1   Key Laboratory of South China Agricultural Plant Molecular Analysis and Genetic Improvement & Guangdong Provincial Key Laboratory of Applied Botany, South China Botanical Garden, Chinese Academy of Sciences, Guangzhou 510650, China; yizhang@scbg.ac.cn
2   Guangdong Provincial Key Laboratory for Plant Epigenetics, College of Life Sciences and Oceanography, Shenzhen University, Shenzhen 518060, China; biohjz@szu.edu.cn
3   College of Life Sciences, University of Chinese Academy of Sciences, Beijing 100049, China; houqiongzhao@scbg.ac.cn (Q.H.); liuyujuan@scbg.ac.cn (Y.L.)
4   CAS Engineering Laboratory for Vegetation Ecosystem Restoration on Islands and Costal Zones, South China Botanical Garden, Chinese Academy of Sciences, Guangzhou 510650, China; wxj@scbg.ac.cn
5   Xiaoliang Research Station for Tropical Coastal Ecosystems, South China Botanical Garden, Chinese Academy of Sciences, Guangzhou 510650, China
6   Center of Economic Botany, Core Botanical Gardens, Chinese Academy of Sciences, Guangzhou 510275, China
*   Correspondence: sldeng@scbg.ac.cn
†   These authors contributed equally to this work.

Citation: Zhang, Y.; Huang, J.; Hou, Q.; Liu, Y.; Wang, J.; Deng, S. Isolation and Functional Characterization of a Salt-Responsive Calmodulin-Like Gene *MpCML40* from Semi-Mangrove *Millettia pinnata. Int. J. Mol. Sci.* **2021**, 22, 3475. https://doi.org/10.3390/ijms22073475

Academic Editors: Jen-Tsung Chen, Ricardo Aroca and Daniela Romano

Received: 24 February 2021
Accepted: 24 March 2021
Published: 27 March 2021

Publisher's Note: MDPI stays neutral with regard to jurisdictional claims in published maps and institutional affiliations.

Abstract: Salt stress is a major increasing threat to global agriculture. Pongamia (*Millettia pinnata*), a semi-mangrove, is a good model to study the molecular mechanism of plant adaptation to the saline environment. Calcium signaling pathways play critical roles in the model plants such as Arabidopsis in responding to salt stress, but little is known about their function in Pongamia. Here, we have isolated and characterized a salt-responsive *MpCML40*, a calmodulin-like (CML) gene from Pongamia. MpCML40 protein has 140 amino acids and is homologous with Arabidopsis AtCML40. MpCML40 contains four EF-hand motifs and a bipartite NLS (Nuclear Localization Signal) and localizes both at the plasma membrane and in the nucleus. *MpCML40* was highly induced after salt treatment, especially in Pongamia roots. Heterologous expression of *MpCML40* in yeast cells improved their salt tolerance. The *35S::MpCML40* transgenic Arabidopsis highly enhanced seed germination rate and root length under salt and osmotic stresses. The transgenic plants had a higher level of proline and a lower level of MDA (malondialdehyde) under normal and stress conditions, which suggested that heterologous expression of *MpCML40* contributed to proline accumulation to improve salt tolerance and protect plants from the ROS (reactive oxygen species) destructive effects. Furthermore, we did not observe any measurable discrepancies in the development and growth between the transgenic plants and wild-type plants under normal growth conditions. Our results suggest that MpCML40 is an important positive regulator in response to salt stress and of potential application in producing salt-tolerant crops.

Keywords: *Millettia pinnata*; calmodulin-like; salt tolerance; heterologous expression

1. Introduction

Salt stress is one of the significant environmental factors affecting plant growth and productivity. Soil salinization is a fast-growing global problem, especially in the arid and semi-arid areas of the world [1]. Moreover, lots of arable lands are changing to salinized land now due to the rising sea level, so improving salt tolerance of crops by genetic modification is a critical aspect of crop breeding. Halophytes are kinds of plants growing in high salinity (salt concentration is around 200 mM NaCl or more) conditions where most crops cannot survive [2]. Mangroves are trees or large shrubs which grow within the intertidal zone in tropical and subtropical regions. Mangrove species include true mangroves and semi-mangroves (also called mangrove associates). True mangroves have morphological specialization, such as aerial roots and vivipary; physiological mechanism for salt exclusion and/or salt excretion [3]. Pongamia (*Millettia pinnata* syn. *Pongamia pinnata*) is a semi-mangrove plant that can grow in either freshwater or moderate salinity water [3]. Unlike true mangroves, Pongamia does not have the salty glands or other specialized morphological traits to endure salinity stress, which suggested that its salt tolerance may be more attributed to gene regulation and protein function [3]. Therefore, investigating the molecular mechanisms of Pongamia salt tolerance may provide promising strategies for crop breeding by genetic modification.

Pongamia is a diploid legume (2n = 22) with a genome size of ~1300 Mb, which is ten times that of Arabidopsis [4,5]. Several transcriptome analyses of Pongamia root, leaf, flower, pod, and seedling have been reported in recent years [6–11]. However, only a few Pongamia functional genes have been identified and characterized. Four circadian clock genes (*ELF4, LCL1, PRR7,* and *TOC1*) were identified from Pongamia using soybean as the reference [12]. A stearoyl-ACP desaturase (SAD) was isolated from Pongamia seeds and suggested a seed development function [13]. Furthermore, two other desaturase genes, *MpFAD2-1* and *MpFAD2-2* (Fatty Acid Desaturase 2), were also isolated and characterized [14]. Although 23,815 candidate salt-responsive genes were identified from Pongamia by comparing the expression pattern under seawater and freshwater treatments using Illumina sequencing, so far, only one report showed that a chalcone isomerase gene, *MpCHI*, enhanced the salt tolerance of yeast (*Saccharomyces cerevisiae*) salt-sensitive mutants [7,15].

The calcium ion Ca^{2+} is a second messenger in all eukaryotes. It is perceived by several Ca^{2+} binding proteins, including calmodulin (CaM), calmodulin-like protein (CML), calcineurin B-like (CBL), and calcium-dependent protein kinase (CPK or CDPK) [16–18]. CaMs are present in all eukaryotes, but CMLs, CBLs, and CPKs are only identified in plants and some protists [17,19]. The most common Ca^{2+} binding motif is the EF-hand motif present in most Ca^{2+} binding proteins, including CaMs and CMLs [20,21]. In Arabidopsis, there are only six typical *CaM* genes, while over 50 *CML* genes have been identified [22]. All six AtCaMs have very similar protein sequences with 149 amino acids containing four EF-hand motifs. In contrast, AtCMLs have 80–330 amino acids with two to four EF-hand motifs [22]. Many *CML* genes have been reported to be involved in abiotic stress signaling [23,24]. In Arabidopsis, it was reported that the expression levels of *AtCML12* (also called *TCH3*) and *AtCML24* (also called *TCH2*) were highly enhanced under heat stress [24,25]. The CML::GUS report gene data showed that *AtCML37, AtCML38,* and *AtCML39* genes were induced by several stimuli, including salt and drought stress [26]. AtCML15 (also called CaM15) could interact with Na^+/H^+ exchanger 1 (AtNHX1) and played roles in maintaining cellular pH and ion homeostasis [27]. *AtCML9* and *AtCML24* genes were induced by abiotic stress and abscisic acid (ABA) and functioned in response to ABA and salt or ion stress [28,29]. *AtCML20* could negatively modulate ABA signaling and drought response [30]. *AtCML42* was reported to function in both herbivory defense and ABA-mediated drought stress response [31]. A rice CML gene, *OsMSR2*, could increase drought or salt tolerance and ABA sensitivity in Arabidopsis [32]. In addition to Arabidopsis and rice, several studies showed that CMLs in other plants were also involved in abiotic stress signaling. Expression of 32 CML genes in wild-growing grapevine (*Vitis amurensis*) was shown to be responsive to abiotic stresses, including drought, salt, heat, and cold [33]. In *Glycine soja*, *GsCML27* participated in salt and osmotic stresses [34]. In *Medicago truncatula*, *MtCML40* was involved in salt stress [35]. In *Camellia sinensis*, *CsCML16, CsCML18-2,* and *CsCML42* were induced by cold and salt conditions, while *CsCML38* was induced by drought and ABA treatments [36].

In the present study, we identified a *CML* gene, *MpCML40*, from Pongamia. The MpCML40 protein contained a typical EF-hand motif. Under salt treatment, the *MpCML40* gene was highly induced in roots. Heterologous expression of *MpCML40* in Arabidopsis strongly enhanced the salt tolerance of transgenic plants.

2. Results

2.1. MpCML40 Is an EF-Hand Motif-Containing Calmodulin-Like Protein

The full-length cDNA of the *MpCML* gene was obtained by 5′ and 3′ rapid amplification of cDNA ends (RACE) assay with four specific internal primers based on the sequence of an EST from our previous study [7]. The full-length sequence of cDNA, which was deposited in GenBank under accession number MW650864, comprised a 423 bp open reading frame (ORF), 135 bp 5′ untranslated region (UTR), and 258 bp 3′ UTR followed by a polyA tail (Table 1). The corresponding protein contained 140 amino acids. A phylogenetic tree based on the amino acid sequences of MpCML and 50 Arabidopsis CML proteins revealed that this MpCML protein was homologous with AtCML40 (Figure 1). Therefore, we named it MpCML40. In addition to AtCML40, the sequence of MpCML40 was also similar to some other members in this subfamily, such as AtCML37, AtCML38, and AtCML39 (Figure 1). Based on SMART (Simple Modular Architecture Research Tool) analysis, MpCML40 also had four EF-hand motifs (Figure 2), the same as AtCML38 and AtCML39. However, the third EF-hand motif lacked the conserved 12 residues with the pattern X•Y•Z•–Y•–X••–Z (also called DxDxDG loop), which might be a pseudo EF-hand motif as previously reported [21,37,38]. Moreover, a predicted bipartite NLS (Nuclear Localization Signal) was found in the third EF-hand motif (Figure 2), which indicated MpCML40 might at least partially localize in the nucleus.

Table 1. The nucleotide sequence of *MpCML40* cDNA. The open reading frame (ORF) of *MpCML40* is highlighted in yellow.

```
   1   CCACAACATA TAATAACAAC TCAATTTTCC ATTTGCATAC AAGTTACATT TTCTCCTTCT
  61   TATTCTTCTT GTTATTGTGT ACATTAAAGA TTTGAACAAA TTACACTACA CCTTTAAGAT
 121   AAGGAGCATT GTAAC ATGAT  GAAGCATGCG GGTTTTGAGG GTGTTCTTCG ATATTTTGAT
 181   GAAGATGGGG ATGGAAAGGT TTCACCTTCA GAGTTAAGGC ATGGATTGGG AATGATGGGT
 241   GGGGAGCTTT TGATGAAAGA AGCAGAGATG GCAATTGAGG CACTGGATTC TGATGGTGAT
 301   GGGTTGTTGA GTTTGGAGGA TTTGATTGCT TTGATGGAAG CAGGGGGAGA GGAACAAAAG
 361   TTGAAGGATT TGAGAGAAGC TTTTGAGATG TATGACACTC AAGGGTGCGG ATTTATAACC
 421   CCAAAGAGCT TGAAGAGGAT GCTTAAGAAG ATGGGAGAGT CCAAGTCCAT TGATGAATGC
 481   AAATTGATGA TTAGTCAATT TGATTTGAAT GGGGATGGGA TGCTTAGCTT TGAAGAATTC
 541   AGAATTATGA TGGAGTGA GG CCAGTATATT TGTTGATGAT ATTGTTTAGT TTGTTTGTTT
 601   GTTTGGGAGG AAGAGGGGTA TAGTTAAGTG GATTTGATTT ATTTGTTTGC AGTTTGCACA
 661   TGTATAAATA ACTCCTTTTG TGCTTTGCAA TACTTTTGAC AATTGATTAA CTGTTAGATT
 721   TCTCCCAAGT TCCTACATAA AAAATTATTC AAATTTTCTT AATGGGAGTT GTATTATGAC
 781   TATTATCATG GTTAAATATA TTTTTTATTT CGTTCCAAAA AAAAAAAAAA AAAAAAAAAA
```

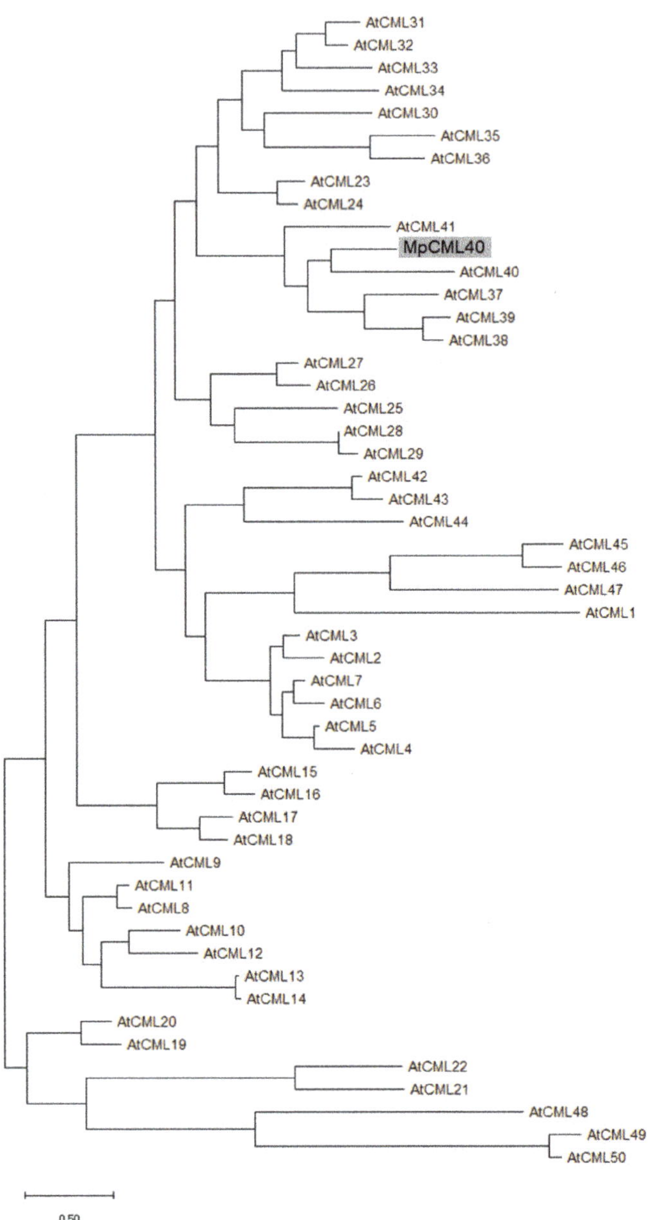

Figure 1. Phylogenetic tree of MpCML40 with Arabidopsis CML proteins. MpCML40 protein and 50 Arabidopsis calmodulin-like (CML) proteins were used for phylogenetic analysis [22]. MpCML40 was marked by a grey box.

```
MpCML40    --------------------------------------------MMKHAGFEGVLRYFDED   17
AtCML40    -----------------------------------------MKSENVNKRDEYQRVFSCFDKS   22
AtCML37    MTLAKNQKSSLSRLYKKVSSKRSESSRNLEDESRTSSNSSGSSSLNVNELRTVFDYMDAN   60
AtCML38    MKNNTQPQSSFKKLCRKLSPKREDSAGEIQQH------NSSNGEDKNRELEAVFSYMDAN   54
AtCML39    --------------------MKNTQRQLSSS------FMKFLEEKNRDLEAVFAYMDAN   33
                                              :    . *:  :*  .

MpCML40    GDGKVSPSELRHGLGMMGGE---LLMKEAEMAIEALDSDGDGLLSLEDLIALMEAGGE--   72
AtCML40    HQGKVSVSTIERCVDAIKSGKRAVVDQEDTTNPNPEESTDDKSLELEDFVKLVEEGEE--   80
AtCML37    SDGKISGEELQSCVSLLGGA---LSSREVEEVVKTSDVDGDGFIDFEEFLKLMEGEDGS-   116
AtCML38    RDGRISPEELQKSFMTLGEQ---LSDEEAVAAVRLSDTDGDGMLDFEEFSQLIKVDDE--   109
AtCML39    RDGRISAEELKKSFKTLGEQ---MSDEEAEAAVKLSDIDGDGMLDINEFALLIKGNDEFT   90
           :*::*  . :.  .  :       :  .*      .  :  .*  :.:::: *::

MpCML40    -EQKLKDLREAFEMYDTQGCGFITPKSLKRMLKKMGESKSIDECKLMISQFDLNGDGMLS   131
AtCML40    -ADKEKDLKEAFKLYEE--SEGITPKSLKRMLSLLGESKSLKDCEVMISQFDINRDGIIN   137
AtCML37    DEERRKELKEAFGMYVMEGEEFITAASLRRTLSRLGESCTVDACKVMIRGFDQNDDGVLS   176
AtCML38    -EEKKMELKGAFRLYIAEGEDCITPRSLKMMLKKLGESRTTDDCRVMISAFDLNADGVLS   168
AtCML39    EEEKKRKIMEAFRMYIADGEDCITPGSLKMMLMKLGESRTTDDCKVMIQAFDLNADGVLS   150
           ::   .:   **  :*      **   **:   *   :***  .  *.:**   **  *  **::.

MpCML40    FEEFRIMME   140
AtCML40    FDEFRAMMQ   146
AtCML37    FDEFVLMMR   185
AtCML38    FDEFALMMR   177
AtCML39    FDEFALMMR   159
```

Figure 2. Amino acid sequence alignment of MpCML40, AtCML37, AtCML38, AtCML39, and AtCML40. The EF-hand motifs were underlined, and the highly conserved amino acids in EF-hand motifs were highlight by yellow color. A predicted bipartite Nuclear Localization Signal (NLS) was highlighted by green color.

2.2. MpCML40 Localizes at the Plasma Membrane and in the Nucleus

MpCML40 was predicted to have a bipartite NLS by the cNLS Mapper online tool with a score of 6.7, which indicated that this protein might partially localize in the nucleus [39–41]. To verify the subcellular localization of this protein, MpCML40-GFP was expressed in tobacco (*Nicotiana benthamiana*) leaves by *Agrobacterium tumefaciens*-mediated transient expression system. The results showed that MpCML40 was indeed partially localized in the nucleus in tobacco epidermal cells, colocalized with free mCherry (Figure 3). Besides, MpCML40-GFP was also colocalized with the specific plasma membrane (PM) mCherry marker, with free GFP as a negative control (Figure 3). These results illustrated MpCML40 localized both at the plasma membrane and in the nucleus in tobacco epidermal cells.

2.3. MpCML40 Gene Is Highly Induced by Salt Stress in Pongamia Roots

The *AtCML37*, *AtCML38*, and *AtCML39* genes were all expressed in Arabidopsis root and highly accumulated after salt treatments [26]. To analyze the spatial and temporal expression pattern of *MpCML40* under salt stress, one-month-old Pongamia seedlings were subjected to qRT-PCR analyses. The relative expression level of *MpCML40* was significantly increased in roots and leaves after salt treatments (Figure 4). Especially in the root, *MpCML40* was strongly up-regulated at three and six hours after salt treatment (Figure 4). These results demonstrated that *MpCML40* was a salt-responsive gene.

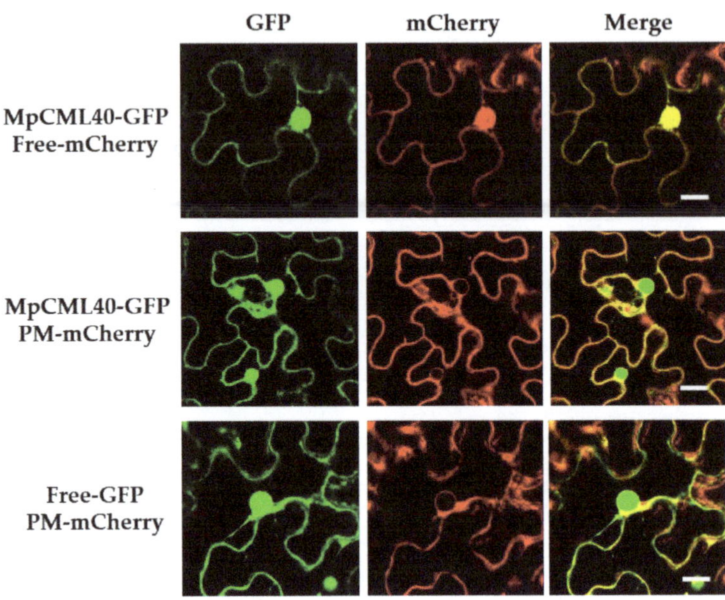

Figure 3. Subcellular localization of MpCML40-GFP. Subcellular localization of MpCML40-GFP was assayed with plasma membrane (PM-mcherry) marker or free mcherry in tobacco leaf epidermal cells. The fluorescence signals were detected 48 h after infiltration. Bar = 20 μM. The experiments were repeated two times with similar results.

Table 2. The list of primer sequences.

Primer	Primer Sequence (5′ → 3′)
MpActin.RtF	AGAGCAGTTCTTCAGTTGAG
MpActin.RtR	TCCTCCAATCCAGACACTAT
MpCML40.RtF	GCACTGGATTCTGATGGTGATGGG
MpCML40.RtR	GCTCTTTGGGGGTTATAAATCCGCA
MpCML40.3′GSP	TGGCAATTGAGGCACTGGATTCTGATGG
MpCML40.3′NGSP	ATGGAAGCAGGGGGAGAGGAACAA
MpCML40.5′GSP	TCCGCACCCTTGAGTGTCATACATCTCA
MpCML40.5′NGSP	CCTCCAAACTCAACAACCCATCACCATC
MpCML40.yeastF	CGCGGATCCATGATGAAGCATGCGGGTTTTGAGG
MpCML40.yeastR	CCGCTCGAGTCACTCCATCATAATTCTGAA
MpCML40.plantF	TATGGCGCGCCCATGATGAAGCATGCGGGTTTTGA
MpCML40.plantR	CTCCATCATAATTCTGAATTCTT
AtACT2.RtF	GACCTTTAACTCTCCCGCTATG
AtACT2.RtR	GAGACACACCATCACCAGAAT

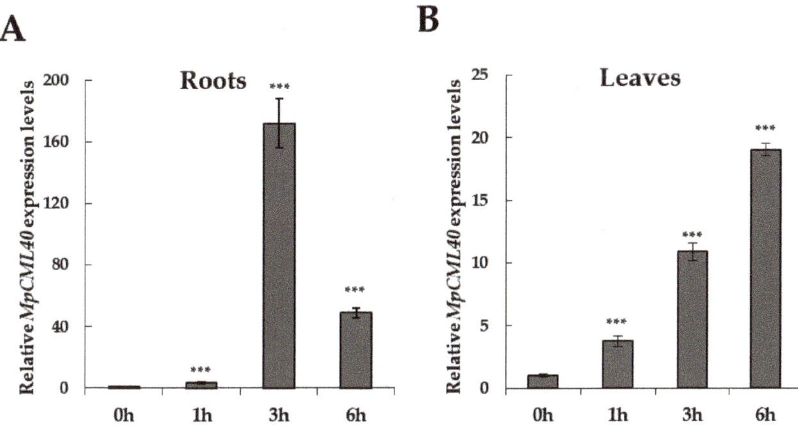

Figure 4. Gene expression changes of *MpCML40* in Pongamia roots and leaves upon salt stress. Relative expression levels of the *MpCML40* gene in roots (**A**) and leaves (**B**) after 500 mM NaCl treatment were analyzed by qRT-PCR using primers listed in Table 2. *MpActin* gene was used as an internal reference. Error bars show mean values (±SD) of three independent samples. *** $p < 0.001$ (Student's *t*-test).

2.4. The pYES22-MpCML40-Transformed Yeast Has Enhanced Salt Tolerance

We used yeast to preliminarily study the function of *MpCML40* in salt stress response. The *S. cerevisiae* strain W303 was transformed with either *pYES2-MpCML40* plasmid or *pYES2* empty vector. The positive colonies were transferred to the SD/-Ura agar plates with different concentrations of NaCl. The yeast strains showed resistance to low concentrations of NaCl. However, when the concentration increased to 1.5 M, the yeasts with *pYES2-MpCML40* plasmid grew faster than those with empty vectors (Figure 5), which suggested that *MpCML40* could substantially improve the salt tolerance of yeast.

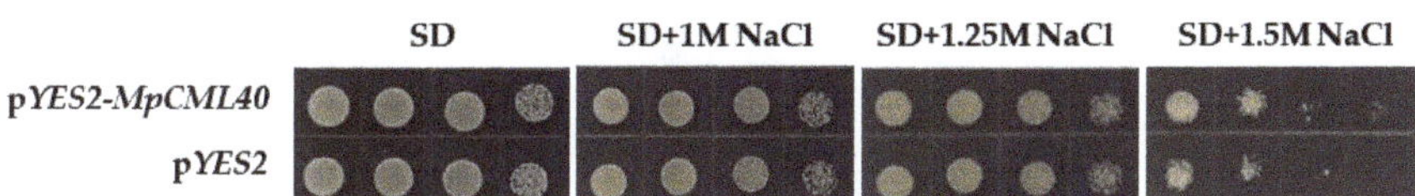

Figure 5. Salt tolerance of the *pYES2-MpCML40* transformed yeast. Series dilutions (1, 10, 100, 1000 folds) of the *pYES2-MpCML40* transformed yeast and *pYES2* (empty control) transformed yeast were grown on SD/-Ura agar plate containing different concentrations of NaCl (1 M, 1.25 M, 1.5 M).

2.5. Heterologous Expression of MpCML40 in Arabidopsis Strongly Enhances Salt and Osmotic Tolerance

To investigate the possible functions of *MpCML40* in salt stress response, we generated the transgenic Arabidopsis plants carrying *35S-Pro::MpCML40* (Figure 6A). Two independent transgenic lines were identified and validated by RT-PCR (Figure 6B). The growth and development phenotype of the transgenic plants were very similar to wild-type plants. We first checked the seeds; germination rate for the *35S::MpCML40* transgenic and wild-type Arabidopsis under salt stress. High concentrations (200 mM and 250 mM) of NaCl strongly inhibited seed germination of wild-type plants, whereas 200 mM of NaCl had no significant effects on the transgenic plants (Figure 6C,D). Moreover, the seeds of the transgenic plants showed obviously higher germination rate on the medium containing 250 mM of NaCl compared with wild-type plants (Figure 6C,D). We also assayed the seed germination rate under osmotic stress. Under low concentrations (200 mM and 300 mM) of sorbitol, the seeds of both wild-type and transgenic plants had high germination rates (Figure 6E,F).

However, the seed germination rate was significantly higher in transgenic plants than in wild-type plants under high concentration (400 mM) of sorbitol (Figure 6E,F).

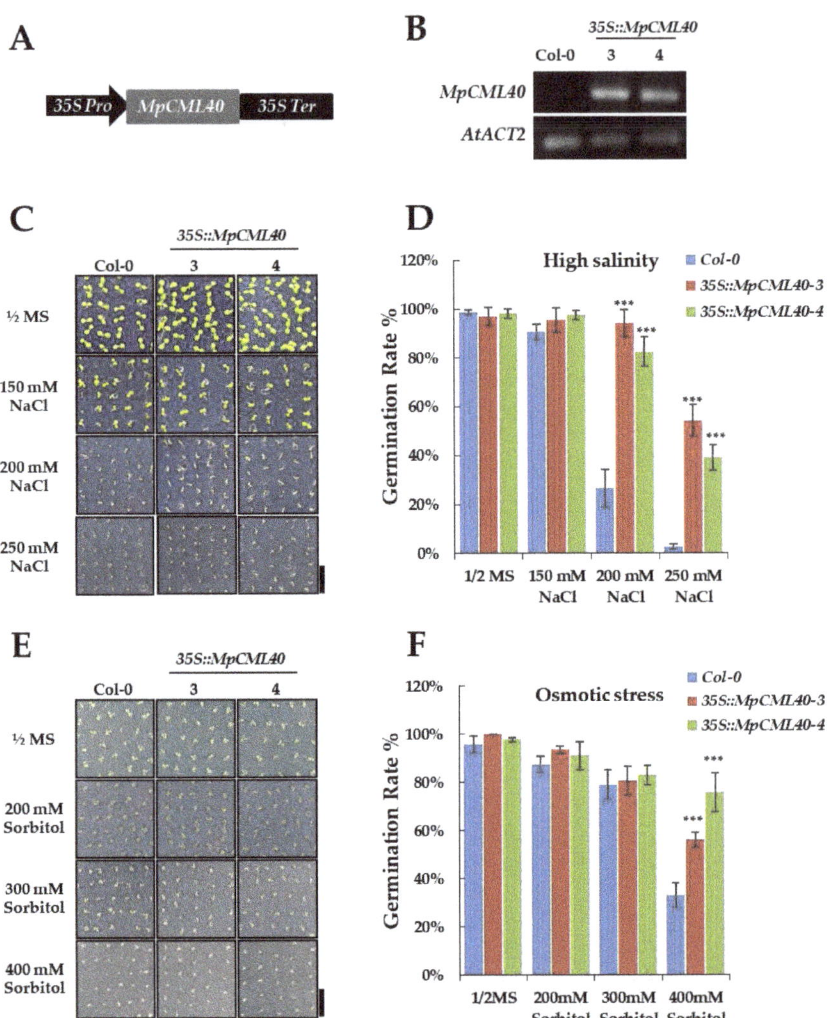

Figure 6. Effects of *MpCML40* on seed germination rate under salt and osmotic stress. (**A**) Schematic structure of the *MpCML40* expression construct. *35S Pro*, cauliflower mosaic virus 35S promoter. *35S Ter*, cauliflower mosaic virus 35S terminator. (**B**) Expression levels of the *MpCML40* gene in the wild-type and transgenic Arabidopsis plants were analyzed by RT-PCRs using the *MpCML40* full-length primers listed in Table 2. *AtACT2* (*ACTIN2*) was used as a control. (**C–F**) Typical phenotype and germination rate of wild-type and transgenic Arabidopsis seeds germinated on $\frac{1}{2}$ MS medium containing different concentrations of NaCl (**C,D**) and sorbitol (**E,F**). Bar = 1 cm. Error bars show mean values ($\pm$SD) of germination rate of three independent plates. *** $p < 0.001$ (Student's *t*-test).

Secondly, the root lengths of wild-type and transgenic seedlings were measured under salt and osmotic stress. The seedlings were germinated and grown for three days on a normal $\frac{1}{2}$ MS agar medium and then transferred to the medium containing different NaCl or sorbitol concentrations. After ten days of treatments, the root lengths of both wild-type and transgenic seedlings were inhibited (Figure 7A–C). However, the roots of the transgenic seedlings were significantly longer than those of wild-type seedlings under salt and osmotic stress (Figure 7B,C).

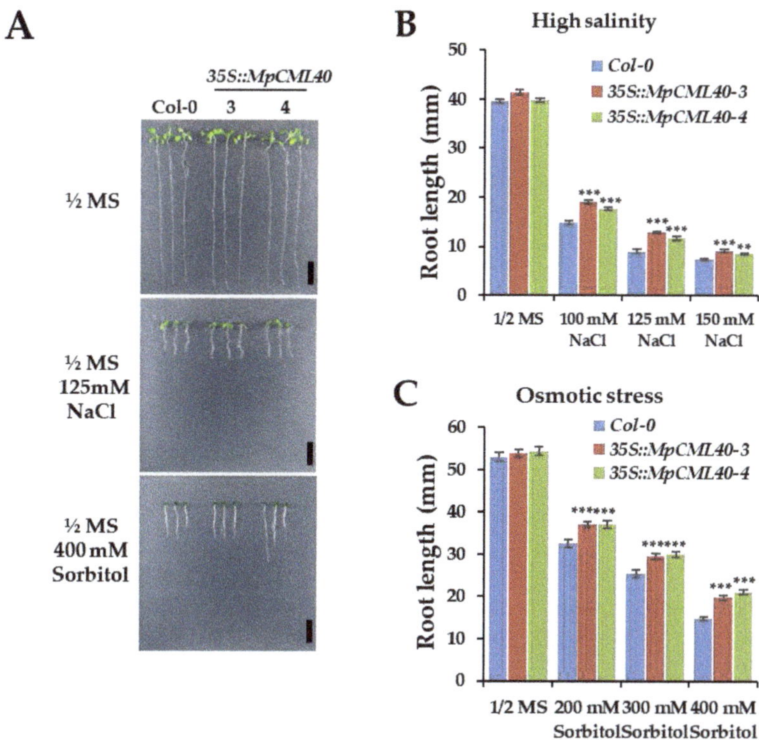

Figure 7. Effects of *MpCML40* on root growth under salt and osmotic stress. (**A**) Typical root length phenotype of two-week wild-type and *MpCML40* heterologously expressing Arabidopsis seedlings grown on $\frac{1}{2}$ MS medium containing 125 mM of NaCl and 400 mM of sorbitol. (**B,C**) The root length of two-week wild-type and transgenic Arabidopsis seedlings grown on $\frac{1}{2}$ MS medium containing different NaCl and sorbitol concentrations. Bar = 1 cm. Error bars show mean values ($\pm$SE) of germination rate from 50 independent seedlings. ** $p < 0.01$, *** $p < 0.001$ (Student's *t*-test).

Lastly, we measured the levels of two critical stress-associated metabolites, malondialdehyde (MDA) and proline, under mock or 200 mM NaCl conditions. Compared with wild-type plants, the transgenic ones showed a significantly higher level of proline and a slightly lower level of MDA (Figure 8A,B). To analyze the reactive oxygen species (ROS) induced by salt stress, nitroblue tetrazolium (NBT) and 3,3'-diaminobenzidine (DAB) staining were performed to detect the contents of H_2O_2 and O_2^- under mock or 200 mM NaCl condition. The leaves of transgenic plants showed lighter staining color than those of wild-type plants did (Figure 8C,D), indicating lower contents of H_2O_2 and O_2^- in transgenic plants under salt stress.

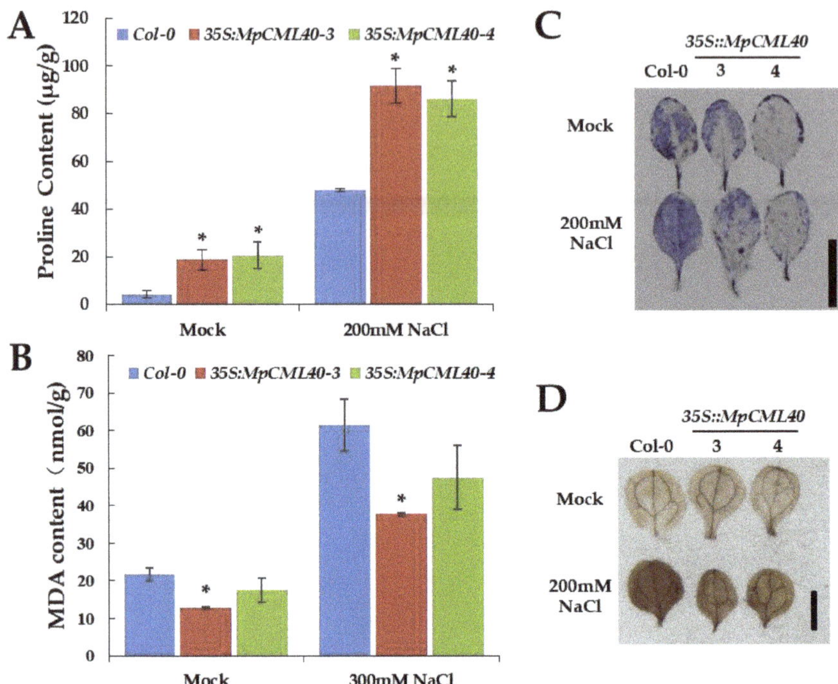

Figure 8. Effects of *MpCML40* heterologous expression on salt-stress-related metabolites. (**A**) Proline content of wild-type and *35S::MpCML40* Arabidopsis plants. (**B**) Micro Malondialdehyde (MDA) content of wild-type and *35S::MpCML40* plants. Error bars show mean values (±SD) of germination rate from three independent samples. * $p < 0.05$ (Student's *t*-test). (**C**) NBT staining of wild-type and *35S::MpCML40* four-week-old Arabidopsis rosette leaves. Bar = 1 cm. (**D**) DAB staining of wild-type and *35S::MpCML40* two-week-old Arabidopsis cotyledons grown on $\frac{1}{2}$ MS agar plates. Bar = 1 mm.

3. Discussion

As a semi-mangrove, Pongamia usually grows under high salinity conditions, which confer this species a unique genetic resource for exploring salt-responsive genes and investigating molecular mechanisms of plant salt tolerance. From 23,815 candidate salt-responsive genes formerly identified in Pongamia using Illumina sequencing [7], we identified a salt-induced CML gene, *MpCML40*, and characterized its function in the process of the salt stress response by heterologous expressing this gene in yeast and Arabidopsis. Our results showed that heterologous expression of *MpCML40* in yeast could improve its salt tolerance. The *35S::MpCML40* transgenic Arabidopsis did not show any growth and development phenotype compared with wild-type plants under normal growth conditions but had a higher germination rate and root length under salt stress. Our findings supported that *MpCML40* played a critical role in salt stress response.

3.1. MpCML40 Is a Salt-Induced CML Gene

CMLs are novel plant-specific Ca^{2+} sensors. There were 50 and 32 CML genes in Arabidopsis and rice, respectively [22,42]. CML genes were recently also identified in tomato, Medicago, grapevine, and apple [33,43–47]. *MpCML40* is the first CML gene identified in Pongamia. The phylogenetic tree showed that *MpCML40* was homologous with AtCML40 (Figure 1). Meanwhile, *MpCML40* was predicted to contain four EF-hand motifs, which were similar to AtCML38 and AtCML39 (Figure 2). Judging from the bipartite NLS, *MpCML40* might be present in the nucleus and also at the plasma membrane (Figures 2 and 3). CML genes in Arabidopsis, rice, and other plants were reported to be induced by several abiotic stresses. In our study, the expression level of *MpCML40* was highly enhanced upon salt stress, especially in roots at three and six hours after treatment (Figure 4). These results were consistent with the previous report that the leaves and roots of Pongamia had differential responses to salt stress while the roots were more efficient than the leaves [48].

3.2. MpCML40 Improves Salt Tolerance in Both Yeast and Arabidopsis

A NaCl-induced gene from Pongamia, *MpCHI*, was formerly reported to enhance salt tolerance of yeast [15]. Accordingly, we transformed *MpCML40* into yeast cells and assessed their salt responses. The wild-type yeast strain could grow in a high salt concentration medium, but the *MpCML40*-transformed yeasts grew obviously better than wild-type yeasts under 1.5 M of NaCl (Figure 5). Furthermore, we heterologously expressed *MpCML40* in Arabidopsis and conducted phenotype analysis. The *35S::MpCML40* transgenic plants had nearly normal germination rate under 200 mM NaCl condition. In contrast, the wild-type plants had significantly lower germination rates (Figure 6C,D). As NaCl concentration increased to 250 mM, nearly no seeds of wild-type plants germinated, while the seeds of transgenic plants still had a germination rate of about 50% (Figure 6C,D). In addition, the transgenic plants also showed a better germination rate than wild-type plants did under high concentrations of sorbitol (Figure 6E,F). Under both salt and osmotic stresses, the transgenic seedlings exhibited longer roots (Figure 7).

Halophytes are considered as one of the best germplasms for identifying salt-responsive genes, but only a few genes that could improve salt tolerance were isolated and characterized from mangroves, especially semi-mangroves [49]. Here, we found *MpCML40* could improve the salt tolerance of transgenic Arabidopsis. Moreover, to evaluate a genetic modification plant, it is essential to check whether or not the growth and development of the transgenic plant have been affected without stress conditions [2]. The *35S::MpCML40* transgenic plants did not show any visible growth and development retardation under normal conditions.

Proline is an important osmolyte for stabilizing macromolecules and membranes in the cell, and a higher level of proline can protect plant cells under salt and osmotic stresses [50–52]. Our results showed that the *35S::MpCML40* transgenic Arabidopsis had a higher level of proline even before salt treatment (Figure 8A), which revealed that MpCML40 could contribute to proline accumulation for salt tolerance. MDA is the main product of membrane lipid peroxidation under salt stress. Hence, the MDA level was regarded as an indicator of cell membrane damage [52]. The lower levels of MDA in the *MpCML40*-heterologous expressing plants supported the potential roles of *MpCML40* in salt stress response (Figure 8B). Besides, the light NBT and DAB staining color in transgenic plants revealed reduced contents of ROS under salt stress (Figure 8C,D). Taken together, our results uncovered that heterologous expression of *MpCML40* in Arabidopsis might contribute to proline accumulation to enhance salt tolerance of the transgenic plants and protect them from the ROS destructive effects.

4. Materials and Methods

4.1. Plant Materials and Growth Conditions

Arabidopsis thaliana ecotype Columbia (Col-0) was used as the genetic background of the *35S::MpCML40* transgenic plants. The transgenic plants were generated by agrobacterium-mediated floral dipping method and selected by BASTA (glufosinate ammonium) resistance. The T3 generation of homozygous plants were used for phenotype analysis. The plants were grown on soils and cultivated in the reach-in growth chamber with 16 h light at 22 °C and 8 h dark at 22 °C with approximately 120 $\mu mol \cdot m^{-2} \cdot s^{-1}$ of fluorescent white light. Arabidopsis seedlings were germinated and cultivated on $\frac{1}{2}$ MS agar plates ($\frac{1}{2} \times$ MS basal salts including 1% sucrose and 0.8% agar) with 16 h light at 22 °C and 8 h dark at 22 °C. Pongamia seeds were soaked in tap water at 28 °C in a growth cabinet until radicle appeared. These germinated seeds were then planted in soil for further growth.

4.2. Full-Length cDNA Cloning, Phylogenetic Analysis, and Motif Prediction

The highly conserved region of the unigene from the Pongamia transcriptome was used as a template for designing gene-specific internal primers for 5' and 3' RACE assay using SMARTer™ RACE cDNA Amplification Kit (Takara Bio, Madison, WI, USA). The total RNA was isolated from one-month-old Pongamia leaves. All primers used were listed in Table 2. The 5' and 3' ends of cDNA were sequenced and assembled into full-length cDNA. MpCML40 protein and 50 Arabidopsis CML proteins were used for phylogenetic analysis [22]. The phylogenetic tree was constructed using the Maximum Likelihood method implemented in the MEGA X program [53,54]. Alignment of MpCML40, AtCML37, AtCML38, AtCML39 and AtCML40 was conducted with Clustal Omega [55]. The conserved EF-hand motifs were analyzed by SMART (Simple Modular Architecture Research Tool) [56,57]. The NLS was searched by cNLS Mapper [39,40].

4.3. Subcellular Localization

PIP2A (plasma membrane intrinsic protein 2A) was used as the mCherry tagged plasma membrane marker (PM-mCherry). The agrobacterial strain GV3101 containing PM marker-mCherry or free mCherry constructs was used at OD600 = 0.5, and GV3101 containing MpCML40-GFP or free-GFP constructs was used at OD600 = 0.25. The four-week-old *N. benthamiana* leaves were used for agrobacteria-mediated transient expression. The images were taken at 48 h after infiltration using LEICA SP8 STED 3X fluorescence microscope confocal system.

4.4. Quantitative Real-Time PCR

One-month-old Pongamia seedlings were transferred from soil into $\frac{1}{2}$ MS liquid medium. After overnight culture, the normal $\frac{1}{2}$ MS liquid medium was replaced by $\frac{1}{2}$ MS liquid medium containing 500 mM NaCl. The seedling samples were collected at 0, 1, 3, 6 h after treatments for total RNA extraction using TRIzol™ Plus (Takara Bio, Madison, Wisconsin, USA) following the manufacturer's protocol. About 1000 ng of total RNA was digested by DNase I for 30 min at 37 °C before reverse transcription. DNase digestion was terminated by addition of 25 mM EDTA and followed by incubation at 70 °C for 10 min. First strand cDNA synthesis was performed using an oligo(dT) 18 primer and the GoScript™ Reverse Transcriptase (Promega, Madison, Wisconsin, USA). Subsequently, qRT-PCR was performed on Roche LightCyler 480 with gene-specific primers and SYBR Green (Life Technologies, Rockville, Maryland, USA). All primers used in qRT-PCR were listed in Table 2.

4.5. Yeast Transformation and Growth Assay

The yeast strain W303 was used for growth assay. The yeast cells were cultured at 30 °C with shaking at 230 rpm and collected at the OD600 0.4–0.6. The yeast cells were centrifuged at 1000 rpm for 5 min at room temperature. The supernatants were discarded, and the cell pellets were suspended in sterile water. The cells were pooled into one tube (final volume 25–50 mL) and centrifuged at 1000 rpm for 5 min at room temperature. The cell pellets were suspended in 1.5 mL freshly prepared and sterile 1× TE/1× LiAc solutions (10 mM Tris-HCl; 1 mM EDTA; pH 7.5; 100 mM LiAc). Next, 0.5 µg plasmid DNA and 0.1 mg salmon sperm carrier DNA were added into 100 µL competent cells in a fresh 1.5 mL tube and mixed by vortexing. The 600 µL freshly prepared and sterile PEG/LiAc solutions (40% PEG 4000; 10 mM Tris-HCl; 1 mM EDTA; pH 7.5; 100 mM LiAc) were added to each tube and vortexed at high speed for 10 s. The mixture was incubated at 30 °C for 30 min with shaking at 200 rpm. Then, 70 µL DMSO was added and mixed well by gentle inversion (Do not vortex). The mixture was heat-shocked for 15 min in a 42 °C water bath and then transferred on ice for 1–2 min. The cells were centrifuged for 30 s at 6000 rpm at room temperature. The supernatants were removed, and the cell pellets were re-suspended in 0.5 mL sterile 1× TE buffer (10 mM Tris-HCl; 1 mM EDTA; pH 7.5). Later, 100–500 µL of suspended cells were spread on each SD/-Ura selection agar plate and incubated at 30 °C until colonies appeared. The positive colonies were cultured overnight and then diluted to 1, 10, 100, 1000 folds. 5 µL of diluted yeast cells were transferred to SD/-Ura agar plate containing different concentrations of NaCl and incubated at 30 °C until colonies appeared.

4.6. Phenotype Analysis of Wild-Type and 35S::MpCML40 Transgenic Arabidopsis Plants

For germination rate assays, at least 100 seeds of wild-type and transgenic plants were sowed on $\frac{1}{2}$ MS medium containing different concentrations of NaCl (150, 200, and 250 mM) and sorbitol (200, 300, and 400 mM). After three days of vernalization at 4 °C in the dark, the seeds were transferred to light for the assessments of germination rates. A seed was considered as germinated when the radical protruded through its envelope.

For root length assay, the seedlings were germinated and grown for three days on normal $\frac{1}{2}$ MS agar medium and then transferred to the medium containing different concentrations of NaCl (100, 125, 150 mM) and sorbitol (200, 300, 400 mM). The root length of at least 50 seedlings was measured by ImageJ software after 10 days of treatments.

4.7. Proline Content Measurement

Proline content was measured using Proline (PRO) Content Assay Kit (BC0290, Solarbio, Beijing, China) following the manufacturer's protocol. Briefly, 10-day-old seedlings from $\frac{1}{2}$ MS agar plates were transferred into $\frac{1}{2}$ MS liquid medium at 12 h before treatments and then treated overnight with 200 mM of NaCl. 100 mg of seedlings were weighted and homogenized with 1 mL of extraction buffer in mortar on ice. Then, the extraction procedure was followed the manufacturer's protocol. Proline content was determined by reading the optical density of the sample at 520 nm using a spectrometer.

4.8. MDA Content Measurement

MDA content was measured using Micro Malondialdehyde (MDA) Assay Kit (BC0020, Solarbio, Beijing, China) following the manufacturer's protocol. Briefly, four-week-old plants in soil were treated with 300 mM of NaCl for one week. 100 mg of seedlings were weighted and homogenized with 1 mL of extraction buffer in mortar on ice. Then, the extraction procedure was followed the manufacturer's protocol. MDA content was determined by reading the optical density of the sample at 600 nm, 532 nm, 450 nm using a spectrometer.

4.9. DAB and NBT Staining

Two-week-old Arabidopsis cotyledons grown on $\frac{1}{2}$ MS agar plate were used for DAB staining, and four-week-old rosette leaves were used for NBT staining. The leaves were vacuumed in $\frac{1}{2}$ MS liquid medium containing 200 mM of NaCl for 5 min and then soaked for another 4 h. Staining was performed by vacuuming the leaves in 1 mg/mL DAB solution or 0.2% NBT solution for 5 min and then staining for another 4 h.

5. Conclusions

In conclusion, we isolated and functionally characterized a gene encoding a calmodulin-like protein, *MpCML40*, from Pongamia. The *35S::MpCML40* transgenic Arabidopsis accumulated high levels of prolines and was more tolerant to salt and osmotic stress than wild-type Arabidopsis, suggesting that MpCML40 was a positive regulator in response to salt stress. Importantly, the transgenic plants grew and developed as well as wild-type plants under normal conditions. Our findings improved the understandings of salt-responsive mechanisms in Pongamia and also provided a potential candidate for crop breeding by genetic modification.

Author Contributions: Conceptualization, J.H. and S.D.; Data curation, Y.Z. and Y.L.; Formal analysis, Y.Z., J.H. and S.D.; Funding acquisition, J.W.; Investigation, Y.Z., J.H., Q.H. and Y.L.; Resources, J.H. and J.W.; Supervision, S.D.; Writing—original draft, Y.Z.; Writing—review & editing, J.H. and S.D. All authors have read and agreed to the published version of the manuscript.

Funding: This research was funded by NSFC-Guangdong Province Union Funds (U1701246), the Guangdong Science and Technology Program (2019B121201005) the Key Research Program of the Chinese Academy of Sciences Grant (KGFZD-135-19-08) and the Shenzhen Fundamental Research Fund (JCYJ20170818142241972).

Institutional Review Board Statement: Not applicable.

Informed Consent Statement: Not applicable.

Data Availability Statement: MpCML40 sequence data is available in GenBank.

Conflicts of Interest: The authors declare no conflict of interest.

References

1. Munns, R.; Tester, M. Mechanisms of salinity tolerance. *Annu. Rev. Plant Biol.* **2008**, *59*, 651–681. [CrossRef]
2. Flowers, T.J. Improving crop salt tolerance. *J. Exp. Bot.* **2004**, *55*, 307–319. [CrossRef]
3. Wang, L.; Mu, M.; Li, X.; Lin, P.; Wang, W. Differentiation between true mangroves and mangrove associates based on leaf traits and salt contents. *J. Plant Ecol.* **2010**, *4*, 292–301. [CrossRef]
4. Choudhury, R.R.; Basak, S.; Ramesh, A.M.; Rangan, L. Nuclear DNA content of Pongamia pinnata L. and genome size stability of in vitro-regenerated plantlets. *Protoplasma* **2014**, *251*, 703–709. [CrossRef]
5. Kaul, S.; Koo, H.L.; Jenkins, J.; Rizzo, M.; Rooney, T.; Tallon, L.J.; Feldblyum, T.; Nierman, W.; Benito, M.I.; Lin, X.; et al. Analysis of the genome sequence of the flowering plant Arabidopsis thaliana. *Nature* **2000**, *408*, 796–815.
6. Wegrzyn, J.L.; Whalen, J.; Kinlaw, C.S.; Harry, D.E.; Puryear, J.; Loopstra, C.A.; Gonzalez-Ibeas, D.; Vasquez-Gross, H.A.; Famula, R.A.; Neale, D.B. Transcriptomic profile of leaf tissue from the leguminous tree, *Millettia pinnata*. *Tree Genet. Genomes* **2016**, *12*, 44. [CrossRef]
7. Huang, J.; Lu, X.; Yan, H.; Chen, S.; Zhang, W.; Huang, R.; Zheng, Y. Transcriptome characterization and sequencing-based identification of salt-responsive genes in Millettia pinnata, a semi-mangrove plant. *DNA Res.* **2012**, *19*, 195–207. [CrossRef] [PubMed]
8. Sreeharsha, R.V.; Mudalkar, S.; Singha, K.T.; Reddy, A.R. Unravelling molecular mechanisms from floral initiation to lipid biosynthesis in a promising biofuel tree species, Pongamia pinnata using transcriptome analysis. *Sci. Rep.* **2016**, *6*, 34315. [CrossRef]
9. Liu, S.J.; Song, S.H.; Wang, W.Q.; Song, S.Q. De novo assembly and characterization of germinating lettuce seed transcriptome using Illumina paired-end sequencing. *Plant Physiol. Biochem.* **2015**, *96*, 154–162. [CrossRef]
10. Huang, J.; Hao, X.; Jin, Y.; Guo, X.; Shao, Q.; Kumar, K.S.; Ahlawat, Y.K.; Harry, D.E.; Joshi, C.P.; Zheng, Y. Temporal transcriptome profiling of developing seeds reveals a concerted gene regulation in relation to oil accumulation in Pongamia (*Millettia pinnata*). *BMC Plant Biol.* **2018**, *18*, 140. [CrossRef] [PubMed]
11. Huang, J.Z.; Guo, X.H.; Hao, X.H.; Zhang, W.K.; Chen, S.Y.; Huang, R.F.; Gresshoff, P.M.; Zheng, Y.Z. De novo sequencing and characterization of seed transcriptome of the tree legume *Millettia pinnata* for gene discovery and SSR marker development. *Mol. Breed.* **2016**, *36*, 75. [CrossRef]
12. Winarto, H.P.; Liew, L.C.; Gresshoff, P.M.; Scott, P.T.; Singh, M.B.; Bhalla, P.L. Isolation and Characterization of Circadian Clock Genes in the Biofuel Plant Pongamia (*Millettia pinnata*). *Bioenergy Res.* **2015**, *8*, 760–774. [CrossRef]
13. Ramesh, A.M.; Kesari, V.; Rangan, L. Characterization of a stearoyl-acyl carrier protein desaturase gene from potential biofuel plant, *Pongamia pinnata* L. *Gene* **2014**, *542*, 113–121. [CrossRef] [PubMed]
14. Moolam, R.A.; Singh, A.; Shelke, R.G.; Scott, P.T.; Gresshoff, P.M.; Rangan, L. Identification of two genes encoding microsomal oleate desaturases (FAD2) from the biodiesel plant *Pongamia pinnata* L. *Trees Struct. Funct.* **2016**, *30*, 1351–1360. [CrossRef]
15. Wang, H.; Hu, T.; Huang, J.; Lu, X.; Huang, B.; Zheng, Y. The expression of Millettia pinnata chalcone isomerase in Saccharomyces cerevisiae salt-sensitive mutants enhances salt-tolerance. *Int. J. Mol. Sci.* **2013**, *14*, 8775–8786. [CrossRef]
16. Zielinski, R.E. Calmodulin and Calmodulin-Binding Proteins in Plants. *Annu. Rev. Plant Physiol. Plant Mol. Biol.* **1998**, *49*, 697–725. [CrossRef]
17. DeFalco, T.A.; Bender, K.W.; Snedden, W.A. Breaking the code: Ca^{2+} sensors in plant signalling. *Biochem. J.* **2010**, *425*, 27–40. [CrossRef] [PubMed]
18. Bender, K.W.; Snedden, W.A. Calmodulin-Related Proteins Step Out from the Shadow of Their Namesake. *Plant Physiol.* **2013**, *163*, 486–495. [CrossRef]
19. Beckmann, L.; Edel, K.H.; Batistic, O.; Kudla, J. A calcium sensor—Protein kinase signaling module diversified in plants and is retained in all lineages of Bikonta species. *Sci. Rep.* **2016**, *6*, 31645. [CrossRef]
20. Kretsinger, R.H.; Nockolds, C.E. Carp muscle calcium-binding protein. II. Structure determination and general description. *J. Biol. Chem.* **1973**, *248*, 3313–3326. [CrossRef]
21. Lewit-Bentley, A.; Rety, S. EF-hand calcium-binding proteins. *Curr. Opin. Struct. Biol.* **2000**, *10*, 637–643. [CrossRef]
22. McCormack, E.; Braam, J. Calmodulins and related potential calcium sensors of Arabidopsis. *New Phytol.* **2003**, *159*, 585–598. [CrossRef]

23. Perochon, A.; Aldon, D.; Galaud, J.P.; Ranty, B. Calmodulin and calmodulin-like proteins in plant calcium signaling. *Biochimie* **2011**, *93*, 2048–2053. [CrossRef]

24. Zeng, H.; Xu, L.; Singh, A.; Wang, H.; Du, L.; Poovaiah, B.W. Involvement of calmodulin and calmodulin-like proteins in plant responses to abiotic stresses. *Front. Plant Sci.* **2015**, *6*, 600. [CrossRef] [PubMed]

25. Braam, J. Regulated expression of the calmodulin-related TCH genes in cultured Arabidopsis cells: Induction by calcium and heat shock. *Proc. Natl. Acad. Sci. USA* **1992**, *89*, 3213–3216. [CrossRef]

26. Vanderbeld, B.; Snedden, W.A. Developmental and stimulus-induced expression patterns of Arabidopsis calmodulin-like genes CML37, CML38 and CML39. *Plant Mol. Biol.* **2007**, *64*, 683–697. [CrossRef] [PubMed]

27. Yamaguchi, T.; Aharon, G.S.; Sottosanto, J.B.; Blumwald, E. Vacuolar Na+/H+ antiporter cation selectivity is regulated by calmodulin from within the vacuole in a Ca2+- and pH-dependent manner. *Proc. Natl. Acad. Sci. USA* **2005**, *102*, 16107–16112. [CrossRef]

28. Magnan, F.; Ranty, B.; Charpenteau, M.; Sotta, B.; Galaud, J.P.; Aldon, D. Mutations in AtCML9, a calmodulin-like protein from Arabidopsis thaliana, alter plant responses to abiotic stress and abscisic acid. *Plant J.* **2008**, *56*, 575–589. [CrossRef] [PubMed]

29. Delk, N.A.; Johnson, K.A.; Chowdhury, N.I.; Braam, J. CML24, regulated in expression by diverse stimuli, encodes a potential Ca^{2+} sensor that functions in responses to abscisic acid, daylength, and ion stress. *Plant Physiol.* **2005**, *139*, 240–253. [CrossRef]

30. Wu, X.; Qiao, Z.; Liu, H.; Acharya, B.R.; Li, C.; Zhang, W. CML20, an Arabidopsis Calmodulin-like Protein, Negatively Regulates Guard Cell ABA Signaling and Drought Stress Tolerance. *Front. Plant Sci.* **2017**, *8*, 824. [CrossRef] [PubMed]

31. Vadassery, J.; Reichelt, M.; Hause, B.; Gershenzon, J.; Boland, W.; Mithofer, A. CML42-mediated calcium signaling coordinates responses to Spodoptera herbivory and abiotic stresses in Arabidopsis. *Plant Physiol.* **2012**, *159*, 1159–1175. [CrossRef]

32. Xu, G.Y.; Rocha, P.S.; Wang, M.L.; Xu, M.L.; Cui, Y.C.; Li, L.Y.; Zhu, Y.X.; Xia, X. A novel rice calmodulin-like gene, OsMSR2, enhances drought and salt tolerance and increases ABA sensitivity in Arabidopsis. *Planta* **2011**, *234*, 47–59. [CrossRef] [PubMed]

33. Dubrovina, A.S.; Aleynova, O.A.; Ogneva, Z.V.; Suprun, A.R.; Ananev, A.A.; Kiselev, K.V. The Effect of Abiotic Stress Conditions on Expression of Calmodulin (CaM) and Calmodulin-Like (CML) Genes in Wild-Growing Grapevine *Vitis amurensis*. *Plants* **2019**, *8*, 602. [CrossRef] [PubMed]

34. Chen, C.; Sun, X.; Duanmu, H.; Zhu, D.; Yu, Y.; Cao, L.; Liu, A.; Jia, B.; Xiao, J.; Zhu, Y. *GsCML27*, a Gene Encoding a Calcium-Binding Ef-Hand Protein from *Glycine soja*, Plays Differential Roles in Plant Responses to Bicarbonate, Salt and Osmotic Stresses. *PLoS ONE* **2015**, *10*, e0141888. [CrossRef]

35. Zhang, X.; Wang, T.; Liu, M.; Sun, W.; Zhang, W.-H. Calmodulin-like gene *MtCML40* is involved in salt tolerance by regulating MtHKTs transporters in *Medicago truncatula*. *Environ. Exp. Bot.* **2019**, *157*, 79–90. [CrossRef]

36. Ma, Q.; Zhou, Q.; Chen, C.; Cui, Q.; Zhao, Y.; Wang, K.; Arkorful, E.; Chen, X.; Sun, K.; Li, X. Isolation and expression analysis of CsCML genes in response to abiotic stresses in the tea plant (*Camellia sinensis*). *Sci. Rep.* **2019**, *9*, 8211. [CrossRef]

37. Rigden, D.J.; Galperin, M.Y. The DxDxDG motif for calcium binding: Multiple structural contexts and implications for evolution. *J. Mol. Biol.* **2004**, *343*, 971–984. [CrossRef]

38. Zhou, Y.; Xue, S.; Yang, J.J. Calciomics: Integrative studies of Ca2+-binding proteins and their interactomes in biological systems. *Metallomics* **2013**, *5*, 29–42. [CrossRef] [PubMed]

39. Kosugi, S.; Hasebe, M.; Tomita, M.; Yanagawa, H. Systematic identification of cell cycle-dependent yeast nucleocytoplasmic shuttling proteins by prediction of composite motifs. *Proc. Natl. Acad. Sci. USA* **2009**, *106*, 10171–10176. [CrossRef] [PubMed]

40. Kosugi, S.; Hasebe, M.; Matsumura, N.; Takashima, H.; Miyamoto-Sato, E.; Tomita, M.; Yanagawa, H. Six classes of nuclear localization signals specific to different binding grooves of importin alpha. *J. Biol. Chem.* **2009**, *284*, 478–485. [CrossRef] [PubMed]

41. Kosugi, S.; Hasebe, M.; Entani, T.; Takayama, S.; Tomita, M.; Yanagawa, H. Design of peptide inhibitors for the importin alpha/beta nuclear import pathway by activity-based profiling. *Chem. Biol.* **2008**, *15*, 940–949. [CrossRef]

42. Boonburapong, B.; Buaboocha, T. Genome-wide identification and analyses of the rice calmodulin and related potential calcium sensor proteins. *BMC Plant Biol.* **2007**, *7*, 4. [CrossRef] [PubMed]

43. Sun, Q.; Yu, S.; Guo, Z. Calmodulin-Like (CML) Gene Family in Medicago truncatula: Genome-Wide Identification, Characterization and Expression Analysis. *Int. J. Mol. Sci.* **2020**, *21*, 7142. [CrossRef] [PubMed]

44. Munir, S.; Khan, M.R.; Song, J.; Munir, S.; Zhang, Y.; Ye, Z.; Wang, T. Genome-wide identification, characterization and expression analysis of calmodulin-like (CML) proteins in tomato (*Solanum lycopersicum*). *Plant Physiol. Biochem.* **2016**, *102*, 167–179. [CrossRef] [PubMed]

45. Shi, J.; Du, X. Identification, characterization and expression analysis of calmodulin and calmodulin-like proteins in *Solanum pennellii*. *Sci. Rep.* **2020**, *10*, 7474. [CrossRef]

46. Zhang, Q.; Liu, X.; Liu, X.; Wang, J.; Yu, J.; Hu, D.; Hao, Y. Genome-Wide Identification, Characterization, and Expression Analysis of Calmodulin-Like Proteins (CMLs) in Apple. *Hortic. Plant J.* **2017**, *3*, 219–231. [CrossRef]

47. Zhu, X.; Dunand, C.; Snedden, W.; Galaud, J.P. CaM and CML emergence in the green lineage. *Trends Plant Sci.* **2015**, *20*, 483–489. [CrossRef]

48. Marriboina, S.; Sengupta, D.; Kumar, S.; Reddy, A.R. Physiological and molecular insights into the high salinity tolerance of *Pongamia pinnata* (L.) pierre, a potential biofuel tree species. *Plant Sci.* **2017**, *258*, 102–111. [CrossRef]

49. Mishra, A.; Tanna, B. Halophytes: Potential resources for salt stress tolerance genes and promoters. *Front. Plant Sci.* **2017**, *8*, 829. [CrossRef]

50. Wang, W.; Vinocur, B.; Altman, A. Plant responses to drought, salinity and extreme temperatures: Towards genetic engineering for stress tolerance. *Planta* **2003**, *218*, 1–14. [CrossRef]
51. Mahajan, S.; Tuteja, N. Cold, salinity and drought stresses: An overview. *Arch. Biochem. Biophys.* **2005**, *444*, 139–158. [CrossRef] [PubMed]
52. Liang, W.; Ma, X.; Wan, P.; Liu, L. Plant salt-tolerance mechanism: A review. *Biochem. Biophys. Res. Commun.* **2018**, *495*, 286–291. [CrossRef] [PubMed]
53. Jones, D.T.; Taylor, W.R.; Thornton, J.M. The rapid generation of mutation data matrices from protein sequences. *Comput. Appl. Biosci.* **1992**, *8*, 275–282. [CrossRef] [PubMed]
54. Kumar, S.; Stecher, G.; Li, M.; Knyaz, C.; Tamura, K. MEGA X: Molecular Evolutionary Genetics Analysis across Computing Platforms. *Mol. Biol. Evol.* **2018**, *35*, 1547–1549. [CrossRef] [PubMed]
55. Madeira, F.; Park, Y.M.; Lee, J.; Buso, N.; Gur, T.; Madhusoodanan, N.; Basutkar, P.; Tivey, A.R.N.; Potter, S.C.; Finn, R.D.; et al. The EMBL-EBI search and sequence analysis tools APIs in 2019. *Nucleic Acids Res.* **2019**, *47*, W636–W641. [CrossRef]
56. Letunic, I.; Khedkar, S.; Bork, P. SMART: Recent updates, new developments and status in 2020. *Nucleic Acids Res.* **2020**, *49*, D458–D460. [CrossRef]
57. Letunic, I.; Bork, P. 20 years of the SMART protein domain annotation resource. *Nucleic Acids Res.* **2018**, *46*, D493–D496. [CrossRef]

International Journal of
Molecular Sciences

Review

Maintenance of Cell Wall Integrity under High Salinity

Jianwei Liu [1], Wei Zhang [1,2], Shujie Long [1,2] and Chunzhao Zhao [1,*]

[1]   Shanghai Center for Plant Stress Biology, CAS Center for Excellence in Molecular Plant Sciences,
      Chinese Academy of Sciences, Shanghai 200032, China; jwliu@psc.ac.cn (J.L.); weizhang@psc.ac.cn (W.Z.);
      shjlong@psc.ac.cn (S.L.)
[2]   University of the Chinese Academy of Sciences, Beijing 100049, China
*   Correspondence: czzhao@psc.ac.cn; Tel.: +86-021-5707-8274

Abstract: Cell wall biosynthesis is a complex biological process in plants. In the rapidly growing cells or in the plants that encounter a variety of environmental stresses, the compositions and the structure of cell wall can be dynamically changed. To constantly monitor cell wall status, plants have evolved cell wall integrity (CWI) maintenance system, which allows rapid cell growth and improved adaptation of plants to adverse environmental conditions without the perturbation of cell wall organization. Salt stress is one of the abiotic stresses that can severely disrupt CWI, and studies have shown that the ability of plants to sense and maintain CWI is important for salt tolerance. In this review, we highlight the roles of CWI in salt tolerance and the mechanisms underlying the maintenance of CWI under salt stress. The unsolved questions regarding the association between the CWI and salt tolerance are discussed.

Keywords: cell wall integrity; cell wall sensor; salt stress; salt tolerance; LRXs; *CrRLK1Ls*

check for
updates

Citation: Liu, J.; Zhang, W.; Long, S.; Zhao, C. Maintenance of Cell Wall Integrity under High Salinity. *Int. J. Mol. Sci.* **2021**, 22, 3260. https://doi.org/10.3390/ijms22063260

Academic Editor: Raffaella Maria Balestrini

Received: 27 February 2021
Accepted: 19 March 2021
Published: 23 March 2021

Publisher's Note: MDPI stays neutral with regard to jurisdictional claims in published maps and institutional affiliations.

1. Introduction

High salinity is an adverse environmental stress that severely affects the growth and yield of crops. Excessive accumulation of sodium in plants confers both ion toxicity and osmotic stress, which in turn dramatically affect the morphological, physiological, biochemical, and metabolic status of plants [1]. Currently, more than 20% of the irrigated lands in the world are threatened by high salinity, and the area of saline soils is increasing gradually every year accompanied by the global climate change and poor irrigation practices [2–4]. It is expected the global population will reach to nearly 10 billion in 2050, and to meet the increasing food demand in future, the utilization of saline soils to grow major crops tends to be inevitable. Therefore, the cultivation of crops with increased salt tolerance is a major objective in salt stress community.

To avoid the damage caused by excessive salts in soil, plants have evolved various strategies to overcome the problems caused by high salinity. Ion homeostasis, osmotic adjustment, ROS balance, and metabolic adjustment are the major factors that are associated with the tolerance of plants to salt stress. Based on the capacity of plants to adapt to salt stress, plants can be classified into glycophytes and halophytes. Our major crops, such as rice, maize, and wheat, are glycophytes that are unable to complete their life cycle when they are being exposed to high salinity. Halophytes, however, have developed various strategies to adapt to the environments with a high concentration of sodium. For example, halophytes are able to extrude salts via glands or store excessive Na^+ in the vacuoles of epidermal bladder cells [5,6].

More and more studies point out that maintenance CWI is also critical for the adaptation of plants to high salinity. Plant cell walls, which mainly consist of polysaccharides and structural proteins, are essential for the establishment of plant morphology and protection of plants against adverse environmental changes [7]. During plant growth and development or in response to environmental stresses, the cell wall compositions and structures are dynamically modulated, allowing rapid cell elongation and increased stress tolerance [8]. To maintain CWI during the reorganization of cell wall, plants need to constantly monitor the chemical and mechanical properties of the cell walls and also need to process an ability to repair cell wall once they are seriously disrupted. It has been shown that CWI maintenance mechanism exists in plants and is essential for the regulation of growth and development and in response to stress conditions [9,10]. The progresses about CWI sensing and maintenance system in plants have been summarized in several outstanding review papers [8,11,12]. In this review, we focus on the elucidation of the associations between CWI and salt tolerance in plants.

2. Importance of Cell Wall Biosynthesis in Salt Tolerance

The plant cell wall is a dynamic network composed of cellulose, hemicellulose, pectin, lignin, and multiple types of structural proteins [13,14]. Moreover, cell wall-remodeling enzymes, various ions, and reactive oxygen species (ROS) also exist in the apoplast and are involved in the regulation of CWI. Upon exposure to high salinity, several changes in the cell wall have been identified, including the reduction of cellulose content [15,16], disruption of the cross-linking of pectins [9], and accumulation of lignin [17]. Studies have shown that the plants that are defective in cell wall biosynthesis are hypersensitive to salt stress, suggesting that maintenance of CWI is important for the adaptation of plants to high salinity.

2.1. Cellulose

Cellulose is the most abundant organic component in the cell wall of terrestrial vascular plants. Cellulose micro-fibrils are composed of β-1,4-linked glucan chains, which are synthesized at the cell surface by cellulose synthase (CesA) complexes (CSCs) [18,19]. Each CSC is assembled into a hexameric rosette structure, harboring CesA catalytic subunits and several accessory proteins. In *Arabidopsis*, there are ten CesA proteins [18]. It is well known that CesA1, CesA3, and CesA6 are assembled in a CSC to synthesize cellulose in the primary cell wall, while CesA4, CesA7, and CesA8 are mainly involved in the synthesis of cellulose in the secondary cell wall [20]. Experimental data have shown that the cellulose contents are significantly reduced after salt treatment and the plants with a loss of function of CESA1 and CESA6 gene display reduced root elongation and severe root tip swelling under salt stress, indicating that cellulose biosynthesis is important for salt tolerance in plants [21,22]. Clear evidences have indicated that the CSCs are dissociated from plasma membrane within 30 min after exposure to high salinity. However, during the growth recovery phase after salt treatment, the CSCs can be reassembled at the plasma membrane to synthesize new cellulose, and the capacity to reassemble CSCs during the growth recovery stage is critical for plants to maintain root and hypocotyl growth under salt stress [16].

Apart from the CesAs, several cellulose biosynthesis-related proteins have also been reported involved in salt tolerance. For example, KORRIGAN1 (KOR1), a putative endo-1,4-β-D-glucanase, is an integral part of the primary cell wall CSC and is required for root elongation under salt stress [22,23]. Cellulose synthase interacting protein 1 (CSI1) and companion of cellulose synthase 1 (CC1 and CC2) proteins, acting as companions of CesAs, are both required for cellulose biosynthesis [16,21]. Mutations in *CSI1* or *CC1* and *CC2* lead to reduced root or hypocotyl elongation under salt stress. *CTL1* encodes a chitinase-like protein that participates in the deposition of the ordered cellulose, and mutation of this gene results in increased sensitivity to high salinity [24] (Table 1).

Table 1. List of the cell wall biosynthesis-related genes that are involved in salt stress response.

Name	Gene ID	Annotation	Function	Reference(s)
AtCesA1/RSW1	At4g32410	Cellulose synthase catalytic subunit	Cellulose synthesis in the primary cell wall	[22]
AtCesA8/IRX1	At4g18780		Cellulose synthesis in the secondary cell wall	[25]
AtCC1	At1g45688	Cellulose synthase companion protein	Cortical microtubules assembly and cellulose biosynthesis under salt stress	[16]
AtCC2	At5g42860			
AtCTL/POM1	At1g05850	Chitinase-like protein 1	Involved in the assembly of glucan chains	[24,26]
AtCSI1/POM2	At2g22125	Cellulose synthase-interactive protein 1	Companion of CesAs; required for cell elongation in root	[21]
AtCCoAOMT1	At4g34050	Caffeoyl-CoA 3-*O*-methyltransferase	Involved in lignin synthesis	[17]
AtKOR/RSW2	At5g49720	Endo-β-1,4-glucanase	Integral component of CSC; required for cell elongation in root	[23]
AtHSR8/MUR4	At1g30620	Golgi-localized UDP-D-xylose 4-epimerase	Arabinose biosynthesis; related to the modification of polysaccharides and glycoproteins	[27]
AtGALS1	At2g33570	β-1,4-galactan synthase	Transfer of galactose from UDP-α-d-Gal or arabinopyranose from UDP-β-l-Ara*p* to growing β-1,4-galactan chains	[28,29]
AtXTH30	At1g32170	Xyloglucan endotrans glucosylase-hydrolase	Cleave or rejoin the xyloglucan; *xth30* mutation decreases crystalline cellulose content and affects the depolymerization of microtubules under salt stress	[30]
AtPMEI13	At5g62360	Pectin methyl-esterase inhibitor 13	Inhibits the activity of PMEs	[31]
AtBPC1	At2g01930	BPC-type transcription factor	Regulation of the expression of *AtGALS1*	[28]
AtBPC2	At1g14685			
AtGCN5	At3g54610	Histone acetyltransferase	Epigenetic regulation of cell wall-related genes	[32]
OsTSD2	Os02g51860	Pectin methyltransferase	Regulation of pectin metabolism	[33]
OsBURP16	Os10g26940	β subunit precursor of polygalacturonase 1	Involved in cell wall pectin degradation	[34]

2.2. Hemicellulose

Hemicelluloses are grouped into xyloglucans (XyG), xylans, mannans, and β-(1,3;1,4)-glucans, and the abundance and structure of these polysaccharides vary greatly in different plants species [35]. Xylan is considered as a cross-linking polysaccharide in the establishment of cell wall architecture [35,36]. XyG contributes to the strengthening of cell wall during cell elongation by binding to cellulose micro-fibrils with hydrogen bonds [37,38]. XyG can be cleaved by the cell wall remodeling enzymes xyloglucan endotransglucosylase/hydrolases (XTHs) [39]. After cleavage, the reducing end of the XyG is attached to the non-reducing end of another XyG oligomer or polymer to produce chimeric XyG molecules [39]. The XTHs-mediated modification of XyG is considered to be important for controlling cell wall extensibility. Studies have reported that XTHs are involved in salt stress response in plants. *Arabidopsis XTH30*, encoding a xyloglucan endotransglucosylase/hydrolase 30, is strongly upregulated under salt stress [30]. Loss of function of the *XTH30* gene leads to increased salt tolerance, which is mainly caused by the slower reduction of crystalline cellulose content and alleviated depolymerization of microtubules in response to salt stress [30]. This result suggests that XTH30 plays a negative role in salt tolerance. However, the positive roles of XTHs in salt tolerance have also been reported. Constitutive expression of *CaXTH3* in hot pepper [40,41] and *PeXTH* in *Populus euphratica* [42] enhance tolerance to salt stress, and disruption of *XTH19* and *XHT23* genes in *Arabidopsis* results in decreased salt tolerance [43].

2.3. Pectin

Pectin is a group of acidic polysaccharides that are enriched with α-(1, 4)-linked galacturonic acids in the backbone [44]. Pectin accounts for up to 40% of the dry weight of higher plant cell walls [44] and plays critical roles in plant growth and development [45], leaf senescence [46], biotic [47] and abiotic stress responses [48]. Pectin is composed of three major types: homogalacturonan (HG), rhamnogalacturonan-I (RG-I), and rhamnogalacturonan-II (RG-II) [7,44]. HG is synthesized in the Golgi apparatus and secreted to the apoplast in a highly methy-esterified form and later it is selectively de-esterified by pectin methyl esterases (PMEs) during cell growth and in response to environmental stimuli [7,44]. The degree and pattern of the methyl-esterification of pectin in some extent determines the stiffness of cell walls [49]. In *Arabidopsis,* there are around 66 members of PME family protein, and for most of PMEs, their activities can be inhibited by endogenous PME inhibitors (PMEIs) or a natural inhibitor epigallocatechin gallate (EGCG) [50,51]. High salinity triggers the demethyl-esterification of loosely bound pectins to inhibit cell swelling [52] and previous studies showed that the activity of PMEs is either positively or negatively associated with salt tolerance in plants [53]. For instance, null *Arabidopsis* function mutant *pme13* is hypersensitive to Na^+ toxicity in seed germination and seedling growth [53]. In contrast, overexpression of *Chorispora bungeana PMEI1* or *AtPMEI13* in *Arabidopsis* causes decreased PMEs activity and enhanced methyl-esterification level of pectins, which subsequently improves seeds germination and survival rate under salt stress [31]. The de-esterified HG molecules can be cross-linked to form the so called egg-box structure, the process of which is mediated by divalent cations, such as Ca^{2+}, and the formation of egg-box structure promotes cell wall stiffening [54]. In the presence of high concentration of Na^+, the ratio of Na^+/Ca^{2+} in the apoplast is increased, and Na^+ is supposed to replace Ca^{2+} to bind pectins and thus disturbs the cross-linking of pectins, leading to reduced cell elongation [55]. Besides, the borate-mediated cross-linking of RG-II contributes to the strength of cell wall and is required for the regulation of growth recovery after exposure to high salinity [56,57].

The roles of pectin in salt tolerance have also been reported in rice. Polygalacturonase 1 (PG1) is a cell wall hydrolase that is responsible for the degradation of cell wall pectin. Overexpression of *OsBURP16*, which encodes a non-catalytic β subunit of PG1, results in an increased pectin degradation and increased salt-hypersensitivity in rice [34]. *OsTSD2* encodes a pectin methyltransferase in rice, and mutation in *OsTSD2* leads to a higher content of Na^+ and a lower level of K^+ in rice shoot under high salinity, which is mainly caused by the reduced expression of genes that are responsible for the maintenance of ion homeostasis, such as *OsHKT1;5*, *OsSOS1*, and *OsKAT1* [33] (Table 1).

2.4. Lignin

As one of the most abundant organic compound in plants, lignin is composed of phenylalanine-derived [58] or tyrosine-derived [59] aromatic monomer substances and is important for the secondary cell wall formation and the responses to a variety of environmental stresses [60]. High salinity induces the accumulation of lignin content and cell wall thickening via the activation of lignin biosynthesis pathway [60]. The accumulation of lignin contributes to the mechanical strengthening of cell wall and protection of membrane integrity under salt stress [61]. The effects of lignin accumulation on salt tolerance have been reported in different crops, including soybean [62], wheat [63], and tomato [64]. *CCoAOMT* encodes a caffeoyl CoA O-methyltransferase (CCoAOMT), which catalyzes caffeoyl CoA to feruloyl CoA in lignin biosynthesis pathway. The expression of *CCoAOMT* is induced in salt-adapted cell, and the plants with a loss-of-function of *CCoAOMT* are hypersensitive to salt stress [17]. *BpMYB46* and *BpNAC012*, encoding two transcription factors in white birch (*Betula platyphylla*), are required for the up-regulation of lignin biosynthetic genes and salt stress-responsive genes, and overexpression of these two genes enhances salt tolerance in *B. platyphylla* [65,66]. AgNAC1, a nuclear-localized protein in celery, acts as a positive regulator in inducing the expression of lignin-related and salt stress-responsive genes, and overexpression of *AgNAC1* enhances the formation of secondary walls and plant salt tolerance [67].

3. The Roles of the Cell Wall-Localized Glycoproteins in Salt Stress Response

In addition to dynamic and complex polysaccharide networks, several types of cell wall proteins (CWPs) have been identified in the apoplast. CWPs play critical roles in cell wall modifications and cell wall stress signals transduction. Hydroxyproline (Hyp)-rich glycoproteins (HRGPs), proline-rich proteins (PRPs), glycine-rich proteins (GRPs), and arabinogalactan proteins AGPs are the major types of CWPs [68]. For most of CWPs, they are secreted into the apoplast in a glycosylation-modified form [69–71].

Extensins (EXTs) are a group of cell wall glycoproteins that belong to the HRGPs family. EXTs are typically characterized for the enrichment of Ser-(Hyp)$_{3-5}$ repeats in their protein sequences [72], and each Hyp residue is decorated with up to five arabinose units by several different arabinosyltransferases, including HPAT1-HPAT3 [73], RRA1-RRA3 [74], XEG113 [75], and ExAD [76]. The arabinosylation of EXTs is suggested to be important for the fulfillment of their biological functions. Our recent study showed that the mutation of *MUR4*, which encodes an UDP-Xyl 4-epimerase that is essential for the conversion of UDP-Xyl to UDP-Ara*p* in Golgi, results in reduced root elongation under salt stress, suggesting that arabinose biosynthesis and subsequently the modification of polysaccharides and glycoproteins by arabinose are important for salt tolerance in plants [27].

Leucine-rich repeat extensins (LRXs) are chimeric proteins that contain an N-terminal leucine-rich repeat (LRR) domain that binds with interacting partners and a C-terminal extensin domain that is likely linked with the EXT network or polysaccharides in the apoplast [77]. *LRXs* gene family consists of 11 members in *Arabidopsis*, among of which *LRX3*, *LRX4*, and *LRX5* are dominantly expressed in vegetative tissues [77]. The biological functions of these three LRX proteins are redundant, as mutation of each single gene does not cause any obvious phenotypes, but *lrx34* double and *lrx345* triple mutants both exhibit dwarfism, increased accumulation of anthocyanin, and increased sensitivity to high salinity [10]. It is worth noting that all these phenotypes are more severe in the *lrx345* triple mutant than that in the *lrx34* double mutant. Our study indicated that *fer-4* mutant as well as the transgenic plants overexpressing *RALF22* and *RALF23* exhibit similar phenotypes as *lrx345* in terms of plant growth and salt sensitivity, and biochemical data show that RALF22 and RALF23 are physically associated with LRX3/4/5 [10]. Combining the data showing that FER is the receptor of RALFs [9], we can conclude that the LRX3/4/5, the secreted peptide RALFs, and the receptor-like kinase FER function as a module to mediate salt stress response in the apoplast. It is supposed that the extensin domain of LRXs is able to anchor polysaccharides in the cell wall [77,78], but it is still unknown whether LRXs directly participate in the sensing of CWI or coordinate with FER to perceive CWI (Figure 1).

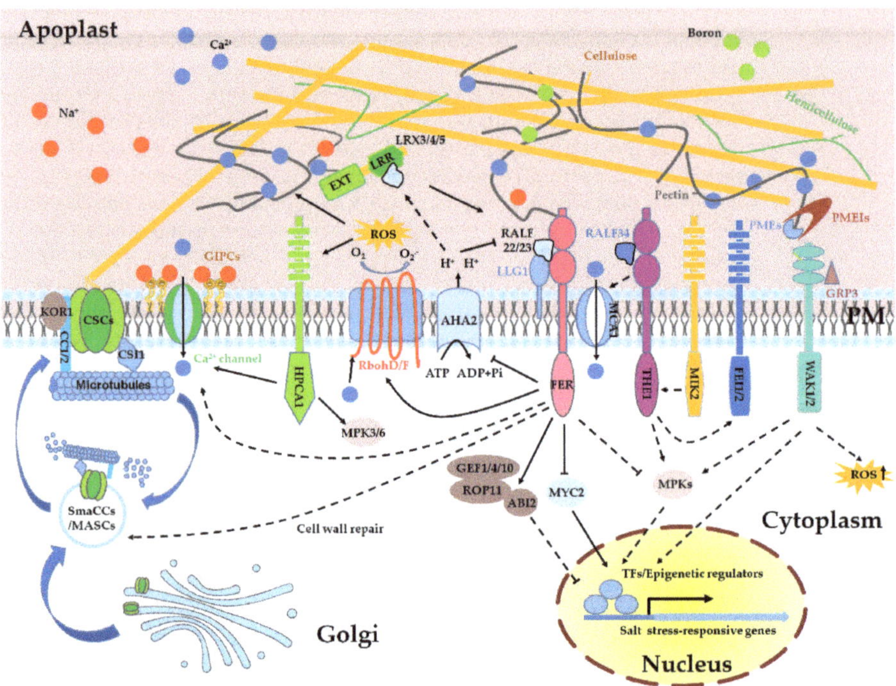

Figure 1. Sensing and maintenance of cell wall integrity under salt stress. Salt stress-induced cell wall changes are proposed to be sensed by multiple receptor-like kinases, including FER, THE1, MIK2, FEI1/2, and WAK1/2. As one of the most important cell wall integrity (CWI) sensors, FER may function alone or together with LRX3/4/5-RALF22/23 module to perceive the perturbation of CWI caused by high salinity. The AHA2-mediated acidification of the apoplastic pH increases the affinity of LRXs with RALFs, while the alkaline state in the apoplast promotes the binding of RALFs with FER. FER and probably also other cell wall sensors convert salt-triggered cell wall signals to multiple intracellular signals, including Ca^{2+}, ROS, abscisic acid (ABA), jasmonic acid (JA), and MPKs, which in turn regulate the expression of salt stress-responsive genes in the nucleus. Salt stress can alter the redox status in the apoplast, and RbohD/F-mediated production of the apoplastic H_2O_2 may affect the cross-linking of cell wall polymers and activate H_2O_2 sensor HPCA1. Glycosyl inositol phosphorylceramide (GIPC) sphingolipids participate in the sensing of extracellular salt by directly binding to sodium ions. Cell wall biosynthesis- and modification-related components, including pectin methyl esterases (PMEs), PME inhibitors (PMEIs), and cellulose synthase (CesA), are involved in the regulation of salt tolerance in plants. Upon initial exposure to salt stress, cortical microtubules are depolymerized and cellulose synthase complex (CSC) together with its companions CSI1 and CC1/2 are internalized into small CesA compartments/microtubule-associated CesA compartments (smaCCs/MASCs). At the growth recovery stage after salt application, FER is probably required for the regulation of the reassembly of cortical microtubules and the relocation of CSCs to the plasma membrane to synthesize cellulose, which subsequently enhances the adaptation of plants to salt stress. Solid lines represent direct regulations, and dashed lines represent in-direct or potential regulations.

AGPs are highly glycosylated with arabinogalactan chains and are proposed to play important roles in salt stress response [69]. Our study showed that the reduced root elongation of the *mur4* mutant under high salinity is partially caused by the decreased AGPs, as application of gum arabic, a commercial source of *Acacia senegal* AGPs, restores the root elongation of the *mur4* mutant under salt stress [27]. As a glycosylphosphatidyli-nositol (GPI)-anchored fasciclin-like AGP, salt overly sensitive 5 (SOS5)/fasciclin-like arabinogalactan-protein 4 (FLA4) was identified based on a screening of mutants with increased sensitivity to salt stress. The *sos5/fla4* mutant exhibits reduced root elongation and severe root tip swelling under salt stress [79,80]. SOS5 is glycosylated by galactosyltrans-ferase 2 (GALT2) and GALT5, both of which belong to AGP-specific galactosyltransferases. The *galt2 galt5* double mutant displays a similar phenotype as the *sos5/fla4* mutant in the presence of high concentration of NaCl [80]. Recently, studies showed that AGPs are able to cross-link with cell wall components. For instance, arabinoxylan pectin arabinogalactan protein 1 (APAP1) is covalently linked to pectins [81], and arabinogalactan protein 31 (AGP31) physically associates with methyl-esterified polygalacturonic acid and galactans, which are the branches of RG-I [82].

Expansins, first isolated from growing cucumber hypocotyls, consist of four subfam-ilies: α-expansin, β-expansin, expansin-like A, and expansin-like B [83,84]. Expansins are key regulators of cell-wall loosening and are required for cell enlargement under a variety of environmental stresses [85]. Several studies have shown that the expression of expansin-encoding genes is induced by high salt and the elevation of the protein levels of expansins tends to promote salt tolerance in plants. *ZmEXPB2*, *ZmEXPB6*, and *ZmEXPB8* genes in maize [86], *AsEXP1* gene in turf grass [87], and *OsEXPA3* gene in rice [88], are induced upon exposure to high salinity. Down-regulation of *ZmEXPB6* is correlated with the reduced leaf growth of maize under salt stress [89]. Overexpression of rice *expansin 7* (*OsEXPA7*) confers substantially enhanced tolerance to salt stress by lowering reactive oxygen species (ROS) accumulation and increasing antioxidant activity in rice [90]. Ec-topic expression of wheat *expansin 2* (*TaEXPA2*) or *TaEXPB23* improves salt tolerance in tobacco [91,92]. Although expansins have been known to positively regulate salt stress response in multiple species, few studies have revealed the mechanisms underlying the expansins-mediated regulation of salt tolerance.

4. Salt Stress Alters the Redox Status in the Apoplast

Reactive oxygen species (ROS) are a class of metabolites, including hydrogen peroxide, singlet oxygen, superoxide, and hydroxyl radicals, which are produced in chloroplasts, mitochondria, peroxisomes, and apoplast [93]. The salt stress-triggered production of ROS and their effects on CWI have been widely reported in plants [93–95]. ROS trig-gers the cross-linking of cell wall compounds and enhances the mechanical strength of cell wall under a short-term stress exposure. Under a prolonged stress treatment, the formation of hydroxyl radicals (•OH) cleave plant polysaccharides, leading to cell wall loosening [96]. The ROS-induced lignin biosynthesis facilitates the adaptation of plants to high salt environment [95,97].

The production of ROS in the apoplast is mainly mediated by respiratory burst oxidase homolog D (RbohD) and RbohF [98], two NADPH oxidases that are localized at the plasma membrane. NADPH oxidases transfer electrons from cytosolic NADPH or NADH to apoplastic oxygen, leading to the production of superoxide (O_2^-), which is then catalyzed to hydrogen peroxide (H_2O_2) by superoxide dismutases [99]. The expression of *RbohD* and *RbohF* is induced under salt stress and *rbohD rbohF* double mutant is hypersensitive to salt stress [100], suggesting that the ROS production in the apoplast is required for salt tolerance. Salt-induced production of ROS by RbohD/F is able to activate Ca^{2+} channel to increase the influx of Ca^{2+} into cytosol, which mediates the modulation of Na^+/K^+ homeostasis [100]. The H_2O_2 generated by RbohD/F during the early stage of stress treatment also acts as a signal molecule to activate antioxidant system to attenuate salt stress-induced oxidative damages [101]. Recent studies showed that RbohD/F form nanoclusters at the plasma membrane in response to osmotic stress and later they are internalized into the cytoplasm via membrane microdomains [102–104]. As high salt conditions are accompanied by osmotic stress, the formation of RbohD/F as nanoclusters at the plasma membrane is perhaps also the case in the plants being exposed to high salinity (Figure 1).

Class III peroxidases are heme-containing enzymes, which are mainly localized in the apoplast and vacuole. Class III peroxidases either positively or negatively modulate apoplastic ROS levels [105]. Class III peroxidases explore H_2O_2 and O_2^- to generate •OH, which leads to the cleavage of polysaccharides and promotes cell wall loosening [106]. Class III *peroxidase 71 (PRX71)*, which is strongly up-regulated in response to cell wall damage (CWD), negatively regulates growth and cell size and positively regulates ROS accumulation [94]. *GsPRX9*, encoding a Class III peroxidase, is induced by salt treatment in soybean root, and the soybean transgenic plants overexpressing *GsPRX9* exhibit increased root growth and decreased H_2O_2 content under salt stress [107].

The biological significance of the salt stress-induced redox change in the apoplast is still far from being fully understood. One of the outputs of the redox change is to affect the formation of intra- and inter-molecular disulfide bond. A large number of cell wall-localized glycoproteins and secreted peptides are characterized with the enrichment of cysteines, which are potentially involved in the formation of disulfide bonds. Therefore, we can speculate that the salt stress-induced redox change can affect the intra- and inter-molecular disulfide bridges of cell wall glycoproteins, which in turn transduce cell wall signals to the cellular interior. LRX8 and RALF4, which are both required for the regulation of pollen tube growth, process cysteines that are involved in the formation of disulfide bridges. A recent structural study showed that the formation of LRX8 homodimer and also the physical association of RALF4 with LRX8 require oxidative environment. Abolishment of the disulfide bonds via sites mutation or treatment of proteins with dithiothreitol (DTT) largely prevents the formation of LRX8 homodimer and affects the affinity of LRX8 with RALF4 [108]. These results suggest that the redox status in the cell wall is required for the regulation of the formation of LRXs-RALFs complex. Based on this hypothesis, we propose that the salt stress-induced change of apoplastic redox status may affect the formation of homo- and hetero-dimers of LRX3/4/5 and also affect the affinity of LRX3/4/5 proteins with RALFs, which finally transduce salt stress signals to the intracellular signaling pathways.

5. The Impact of Apoplastic pH on Salt Tolerance

In the early 1970s, the acid growth theory was proposed, which states that acidification of the apoplast promotes cell elongation, whereas alkaline state in the apoplast prevents cell growth [109]. The reduction of apoplastic pH ($_{apo}$pH) activates several cell wall proteins, including expansins and other remodeling enzymes, resulting in the loosening of cell wall [110]. $_{apo}$pH in linear growing cells is regulated by plasma membrane-localized H^+-ATPases (AHAs) [111]. RALFs are a class of peptides that cause the alkalinization of the apoplast by regulating H^+-ATPases via *Catharanthus roseus* RLK1-like kinases (*Cr*RLK1Ls). FER is one of the *Cr*RLK1L family proteins that consist of two carbohydrate-binding malectin-like domains, a transmembrane domain, and an intracellular serine/threonine-kinase domain [112,113]. FER inhibits the proton transport activity of AHA2 likely via direct phosphorylation [114]. It is known that salinity triggers the transient alkalization in the apoplast and inhibits plant growth [115], and our study showed that salt stress can induce the formation of mature RALFs [10]. These data suggest that salt stress-induced alkalinization of the apoplast is probably mediated by RALFs-FER-AHA2 module and the acidification of the extracellular environment is important for salt tolerance. Two halophyte species, *Atriplex lentiformis* and *Chenopodium quinoa*, which have a capacity to tolerate a high concentration of sodium ion, display a high H^+-ATPase activity under salt stress, which contributes to a low $_{apo}$pH and fast Na^+ efflux [116]. *SOS1*, encoding a plasma membrane membrane-localized Na^+/H^+ antiporter, is required for the extrusion of excessive Na^+ from the cytosol [117]. The Na^+/H^+ exchange activity of SOS1 is absent under normal growth conditions. Upon salt stress, however, Na^+-induces induced formation of an ATP-dependent pH gradient can enhance the Na^+/H^+ transport activity of SOS1 [118].

Altogether, low $_{apo}$pH facilitates plant growth under salt stress, but the direct effects of low $_{apo}$pH on cell wall networks need more detailed studies.

6. Cell Wall Integrity Sensing and Signal Transduction under High Salinity

Unlike the traditional activation of plant receptor-like kinases by the corresponding ligands, the sensing of CWI is not limited by ligand-receptor pattern, e.g., recognition of wall fragments released from the damaged cell walls by receptor-like kinases, and is probably also achieved via the recognition of the cell wall modifications and the alteration of redox and $_{apo}$pH status. Currently, a series of plasma membrane-localized receptor-like kinases and cell wall glycoproteins have been identified that are involved in the sensing and maintenance of CWI. As a universal signal molecule, Ca^{2+} is also involved in the transduction of CWI signaling signals in plants.

The cell wall appears to be the largest source of Ca^{2+} in plant cell [119]. Under normal conditions, Ca^{2+} is used to stabilize pectins via the dimerization of HG chains [120]. AGPs have been shown to bind abundant Ca^{2+} [121]. Under salt stress, the excessive accumulation of Na^+ in the apoplast disrupts ion homeostasis, leading to rapid sodium-specific calcium waves occurred in roots [122]. The imported calcium ions directly bind the EF hands of RbohD/F and improve their catalytic activity [123,124]. Ca^{2+} is also an initial signal to activate the SOS signaling pathway, which promotes the extrusion of Na^+ from the cytosol [125,126].

In addition to high salinity, other abiotic stresses, such as drought, cold, and osmotic stress, can also induce the cytosolic Ca^{2+} influx within a few seconds to minutes. Although the induction of Ca^{2+} signaling is a common event for these different abiotic stresses, studies have shown that the different stresses-triggered Ca^{2+} influx is mediated by different components. Reduced hyperosmolality-induced $[Ca^{2+}]_i$ increase 1 (OSCA1) is specifically required for the osmotic stress-triggered uptake of Ca^{2+} [127], and hydrogen-peroxide-induced Ca^{2+} increases 1 (HPCA1) is required for H_2O_2-, but not for salt- and osmotic stress-, induced influx of Ca^{2+} [128]. Glycosyl inositol phosphorylceramide (GIPC) sphingolipids, which are glycosylated via glucuronosyltransferase MOCA1, was discovered as a sensor of extracellular salt by directly binding to sodium ions [129]. The *moca1* mutant lacking functional GIPCs is defective in the activation of Ca^{2+} waves when being exposed to high concentration of Na^+, K^+, or Li^+ ion. GIPCs can bind Na^+ to gate Ca^{2+} influx channels and trigger the activation of SOS signaling pathway. However, which Ca^{2+} channels are activated by GIPCs and the mechanism underlying the activation need further study (Figure 1).

FER is considered as a CWI sensor and required for the activation of Ca^{2+} influx and maintenance of CWI under salt stress [9]. Mutation of *FER* reduces salt-induced Ca^{2+} influx in the root epidermis and increases sensitivity to high salinity. FER contains two malectin domains that have been experimentally demonstrated to directly bind with de-methyl-esterified HG in vitro and in vivo [9,130], suggesting that FER probably senses the cell wall changes directly via its extracellular domain and then transduces the cell wall signals to cellular interior via its cytoplasmic kinase domain. However, how the modification of pectin affects the activity of FER is still elusive. Our recent study showed that LRX3/4/5, RALFs, and FER function as a module to regulate salt stress response, which implies that FER-mediated perception of CWI probably needs the aid of LRX3/4/5-RALFs regulatory module [10]. Salt stress may dissociate the LRX3/4/5-RALFs complex via the salt stress-induced redox and pH changes in the apoplast, and the released RALFs bind to LLG1-FER complex and thereby allow the transduction of cell wall signals. The mechanism behind the dissociation of LRX3/4/5 and RALFs under salt stress needs to be further investigated.

THESEUS1 (THE1) is a *Cr*RLK1L family protein that was first identified in a screening for the suppressors of *prc1-1* [131]. The null mutation of *the1* partially suppresses the stunted growth and lignin deposition of the *prc1-1* mutant, despite the reduced cellulose content in the *prc1-1* is not restored [131]. HERKULES1 (HERK1) is another *Cr*RLK1L protein that is phylogenetically closely related to FER and THE1. Double mutant *herk1* *the1-4* displays similar phenotypes as *fer-4* in terms of growth and salt stress response [52]. A recent study indicates that THE1 acts as the receptor of RALF34 to fine-tune lateral root initiation [132]. These results suggest that FER, THE, and HERK1 may work together to replay RALFs-mediated cell wall signals, but the biochemical associations among these three *CrRLK1L* proteins are still largely unknown.

Male discoverer 1-interacting receptor like kinase 2 (MIK2) is a leucine-rich repeat receptor-like kinase (LRR-RLK) that was identified by a genome-wide association study (GWAS) based on the natural variations in response to salinity stress [133]. MIK2 controls root growth direction under salt stress in a THE1-dependent manner [134]. The salt-hypersensitive phenotype of *mik2* mutant can be suppressed by *the1-1*, a null mutation of *THE1* [134]. Recently, the serine rich endogenous peptide (SCOOP) phytocytokines were identified as the ligands of MIK2 to trigger immune responses [135], but whether the SCOOP peptides participate in MIK2-mediated regulation of salt tolerance is still unknown. FEI1 and FEI2 are two LRR-RLKs that are associated with cellulose synthesis and anisotropic cell expansion and are involved in CWI sensing [136]. Double mutant *fei1 fei2* displays root swelling and reduced cellulose biosynthesis under high sucrose or high salt conditions [137]. Genetic analysis indicated that FEI2 functions downstream of THE1 in mediating CWI perception [138]. Mid1-complementing activity 1 (MCA1) is a plasma membrane–localized stretch-activated Ca^{2+} channel and functions downstream of THE1 in *Arabidopsis* [95,139]. Like *the1-1* mutant, *mca1* seedlings exhibit reduced deposition of lignin and decreased jasmonic acid and salicylic acid biosynthesis in response to isoxaben-induced CWD [138].

Wall-associated kinases (WAKs) are a family of receptor-like Ser/Thr kinases whose extracellular domains are cross-linked with pectin fraction in a high affinity [140,141]. The EGF-like domain of WAK1/2 preferentially binds to de-methyl-esterified HG over methyl-esterified HG, and WAK1 also exhibits a high affinity with oligogalacturonides (OGs) in vitro [140,142]. The binding of WAKs to pectin and OGs occurs only in the presence of Ca^{2+} [140]. GRP-3, a glycine-rich cell wall protein, also acts as a potential switch for the kinase activity of WAK1 and negatively regulates the defense responses elicited by OGs [143]. A dominant allele of *wak2* mutant exhibits constitutive activation of stress responses, including increased ROS accumulation and cell wall biogenesis [142,144]. Under long-term salt stress, tomato *WAK1* mutant *slwak1* exhibits disrupted osmotic homeostasis and elevated sucrose content in roots, which in turn negatively affects fruit yield [145]. Similarly, *Ds* transposon insertion mutant of *HvWAK1* in barley displays decreased salt tolerance [146]. Although the WAKs have been shown to participate in the salt stress response, the existing experimental evidences to elaborate the roles of WAKs in sensing the CWI under salinity are still lacking. Recently, Gigli-Bisceglia et al. indicated that salinity stress-induced de-methyl-esterification of pectin activates stress signaling pathways, which may provide a direction to study the roles of WAKs in salt stress response [52] (Figure 1).

The CWD caused by salinity stress, isoxaben, an inhibitor of cellulose biosynthesis, or driselase, a cell wall-degrading enzyme, can increase the protein levels of hormone-like peptides PROPEP1/3, the precursors of plant elicitor peptide 1/3 (Pep1/3) [138,147]. The *Pep3* knockdown plants and the null mutant of *Pep1 receptor 1* (*PEPR1*) both exhibit salt-hypersensitivity [148]. These results suggest that the activation of PEPR1 by PROPEP3 positively regulates salt tolerance in *Arabidopsis*. Currently, the majority of studies on Peps-PEPRs complexes focus on their roles in plant immunity, and in future the roles of the Peps-PEPRs complexes-mediated signaling in abiotic stress responses need more investigations.

HPCA1 is a LRR-RLK required for the sense of extracellular H_2O_2 [128]. The two pairs of cysteine residues in the extracellular domain of HPCA1 are covalently modified by extracellular H_2O_2, which leads to the activation of HPCA1 and elevation of Ca^{2+} influx. In *hpca1* mutant seedlings, the extracellular H_2O_2-induced Ca^{2+} influx, the activation of ABA signaling, and the phosphorylation of MPK3/6 are all inhibited [128]. It was shown that HPCA1 is not required for the salt stress-induced influx of Ca^{2+}, but considering that high salinity can affect the redox status in the apoplast, so whether HPCA1 is also required for the sense of salt stress-induced redox changes worth further investigations.

7. Salt Stress-Triggered Intracellular Signaling Pathway Regulated by Cell Wall Sensors

Although several plasma membrane-localized cell wall integrity sensors have been identified that perceive cell wall changes, the intracellular signaling pathways that relay cell wall signals are still largely unknown. The phosphorylation of MPK6 is a marker of the environmental stimuli, and the transient activation of MPK6 under abiotic stress conditions, including high salinity and cold, has been reported [149]. As a major signaling transducer, the activity of MPK6 is regulated by multiple CWI sensors, such as FER, THE1, HERK1, and HPCA1 [52,128]. In future, the regulatory mechanisms of these CWI sensors on the activity of MPK6 need to be addressed.

After perception of CWD by cell wall sensors, plants can integrate and balance multiple hormone signals to improve salt tolerance. ABA and JA are the major hormones involved in the response to diverse environmental stresses. In the *lrx345* and *fer-4* mutants, the ABA and JA contents are constitutively increased and the salt-hypersensitivity of these two mutants is largely caused by the disrupted homeostasis of phytohormones [150]. Phosphatase ABA insensitive 2 (ABI2) is a negative regulator of ABA signaling pathway, and FER activates the guanine nucleotide exchange factor (GEF) 1/4/10/Rho of plant 11 (ROP11) pathway to positively regulate the activity of ABI2 phosphatase, and thereby modulating ABA signaling pathway [151,152]. MYC2, a master transcription factor in JA signaling pathway, is also regulated by FER. FER positively regulates immunity by inhibiting JA signaling via the phosphorylation-mediated destabilization of MYC2 [153]. It has also been shown that MYC2 negatively regulates salt tolerance via the inhibition of proline biosynthesis [154]. In brief, these results suggest that FER controls the environmental stress responses via the modulation of the homeostasis of phytohormones (Figure 1).

8. Cell Wall Repair under High Salinity

Upon exposure to salt stress, the cortical microtubules in the hypocotyl of seedling are rapidly depolymerized, the process of which usually occurs within 2 h of salt application. However, at the growth recovery stage (after salt treatment for ~8 h), the cortical microtubules are reassembled into stable cortical arrays [16]. Evidences have shown that the rapid depolymerization of the cortical microtubules network is important for salt tolerance. For instance, stabilization of microtubules with paclitaxel leads to increased salt-hypersensitivity, whereas constitutive disruption of microtubules with oryzalin or propyzamide improves salt tolerance [155].

The depolymerization of cortical microtubules requires the alteration of the activities of the atypical microtubule-associated protein kinase propyzamide hypersensitive 1 (PHS1) and microtubule-associated protein SPIRAL1 (SPR1). Under normal growth conditions, the kinase activity of PHS1 is inhibited by its own phosphatase domain, while salt or osmotic stress blocks this inhibition and then enhances the phosphorylation and depolymerization of α-tubulin [156]. SPR1 binds to the microtubules and antagonizes stress-induced cortical microtubule depolymerization. Under salt stress, SPR1 is rapidly degraded by the 26S proteasome and the inhibition of microtubule depolymerization is relieved [157]. Histone H2B monoubiquitination (H2Bub1) participates in the regulation of the expression of *protein tyrosine phosphatase 1* (*PTP1*) and *MAP kinase phosphatase* (*MKP*) genes, which in turn modulate the phosphorylation and dephosphorylation of microtubule-binding proteins via a PTP1/MKP-MPK3/6 signal mode, and finally promotes the rapid microtubule depolymerization under salt stress [158].

CSCs synthesize cellulose via the binding with cortical microtubules, and the polymerization status of cortical microtubules determines the movement of CSCs at the cell surface. CSCs are assembled in the Golgi apparatus and translocated to the plasma membrane via vesicle trafficking. Salt-induced depolymerization of microtubules is accompanied by the internalization of CSCs into small CesA compartments/microtubule-associated CesA compartments (smaCCs/MASCs) [15]. At the growth recovery stage after salt treatment, cortical microtubule is reassembled and CSCs is relocated to the plasma membrane to synthesize cellulose. Increasing evidences have shown that the efficiency of plants to reassembly cortical microtubule and cellulose during the growth recovery stage is critical for salt tolerance. CC1 and its paralog CC2 were identified as companions of CSCs and are required for the reassembly of cortical microtubule and subsequently cellulose biosynthesis during the growth recovery stage [16]. In *cc1 cc2* double mutants, CSCs dissociate from the microtubules after salt treatment, but a stress-tolerant microtubule complex cannot be reproduced, resulting in the abolishment of the localization of CSCs at the plasma membrane and decreased cellulose synthesis. Microtubules-associated proteins 65-1 (MAP65-1) is a plant microtubule-bundling protein, which participates in the polymerization and bundling of cortical microtubules [159]. Phosphatidic acid (PA), a product of phospholipase D (PLD), binds to MAP65-1 and increases its activity to enhance microtubule polymerization and bundling [160]. The *pldα1* mutant exhibits a defect in microtubule organization under salt stress and increased salt-hypersensitivity. Moreover, 16:0–18:2 PA can activate MPK6 via directly binding to MPK6 and the salt-induced transient activation of MPK6 is abolished in the pldα1 mutant [149].

Brassinosteroid insensitive 2 (BIN2), a master negative factor in brassinosteroid signal pathway, regulates the balance between salt stress response and growth recovery [161]. BIN2 is required for the negative regulation of cellulose biosynthesis. BIN2 phosphorylates CESA1 to inhibit the activity of CSCs [162]. By exploring turboID-mediated proximity labeling technology, Kim et al. found that BIN2 interacts with FER, but the biological significance of this interaction has not yet been resolved [163]. It is possible that FER regulates the activity of BIN2 via phosphorylation, and then modulates CesAs activity and cellulose biosynthesis under salt stress.

9. Transcriptional Regulation of Cell Wall-Associated Genes under Salt Stress

Under salinity stress, plant cells sense salt signals via receptors or sensors and then transmit the signals to the downstream regulatory networks to trigger the transcription of salt stress-responsive genes, which in turn promote the adjustment of the physiological, biochemical, and metabolic properties of plant cells to adapt to high salinity.

The transcriptional regulation of genes largely depends on the activity of the corresponding transcription factors. Some transcription factors have been identified that are required for the regulation of cell wall-associated genes in response to salt stress. For example, salt stress induces the accumulation of β-1,4-galactan in root cell walls through the up-regulation the of *galactan synthase 1* (*GALS1*) gene. Based on a genetic screening, two transcription factors basic pentacysteine 1 (BPC1) and BPC2 were identified that directly bind to the promoter of the *GALS1* gene and repress its expression [28]. The expression of *BPC1* and *BPC2* genes is significantly reduced under salt stress. The *bpc1 bpc2* double mutant, in which the accumulation of β-1,4-galactan is elevated under salt stress compared with the wild type, exhibits increased salt tolerance [28]. *Oryza sativa MULTIPASS* (*OsMPS*) encodes an R2R3-type MYB transcription factor in rice. Expression profiling revealed that, upon ABA or salt stress treatment, the expression of expansins, such as *OsEXPA4*, *OsEXPA8*, *OsEXPB2*, *OsEXPB3*, and *OsEXPB6*, and the expression of cell wall biosynthesis genes, such as endoglucanase genes *OsGLU5* and *OsGLU14*, are negatively regulated by OsMPS [164]. *XTH19* and *XTH23*, belonging to xyloglucan endotransglucosylase/hydrolase group II, are up-regulated by salt stress and BR [43]. In the *xth23* single or *xth19 xth23* double mutant, lateral root growth is disrupted under salt stress, whereas overexpression of *XTH19* or *XTH23* enhances salt tolerance and increases lateral root initiation [43]. BRI1-EMS-SUPPRESSOR 1 (BES1) is a transcription factor that is involved in BR signaling pathway. BES1 directly binds the promoter of *XTH19* and *XTH23* and positively regulates their expression under salt stress [43] (Table 1).

Gene expression is also influenced by epigenetic regulation, such as histone modification and DNA methylation. Salt stress triggers the histone H3K9/K14 acetylation of some abiotic stress-responsive genes to crease their transcript levels [165]. *General control nonderepressible 5* (*GCN5*), encoding a histone acetyltransferase, is induced by salt stress and acts as a maintainer of CWI. GCN5 mediates the acetylation of H3K9 and H3K14 in the promoters of *CTL1*, *PGX3* (*polygalacturonase involved in expansion-3*), and *MYB54* under salt stress, and thus fine-tunes their gene expression [32]. Constitutive expression of *CTL1* partially restores the salt-hypersensitivity and CWD of the *gcn5* mutant [32]. Similarly, the H3K9 acetylation level in the genome of maize is also elevated after salt treatment, and the increased acetylation level enhances the expression of *ZmGCN5*, which in turn promotes the expression of *ZmEXPB2* and *ZmXET1* genes [166].

10. Conclusions and Future Perspective

Cell wall is not just a mechanical support for plant cells, but is also the frontline to sense and transduce environmental stress signals. High salinity, as one of the globally distributed abiotic stresses, can disrupt the CWI, and the severity of the salt-triggered CWD largely depends on the concentration of the surrounding sodium ion combined with other environmental conditions, such as light intensity and water availability. Study of the mechanisms underlying the sensing and maintenance of CWI under salt stress not only strengthens our understanding of salt stress responses in plants but also provides new strategies for the cultivation of crops with improved salt tolerance. Regarding the associations between CWI and salt tolerance, there are still many questions remain to be addressed, and the most important ones could be that how the excessive accumulation of Na^+ in the apoplast affects the CWI, and how the salt-induced cell wall changes are sensed by the cell wall sensors. Moreover, the Ca^{2+} channels that are required for the relay of salt-triggered cell wall stress signals need to be identified and the cell wall repair mechanisms under stress conditions need to be further investigated. With the development of gene editing technologies and improved transformation efficiency, editing of CWI-related genes in crops to generate salt-tolerant varieties can be applied in future.

Author Contributions: J.L., S.L. and W.Z. compiled the materials and wrote the first draft; J.L. and C.Z. edited and finalized the manuscript. All authors have read and agreed to the published version of the manuscript.

Funding: This work was supported by the National Natural Science Foundation of China (NSFC) (Grant No. 32070295), and the Shanghai Pujiang Program (Grant No. 20PJ1414800), and the Strategic Priority Research Program from the Chinese Academy of Sciences (Grant No. XDA27040104).

Data Availability Statement: Not applied in this study.

Conflicts of Interest: The authors declare no conflict of interest.

References

1. Munns, R.; Tester, M. Mechanisms of salinity tolerance. *Annu. Rev. Plant Biol.* **2008**, *59*, 651–681. [CrossRef] [PubMed]
2. Singh, A. Soil salinization management for sustainable development: A review. *J. Environ. Manag.* **2021**, *277*, 111383. [CrossRef]
3. Jesus, J.M.; Danko, A.S.; Fiúza, A.; Borges, M.T. Phytoremediation of salt-affected soils: A review of processes, applicability, and the impact of climate change. *Environ. Sci. Pollut. Res. Int.* **2015**, *22*, 6511–6525. [CrossRef] [PubMed]
4. Rengasamy, P. World salinization with emphasis on Australia. *J. Exp. Bot.* **2006**, *57*, 1017–1023. [CrossRef] [PubMed]
5. Van Zelm, E.; Zhang, Y.; Testerink, C. Salt tolerance mechanisms of plants. *Annu. Rev. Plant Biol.* **2020**, *71*, 403–433. [CrossRef] [PubMed]
6. Zhao, C.; Zhang, H.; Song, C.; Zhu, J.; Shabala, S. Mechanisms of plant responses and adaptation to soil salinity. *Innovation* **2020**, *1*, 100017. [CrossRef]
7. Caffall, K.H.; Mohnen, D. The structure, function, and biosynthesis of plant cell wall pectic polysaccharides. *Carbohydr. Res.* **2009**, *344*, 1879–1900. [CrossRef]
8. Voxeur, A.; Höfte, H. Cell wall integrity signaling in plants: "To grow or not to grow that's the question". *Glycobiology* **2016**, *26*, 950–960. [CrossRef]
9. Feng, W.; Kita, D.; Peaucelle, A.; Cartwright, H.N.; Doan, V.; Duan, Q.; Liu, M.C.; Maman, J.; Steinhorst, L.; Schmitz-Thom, I.; et al. The FERONIA receptor kinase maintains cell-wall integrity during salt stress through Ca^{2+} Signaling. *Curr. Biol.* **2018**, *28*, 666–675.e5. [CrossRef]

10. Zhao, C.; Zayed, O.; Yu, Z.; Jiang, W.; Zhu, P.; Hsu, C.C.; Zhang, L.; Tao, W.A.; Lozano-Durán, R.; Zhu, J.K. Leucine-rich repeat extensin proteins regulate plant salt tolerance in *Arabidopsis*. *Proc. Natl. Acad. Sci. USA* **2018**, *115*, 13123–13128. [CrossRef] [PubMed]

11. Rui, Y.; Dinneny, J.R. A wall with integrity: Surveillance and maintenance of the plant cell wall under stress. *New Phytol.* **2020**, *225*, 1428–1439. [CrossRef] [PubMed]

12. Bacete, L.; Hamann, T. The role of mechanoperception in plant cell wall integrity maintenance. *Plants* **2020**, *9*, 574. [CrossRef] [PubMed]

13. Lampugnani, E.R.; Khan, G.A.; Somssich, M.; Persson, S. Building a plant cell wall at a glance. *J. Cell Sci.* **2018**, *131*. [CrossRef]

14. Somerville, C.; Bauer, S.; Brininstool, G.; Facette, M.; Hamann, T.; Milne, J.; Osborne, E.; Paredez, A.; Persson, S.; Raab, T.; et al. Toward a systems approach to understanding plant cell walls. *Science* **2004**, *306*, 2206–2211. [CrossRef]

15. Kesten, C.; Wallmann, A.; Schneider, R.; McFarlane, H.E.; Diehl, A.; Khan, G.A.; van Rossum, B.J.; Lampugnani, E.R.; Szymanski, W.G.; Cremer, N.; et al. The companion of cellulose synthase 1 confers salt tolerance through a Tau-like mechanism in plants. *Nat. Commun.* **2019**, *10*, 857. [CrossRef]

16. Endler, A.; Kesten, C.; Schneider, R.; Zhang, Y.; Ivakov, A.; Froehlich, A.; Funke, N.; Persson, S. A mechanism for sustained cellulose synthesis during salt stress. *Cell* **2015**, *162*, 1353–1364. [CrossRef] [PubMed]

17. Chun, H.J.; Baek, D.; Cho, H.M.; Lee, S.H.; Jin, B.J.; Yun, D.J.; Hong, Y.S.; Kim, M.C. Lignin biosynthesis genes play critical roles in the adaptation of *Arabidopsis* plants to high-salt stress. *Plant Signal. Behav.* **2019**, *14*, 1625697. [CrossRef] [PubMed]

18. McFarlane, H.E.; Döring, A.; Persson, S. The cell biology of cellulose synthesis. *Annu. Rev. Plant Biol.* **2014**, *65*, 69–94. [CrossRef]

19. Paredez, A.R.; Somerville, C.R.; Ehrhardt, D.W. Visualization of cellulose synthase demonstrates functional association with microtubules. *Science* **2006**, *312*, 1491–1495. [CrossRef] [PubMed]

20. Endler, A.; Persson, S. Cellulose synthases and synthesis in *Arabidopsis*. *Mol. Plant* **2011**, *4*, 199–211. [CrossRef] [PubMed]

21. Zhang, S.S.; Sun, L.; Dong, X.; Lu, S.J.; Tian, W.; Liu, J.X. Cellulose synthesis genes *CESA6* and *CSI1* are important for salt stress tolerance in *Arabidopsis*. *J. Integr. Plant Biol.* **2016**, *58*, 623–626. [CrossRef]

22. Kang, J.S.; Frank, J.; Kang, C.H.; Kajiura, H.; Vikram, M.; Ueda, A.; Kim, S.; Bahk, J.D.; Triplett, B.; Fujiyama, K.; et al. Salt tolerance of *Arabidopsis thaliana* requires maturation of N-glycosylated proteins in the Golgi apparatus. *Proc. Natl. Acad. Sci. USA* **2008**, *105*, 5933–5938. [CrossRef] [PubMed]

23. Vain, T.; Crowell, E.F.; Timpano, H.; Biot, E.; Desprez, T.; Mansoori, N.; Trindade, L.M.; Pagant, S.; Robert, S.; Höfte, H.; et al. The cellulase KORRIGAN is part of the cellulose synthase complex. *Plant Physiol.* **2014**, *165*, 1521–1532. [CrossRef] [PubMed]

24. Kwon, Y.; Kim, S.H.; Jung, M.S.; Kim, M.S.; Oh, J.E.; Ju, H.W.; Kim, K.I.; Vierling, E.; Lee, H.; Hong, S.W. Arabidopsis *hot2* encodes an endochitinase-like protein that is essential for tolerance to heat, salt and drought stresses. *Plant J.* **2007**, *49*, 184–193. [CrossRef] [PubMed]

25. Chen, Z.; Hong, X.; Zhang, H.; Wang, Y.; Li, X.; Zhu, J.K.; Gong, Z. Disruption of the cellulose synthase gene, *AtCesA8/IRX1*, enhances drought and osmotic stress tolerance in *Arabidopsis*. *Plant J.* **2005**, *43*, 273–283. [CrossRef]

26. Sánchez-Rodríguez, C.; Bauer, S.; Hématy, K.; Saxe, F.; Ibáñez, A.B.; Vodermaier, V.; Konlechner, C.; Sampathkumar, A.; Rüggeberg, M.; Aichinger, E.; et al. CHITINASE-LIKE1/POM-POM1 and its homolog CTL2 are glucan-interacting proteins important for cellulose biosynthesis in *Arabidopsis*. *Plant Cell* **2012**, *24*, 589–607. [CrossRef]

27. Zhao, C.; Zayed, O.; Zeng, F.; Liu, C.; Zhang, L.; Zhu, P.; Hsu, C.C.; Tuncil, Y.E.; Tao, W.A.; Carpita, N.C.; et al. Arabinose biosynthesis is critical for salt stress tolerance in *Arabidopsis*. *New Phytol.* **2019**, *224*, 274–290. [CrossRef]

28. Yan, J.; Liu, Y.; Yang, L.; He, H.; Huang, Y.; Fang, L.; Scheller, H.V.; Jiang, M.; Zhang, A. Cell wall β-1,4-galactan regulated by the BPC1/BPC2-GALS1 module aggravates salt sensitivity in *Arabidopsis thaliana*. *Mol. Plant* **2020**, *14*, 411–425. [CrossRef]

29. Laursen, T.; Stonebloom, S.H.; Pidatala, V.R.; Birdseye, D.S.; Clausen, M.H.; Mortimer, J.C.; Scheller, H.V. Bifunctional glycosyltransferases catalyze both extension and termination of pectic galactan oligosaccharides. *Plant J.* **2018**, *94*, 340–351. [CrossRef]

30. Yan, J.; Huang, Y.; He, H.; Han, T.; Di, P.; Sechet, J.; Fang, L.; Liang, Y.; Scheller, H.V.; Mortimer, J.C.; et al. Xyloglucan endotransglucosylase-hydrolase 30 negatively affects salt tolerance in *Arabidopsis*. *J. Exp. Bot.* **2019**, *70*, 5495–5506. [CrossRef]

31. Chen, J.; Chen, X.; Zhang, Q.; Zhang, Y.; Ou, X.; An, L.; Feng, H.; Zhao, Z. A cold-induced pectin methyl-esterase inhibitor gene contributes negatively to freezing tolerance but positively to salt tolerance in *Arabidopsis*. *J. Plant Physiol.* **2018**, *222*, 67–78. [CrossRef]

32. Zheng, M.; Liu, X.; Lin, J.; Liu, X.; Wang, Z.; Xin, M.; Yao, Y.; Peng, H.; Zhou, D.X.; Ni, Z.; et al. Histone acetyltransferase GCN5 contributes to cell wall integrity and salt stress tolerance by altering the expression of cellulose synthesis genes. *Plant J.* **2019**, *97*, 587–602.

33. Fang, C.; Li, K.; Wu, Y.; Wang, D.; Zhou, J.; Liu, X.; Li, Y.; Jin, C.; Liu, X.; Mur, L.; et al. OsTSD2-mediated cell wall modification affects ion homeostasis and salt tolerance. *Plant Cell Environ.* **2019**, *42*, 1503–1512. [CrossRef]

34. Liu, H.; Ma, Y.; Chen, N.; Guo, S.; Liu, H.; Guo, X.; Chong, K.; Xu, Y. Overexpression of stress-inducible *OsBURP16*, the β subunit of polygalacturonase 1, decreases pectin content and cell adhesion and increases abiotic stress sensitivity in rice. *Plant Cell Environ.* **2014**, *37*, 1144–1158. [CrossRef]

35. Scheller, H.V.; Ulvskov, P. Hemicelluloses. *Annu. Rev. Plant Biol.* **2010**, *61*, 263–289. [CrossRef]

36. Zhang, B.; Gao, Y.; Zhang, L.; Zhou, Y. The plant cell wall: Biosynthesis, construction, and functions. *J. Integr. Plant Biol.* **2020**, *63*, 251–272. [CrossRef]

37. Park, Y.B.; Cosgrove, D.J. Xyloglucan and its interactions with other components of the growing cell wall. *Plant Cell Physiol.* **2015**, *56*, 180–194. [CrossRef] [PubMed]

38. Hayashi, T.; Kaida, R. Functions of xyloglucan in plant cells. *Mol. Plant* **2011**, *4*, 17–24. [CrossRef] [PubMed]

39. Nishitani, K.; Tominaga, R. Endo-xyloglucan transferase, a novel class of glycosyltransferase that catalyzes transfer of a segment of xyloglucan molecule to another xyloglucan molecule. *J. Biol. Chem.* **1992**, *267*, 21058–21064. [CrossRef]

40. Choi, J.Y.; Seo, Y.S.; Kim, S.J.; Kim, W.T.; Shin, J.S. Constitutive expression of *CaXTH3*, a hot pepper xyloglucan endotransglucosylase/hydrolase, enhanced tolerance to salt and drought stresses without phenotypic defects in tomato plants (*Solanum lycopersicum cv.* Dotaerang). *Plant Cell Rep.* **2011**, *30*, 867–877. [CrossRef] [PubMed]

41. Cho, S.K.; Kim, J.E.; Park, J.A.; Eom, T.J.; Kim, W.T. Constitutive expression of abiotic stress-inducible hot pepper *CaXTH3*, which encodes a xyloglucan endotransglucosylase/hydrolase homolog, improves drought and salt tolerance in transgenic *Arabidopsis* plants. *FEBS Lett.* **2006**, *580*, 3136–3144. [CrossRef]

42. Han, Y.; Wang, W.; Sun, J.; Ding, M.; Zhao, R.; Deng, S.; Wang, F.; Hu, Y.; Wang, Y.; Lu, Y.; et al. *Populus euphratica XTH* overexpression enhances salinity tolerance by the development of leaf succulence in transgenic tobacco plants. *J. Exp. Bot.* **2013**, *64*, 4225–4238. [CrossRef]

43. Xu, P.; Fang, S.; Chen, H.; Cai, W. The brassinosteroid-responsive xyloglucan endotransglucosylase/hydrolase 19 (*XTH19*) and *XTH23* genes are involved in lateral root development under salt stress in *Arabidopsis*. *Plant J.* **2020**, *104*, 59–75. [CrossRef]

44. Atmodjo, M.A.; Hao, Z.; Mohnen, D. Evolving views of pectin biosynthesis. *Annu Rev. Plant Biol.* **2013**, *64*, 747–779. [CrossRef]

45. Wolf, S.; Mouille, G.; Pelloux, J. Homogalacturonan methyl-esterification and plant development. *Mol. Plant* **2009**, *2*, 851–860. [CrossRef] [PubMed]

46. Fang, C.; Zhang, H.; Wan, J.; Wu, Y.; Li, K.; Jin, C.; Chen, W.; Wang, S.; Wang, W.; Zhang, H.; et al. Control of leaf senescence by an MeOH-Jasmonates cascade that is epigenetically regulated by OsSRT1 in rice. *Mol. Plant* **2016**, *9*, 1366–1378. [CrossRef]

47. Lionetti, V.; Cervone, F.; Bellincampi, D. Methyl esterification of pectin plays a role during plant-pathogen interactions and affects plant resistance to diseases. *J. Plant Physiol.* **2012**, *169*, 1623–1630. [CrossRef] [PubMed]

48. Wormit, A.; Usadel, B. The multifaceted role of pectin methylesterase inhibitors (PMEIs). *Int. J. Mol. Sci.* **2018**, *19*, 2878. [CrossRef] [PubMed]

49. Wachsman, G.; Zhang, J.; Moreno-Risueno, M.A.; Anderson, C.T.; Benfey, P.N. Cell wall remodeling and vesicle trafficking mediate the root clock in *Arabidopsis*. *Science* **2020**, *370*, 819. [CrossRef]

50. Sénéchal, F.; Wattier, C.; Rustérucci, C.; Pelloux, J. Homogalacturonan-modifying enzymes: Structure, expression, and roles in plants. *J. Exp. Bot.* **2014**, *65*, 5125–5160. [CrossRef]

51. Lewis, K.C.; Selzer, T.; Shahar, C.; Udi, Y.; Tworowski, D.; Sagi, I. Inhibition of pectin methyl esterase activity by green tea catechins. *Phytochemistry* **2008**, *69*, 2586–2592. [CrossRef]

52. Gigli-Bisceglia, N.; Van Zelm, E.; Huo, W.; Lamers, J.; Testerink, C. Salinity stress-induced modification of pectin activates stress signaling pathways and requires HERK/THE and FER to attenuate the response. *bioRxiv* **2020**. [CrossRef]

53. Yan, J.; He, H.; Fang, L.; Zhang, A. Pectin methylesterase 31 positively regulates salt stress tolerance in *Arabidopsis*. *Biochem. Biophys. Res. Commun.* **2018**, *496*, 497–501. [CrossRef]

54. Hocq, L.; Pelloux, J.; Lefebvre, V. Connecting homogalacturonan-type pectin remodeling to acid growth. *Trends Plant Sci.* **2017**, *22*, 20–29. [CrossRef]

55. Manunza, B.; Deiana, S.; Pintore, M.; Gessa, C. Interaction of Ca^{2+} and Na^+ ions with polygalacturonate chains: A molecular dynamics study. *Glycoconj. J.* **1998**, *15*, 297–300. [CrossRef] [PubMed]

56. Sechet, J.; Htwe, S.; Urbanowicz, B.; Agyeman, A.; Feng, W.; Ishikawa, T.; Colomes, M.; Kumar, K.S.; Kawai-Yamada, M.; Dinneny, J.R.; et al. Suppression of *Arabidopsis* GGLT1 affects growth by reducing the L-galactose content and borate cross-linking of rhamnogalacturonan-II. *Plant J.* **2018**, *96*, 1036–1050. [CrossRef]

57. O'Neill, M.A.; Eberhard, S.; Albersheim, P.; Darvill, A.G. Requirement of borate cross-linking of cell wall rhamnogalacturonan II for *Arabidopsis* growth. *Science* **2001**, *294*, 846–849. [CrossRef]

58. Vanholme, R.; De Meester, B.; Ralph, J.; Boerjan, W. Lignin biosynthesis and its integration into metabolism. *Curr. Opin. Biotechnol.* **2019**, *56*, 230–239. [CrossRef] [PubMed]

59. Barros, J.; Serrani-Yarce, J.C.; Chen, F.; Baxter, D.; Venables, B.J.; Dixon, R.A. Role of bifunctional ammonia-lyase in grass cell wall biosynthesis. *Nat. Plants* **2016**, *2*, 16050. [CrossRef] [PubMed]

60. Moura, J.C.; Bonine, C.A.; de Oliveira, F.V.J.; Dornelas, M.C.; Mazzafera, P. Abiotic and biotic stresses and changes in the lignin content and composition in plants. *J. Integr. Plant Biol.* **2010**, *52*, 360–376. [CrossRef]

61. Naseer, S.; Lee, Y.; Lapierre, C.; Franke, R.; Nawrath, C.; Geldner, N. Casparian strip diffusion barrier in *Arabidopsis* is made of a lignin polymer without suberin. *Proc. Natl. Acad. Sci. USA* **2012**, *109*, 10101–10106. [CrossRef]

62. Hilal, M.; Zenoff, A.M.; Ponessa, G.; Moreno, H.; Massa, E.M. Saline stress alters the temporal patterns of xylem differentiation and alternative oxidase expression in developing soybean roots. *Plant Physiol.* **1998**, *117*, 695–701. [CrossRef]

63. Jbir, N.; Chaïbi, W.; Ammar, S.; Jemmali, A.; Ayadi, A. Root growth and lignification of two wheat species differing in their sensitivity to NaCl, in response to salt stress. *C. R. Acad. Sci. III* **2001**, *324*, 863–868. [CrossRef]

64. Sánchez-Aguayo, I.; Rodríguez-Galán, J.M.; García, R.; Torreblanca, J.; Pardo, J.M. Salt stress enhances xylem development and expression of S-adenosyl-L-methionine synthase in lignifying tissues of tomato plants. *Planta* **2004**, *220*, 278–285. [CrossRef]

65. Hu, P.; Zhang, K.; Yang, C. BpNAC012 positively regulates abiotic stress responses and secondary wall biosynthesis. *Plant Physiol.* **2019**, *179*, 700–717. [CrossRef]

66. Guo, H.; Wang, Y.; Wang, L.; Hu, P.; Wang, Y.; Jia, Y.; Zhang, C.; Zhang, Y.; Zhang, Y.; Wang, C.; et al. Expression of the MYB transcription factor gene *BplMYB46* affects abiotic stress tolerance and secondary cell wall deposition in *Betula platyphylla*. *Plant Biotechnol. J.* **2017**, *15*, 107–121. [CrossRef]

67. Duan, A.Q.; Tao, J.P.; Jia, L.L.; Tan, G.F.; Liu, J.X.; Li, T.; Chen, L.Z.; Su, X.J.; Feng, K.; Xu, Z.S.; et al. AgNAC1, a celery transcription factor, related to regulation on lignin biosynthesis and salt tolerance. *Genomics* **2020**, *112*, 5254–5264. [CrossRef] [PubMed]

68. Showalter, A.M. Structure and function of plant cell wall proteins. *Plant Cell* **1993**, *5*, 9. [PubMed]

69. Hervé, C.; Siméon, A.; Jam, M.; Cassin, A.; Johnson, K.L.; Salmeán, A.A.; Willats, W.G.; Doblin, M.S.; Bacic, A.; Kloareg, B. Arabinogalactan proteins have deep roots in eukaryotes: Identification of genes and epitopes in brown algae and their role in *Fucus serratus* embryo development. *New Phytol.* **2016**, *209*, 1428–1441. [CrossRef] [PubMed]

70. Sampedro, J.; Cosgrove, D.J. The expansin superfamily. *Genome Biol.* **2005**, *6*, 242. [CrossRef]

71. Ding, J.L.C.; Hsu, J.S.F.; Wang, M.M.C.; Tzen, J.T.C. Purification and glycosylation analysis of an acidic pectin methylesterase in jelly fig (*Ficus awkeotsang*) achenes. *J. Agric. Food Chem.* **2002**, *50*, 2920–2925. [CrossRef]

72. Castilleux, R.; Plancot, B.; Gügi, B.; Attard, A.; Loutelier-Bourhis, C.; Lefranc, B.; Nguema-Ona, E.; Arkoun, M.; Yvin, J.; Driouich, A.; et al. Extensin arabinosylation is involved in root response to elicitors and limits oomycete colonization. *Ann. Bot. Lond.* **2020**, *125*, 751–763. [CrossRef] [PubMed]

73. Ogawa-Ohnishi, M.; Matsushita, W.; Matsubayashi, Y. Identification of three hydroxyproline *O*-arabinosyltransferases in *Arabidopsis thaliana*. *Nat. Chem. Biol.* **2013**, *9*, 726–730. [CrossRef] [PubMed]

74. Egelund, J.; Obel, N.; Ulvskov, P.; Geshi, N.; Pauly, M.; Bacic, A.; Petersen, B.L. Molecular characterization of two *Arabidopsis thaliana* glycosyltransferase mutants, *rra1* and *rra2*, which have a reduced residual arabinose content in a polymer tightly associated with the cellulosic wall residue. *Plant Mol. Biol.* **2007**, *64*, 439–451. [CrossRef] [PubMed]

75. Gille, S.; Hänsel, U.; Ziemann, M.; Pauly, M. Identification of plant cell wall mutants by means of a forward chemical genetic approach using hydrolases. *Proc. Natl. Acad. Sci. USA* **2009**, *106*, 14699. [CrossRef]

76. Velasquez, M.; Salter, J.S.; Dorosz, J.G.; Petersen, B.L.; Estevez, J.M. Recent advances on the posttranslational modifications of EXTs and their roles in plant cell walls. *Front. Plant Sci.* **2012**, *3*, 93. [CrossRef]

77. Draeger, C.; Ndinyanka, F.T.; Gineau, E.; Mouille, G.; Kuhn, B.M.; Moller, I.; Abdou, M.T.; Frey, B.; Pauly, M.; Bacic, A.; et al. Arabidopsis leucine-rich repeat extensin (LRX) proteins modify cell wall composition and influence plant growth. *BMC Plant Biol.* **2015**, *15*, 1–11. [CrossRef] [PubMed]

78. Ringli, C. The hydroxyproline-rich glycoprotein domain of the *Arabidopsis* LRX1 requires Tyr for function but not for insolubilization in the cell wall. *Plant J.* **2010**, *63*, 662–669. [CrossRef]

79. Shi, H.; Kim, Y.; Guo, Y.; Stevenson, B.; Zhu, J.K. The *Arabidopsis SOS5* locus encodes a putative cell surface adhesion protein and is required for normal cell expansion. *Plant Cell* **2003**, *15*, 19–32. [CrossRef]

80. Basu, D.; Tian, L.; Debrosse, T.; Poirier, E.; Emch, K.; Herock, H.; Travers, A.; Showalter, A.M. Glycosylation of a fasciclin-like arabinogalactan-protein (SOS5) mediates root growth and seed mucilage adherence via a cell wall receptor-like kinase (FEI1/FEI2) pathway in *Arabidopsis*. *PLoS ONE* **2016**, *11*, e0145092. [CrossRef]

81. Tan, L.; Eberhard, S.; Pattathil, S.; Warder, C.; Glushka, J.; Yuan, C.; Hao, Z.; Zhu, X.; Avci, U.; Miller, J.S.; et al. An *Arabidopsis* cell wall proteoglycan consists of pectin and arabinoxylan covalently linked to an arabinogalactan protein. *Plant Cell* **2013**, *25*, 270–287. [CrossRef]

82. Hijazi, M.; Roujol, D.; Nguyen-Kim, H.; Del, R.C.C.L.; Saland, E.; Jamet, E.; Albenne, C. Arabinogalactan protein 31 (AGP31), a putative network-forming protein in *Arabidopsis thaliana* cell walls? *Ann. Bot.* **2014**, *114*, 1087–1097. [CrossRef]

83. Marowa, P.; Ding, A.; Kong, Y. Expansins: Roles in plant growth and potential applications in crop improvement. *Plant Cell Rep.* **2016**, *35*, 949–965. [CrossRef] [PubMed]

84. Cosgrove, D.J. Characterization of long-term extension of isolated cell walls from growing cucumber hypocotyls. *Planta* **1989**, *177*, 121–130. [CrossRef] [PubMed]

85. Wu, Y.; Cosgrove, D.J. Adaptation of roots to low water potentials by changes in cell wall extensibility and cell wall proteins. *J. Exp. Bot.* **2000**, *51*, 1543–1553. [CrossRef] [PubMed]

86. Geilfus, C.M.; Zörb, C.; Mühling, K.H. Salt stress differentially affects growth-mediating β-expansins in resistant and sensitive maize (*Zea mays* L.). *Plant Physiol. Biochem.* **2010**, *48*, 993–998. [CrossRef]

87. Xu, J.; Tian, J.; Belanger, F.C.; Huang, B. Identification and characterization of an expansin gene *AsEXP1* associated with heat tolerance in C3 Agrostis grass species. *J. Exp. Bot.* **2007**, *58*, 3789–3796. [CrossRef]

88. Qiu, S.; Ma, N.; Che, S.; Wang, Y.; Peng, X.; Zhang, G.; Wang, G.; Huang, J. Repression of *OsEXPA3* expression leads to root system growth suppression in rice. *Crop. Sci.* **2014**, *54*, 2201–2213. [CrossRef]

89. Geilfus, C.; Ober, D.; Eichacker, L.A.; Mühling, K.H.; Zörb, C. Down-regulation of ZmEXPB6 (*Zea mays* β-expansin 6) protein is correlated with salt-mediated growth reduction in the leaves of *Z. mays* L. *J. Biol. Chem.* **2015**, *290*, 11235–11245. [CrossRef]

90. Jadamba, C.; Kang, K.; Paek, N.; Lee, S.I.; Yoo, S. Overexpression of rice *expansin 7* (*Osexpa7*) confers enhanced tolerance to salt stress in rice. *Int. J. Mol. Sci.* **2020**, *21*, 454. [CrossRef]

91. Chen, Y.; Han, Y.; Kong, X.; Kang, H.; Ren, Y.; Wang, W. Ectopic expression of wheat expansin gene *TaEXPA2* improved the salt tolerance of transgenic tobacco by regulating Na$^+$/K$^+$ and antioxidant competence. *Physiol. Plant.* **2017**, *159*, 161–177. [CrossRef] [PubMed]

92. Han, Y.; Li, A.; Li, F.; Zhao, M.; Wang, W. Characterization of a wheat (*Triticum aestivum* L.) expansin gene, *TaEXPB23*, involved in the abiotic stress response and phytohormone regulation. *Plant Physiol. Biochem.* **2012**, *54*, 49–58. [CrossRef] [PubMed]

93. Miller, G.; Suzuki, N.; Ciftci-Yilmaz, S.; Mittler, R. Reactive oxygen species homeostasis and signalling during drought and salinity stresses. *Plant Cell Environ.* **2010**, *33*, 453–467. [CrossRef]

94. Raggi, S.; Ferrarini, A.; Delledonne, M.; Dunand, C.; Ranocha, P.; De Lorenzo, G.; Cervone, F.; Ferrari, S. The arabidopsis class III peroxidase AtPRX71 negatively regulates growth under physiological conditions and in response to cell wall damage. *Plant Physiol.* **2015**, *169*, 2513–2525. [CrossRef] [PubMed]

95. Denness, L.; McKenna, J.F.; Segonzac, C.; Wormit, A.; Madhou, P.; Bennett, M.; Mansfield, J.; Zipfel, C.; Hamann, T. Cell wall damage-induced lignin biosynthesis is regulated by a reactive oxygen species- and jasmonic acid-dependent process in *Arabidopsis*. *Plant Physiol.* **2011**, *156*, 1364–1374. [CrossRef]

96. Tenhaken, R. Cell wall remodeling under abiotic stress. *Front. Plant Sci.* **2015**, *5*, 771. [CrossRef] [PubMed]

97. Lee, Y.; Rubio, M.C.; Alassimone, J.; Geldner, N. A mechanism for localized lignin deposition in the endodermis. *Cell* **2013**, *153*, 402–412. [CrossRef]

98. Ozgur, R.; Turkan, I.; Uzilday, B.; Sekmen, A.H. Endoplasmic reticulum stress triggers ROS signalling, changes the redox state, and regulates the antioxidant defence of *Arabidopsis thaliana*. *J. Exp. Bot.* **2014**, *65*, 1377–1390. [CrossRef]

99. Torres, M.A.; Dangl, J.L. Functions of the respiratory burst oxidase in biotic interactions, abiotic stress and development. *Curr. Opin. Plant Biol.* **2005**, *8*, 397–403. [CrossRef]

100. Ma, L.; Zhang, H.; Sun, L.; Jiao, Y.; Zhang, G.; Miao, C.; Hao, F. NADPH oxidase AtrbohD and AtrbohF function in ROS-dependent regulation of Na$_+$/K$_+$ homeostasis in *Arabidopsis* under salt stress. *J. Exp. Bot.* **2012**, *63*, 305–317. [CrossRef]

101. Ben, R.K.; Benzarti, M.; Debez, A.; Bailly, C.; Savouré, A.; Abdelly, C. NADPH oxidase-dependent H$_2$O$_2$ production is required for salt-induced antioxidant defense in *Arabidopsis thaliana*. *J. Plant Physiol.* **2015**, *174*, 5–15.

102. Smokvarska, M.; Francis, C.; Platre, M.P.; Fiche, J.B.; Alcon, C.; Dumont, X.; Nacry, P.; Bayle, V.; Nollmann, M.; Maurel, C.; et al. A plasma membrane nanodomain ensures signal specificity during osmotic signaling in plants. *Curr. Biol.* **2020**, *30*, 4654–4664.e4. [CrossRef] [PubMed]

103. Martinière, A.; Fiche, J.B.; Smokvarska, M.; Mari, S.; Alcon, C.; Dumont, X.; Hematy, K.; Jaillais, Y.; Nollmann, M.; Maurel, C. Osmotic stress activates two reactive oxygen species pathways with distinct effects on protein nanodomains and diffusion. *Plant Physiol.* **2019**, *179*, 1581–1593. [CrossRef]

104. Hao, H.; Fan, L.; Chen, T.; Li, R.; Li, X.; He, Q.; Botella, M.A.; Lin, J. Clathrin and membrane microdomains cooperatively regulate RbohD dynamics and activity in *Arabidopsis*. *Plant Cell* **2014**, *26*, 1729–1745. [CrossRef] [PubMed]

105. Passardi, F.; Penel, C.; Dunand, C. Performing the paradoxical: How plant peroxidases modify the cell wall. *Trends Plant Sci.* **2004**, *9*, 534–540. [CrossRef] [PubMed]

106. Schopfer, P. Hydroxyl radical-induced cell-wall loosening *in vitro* and *in vivo*: Implications for the control of elongation growth. *Plant J.* **2001**, *28*, 679–688. [CrossRef]

107. Jin, T.; Sun, Y.; Zhao, R.; Shan, Z.; Gai, J.; Li, Y. Overexpression of peroxidase gene *GsPRX9* confers salt tolerance in soybean. *Int. J. Mol. Sci.* **2019**, *20*, 3745. [CrossRef]

108. Moussu, S.; Broyart, C.; Santos-Fernandez, G.; Augustin, S.; Wehrle, S.; Grossniklaus, U.; Santiago, J. Structural basis for recognition of RALF peptides by LRX proteins during pollen tube growth. *Proc. Natl. Acad. Sci. USA* **2020**, *117*, 7494–7503. [CrossRef]

109. Rayle, D.L.; Cleland, R. Enhancement of wall loosening and elongation by acid solutions. *Plant Physiol.* **1970**, *46*, 250–253. [CrossRef]

110. Hager, A. Role of the plasma membrane H$^+$-ATPase in auxin-induced elongation growth: Historical and new aspects. *J. Plant Res.* **2003**, *116*, 483–505. [CrossRef]

111. Mangano, S.; Martínez, P.J.; Marino-Buslje, C.; Estevez, J.M. How does pH fit in with oscillating polar growth? *Trends Plant Sci.* **2018**, *23*, 479–489. [CrossRef]

112. Lindner, H.; Müller, L.M.; Boisson-Dernier, A.; Grossniklaus, U. CrRLK1L receptor-like kinases: Not just another brick in the wall. *Curr. Opin. Plant Biol.* **2012**, *15*, 659–669. [CrossRef]

113. Schallus, T.; Jaeckh, C.; Fehér, K.; Palma, A.S.; Liu, Y.; Simpson, J.C.; Mackeen, M.; Stier, G.; Gibson, T.J.; Feizi, T.; et al. Malectin: A novel carbohydrate-binding protein of the endoplasmic reticulum and a candidate player in the early steps of protein *N*-glycosylation. *Mol. Biol. Cell* **2008**, *19*, 3404–3414. [CrossRef]

114. Haruta, M.; Sabat, G.; Stecker, K.; Minkoff, B.B.; Sussman, M.R. A peptide hormone and its receptor protein kinase regulate plant cell expansion. *Science* **2014**, *343*, 408–411. [CrossRef] [PubMed]

115. Geilfus, C.M. The pH of the apoplast: Dynamic factor with functional impact under stress. *Mol. Plant* **2017**, *10*, 1371–1386. [CrossRef]

116. Bose, J.; Rodrigo-Moreno, A.; Lai, D.; Xie, Y.; Shen, W.; Shabala, S. Rapid regulation of the plasma membrane H$_+$-ATPase activity is essential to salinity tolerance in two halophyte species, *Atriplex lentiformis* and *Chenopodium quinoa*. *Ann. Bot.* **2015**, *115*, 481–494. [CrossRef] [PubMed]

117. Shi, H.; Quintero, F.J.; Pardo, J.M.; Zhu, J.K. The putative plasma membrane Na^+/H^+ antiporter SOS1 controls long-distance Na^+ transport in plants. *Plant Cell* **2002**, *14*, 465–477. [CrossRef] [PubMed]

118. Qiu, Q.; Barkla, B.J.; Vera-Estrella, R.; Zhu, J.; Schumaker, K.S. Na^+/H^+ exchange activity in the plasma membrane of *Arabidopsis*. *Plant Physiol.* **2003**, *132*, 1041–1052. [CrossRef]

119. Moore, A.L.; Åkerman, K.E.O. Calcium and plant organelles. *Plant Cell Environ.* **1984**, *7*, 423–429. [CrossRef]

120. Liners, F.; Letesson, J.J.; Didembourg, C.; Van Cutsem, P. Monoclonal antibodies against pectin: Recognition of a conformation induced by calcium. *Plant Physiol.* **1989**, *91*, 1419–1424. [CrossRef]

121. Lopez-Hernandez, F.; Tryfona, T.; Rizza, A.; Yu, X.L.; Harris, M.O.B.; Webb, A.A.R.; Kotake, T.; Dupree, P. Calcium binding by arabinogalactan polysaccharides is important for normal plant development. *Plant Cell* **2020**, *32*, 3346. [CrossRef]

122. Choi, W.G.; Toyota, M.; Kim, S.H.; Hilleary, R.; Gilroy, S. Salt stress-induced Ca^{2+} waves are associated with rapid, long-distance root-to-shoot signaling in plants. *Proc. Natl. Acad. Sci. USA* **2014**, *111*, 6497–6502. [CrossRef]

123. Ogasawara, Y.; Kaya, H.; Hiraoka, G.; Yumoto, F.; Kimura, S.; Kadota, Y.; Hishinuma, H.; Senzaki, E.; Yamagoe, S.; Nagata, K.; et al. Synergistic activation of the *Arabidopsis* NADPH oxidase AtrbohD by Ca^{2+} and phosphorylation. *J. Biol. Chem.* **2008**, *283*, 8885–8892. [CrossRef] [PubMed]

124. Keller, T.; Damude, H.G.; Werner, D.; Doerner, P.; Dixon, R.A.; Lamb, C. A plant homolog of the neutrophil NADPH oxidase $gp91^{phox}$ subunit gene encodes a plasma membrane protein with Ca^{2+} binding motifs. *Plant Cell* **1998**, *10*, 255–266. [CrossRef]

125. Halfter, U.; Ishitani, M.; Zhu, J.K. The *Arabidopsis* SOS2 protein kinase physically interacts with and is activated by the calcium-binding protein SOS3. *Proc. Natl. Acad. Sci. USA* **2000**, *97*, 3735–3740. [CrossRef]

126. Liu, J.; Zhu, J.K. A calcium sensor homolog required for plant salt tolerance. *Science* **1998**, *280*, 1943–1945. [CrossRef]

127. Yuan, F.; Yang, H.; Xue, Y.; Kong, D.; Ye, R.; Li, C.; Zhang, J.; Theprungsirikul, L.; Shrift, T.; Krichilsky, B.; et al. OSCA1 mediates osmotic-stress-evoked Ca^{2+} increases vital for osmosensing in *Arabidopsis*. *Nature* **2014**, *514*, 367–371. [CrossRef] [PubMed]

128. Wu, F.; Chi, Y.; Jiang, Z.; Xu, Y.; Xie, L.; Huang, F.; Wan, D.; Ni, J.; Yuan, F.; Wu, X.; et al. Hydrogen peroxide sensor HPCA1 is an LRR receptor kinase in *Arabidopsis*. *Nature* **2020**, *578*, 577–581. [CrossRef] [PubMed]

129. Jiang, Z.; Zhou, X.; Tao, M.; Yuan, F.; Liu, L.; Wu, F.; Wu, X.; Xiang, Y.; Niu, Y.; Liu, F.; et al. Plant cell-surface GIPC sphingolipids sense salt to trigger Ca^{2+} influx. *Nature* **2019**, *572*, 341–346. [CrossRef]

130. Lin, W.; Tang, W.; Anderson, C.T.; Yang, Z. FERONIA's sensing of cell wall pectin activates ROP GTPase signaling in *Arabidopsis*. *bioRxiv* **2018**, 269647.

131. Hématy, K.; Sado, P.E.; Van Tuinen, A.; Rochange, S.; Desnos, T.; Balzergue, S.; Pelletier, S.; Renou, J.P.; Höfte, H. A receptor-like kinase mediates the response of *Arabidopsis* cells to the inhibition of cellulose synthesis. *Curr. Biol.* **2007**, *17*, 922–931. [CrossRef]

132. Gonneau, M.; Desprez, T.; Martin, M.; Doblas, V.G.; Bacete, L.; Miart, F.; Sormani, R.; Hématy, K.; Renou, J.; Landrein, B.; et al. Receptor kinase THESEUS1 is a rapid alkalinization factor 34 receptor in *Arabidopsis*. *Curr. Biol.* **2018**, *28*, 2452–2458.e4. [CrossRef]

133. Julkowska, M.M.; Klei, K.; Fokkens, L.; Haring, M.A.; Schranz, M.E.; Testerink, C. Natural variation in rosette size under salt stress conditions corresponds to developmental differences between *Arabidopsis* accessions and allelic variation in the *LRR-KISS* gene. *J. Exp. Bot.* **2016**, *67*, 2127–2138. [CrossRef]

134. Van der Does, D.; Boutrot, F.; Engelsdorf, T.; Rhodes, J.; McKenna, J.F.; Vernhettes, S.; Koevoets, I.; Tintor, N.; Veerabagu, M.; Miedes, E.; et al. The *Arabidopsis* leucine-rich repeat receptor kinase MIK2/LRR-KISS connects cell wall integrity sensing, root growth and response to abiotic and biotic stresses. *PLoS Genet.* **2017**, *13*, e1006832. [CrossRef] [PubMed]

135. Rhodes, J.; Yang, H.; Moussu, S.; Boutrot, F.; Santiago, J.; Zipfel, C. Perception of a divergent family of phytocytokines by the *Arabidopsis* receptor kinase MIK2. *Nat. Commun.* **2021**, *12*, 705. [CrossRef]

136. Xu, S.L.; Rahman, A.; Baskin, T.I.; Kieber, J.J. Two leucine-rich repeat receptor kinases mediate signaling, linking cell wall biosynthesis and ACC synthase in *Arabidopsis*. *Plant Cell* **2008**, *20*, 3065–3079. [CrossRef] [PubMed]

137. Basu, D.; Wang, W.; Ma, S.; DeBrosse, T.; Poirier, E.; Emch, K.; Soukup, E.; Tian, L.; Showalter, A.M. Two hydroxyproline galactosyltransferases, GALT5 and GALT2, function in arabinogalactan-protein glycosylation, growth and development in *Arabidopsis*. *PLoS ONE* **2015**, *10*, e0125624. [CrossRef]

138. Engelsdorf, T.; Gigli-Bisceglia, N.; Veerabagu, M.; McKenna, J.F.; Vaahtera, L.; Augstein, F.; Van der Does, D.; Zipfel, C.; Hamann, T. The plant cell wall integrity maintenance and immune signaling systems cooperate to control stress responses in *Arabidopsis thaliana*. *Sci. Signal.* **2018**, *11*, eaao3070. [CrossRef] [PubMed]

139. Nakagawa, Y.; Katagiri, T.; Shinozaki, K.; Qi, Z.; Tatsumi, H.; Furuichi, T.; Kishigami, A.; Sokabe, M.; Kojima, I.; Sato, S.; et al. Arabidopsis plasma membrane protein crucial for Ca^{2+} influx and touch sensing in roots. *Proc. Natl. Acad. Sci. USA* **2007**, *104*, 3639–3644. [CrossRef] [PubMed]

140. Decreux, A.; Messiaen, J. Wall-associated kinase WAK1 interacts with cell wall pectins in a calcium-induced conformation. *Plant Cell Physiol.* **2005**, *46*, 268–278. [CrossRef] [PubMed]

141. He, Z.H.; Fujiki, M.; Kohorn, B.D. A cell wall-associated, receptor-like protein kinase. *J. Biol. Chem.* **1996**, *271*, 19789–19793. [CrossRef] [PubMed]

142. Kohorn, B.D.; Johansen, S.; Shishido, A.; Todorova, T.; Martinez, R.; Defeo, E.; Obregon, P. Pectin activation of MAP kinase and gene expression is WAK2 dependent. *Plant J.* **2009**, *60*, 974–982. [CrossRef] [PubMed]

143. Gramegna, G.; Modesti, V.; Savatin, D.V.; Sicilia, F.; Cervone, F.; De Lorenzo, G. GRP-3 and KAPP, encoding interactors of WAK1, negatively affect defense responses induced by oligogalacturonides and local response to wounding. *J. Exp. Bot.* **2016**, *67*, 1715–1729. [CrossRef] [PubMed]

144. Kohorn, B.D.; Kohorn, S.L.; Todorova, T.; Baptiste, G.; Stansky, K.; McCullough, M. A dominant allele of *Arabidopsis* pectin-binding wall-associated kinase induces a stress response suppressed by MPK6 but not MPK3 mutations. *Mol. Plant* **2012**, *5*, 841–851. [CrossRef]

145. Meco, V.; Egea, I.; Ortíz-Atienza, A.; Drevensek, S.; Esch, E.; Yuste-Lisbona, F.J.; Barneche, F.; Vriezen, W.; Bolarin, M.C.; Lozano, R.; et al. The salt sensitivity induced by disruption of cell wall-associated kinase 1 (*SlWAK1*) tomato gene is linked to altered osmotic and metabolic homeostasis. *Int. J. Mol. Sci.* **2020**, *21*, 6308. [CrossRef] [PubMed]

146. Kaur, R.; Singh, K.; Singh, J. A root-specific wall-associated kinase gene, *HvWAK1*, regulates root growth and is highly divergent in barley and other cereals. *Funct. Integr. Genom.* **2013**, *13*, 167–177. [CrossRef]

147. Safaeizadeh, M.; Boller, T. Differential and tissue-specific activation pattern of the *AtPROPEP* and *AtPEPR* genes in response to biotic and abiotic stress in *Arabidopsis thaliana*. *Plant Signal. Behav.* **2019**, *14*, e1590094. [CrossRef] [PubMed]

148. Nakaminami, K.; Okamoto, M.; Higuchi-Takeuchi, M.; Yoshizumi, T.; Yamaguchi, Y.; Fukao, Y.; Shimizu, M.; Ohashi, C.; Tanaka, M.; Matsui, M.; et al. AtPep3 is a hormone-like peptide that plays a role in the salinity stress tolerance of plants. *Proc. Natl. Acad. Sci. USA* **2018**, *115*, 5810–5815. [CrossRef]

149. Yu, L.; Nie, J.; Cao, C.; Jin, Y.; Yan, M.; Wang, F.; Liu, J.; Xiao, Y.; Liang, Y.; Zhang, W. Phosphatidic acid mediates salt stress response by regulation of MPK6 in *Arabidopsis thaliana*. *New Phytol.* **2010**, *188*, 762–773. [CrossRef]

150. Zhao, C.; Jiang, W.; Zayed, O.; Liu, X.; Tang, K.; Nie, W.; Li, Y.; Xie, S.; Li, Y.; Long, T.; et al. The LRXs-RALFs-FER module controls plant growth and salt stress responses by modulating multiple plant hormones. *Natl. Sci. Rev.* **2020**, *8*, nwaa149. [CrossRef]

151. Chen, J.; Yu, F.; Liu, Y.; Du, C.; Li, X.; Zhu, S.; Wang, X.; Lan, W.; Rodriguez, P.L.; Liu, X.; et al. FERONIA interacts with ABI2-type phosphatases to facilitate signaling cross-talk between abscisic acid and RALF peptide in *Arabidopsis*. *Proc. Natl. Acad. Sci. USA* **2016**, *113*, E5519–E5527. [CrossRef] [PubMed]

152. Yu, F.; Qian, L.; Nibau, C.; Duan, Q.; Kita, D.; Levasseur, K.; Li, X.; Lu, C.; Li, H.; Hou, C.; et al. FERONIA receptor kinase pathway suppresses abscisic acid signaling in *Arabidopsis* by activating ABI2 phosphatase. *Proc. Natl. Acad. Sci. USA* **2012**, *109*, 14693–14698. [CrossRef] [PubMed]

153. Guo, H.; Nolan, T.M.; Song, G.; Liu, S.; Xie, Z.; Chen, J.; Schnable, P.S.; Walley, J.W.; Yin, Y. FERONIA receptor kinase contributes to plant immunity by suppressing jasmonic acid signaling in *Arabidopsis thaliana*. *Curr. Biol.* **2018**, *28*, 3316–3324.e6. [CrossRef] [PubMed]

154. Verma, D.; Jalmi, S.K.; Bhagat, P.K.; Verma, N.; Sinha, A.K. A bHLH transcription factor, MYC2, imparts salt intolerance by regulating proline biosynthesis in *Arabidopsis*. *FEBS J.* **2020**, *287*, 2560–2576. [CrossRef] [PubMed]

155. Wang, C.; Li, J.; Yuan, M. Salt tolerance requires cortical microtubule reorganization in *Arabidopsis*. *Plant Cell Physiol.* **2007**, *48*, 1534–1547. [CrossRef]

156. Fujita, S.; Pytela, J.; Hotta, T.; Kato, T.; Hamada, T.; Akamatsu, R.; Ishida, Y.; Kutsuna, N.; Hasezawa, S.; Nomura, Y.; et al. An atypical tubulin kinase mediates stress-induced microtubule depolymerization in *Arabidopsis*. *Curr. Biol.* **2013**, *23*, 1969–1978. [CrossRef]

157. Wang, S.; Kurepa, J.; Hashimoto, T.; Smalle, J.A. Salt stress-induced disassembly of *Arabidopsis* cortical microtubule arrays involves 26S proteasome-dependent degradation of SPIRAL1. *Plant Cell* **2011**, *23*, 3412–3427. [CrossRef]

158. Zhou, S.; Chen, Q.; Sun, Y.; Li, Y. Histone H2B monoubiquitination regulates salt stress-induced microtubule depolymerization in *Arabidopsis*. *Plant Cell Environ.* **2017**, *40*, 1512–1530. [CrossRef] [PubMed]

159. Zhou, S.; Chen, Q.; Li, X.; Li, Y. MAP65-1 is required for the depolymerization and reorganization of cortical microtubules in the response to salt stress in *Arabidopsis*. *Plant Sci.* **2017**, *264*, 112–121. [CrossRef]

160. Zhang, Q.; Lin, F.; Mao, T.; Nie, J.; Yan, M.; Yuan, M.; Zhang, W. Phosphatidic acid regulates microtubule organization by interacting with MAP65-1 in response to salt stress in *Arabidopsis*. *Plant Cell* **2012**, *24*, 4555–4576. [CrossRef]

161. Li, J.; Zhou, H.; Zhang, Y.; Li, Z.; Yang, Y.; Guo, Y. The GSK3-like kinase BIN2 is a molecular switch between the salt stress response and growth recovery in *Arabidopsis thaliana*. *Dev. Cell* **2020**, *55*, 367–380.e6. [CrossRef] [PubMed]

162. Sánchez-Rodríguez, C.; Ketelaar, K.; Schneider, R.; Villalobos, J.A.; Somerville, C.R.; Persson, S.; Wallace, I.S. BRASSINOSTEROID INSENSITIVE2 negatively regulates cellulose synthesis in *Arabidopsis* by phosphorylating cellulose synthase 1. *Proc. Natl. Acad. Sci. USA* **2017**, *114*, 3533–3538. [CrossRef] [PubMed]

163. Kim, T.; Park, C.H.; Hsu, C.; Zhu, J.; Hsiao, Y.; Branon, T.; Xu, S.; Ting, A.Y.; Wang, Z. Application of TurboID-mediated proximity labeling for mapping a GSK3 kinase signaling network in *Arabidopsis*. *bioRxiv* **2019**, 636324.

164. Schmidt, R.; Schippers, J.H.; Mieulet, D.; Obata, T.; Fernie, A.R.; Guiderdoni, E.; Mueller-Roeber, B. MULTIPASS, a rice R2R3-type MYB transcription factor, regulates adaptive growth by integrating multiple hormonal pathways. *Plant J.* **2013**, *76*, 258–273. [CrossRef]

165. Luo, M.; Wang, Y.; Liu, X.; Yang, S.; Lu, Q.; Cui, Y.; Wu, K. HD2C interacts with HDA6 and is involved in ABA and salt stress response in *Arabidopsis*. *J. Exp. Bot.* **2012**, *63*, 3297–3306. [CrossRef] [PubMed]

166. Hui, L.; Yan, S.; Zhao, L.; Tan, J.; Zhang, Q.; Gao, F.; Wang, P.; Hou, H.; Li, L. Histone acetylation associated up-regulation of the cell wall related genes is involved in salt stress induced maize root swelling. *BMC Plant Biol.* **2014**, *14*, 1–14.

International Journal of
Molecular Sciences

Review

Roles of Plant Growth-Promoting Rhizobacteria (PGPR) in Stimulating Salinity Stress Defense in Plants: A Review

Dung Minh Ha-Tran [1,2,3], Trinh Thi My Nguyen [2], Shih-Hsun Hung [2,4], Eugene Huang [5] and Chieh-Chen Huang [2,6,*,†]

1   Molecular and Biological Agricultural Sciences Program, Taiwan International Graduate Program, Academia Sinica and National Chung Hsing University, Taipei 11529, Taiwan; hatranminhdung@gmail.com
2   Department of Life Sciences, National Chung Hsing University, Taichung 40227, Taiwan; mytrinhnguyen0410@gmail.com (T.T.M.N.); walter030170@gmail.com (S.-H.H.)
3   Graduate Institute of Biotechnology, National Chung Hsing University, Taichung 40227, Taiwan
4   Department of Horticulture, National Chung Hsing University, Taichung 40227, Taiwan
5   College of Agriculture and Natural Resources, National Chung Hsing University, Taichung 40227, Taiwan; eugenehuang@smail.nchu.edu.tw
6   Innovation and Development Center of Sustainable Agriculture, National Chung Hsing University, Taichung 40227, Taiwan
*   Correspondence: cchuang@dragon.nchu.edu.tw
†   Mailing Address: Life Sciences Building 4F. R.403, No.145, Xingda Rd., Taichung 402, Taiwan.

Abstract: To date, soil salinity becomes a huge obstacle for food production worldwide since salt stress is one of the major factors limiting agricultural productivity. It is estimated that a significant loss of crops (20–50%) would be due to drought and salinity. To embark upon this harsh situation, numerous strategies such as plant breeding, plant genetic engineering, and a large variety of agricultural practices including the applications of plant growth-promoting rhizobacteria (PGPR) and seed biopriming technique have been developed to improve plant defense system against salt stress, resulting in higher crop yields to meet human's increasing food demand in the future. In the present review, we update and discuss the advantageous roles of beneficial PGPR as green bioinoculants in mitigating the burden of high saline conditions on morphological parameters and on physio-biochemical attributes of plant crops via diverse mechanisms. In addition, the applications of PGPR as a useful tool in seed biopriming technique are also updated and discussed since this approach exhibits promising potentials in improving seed vigor, rapid seed germination, and seedling growth uniformity. Furthermore, the controversial findings regarding the fluctuation of antioxidants and osmolytes in PGPR-treated plants are also pointed out and discussed.

Keywords: PGPR; salt stress; salinity; abiotic stress; ACC deaminase; seed priming; IAA

check for
updates

Citation: Ha-Tran, D.M.; Nguyen, T.T.M.; Hung, S.-H.; Huang, E.; Huang, C.-C. Roles of Plant Growth-Promoting Rhizobacteria (PGPR) in Stimulating Salinity Stress Defense in Plants: A Review. *Int. J. Mol. Sci.* **2021**, 22, 3154. https:// doi.org/10.3390/ijms22063154

Academic Editor: Daniela Romano

Received: 28 February 2021
Accepted: 18 March 2021
Published: 19 March 2021

Publisher's Note: MDPI stays neutral with regard to jurisdictional claims in published maps and institutional affiliations.

1. Introduction

Soil salinization caused by saline irrigation regimes [1], by water scarcity [2], and by the rise in sea level due to global warming [3]. Another potential source causing soil salinity comes from compost fertilizer since the raw materials for composting operations are food waste and municipal organic waste that contain large quantities of NaCl [4]. Salinity not only hampers crop productivity, but also threatens the sustainability of agro-ecosystems worldwide. The osmotic stress caused by high salinity (100–200 mM) is originated from the reduction in solute potential of soil solution. The reduced solute potential, in turn, leads to the decrease in hydraulic conductance and then in water and solute uptake by plants [5]. This conducts the prevalence of drought-like conditions and makes drought and salinity occur simultaneously in various agricultural systems [6]. Salinity stress also imposes nutrients deficiencies by interfering directly with ion transporters in the root plasma membrane (e.g., K^+-selective ion channels) [7], and by inhibiting root growth [8–10]. Due to the rising severity of salinity on global food production, numerous strategies have been offered to cope with the increasing challenging soil conditions. Along with plant breeding [11], plant genetic engineering [12], and genetic transformation [13], agricultural practices have dramatically contributed to the improvement of plant tolerance to salinity stress. The supplement of calcium (5 mM $CaCl_2$) ameliorated the reduction in shoot and root of salt-treated strawberry plants [14]. The pivotal physio–mechanical property of silicon (Si) has been widely noticed in most plants, especially its alleviating role in improving photosynthetic activity, enhancing essential nutrient uptake, and mitigating negative influence of abiotic stress [15]. In the study of Hassanvand et al. (2019) [16], the reduction of pigment content and essential oil yield in geranium (*Pelargonium graveolens*) plants caused by elevated EC levels was effectively ameliorated by a weekly K_2SiO_3 application. Green leaf volatiles (GLVs) are an important group of volatile organic compounds (VOCs) emitted by plants under stressful conditions [17], and enable plants to activate defense-related genes [18]. Z-3-hexeny-1-yl acetate (Z-3-HAC), a GLV, was used in seed priming to promote a better salt stress tolerance [17]. The Z-3-HAC-primed peanut (*Arachis hypogaea* L.) seedlings exhibited higher antioxidant enzymes (AEs) activities, a higher net photosynthetic rate (Pn), and an increased osmolyte accumulation, while reduced reactive oxygen species (ROS) levels, electrolyte leakage (EL), and lipid peroxidation (LP) as compared to the non-primed plants [17]. Being a metabolic intermediate in higher plants, 5-aminolevulinic acid (ALA) is a common precursor of tetrapyrroles such as chlorophyll (Chl), heme and siroheme, and this small signaling molecule also participates in several physiological processes to counteract salt stress damage [19]. In the study of Wu et al. (2018) [20], an exogenous application of ALA under salinity increased the contents of intermediates and Chl a, Chl b, as well as repaired the damages of photosynthetic apparatus. Melatonin (N-acetyl-5-methoxytryptamine) (Mel), a ubiquitous multifunctional signaling molecule, functions as a stimulator in several physiochemical responses against stresses [21]. The application of exogenous Mel mitigated salt stress by increasing the contents of polyamines (PAs), the ubiquitous cellular components acting as antistress agents, in wheat seedlings [22]. Moreover, salt tolerance in Mel-treated rice plants was improved via the upregulation of K^+ transporter genes, the modulation of K^+ homeostasis and the scavenging of hydroxyl radicals [23].

The bacterization of plant crops with PGPR and the implementation of these useful rhizobacteria in seed biopriming have demonstrated their beneficial properties in enhancing plant growth and development, and in augmenting plant salt stress tolerance through different mechanisms. PGPR aid to alleviate salinity stress in plants by boosting water absorption capability, enhancing essential nutrients uptake, accumulating osmolytes (OS) (e.g., proline (Pro), glutamate (Glu), glycine betaine, soluble sugars, choline, O-sulphate, and polyols), increasing AEs activities (e.g., superoxide dismutase (SOD, EC 1.15.1.1), peroxidase (POD, EC 1.11.1.7), catalase (CAT, EC 1.11.1.6), ascorbate peroxidase (APX, EC 1.11.1.11), monodehydroascorbate reductase (MDAR, EC 1.6.5.4), dehydroascorbate reductase (DHAR, EC 1.8.5.1), glutathione reductase (GR, EC 1.6.4.2), and non-enzymatic antioxidants (NEAs) (e.g., ascorbate (ASC), glutathione (GSH), tocopherols (TCP), carotenoids (Car), and polyphenols (PPs)) in plant tissues [24–28]. In all types of salinity, sodium chloride (NaCl) is the most soluble and widespread salt [17] and Na^+ is the primary cause of ion-specific damage for many plants, especially for graminaceous crops [29]. Consequently, to narrow down the scope of this review, we focus mainly on the negative effects of Na^+ ion on plants, although high concentrations of Cl^- anion are also toxic to plants. In this point of view, three terms "salt", "saline", and "Na^+" were used interchangeably in the review to indicate the salinity.

2. Adverse Effects of Salinity on Plants

2.1. Na⁺ Accumulation, Nutrients Uptake Inhibition, and Plant Growth Reduction

Under salt stress, Na^+ is accumulated at higher concentrations in plant tissues, causing changes in Na^+/K^+ ratio and the inhibition of essential nutrient uptake [14,24,30]. This could be attributed to the competition between similar ionic radii of Na^+ and K^+ in soils [31], causing the dysfunctional ionic selectivity of the cell membranes. In the review of Manishankar et al. [32], high Na^+ concentration in soil can change soil texture, leading to a decrease in soil porosity. This leads to the reduction of soil aeration and water conductance. Also according to Manishankar et al. [32], the zones of low water potential caused by high salt deposition in the soil make difficult for the roots to uptake water and nutrients. With 35 mM NaCl treatment, the Na^+ concentration in strawberry leaves and roots was 3.4-fold higher than that in the control plants [14]. Moreover, salt stress also caused the critical reduction in the fruit yield (FY) with 35% yield loss in the variety Camarosa and 45% in the variety Oso Grande [14]. At 150 mM NaCl, a tremendous increase (50.4-fold) in the Na^+ content, and an increase in the Na^+/K^+ ratio (1.48 vs. 0.02) in the roots of *Broussonetia papyrifera*, a woody plant used in paper industry, was recorded, in harmony with the decrease in K^+ (25.6%), Ca^{2+} (23.3%), Mg^{2+} (21.4%), and P^{3+} (8.4%) contents [33]. In contrast, an upsurge of Na^+ concentration was found in the leaves of canola plants (Brassica napus L.), with approximately 4-fold greater than that in the roots [30]. The Na^+ content in the common bean leaves (*Phaseolus vulgaris* L.) was 5–7–fold higher than that in the control common bean leaves, whereas the K^+ content was decreased by 32–35% relative to the control plants [34]. Likewise, the Na^+ content in the salt treated-chickpea leaves (*Cicer arietinum* L.) was 3.2-fold higher than that in the control leaves, leading changes in Na^+/K^+ ratio from 0.31 in the non-saline condition to 2.24 in the saline condition. The reduction in N, K, Ca, Mg contents was also recorded by 54%, 55%, 60%, and 55%, respectively as compared to those in the control leaves [35].

In general, phytotoxicity caused by high salt concentrations was found under in vitro and greenhouse conditions and the toxic symptoms increase correlatively with the increase in NaCl treatments. High salinity significantly affects plant growth and physio-biochemical aspects, resulting in the decrease in germination rate (GRA), fresh and dry matters, photosynthetic pigments, essential nutrients uptake, and most importantly, in the loss of final crop yields. In contrast, a significant increase in AEs activities, osmoregulators, LP, membrane damage, ROS contents, Na^+ accumulation, and Na^+/K^+ ratio was obviously observed with the increasing NaCl concentrations [36]. The shoot dry weight (SDW) and root dry weight (RDW) of 35 mM NaCl-treated strawberry plants (*Fragaria x ananassa* Duch) were 45.8% and 58.6% lower than those in the control plants, respectively [14]. Salt stress adversely hampers all stages of plant growth, causing the reduction in FY (227 vs. 415 g/plant), fruit weight (FW) (8.4 vs. 9.6 g/fruit), number of fruits per plant (NF) (27 vs. 43), and water-soluble dry matter (SDM) (6.6% vs. 8.4%) of stressed plants relative to the unstressed plants. The root dry weight (RDW) of common bean decreased by 59–61% and the final yield lost by 27–30% under 200 mM NaCl [34]. Similarly, at 200 mM NaCl concentration, salt stress reduced 38% SDW and 50% RDW of chickpea (*Cicer arietinum* cv. Giza 1) compared to the control plants [35]. Regarding the influence of salinity on nutritional values, although moderate saline stress enhanced glucosinolates and antioxidants contents in broccoli (*Brassica oleracea* L. var. *italica* cv. Marathon) (40 mM NaCl) [22], and the application of 6 dS m^{-1} (66 mM NaCl) increased the contents of lycopene, β-carotene, vitamin C and overall phenolic compounds (PCs) of tomato fruits [23], high salinity concentrations (15 dS m^{-1} ~ 200 mM) markedly reduced protein, fat, and crude fiber contents of wheat grains [24]. Moreover, the fruit size of tomato [37] and the FW of pepper [38], which are considered major determinants of price and marketable characteristics, were strongly reduced with the increase of saline levels. However, it is noteworthy that although high salt concentration affects plants in an adverse manner, the definition of "low", "moderate", or "high" salinity depends fundamentally on plant variety, growth stage, nutrient composition in soil, and irrigation regime, etc.

2.2. Impairment of Physio-Biochemical Attributes

2.2.1. Reduction in Photosynthetic Pigments

Salinity stress causes an unrepairable damage to the photosynthetic apparatus at any development stage of plant's life as it alters the chloroplasts structure, degrades chloroplast envelope, and triggers chloroplast protrusions [39]. Numerous studies indicated that high salinity led to a serious degradation of Chl and Car in salt-stressed plants; however, the degrees of reduction in these photosynthetic pigments (PhoPs) depended largely on plant species, plant age, NaCl concentration and the duration of salt stress exposure. Specifically, only 9%, 11%, 13%, and 14% reduction in the total chlorophyll (Tchl) were determined in the rice (*Oryza sativa* L.) [40], soybean (*Glycine max* (L.) Merr.) [41], maize (*Zea mays* L.) [42], and cucumber (*Cucumis sativus* L.) [43] seedlings, respectively. However, contrary to these studies, the reduced contents of Tchl were tremendously varied from 22% in oat (*Avena sativa*) seedlings [44], 41–42% in tomato (*Solanum lycopersicum* L.) seedlings [45,46], 44% in tomato (*Solanum lycopersicum* L.) plants [47], 50% in peanut (*Arachis hypogaea*) seedlings [48], 56% in rice (*Oryza sativa* L.) seedlings [49] to 61% in rapeseed (*Brassica napus* L.) plants [30], respectively. In addition, under salt detriment, 16% of Car decreased in the ginseng plantlets, 19% of Car reduced in the mung bean plants, and 49% of Car decreased in the tomato seedings were reported by Sukweenadhi et al. (2018) [50], Shahid et al. (2021) [36], and Akram et al. (2019) [45], respectively. In addition, NaCl toxicity also declined Pn, stomatal conductance, and transpiration rate in the stressed plants [51].

2.2.2. Increase in LP

Lipids are essential components of cell membranes responsible for structure maintenance and cell functions control [52]. ROS are generated from several life processes and an excess of ROS can damage cell, tissues and organs [53]. Salinity exposure brings about a disturbance, an overflow, or even a disruption of electron transport chains (ETC) in mitochondria and chloroplasts in higher plants, resulting in ROS accumulation [54]. The major site involved in the production of $O_2^{\bullet-}$ is the photosystem I (PSI). In the presence of light, O_2 which is continuously provided by the water autolysis (Reaction 1: $2H_2O \rightarrow 4\,e^- + O_2 + 4\,H^+$) can be reduced to $O_2^{\bullet-}$ (Reaction 2: $2O_2 + 2\,e^- \rightarrow 2\,O_2^{\bullet-}$). The excess amount of reduced ferredoxin (Fd_{red}) and the limited NADP availability induce the autoxidation of Fd_{red} to Fd_{ox} and the generation of $O_2^{\bullet-}$ (Reaction 3: $Fd_{red} + O_2 \rightarrow Fd_{ox} + O_2^{\bullet-}$). In addition, the Fd_{red} can react with $O_2^{\bullet-}$ to form H_2O_2 (Reaction 4: $Fd_{red} + O_2^{\bullet-} + 2\,H^+ \rightarrow Fd_{ox} + H_2O_2$). Lipids are primary targets of ROS attack and the free radicals oxidation of polyunsaturated fatty acids is called LP [55]. As a byproduct of LP, malonaldehyde (MDA) has been largely used as an important indicator to evaluate the extent of damaging effects caused by ROS and oxidative stress combination on membrane lipids to reduce membrane stability [56]. The MDA content was tremendously increased by 36% in ginseng root plantlets [50], 39% in maize [42], 47% in peanut [48], 70% in chickpea [35], 131% in oat [44], 153% in mung bean [36], and 300% in rice seedlings [57], indicating an severe damage to cell membrane integrity and/or membrane permeability during salinity exposure [58].

2.3. Increased Accumulation of ROS and Elevated Production of AEs, NEAs, and OS

On the one hand, ROS function as signaling molecules to mediate a wide range of important biological processes during plant growth and development such as seed germination [59], cell differentiation [60], root primary growth [61], and stem cell activities [62]. On the other hand, an elevated accumulation of ROS in plant tissues also causes oxidative damage to protein, DNA, lipids, and Chl biosynthesis [63,64]. Salinity stress brings about excessive accumulations of ROS including superoxide radical ($O_2^{\bullet-}$), hydrogen peroxide (H_2O_2), singlet oxygen ($^1O_2$), and hydroxyl radicals ($^{\bullet}OH$), which disturb cellular redox homeostasis and lead to oxidative stress [65]. ROS homeostasis, therefore, is essential to maintain a delicate balance for plant growth, especially under environmentally adverse conditions. To deal with salinity-derived oxidative stress, plants possess enzymatic defense system that synthesizes an array of AEs, along with NEAs to neutralize and detoxify

ROS [26,27]. The AEs conduct the scavenging activity by breaking down and removing free radicals, while the NEAs perform their scavenging functions by interrupting free radical chain reactions [66]. Furthermore, the accelerated synthesis and accumulation of OS are also the common responses executed by plants to provide osmotic adjustments and to protect cell membrane integrity [67]. In plants, Pro is synthesized by either glutamate pathway or orinithine pathway [68] and is accumulated in cytosol and vacuole under stressful conditions. Under non-stress conditions, Pro only accounts for less than 5% of the total pool of free amino acids in plants. However, under various stresses, the Pro concentration might increase up to 80% of the total amino acid pool, indicating its vital roles in ROS homeostasis and water balance in plants [69]. Pro was found to exhibit protective roles against damages caused by $^1O_2$ or $^{\bullet}OH$ [70]. Ethylene (C_2H_4), a small volatile phytohormone in higher plants, is involved in all stages of plant growth and development, from seed germination to fruit ripening [71]. Furthermore, ethylene has been considered as a stress hormone since it participates in plant responses to various types of stress such as wounding [72], salinity [73], and drought [74]. Although a small amount of ethylene, which is immediately produced after the onset of a stress, can initiate the systemic resistance in plants, the excess amount of ethylene from the second peak could bring about the inhibition of plant growth or even lead to cell death [75].

3. Plant Growth-Promoting Rhizobacteria as the Promising Bioinoculants for Plant Crops

3.1. Key Criteria for Being Applicable PGPR

The close alliance among soil, plant, and microbes exists during the entire life cycle of plants promotes plant development, induces systemic resistance in the host plant against pathogens and mitigates salinity stress [76]. PGPR have been widely used for decades to control insects pests [77], plant diseases [78], to promote plant growth [79], to manage nutrient [80], and to alleviate abiotic stress [81]. The ameliorative functions of PGPR consist of three aspects, namely, the ability to protect themselves against hyperosmotic conditions and abnormal NaCl concentrations, the capacity to aid plant tolerate better to elevated salinity, and to improve soil quality [82]. Regarding the alleviating roles of PGPR in promoting plant salinity tolerance, PGPR exhibit beneficial traits in mitigating the toxic effects of high salt concentrations on morphological, physiological, and biochemical processes in plants, resulting in the significant rescue of yield loss. According to Fouda et al. [83], the application of PGPR could ameliorate the negative impacts of salinity via two main mechanisms as follows: (1) PGPR activate stress response systems in the host plants soon after the exposure of the plants to salinity, and (2) PGPR synthesize anti-stress biochemicals such as AEs, NEAs, and OS that are responsible for the removal of ROS [84]. Furthermore, PGPR can also mitigate salt stress symptoms by producing Na^+-binding exopolysaccharides (EPS), improving ion homeostasis, decreasing ethylene levels through enzyme 1-aminocyclopropane-1-carboxylate (ACC) deaminase, and synthesizing phytohormones [85–87].

3.1.1. ACC Deaminase-Producing PGPR and Other Plant Growth Promoting Attributes

Enzyme ACC deaminase [EC 4.1.99.4] catalyzes the cleavage of 1-aminocyclopropane-1-carboxylate (ACC), an intermediate precursor of ethylene in higher plants, to produce α-ketobutyrate and ammonia [88]. A proper amount of ethylene derived from the existing pool of ACC, or so called the small peak of ethylene in the biphasic ethylene response model described by Glick et al. [89] and Pierik et al. [90], is thought to be useful to plants in activating plant defensive responses to stress stimuli (e.g., temperature extremes, drought or flooding, insect pest damages, phytopathogens, and mechanical wounding) [91]. However, an elevated ethylene accumulation, also called stress ethylene or the larger peak of ethylene in the biphasic model, may cause harmful effects (e.g., chlorosis, abscission, and senescence) on plant growth [92], even lead to dead when present at high concentrations in plant tissues [93]. Although PGPR possess many different mechanisms to maintain plant growth under salinity detriment, the production of ACC deaminase is extremely important in reducing the elevated levels of ethylene, thereby indirectly support plant growth. The ACC deaminase-producing PGPR that live on plant surfaces or colonize in the plant tissues function as a sink for ACC [30] and the use of ACC as a nitrogen (N) source is beneficial to plant health since N uptake is always suppressed under salt conditions [94]. Up to now, a plethora of PGPR that have been studied to evaluate their roles in mitigating salinity stress in plants. The PGPR, namely, *Pseudomonas putida* UW4 [30], *Arthrobacter protophormiae* [95], *Enterobacter* sp. EJ01 [96], *Enterobacter* sp. UPMR18 [97], *Zhihengliuella halotolerans*, *Bacillus gibsonii*, *Halomonas* sp. [98], *Chryseobacterium gleum* sp. SUK [99], *Pseudomonas fluorescens* 002 [100], *Microbacterium oleivorans* KNUC7074, *Brevibacterium iodinum* KNUC7183, and *Rhizobium massiliae* KNUC7586 [101], *Stenotrophomonas maltophilia* SBP-9 [102], *Enterobacter* sp. P23 [49], *Burkholderia* sp. MTCC 12259 [57], *Paenibacillus yonginensis* DCY84 [50], *Bacillus pumilus* strain FAB10 [51], *Pantoea agglomerans* [103], *Aneurinibacillus aneurinilyticus* and *Paenibacillus* sp. [88], *Leclercia adecarboxylata* MO1 [104], *Pseudomonas argentinensis* and *Pseudomonas azotoformans* [105], *Bacillus subtilis* (NBRI 28B), *B. subtilis* (NBRI 33 N), *Bacillus safensis* (NBRI 12 M) [106], *Bacillus megaterium* NRCB001, *B. subtilis* subsp. *subtilis* NRCB002, *B. subtilis* NRCB003 [107], and *Kosakonia sacchari* [36] can produce ACC deaminase, as well as other important products such as indole-3-acetic acid (IAA), siderophore (Sid), EPS, and Pro. In addition, PGPR can conduct biofilm forming, N fixation, phosphate (P) solubilization, hydrogen cyanide (HCN) and antifungal enzymes production [99]. The capability of PGPR for moderating salinity damage could be considered an indispensable trait for strain selection [108], reflecting in the elevated amounts of ACC deaminase, IAA, EPS, GSH, and Pro produced by themselves during salt exposure to protect their cells against the damaging effects of high NaCl concentrations. For instance, at 500 mM NaCl, *Sphingomonas* sp. LK11 produced more GSH and Pro to counteract the detrimental effects of salinity imposed on its growth [108]. Similarly, the productions of ACC deaminase and Pro by the halotolerant *Burkholderia* sp. MTCC 12259 were highly correlated with the increasing NaCl concentrations in the medium broth, in which ACC deaminase reached the highest at 600 mM NaCl, while the highest Pro level was obtained at 1000 mM NaCl [57]. This result was in accordance with the report of Ilyas et al. [109] when the Pro produced by a consortium consisting of *Bacillus* sp. (KF719179), *Azospirillum brasilense* (KJ194586), *Azospirillum lipoferum* (KJ434039), and *Pseudomonas stutzeri* (KJ685889) reached the maximum value at the highest NaCl concentration (10%, *w/v*). Also, the productions of ROS-quenching enzymes SOD, CAT, POD, PPO, and Pro in *Enterobacter* sp. P23 were increased with the increase in NaCl concentrations [49]. The levels of IAA, Sid, and ACC deaminase produced by *K. sacchari* strain MSK1 were increased with the increasing NaCl concentrations and reached the highest levels at the highest NaCl concentration (400 mM) [36]. Recently, Misra and Chauhan [106] found that two *B. subtilis* strains NBRI 28B, NBRI 33N, and *B. safensis* NBRI 12 M increased the production of ACC deaminase, biofilm, EPS, and Alginate (Alg) in proportion to the increasing NaCl concentrations in nutrient broth. This finding was in corroboration with the previous study of Mukherjee et al. [110], in which *Halomonas* sp. Exo1 could tolerate up to 20% (*w/v*) salt concentration and its EPS yield was directly proportional to the increasing NaCl. These findings indicate that to be selected as potential bioinoculants for improving crop yield in saline soil, the PGPR candidates need to possess an ability to withstand and respond appropriately to high salinity in the environment.

3.1.2. Improvements of Growth Parameters, Nutrients Uptake, and Photosynthetic Pigments in PGPR-Inoculated Plants under Non-Stress Conditions

The halotolerant bacterium *Enterobacter* sp. strain P23 isolated from India's rice fields possesses the abilities to exhibit high ACC deaminase activity, to solubilize P, to produce IAA, Sid, and HCN [49]. In non-tress conditions, the P23-inoculated rice seedlings showed better morphological parameters, namely shoot length (SL), root length (RL), shoot fresh weight (SFW), SDW, root fresh weight (RFW), and RDW, higher Chl content than those in the non-inoculated rice seedlings. This result was consistent with numerous other studies where the PGPR-inoculated plants grew better than the non-inoculated plants in normal environments. Specifically, the values of SFW, RFW, SDW, RDW, Chl a, Chl b, Car, and N, P, and K concentrations in the S20-inoculated maize seedlings were increased by 2%, 6%, 5%, 2%, 4%, 7%, 2%, 16%, 43%, and 2%, respectively, as compared to the control seedlings [111]. Also, in the *Chryseobacterium gleum* sp. SUK + feather lysate inoculum (FLI)-inoculated wheat seedlings, an increase in 24% Tchl, and in 13% amino acids was noticed [99]. Likewise, an increase in SL, RL, SFW, RFW, and Tchl was observed in the *L. adecarboxylata*-inoculated tomato plants with 22%, 16%, 28%, 51%, and 13% higher than those in the control plants, respectively [104]. The same trend in increased vegetative parameters was found in the studies of Li and Jiang [42], Khan et al. [40], Sapre et al. [44], Sarkar et al. [49], Akram et al. [45], and Alexander et al. [48]. The increase in Tchl was widely observed in various studies, however, the extent to which these pigments increased depends on PGPR strains, NaCl treatments, and plant species. For instance, only a 5% Chl increase in maize seedling bacterized with *B. aquimaris* DY-3 was noticed by Li and Jiang (2017) [42], whereas a 12% increase in *P. putida* H-2-3-inoculated soybean seedlings [41], a 17% increase in *S. maltophilia* BJ01-peanut seedlings [48], a 29% increase in *K. sacchari*-treated mung bean seedlings [36], a 41% increase in *Bacillus megaterium* BMA12-bacterized tomato seedlings [45], 46% in *B. pupilus*-inoculated rice seedlings [40], and 60% in *A. brasilense*-treated white clover plants [58].

PGPR can change root-system architecture by producing phytohormones, especially auxins (Aux) [112], volatile compounds [113,114], and by mediating plant ethylene levels via enzyme ACC deaminase [115]. The inoculation of Arabidopsis plants with *Bacillus megaterium* caused a suppression in primary root growth, while induced lateral root growth development, increased lateral root number, and promoted root hair length [116]. Recently, the research group of Chu et al. (2020) [117] also found that *Pseudomonas* PS01 inhibited the elongation of primary roots and triggered the formation of lateral root and the development of root hair. López-Bucio and colleagues [116] suggested that the inhibition of primary root was caused by a decrease in cell elongation and by a reduced cell proliferation in the root meristem. Vegetative parameters of the endophytes-inoculated sorghum plants (*Sorghum bicolor*) were widely varied with different endophytic PGPR species [118]. Intriguingly, although the amounts of IAA produced by *Pseudomonas plecoglossicida*-R382, *Serratia marcescens*-R381, *Pantoea coffeiphila*-R342, *Bacillus cereus*-R8, *Rhodopseudomonas boonkerdii*-R102, and *Nocardioides aromaticivorans*-R21 were comparable, the RDWs of their respective inoculated sorghum plants were significantly different [118]. This finding suggests that besides the effects of the bacterial IAA on root plant architecture, the interactions between plant and microbe are multifaceted and might play a major role in shaping root system development [119].

The positive influences of PGPR treatment on fruit/grain quality, total yield, and marketable grade yield were also investigated. The FY, fruit marketable yield (FMY), FW, fruit length (FL), fruit diameter (FD), and texture of red fruit in *Bacillus subtilis* BEB-13bs-inoculated tomato plants were improved by 21%, 6%, 29%, 9%, and 5%, respectively in comparison with the control plants [120]. The maximum grain yield was recorded in the wheat plants treated with a triple combination of *Bacillus megaterium*, *Enterobacter* sp. and *Arthrobacter chlorophenolicus* [121], as well as the highest nutrient contents (e.g., N, P, Cu, Zn, Mn, and Fe) were observed in the treated wheat grains.

Nevertheless, in some exception cases, the applications of PGPR under normal conditions did not promote plant growth and yield. The PGPR even exhibited negative effects on the growth of eggplant and tomato plants as reported in the studies of Abd El-Azeem et al. [24] and Vaishnav et al., respectively [47]. Specifically, the SFW, SDW, and yield of eggplant were decreased by 8%, 9%, 12%, respectively after inoculated with *X. autotrophicus* BM13, decreased by 12%, 21%, and 30%, respectively when inoculated with *Bacillus brevis* FK2 [24], as well as the SL of *Sphingobacterium* BHU-AV3-inoculated tomato was reduced by 11% [47]. Similarly, the SL, RL, and total plant fresh weight (TPFW) of *C. gleum*-inoculated wheat plants were decreased by 16%, 36%, and 13%, respectively relative to the control [99]. The data in these previous reports were in accordance with our preliminary data (unpublished data) as the SFW and RFW values of the *Curtobacterium* sp. C1-inoculated Arabidopsis plants were lower than those in the uninoculated plants.

Although the suppressive impacts of PGPR on plant growth and yield are scarcely recorded under non-stress conditions, this should be taken into consideration prior to PGPR bacterization practices in field. Furthermore, the response of plant variety to PGPR is genotype-dependent as shown in the report of Nawaz et al. [122] where the salt tolerant wheat genotype Aas-11 responded positively to *Bacillus pumilus* and *Exiguobacterium aurantiacum*, whereas the salt sensitive wheat genotype Galaxy-13 responded better to *Pseudomonas fluorescence*. In this regard, we should agree that the interactions between host plants and microbes are complicated and not always a win–win situation. In addition, the adaptation of plant species to PGPR might markedly vary from case to case due to genetic variation. More investigations at molecular level are required to deeply elucidate the multi-dimensional impacts of microbes on plants.

3.1.3. Improvements of Growth Parameters, Nutrients Uptake, and Photosynthesis in PGPR-Inoculated Plants under Salinity Conditions

Although PGPR can promote plant growth and improve nutrients uptake, as well as stimulate the synthesis of PhoPs in non-stress environments, their ameliorative roles in plant defense responses are fully expressed till plant crops endure harsh environmental conditions. In the reports of Awad et al. [123] and Abd El-Ghany and Attia [124], they found that the bacterization of maize (*Zea mays* L.) plants and faba bean (*Vicia faba* cv. Giza3) seeds with *Azotobacter chroococcum*, an EPS-producing bacterium, had the decreased Na^+ and Cl^- concentrations and the increased N, P, and K concentrations in their plant tissues. PPs, known as potent antioxidants, can eliminate radical species (e.g., $^1O_2$, $O_2^{\bullet-}$, OH^-, H_2O_2), thus preventing the propagation of oxidative chain reactions [125]. In the study of Hichem et al. [126], the amounts of total PPs including phenolic acids, flavonoids, anthocyanins and proanthocyanidins increased accordingly with the increased salinity in young and mature maize leaves and the elevated concentrations of these PCs had an inverse correlation with H_2O_2 content and LP level in leaves, indicating the scavenging activity of endogenous PCs against free radicals [127]. The total PPs in the leaves of *Azotobacter chroococcum*-inoculated maize seedlings were always higher than those in the non-inoculated maize seedlings, regardless of salt concentrations [128]. Moreover, the total PPs reached the highest level at the highest NaCl treatment (5.85 g NaCl/kg soil). Abd_Allah et al. [35], who evaluated the effects of endophytic *B. subtilis* (BERA71) on mitigating saline soil stress in chickpea plants (*Cicer arietinum* cv. Giza 1), found that the *B. subtilis* (BERA71)-inoculated chickpea plants yielded higher plant biomass, achieved higher photosynthetic pigments, while reduced ROS levels, and LP compared to the non-inoculated seedlings. The positive correlation between Pro accumulation and salt stress adaptation has been widely recognized. However, the results are still controversial, and more investigations should be conducted to thoroughly explain the underlying mechanisms that regulate AEs and OS production.

Regarding nutrient acquisition, the PGPR helped to decrease Na^+ accumulation, whereas enhanced the acquisition of N, Ca, Mg, and K contents in the chickpea plants [35]. The increased uptake of Mg^{2+} induced by *Bacillus subtilis* and *Bacillus pumilus* inoculation was associated with the elevated PhoPs contents since Mg^{2+} is the major component of Chl [40,129]. Accordingly, the expression level of *Cab2*, the gene encoding a Chl a/b protein in Arabidopsis plant, was downregulated in Mg-deficient plants before any obvious symptom of chlorophyll deficiency appears [130]. However, Abd_Allah and his colleagues [35] did not investigate the mechanisms that enhanced the uptake of essential nutrients. Therefore, it is unclear whether the increased nutrient acquisition in the *B. subtilis*-inoculated chickpea plants was due to the modulation of root architecture [117,131], the mobilization of P in the soil [132,133], or the N fixation [134,135] induced by *B. subtilis*. Similarly, Khan et al. [40] noticed a limited uptake of Na^+ in *B. pumilus*-inoculated paddy plants, but the fundamental mechanism that suppressed Na^+ uptake was not thoroughly investigated yet. In contrast, an extensive accumulation of Na^+ was observed in the shoots of

Bacillus-inoculated halophyte *Arthrocnemum macrostachyum* under high NaCl concentration (1030 mM) [136]. Up to now, a plenty of studies recognize the roles of PGPR in increasing K^+/Na^+ ratio, in activating K^+-Na^+ selectivity, in maintaining PhoPs, in enhancing nutrient uptakes, thereby alleviating salt stress in saline environments [40]. However, more studies are needed to clearly elucidate the mechanisms underlying these phenomena. The key findings in recent PGPR studies were presented in Table 1.

3.1.4. Improvements of Growth Parameters, Nutrients Uptake, and Photosynthesis in PGPR-Primed Seeds and Their Respective Seedlings under Salinity Conditions

Seed is a dramatically important component of agricultural production since it is considered the primary determinant in establishing a fruitful crop. Moreover, seed germination is the first and the most critical stages of the plant's life cycle [137,138]. The uniformity of seed germination is one of the fundamental criteria that is used to evaluate SV [139]. In the era of climate change, seeds always suffer from the environmental challenges that may cause the reduction in seed GRA, GP, and the dysfunction of seedlings, resulting in a decrease in ultimate crop yields. Germinating seeds and seedlings appear to be more sensible to salinity than the growing plants since the germination stage occurs on saline soil surface where the drought-like condition reduces SV, suppresses protein synthesis, and disturbs structural organization in germinating embryos [140,141]. In addition, seed germination is strongly associated with the seedlings survival rate, as well as the subsequent vegetative growth [142]. α-amylase is a key player in starch hydrolysis during seed germination since it supplies carbon source and energy to germinating seeds in the early stages of development before the initiation of the photosynthetic machinery [137]. A reduced water uptake and a decrease in α-amylase activity caused by NaCl might cause the delay of germination time [143]. Furthermore, the data from Dehnavi et al. (2020) [138] demonstrated that salinity accounted for 98% of the variation in tested parameters including GP, germination index, mean germination time, SVI, SL, and RL of seedlings, fresh and dry weight of seedlings, and salinity tolerance indices.

Seed biopriming with living PGPR inoculum stimulates a speed and an uniformity of gemination, assures a rapid, uniform, and high establishment of crops, thereby improving yield and fruit/grain quality in non-stress and harsh conditions [144]. Under non-stress conditions, the GRAs of two endangered fir plant species *Abies hickelii* and *Abies religiosa* were highly stimulated by a combination of 12 h-hydropriming with PGPR biopriming, resulting an improved GRA up to 91% of *P. fluorescens* JUV8-primed *A. hickelli* seeds vs. 28% of unprimed control and up to 68% of *B. subtilis* BsUV-primed *A. religiosa* seeds vs. 32% of unprimed control [145]. Similarly, the GRA of isolate Ac26-primed wheat seeds was increased to 93.3% and the vigor index was 2830.7, much higher than those of the unprimed control with 53.3% and 1097.5, respectively [146]. The subsequent development of primed plants was also better than the unprimed plants, suggesting the lasting impacts of PGPR treatment on physio–biochemical attributes of the treated plants [145,146].

Table 1. Ameliorative effects of plant growth-promoting rhizobacteria (PGPR) on plant growth and physio-biochemical parameters under salinity conditions.

PGPR	Treatments	GP	Hormones	PhoPs	MDA	AEs	NEAs	Pro	Ion Content	Sources
B. cepacia SE4, *Promicromonospora* SE188, and *A. calcoaceticus* SE370. 7-day-old tomato seedlings inoculated with PGPR.	Control									[43]
	120 mM NaCl + Uninoculated	↓ 17% SFW ↓ 25% SDW	↑ 255% ABA, ↑ 194% SA	↓ 14% Tchl	N/A	↑ 86% CAT ↑ 213% POD ↑ 456% PPO	↑ 79% PP	N/A	In shoot: ↑ 740% Na$^+$ ↓ 4% K$^+$ Na$^+$/K$^+$ ratio ~0.28	
	120 mM NaCl + SE4	↓ 11% SFW ↓ 8% SDW	↑ 10% ABA ↑ 367% SA	↓ 0% Tchl		↑ 27% CAT ↑ 163% POD ↑ 333% PPO	↑ 45% PP		In shoot: ↑ 297% Na$^+$ ↑ 17% K$^+$ Na$^+$/K$^+$ ratio ~0.11	
	120 mM NaCl + SE118	↓ 13% SFW ↓ 6% SDW	↑ 6% ABA ↑ 217% SA	↓ 0% Tchl		↑ 23% CAT ↑ 131% POD ↑ 322% PPO	↑ 35% PP			
	120 mM NaCl + SE370	↓ 10% SFW ↓ 9% SDW	↑ 23% ABA, ↑ 261% SA	↓ 0% Tchl		↑ 46% CAT ↑ 156% POD ↑ 322% PPO	↑ 52% PP			
P. putida H-2-3. 21-day-old soybean seedlings inoculated with *P. putida*.	Control									[41]
	120 mM NaCl + Uninoculated	↓ 18% SL ↓ 12% TPFW	↑ 33% ABA ↑ 114% SA ↓ 11% JA	↓ 11% Tchl	N/A	↑ 301% SOD	↓ 23% total PP	N/A	In whole plant: ↑ 86% Na$^+$, ↑ 55% P	
	0 mM NaCl + *P. putida*	↑ 17% SL ↑ 8% TPFW	↑ 18% ABA ↑ 29% SA ↓ 25% JA	↑ 12% Tchl		↑ 2% SOD	Unchanged total PP		In whole plant: ↓ 17% Na$^+$, ↑ 22% P	
	120 mM NaCl + *P. putida*	↓ 9% SL ↓ 0% TPFW	↓ 6% ABA ↓ 26% SA ↑ 54% JA	↓ 7% Tchl		↑ 4% SOD	↑ 4% total PP		In whole plant: ↑ 45% Na$^+$, ↑ 30% P	
B. pumilus. 14-day-old rice seedlings inoculated with *B. pumilus*.	Control									[40]
	0 mM NaCl + *B. pumilus*	↑ 22% SFW		↑ 46% Tchl	N/A	↑ 22% SOD ↑ 20% POD ↑ 73% CAT	N/A	↑ 7%	In shoot: ↓ 54% Na$^+$, ↑ 57% K$^+$, ↑ 76% Mg^{2+}, ↑ 18% Ca^{2+}, Na$^+$/K$^+$ ratio ~0.27	
	10 ppm Boron + Uninoculated	↓ 0% SFW	N/A	↓ 18% Tchl		↑ 274% SOD ↑ 212% POD ↑ 204% CAT		↑ 41%	In shoot: ↓ 23% Na$^+$, ↑ 7% K$^+$, ↑ 5% Mg^{2+}, ↑ 0% Ca^{2+}, Na$^+$/K$^+$ ratio ~0.67	

Table 1. *Cont.*

PGPR	Treatments	GP	Hormones	PhoPs	MDA	AEs	NEAs	Pro	Ion Content	Sources
	10 ppm Boron + Inoculated	↑ 18% SFW		↑ 59% Tchl		↑ 400% SOD ↑ 272% POD ↑ 254% CAT		↑ 74%	In shoot: ↓ 31% Na$^+$, ↑ 61% K$^+$, ↑ 67% Mg^{2+}, ↑ 18% Ca^{2+}, Na$^+$/K$^+$ ratio ~0.4	
	150 mM NaCl + Uninoculated	↓ 10% SFW		↓ 9% Tchl		↑ 248% SOD ↑ 168% POD ↑ 204% CAT		↑ 56%	In shoot: ↑ 458% Na$^+$, ↓ 50% K$^+$, ↓ 38% Mg^{2+}, ↓ 76% Ca^{2+}, Na$^+$/K$^+$ ratio ~10.4	
	150 mM NaCl + *B. pumilus*	↑ 11% SFW		↑ 86% Tchl		↑ 348% SOD ↑ 220% POD ↑ 273% CAT		↑ 83%	In shoot: ↑ 185% Na$^+$, ↑ 24% Mg^{2+}, ↓ 7% K$^+$, ↓ 18% Ca^{2+}, Na$^+$/K$^+$ ratio ~3	
	10 ppm Boron + 150 mM NaCl + Uninoculated	↓ 10% SFW		↓ 23% Tchl		↑ 300% SOD ↑ 388% POD ↑ 377% CAT		↑ 146%	In shoot: ↑ 531% Na$^+$, ↓ 32% K$^+$, ↓ 33% Mg^{2+}, ↓ 27% Ca^{2+}, Na$^+$/K$^+$ ratio ~8.6	
	10 ppm Boron + 150 mM NaCl + *B. pumilus*	↑ 3% SFW		↓ 5% Tchl		↑ 322% SOD ↑ 316% POD ↑ 254% CAT		↑ 85%	In shoot: ↑ 115% Na$^+$, ↓ 11% K$^+$, ↓ 5% Mg^{2+}, ↓ 4% Ca^{2+}, Na$^+$/K$^+$ ratio ~2.2	
C. gleum SUK. Wheat plantlets inoculated with *C. gleum*.	0 mM NaCl + Uninoculated									[99]
	100 mM NaCl + Uninoculated	↓ 41% SL ↓ 46% RL ↓ 16% TPFW		↓ 36% Tchl			↑ 80% FLA	↑ 31%	In shoot: ↑ 128% Na$^+$, ↓ 30% K$^+$, Na$^+$/K$^+$ ratio ~0.12	
	0 mM NaCl + SUK + FLI	↓ 13% SL ↓ 14% RL ↓ 0% TPFW		↑ 18% Tchl			↑ 96% FLA	↑ 48%	N/A	
	100 mM NaCl + SUK + FLI	↓ 9% SL ↓ 9% RL ↑ 19% TPFW	N/A	↑ 5% Tchl	N/A	N/A	↑ 147% FLA	↑ 63%	In shoot: ↑ 61% Na$^+$, ↓ 19% K$^+$, Na$^+$/K$^+$ ratio ~0.08	
	0 mM NaCl + SUK	↓ 16% SL ↓ 36% RL ↓ 13% TPFW		↓ 11% Tchl			↑ 57% FLA	↑ 25%	N/A	
	100 mM NaCl + SUK	↓ 19% SL, ↓ 9% RL, ↓ 6% TPFW		↓ 23% Tchl			↑ 84% FLA	↑ 47%	In shoot: ↑ 67% Na$^+$, ↓ 19% K$^+$, Na$^+$/K$^+$ ratio 0.08	

Table 1. *Cont.*

PGPR	Treatments	GP	Hormones	PhoPs	MDA	AEs	NEAs	Pro	Ion Content	Sources
B. aquimaris DY-3. Three-day-old maize seedlings inoculated with DY-3	Control									[42]
	1% (*w/v*) NaCl + Uninoculated	↓ 34% TPDW	N/A	↓ 13% Tchl	↑ 39%	↑ 21% SOD ↑ 16% POD ↑18% CAT ↑ 23% APX	↑ 22% PHE	↑ 36%	N/A	
	0% NaCl + DY-3	↑ 12% TPDW		↑ 5% Tchl	↓ 8%	↑ 13% SOD ↑ 9% CAT ↑ 9% APX ↓ 12% POD	↑ 11% PHE	↑ 24%		
	1% (*w/v*) NaCl + DY-3	↓ 13% TPDW		↓ 9% Tchl	↑ 26%	↑ 53% SOD ↑ 42% CAT ↑ 65% APX ↓ 2% POD	↑ 67% PHE	↑ 77%		
Klebsiella IG3. Oat seedlings inoculated with IG3	Control									[44]
	100 mM NaCl + Uninoculated	↓ 22% SL ↓ 31% SFW ↓ 29% RFW ↓ 18% RL	↓ 6% IAA	↓ 22% Tchl	In shoot: ↑ 135% In root: ↑ 231%	↑ 353% SOD ↑ 540% POD	N/A	↑ 230%	N/A	
	0 mM NaCl + IG3	↑ 3% SL ↑ 3% SFW ↑ 1% RFW ↑ 13% RL	↑ 41% IAA	↑ 4% Tchl	In shoot: ↑ 3% In root: ↑ 18%	↑ 0% SOD ↑ 2% POD		↑ 42%		
	100 mM NaCl + IG3	↓ 10% SL, ↓ 18% SFW ↓ 16% RFW ↓ 2% RL	↑ 67% IAA	↓ 13% Tchl	In shoot: ↑ 27% In root: ↑ 45%	↑ 96% SOD ↑ 286% POD		↑ 155%		
P. yonginensis DCY84. Root seedlings of ginseng inoculated with *P. yonginensis* DCY84	Short period of stress (3 days of 300 mM NaCl exposure)									[50]
	Control								In shoot: Na⁺/K⁺ ratio ~13 In root: Na⁺/K⁺ ratio ~11	
	0 mM NaCl + DCY84	↑ 15% SFW ↑ 5% RFW	N/A	↑ 3% Chl a ↑ 2% Chl b ↑ 2% Car	Unchanged	↑ 62% APX ↑ 40% POD ↑ 114% CAT	N/A	↑ 253%	In shoot: Na⁺/K⁺ ratio ~15 In root: Na⁺/K⁺ ratio ~11	

Table 1. *Cont.*

PGPR	Treatments	GP	Hormones	PhoPs	MDA	AEs	NEAs	Pro	Ion Content	Sources
	300 mM NaCl + Uninoculated	↓ 13% SFW ↓ 9% RFW		↓ 15% Chl a ↓ 13% Chl b ↓ 16% Car	↑ 29%	↑ 55% POD ↑ 0% APX ↓ 14% CAT		↑ 20%	In shoot: Na$^+$/K$^+$ ratio ~6.4 In root: Na$^+$/K$^+$ ratio ~7	
	300 mM NaCl + DCY84	↑ 12% SFW ↑ 0% RFW		↑ 3% Chl a ↓ 2% Chl b ↓ 9% Car	↑ 21%	↑ 54% APX ↑ 80% POD ↑ 114% CAT		↑ 233%	In shoot: Na$^+$/K$^+$ ratio ~8.2 In root: Na$^+$/K$^+$ ratio ~9.8	
	Long period of stress (12 days of 300 mM NaCl exposure)								
	Control								In shoot: Na$^+$/K$^+$ ratio ~7.2; In root: Na$^+$/K$^+$ ratio ~9.9	
	0 mM NaCl + DCY84	↑ 17% SFW ↑1% RFW		↑ 3% Chl a ↑ 2% Chl b ↓ 2% Car	Unchanged	↑ 45% APX ↑ 100% POD ↑ 143% CAT		↑ 300%	In shoot: Na$^+$/K$^+$ ratio ~7.8, In root: Na$^+$/K$^+$ ratio ~11	
	300 mM NaCl + Uninoculated	↓ 18% SFW ↓22% RFW	N/A	↓ 53% Chl a ↓ 66% Chl b ↓ 57% Car	↑ 36%	↓ 31% APX ↓ 33% POD ↓ 71% CAT	N/A	↑ 13%	In shoot: Na$^+$/K$^+$ ratio ~4.4, In root: Na$^+$/K$^+$ ratio ~3.8	
	300 mM NaCl + DCY84	↑ 12% SFW ↓ 3% RFW		↓ 3% Chl a ↓ 11% Chl b ↓ 7% Car	↑ 14%	↑ 90% APX ↑ 317% POD ↑ 343% CAT		↑ 287%	In shoot: Na$^+$/K$^+$ ratio ~6.5, In root: Na$^+$/K$^+$ ratio ~7.7	
B. megaterium A12 (BMA12). Ten-day-old tomato seedlings inoculated with BMA12	Control									[45]
	200 mM NaCl + Uninoculated	↓ 37% PH ↓ 50% RL ↓ 59% TPFW ↓ 54% TPDW ↓ 35% TLA	↓ 32% IAA ↓ 43% GA4 ↑ 100% C$_2$H$_4$ ↑ 82% ABA	↓ 32% Chl a ↓ 40% Chl b ↓ 41% TChl ↓ 49% Car	N/A	↑ 24% SOD ↑ 27% CAT ↓ 24% APX ↓ 24% POD ↓ 57% PPO	↑ 74% GSH ↑ 228% ASC	N/A	N/A	
	0 mM NaCl + BMA12	↑ 23% PH ↑ 37% RL ↑ 28% TPFW ↑ 40% TPDW ↑ 25% TLA	↑ 53% IAA, ↑ 170% GA4 ↓ 16% C$_2$H$_4$ ↓ 14% ABA	↑ 53% Chl a ↑ 14% Chl b ↑ 41% TChl ↑ 35% Car		↑ 86% SOD ↑ 54% CAT ↑ 34% APX ↑ 37% POD ↑ 55% PPO	↑ 17% GSH ↑ 5% ASC			

Table 1. *Cont.*

PGPR	Treatments	GP	Hormones	PhoPs	MDA	AEs	NEAs	Pro	Ion Content	Sources
	2000 mM NaCl + BMA12	↓ 13% PH ↓ 28% RL ↓ 33% TPFW ↓ 35% TPDW ↓ 21% TLA	↑ 0% IAA ↑ 11%C_2H_4 ↑186% ABA, ↑ 86% GA4	↑ 5% Chl a ↓ 17% Chl b ↓ 4% TChl ↓ 24% Car		↑ 213% SOD ↑ 91% CAT ↑ 78% APX ↑ 18% POD ↓ 10% PPO	↑ 250% GSH ↑ 100% ASC			
Pseudomonas (wild-type UW4 and mutant strains). Seven-day-old tomato plants inoculated with UW4.	Control									[46]
	200 mM NaCl + Uninoculated	↓ 56% RL ↓ 37% SL ↓ 37% TPDW		↓ 42% Tchl						
	200 mM NaCl + WT UW4	↑ 16% RL ↑ 3% SL ↑ 25% TPDW		↑ 31% Tchl						
	200 mM NaCl + acdS-mutant	↓ 33% RL ↓ 9% SL ↓ 17% TPDW	N/A	↓ 25% Tchl				N/A		
	200 mM NaCl + treS- mutant	↓ 39% RL ↓ 31% SL ↓ 8% TPDW		↓ 13% Tchl						
	200 mM NaCl + acdS-/treS- double mutant	↓ 58% RL ↓ 37% SL ↓ 35% TPDW		↓ 56% Tchl						
	200 mM NaCl + OxtreS	↑ 45% RL ↑ 3% SL ↑ 54% TPDW		↑ 61% Tchl						
S. maltophilia BJ01. Seven-day-old peanut seedlings inoculated with BJ01	Control									[48]
	0 mM NaCl + BJ01	↑ 4% SL, ↑ 11% TPFW, ↓ 15% RL	↑ 19% Aux	↑ 11% Chla, ↑ 0% Chl b, ↑ 17% Tchl	↓ 26%	N/A	N/A	↓ 32%	N/A	

Table 1. *Cont.*

PGPR	Treatments	GP	Hormones	PhoPs	MDA	AEs	NEAs	Pro	Ion Content	Sources
	100 mM NaCl + Uninoculated	↑ 9% RL, ↓ 45% TPFW, ↓ 39% SL	↑ 16% Aux	↓ 56% Chl a, ↓ 42% Chl b, ↓ 50% Tchl	↑ 47%			↑ 1355%		
	100 mM NaCl + BJ01	↓ 26% SL, ↓ 3% RL, ↓ 26% TPFW	↑ 29% Aux	↓ 11% Chl a ↓ 34% Chl b ↓ 23% Tchl	↑ 16%			↑ 1173%		

Note: All calculations in the Table 1 represent the comparisons between the treated plants and the control plants (non-stress conditions and un-inoculation). The up arrowhead (↑) indicates an increase in a tested parameter as compared to the control. The down arrowhead (↓) displays a reduction in a tested parameter relative to the control. Abbreviation in the Table 1: ABA, Abscisic acid; *A. calcoaceticus, Acinetobacter calcoaceticus*; *A. aneurinilyticus, Aneurinibacillus aneurinilyticus*; AEs, antioxidant enzymes; APX, Ascorbate peroxidase; ASC, Ascorbate; Aux, Auxin; *B. aquimaris, Bacillus aquimaris*; *B. brevis, Bacillus brevis*; *B. cepacia, Burkholdera cepacia*; *B. megaterium, Bacillus megaterium*; *B. pumilus, Bacillus pumilus*; C_2H_4, ethylene; Car, Carotenoids; CAT, Catalase; *C. gleum, Chryseobacterium gleum*; Chl, Chlorophyll; *E. aerogenes, Enterobacter aerogenes*; FLA, Flavonoids; FLI, Feather lysate inoculum; GA4, Gibberellins 4; GSH, Glutathione; GP, Growth parameter; IAA, Indole-3-acetic acid; JA, Jasmonates; MDA, Malondialdehyde; N/A, Not available; NEAs, Non-enzymatic antioxidants; *P. fluorescence, Pseudomonas fluorescence*; PH, Plant height; PHE, Phenols; PhoPs, Photosynthetic pigments; POD, Peroxidase; PPs, Polyphenols; Pro, Proline; *P. putida, Pseudomonas putida*; PPO, Polyphenol oxidase; Pro, Proline; *P. yonginensis, Paenibacillus yonginensis*; RDW, Root dry weight; RFW, Root fresh weight; RL, Root length; SA, Salicylic acid; SDW, Shoot dry weight; SFW, Shoot fresh weight; SL, Shoot length; *S. maltophilia, Stenotrophomonas maltophilia*; SOD, Superoxide dismutase; Tchl, Total chlorophyll; TLA, Total leaves area per plant; TPDW, Total plant dry weight; TPFW, Total plant fresh weight; TPP, Total polyphenol; *X. autotrophicus, Xanthobacter autotrophicus*; Y, Yield.

In the study of Sarkar et al. [57], the inoculation of rice seeds with *Enterobacter* sp. strain P23 promoted higher germination percentage (GP) (76% ± 7.03 vs. 48% ± 4.78), and higher seedling vigor index (SVI) (881.6 ± 67 vs. 57.5 ± 12.6) as compared to the non-inoculated seeds. Under salt conditions, the Pro peaked its highest level, the SOD, CAT, POD, PPO, and MDA exhibited their highest contents in uninoculated rice seedlings. However, the activities of these enzymes in P23-inoculated seedlings were significantly reduced relative to those in the non-inoculated seedlings. The productions of ethylene in non-inoculated seedlings and P23 AcdS mutant-inoculated seedlings were comparable, consistent with the study of Cheng et al. [30], while ethylene in the WT P23 strain-treated plants was lower, indicating that the WT P23 succeeded in decreasing stress ethylene production. Under 250 mM NaCl treatment, the SFW and SDW of *P. putida* UW 4-inoculated canola plants (*Brassica napus* L.) were 1.7-fold higher than those of untreated plants [30]. However, the *P. putida* ACC deaminase (AcdS) minus mutant-inoculated canola plants did not show significant difference in SFW and SDW relative to the untreated plants, indicating the critical role of a functional ACC deaminase enzyme in plant growth under salinity stress. The proteins involved in photosynthesis in the WT *P. putida* UW4 plants were downregulated; however, to a lesser extent as compared to that in the uninoculated plants or in the *P. putida* AcdS plants, resulting in the higher chlorophyll contents relative to the uninoculated plants. Surprisingly, both AcdS and WT *P. putida* plants could accumulate large amount of NaCl in their shoots with 3.7–7-fold higher than that in the uninoculated plants, respectively while being able to maintain their normal growth. This could be partly explained by the increase cell permeability caused by IAA that was produced by the WT *P. putida* and the AcdS mutant. This finding is intriguing and controversial since numerous other studies recognized the decreased Na$^+$ uptake in PGPR-bacterized plants [40,42,99,111]. In their another study, Sarkar et al. (2018) [49] primed the rice seeds (*Oryza sativa* cv. Ratna) with *Enterobacter* sp. P23 and achieved greater GRA (76% vs. 48%), as well as SVI (881.6 vs. 57.6) relative to the unprimed seeds. Subsequently, the growth and development of the primed seedlings were better than the unprimed control, representing via greater SFW, RFW, SDW, RDW, SL, RL, amylase, protease, Aux, and Chl values [49,57]. In the study of Zhu et al. (2020) [107], the treatment with 130 mM NaCl severely affected the GRA of the non-primed alfalfa seeds (*Medicago sativa* L.) in comparison with the primed seeds. Specifically, the gemination rate of the non-primed seeds reduced to 29% versus 32% of *B. megaterium* NRCB001-primed seeds, 42% of *B. subtilis* NRCB002, and 40% of *B. subtilis* NRCB003. Also in Zhu et al. [107], the vegetative parameters such as PH, RL, NL, TLA, and TPDW of primed seedlings were always higher than those of unprimed seedlings and the MDA content in their leaves were lower, suggesting a less injured cellular membrane in the primed alfalfa grass.

Regarding the synergy between different PGPR and/or between the microbes and chemicals, the synergistic effects of a consortium (*A. aneurinilyticus* + *Paenibacillus* sp.) were observed via the maximum physio-morphology parameters of primed French bean seedlings (*Phaseolus vulgaris*) in comparison to uninoculated- or individual *A. aneurinilyticus* and *Paenibacillus*-primed seedlings [88]. Two VOCs, namely, 4-nitroguaiacol and quinoline derived from *Pseudomonas simiae* exhibited their ability to induce soybean seed germination under 100 mM NaCl treatment [147]. Furthermore, the combined treatment of sodium nitroprusside (SNP) and *P. simiae* resulted in the higher biomass, the lower MDA content and EL in the treated soybean plants than other treatment plants [147]. Mel exhibits pleiotropic biological activities such as growth regulation [148] and antioxidative property [149] and has been widely used as a promising tool for mitigating salt stress in plants. Abd El-Ghany and Attia (2020) [124] found that the combination of Mel and peat-based inoculants (*Rhizobium leguminosarum*, a N fixing bacterium, and *Azotobacter chroococcum*, an EPS-producing bacterium) synergistically enhanced salt stress tolerance in faba bean plants (*Vicia faba*) as compared to Mel- or inoculants-treated seeds alone. Specifically, in the combined treatment (100 μM Mel + inoculants), the content of Chl a, Chl b, Car, and Pro reached the highest, suggesting the synergistic effects of Mel and beneficial PGPR in improving plant growth and other physiological aspects in salt stress conditions. The combination of Mel and bacterial inoculants, in contrast, helped to boost the faba bean plant growth, to increase PhoPs, Pro, N–P–K uptake, and to reduce the Na^+/K^+ ratio.

In conclusion, seed biopriming using PGPR enhances the GRA and SV index in the primed seeds as compared to the unprimed seeds under saline conditions, thereby supporting plants a vigorous growth and a better salinity tolerance during their whole life [150]. The key findings in recent seed biopriming studies were presented in Table 2.

3.2. The Increase in AEs and/or Osmoregulators in PGPR-Inoculated Plants and PGPR-Primed Seedlings under Salt Stress

The increased activities of AEs and/or the elevated accumulations of osmoregulators in PGPB-inoculated plants were reported by Li and Jiang [42], Akram et al. [45], Vaishnav et al. [47], Khalid et al. [58], Kim et al. [96], Habib et al. [97], Kang et al. [104], Zhu et al. [107], Halo et al. [151], Bharti et al. [152], El-Esawi et al. [153], Vimal et al. [154], El-Nahrawy and Yassin [155], Sun et al. [156]. For instance, the activity of ROS-scavenging enzymes SOD, CAT of *Enterobacter*-treated okra plants was the highest amongst all treatments, in parallel with their highest vegetative parameters SFW, SDW, RFW, and RDW [97]. Likewise, APX activity in *Enterobacter*-inoculated tomato plants was 20% higher and 2,2-diphenyl-1-picryl-hydrazyl-hydrate (DPPH) assay showed 24% increase in scavenging capacity in the inoculated plants relative to the control plants [96]. In the study of Abd_Allah et al. [35], the activities of POD, CAT, GR, and SOD, and the contents of AsA, GSH and proline were always the highest in the inoculated chickpea plants. The Arabidopsis plants inoculated with *Burkholderia phytofirmans* PsJN revealed an elevated Pro accumulation in comparison with the control plants [157]. In the *Leclercia adcarboxylata*-treated tomato plants, Pro, serine (Ser), glycine (Gly), methionine (Met), and threonine (Thr), as well as citric acid (CA) and malic acid (MA) were significantly accumulated [104]. The detailed profiles of AEs were presented in Tables 1 and 2.

Table 2. The use of PGPR in seed biopriming technique for improving salinity stress tolerance in plants.

PGPR	Treatments	GP	PhoPs	AEs	MDA	Ion Content	Pro	Ethylene	Sources
S. maltophilia SBP-9. Bacterized wheat seeds with *S. maltophilia* SBP-9 for 1 h.	Control							N/A	[102]
	0 mM NaCl + SBP-9	↑ 15% SL, ↑ 10% RL, ↑ 12% SFW, ↑ 17% SDW, ↑ 33% RFW, ↑ 9% RDW	↑ 8% Tchl	↑ 33% SOD ↑ 20% CAT ↑ 33% POD	↓ 27%	In shoot: ↓ 4% Na$^+$, ↑ 12% K$^+$, Na$^+$/K$^+$ ratio ~0.54	↓ 10%		
	150 mM NaCl + Unprimed	↓ 11% SL, ↓ 5% RL, ↓ 8% SFW, ↓ 24% SDW, ↓ 11% RFW, ↓ 23% RDW	↓ 15% Tchl	↑ 58% SOD ↑ 7% CAT ↑ 67% POD	↑ 17%	In shoot: ↑ 48% Na$^+$, ↓ 17% K$^+$, Na$^+$/K$^+$ ratio ~1.12	↑ 74%		
	150 mM NaCl + SBP-9	↑ 4% SL, ↑ 15% RL, ↑ 4% SFW, ↑ 7% SDW, ↑ 22% RFW, ↓ 5% RDW	↑ 5% Tchl	↑ 133% SOD ↑ 93% CAT ↑ 133% POD	↓ 13%	In shoot: ↑ 20% Na$^+$, ↑ 3% K$^+$ Na$^+$/K$^+$ ratio ~0.73	↑ 23%		
	200 mM NaCl + Unprimed	↓ 37% SL, ↓ 20% RL, ↓ 32% SFW, ↓ 38% SDW, ↓ 33% RFW, ↓ 64% RDW	↓ 59% Tchl	↑ 217% SOD ↑ 100% CAT ↑ 192% POD	↑ 93%	In shoot: ↑ 107% Na$^+$, ↓ 25% K$^+$, Na$^+$/K$^+$ ratio ~1.73	↑ 165%		
	200 mM NaCl + SBP-9	↓ 11% SL, ↓ 5% RL, ↓ 16% SFW, ↓ 17% SDW, ↓ 6% RFW, ↓ 41% RDW	↓ 39% Tchl	↑ 350% SOD ↑ 180% CAT ↑ 283% POD	↑ 50%	In shoot: ↑ 54% Na$^+$, ↓ 3% K$^+$ Na$^+$/K$^+$ ratio ~1	↑ 110%		
Enterobacter P23. Seeds of *Oryza sativa* cv. Ratna treated with bacterial suspension.	Control							N/A	[49]
	150 mM NaCl + Unprimed	↓ 51% GP, ↓ 97% SVI, ↓ 45% SFW, ↓ 58% SDW, ↓ 33% SL, ↓ 39% RFW, ↓ 63% RDW, ↓ 44% RL	↓ 54% Chl a ↓ 80% Chl b ↓ 56% Tchl	↑ 120% SOD ↑ 112% CAT ↑ 174% POD ↑ 700% PPO	↑ 300%	N/A	↑ 175%		
	150 mM NaCl + P23	↓ 22% GP, ↓ 58% SVI, ↓ 16% SFW, ↓ 11% SL, ↓ 23% SDW, ↓ 15% RFW, ↓ 30% RDW, ↓ 11% RL	↓ 13% Chl a ↓ 10% Chl b ↓ 8% Tchl	↑ 32% SOD ↑ 46% CAT ↑ 70% POD ↑ 300% PPO	↑ 195%		↑ 75%		
B. pumilus FAB10. Wheat seeds cv. 343 treated with FAB10.	Control							N/A	[51]
	75 mM NaCl + Unprimed	↓ 17% SL, ↓ 35% RL, ↓ 49% SDW, ↓ 53% RDW, ↓ 35% SpDW, ↓ 21% GY, ↓ 17% GPr	N/A	↑ 20% SOD ↑ 40% CAT ↑ 50% GR	↑ 189%	N/A	↑ 105%		
	125 mM NaCl + Unprimed	↓ 24% SL, ↓ 49% RL, ↓ 42% SDW, ↓ 67% RDW, ↓ 48% SpDW, ↓ 31% GY, ↓ 22% GPr		↑ 23% SOD ↑ 80% CAT ↑ 75% GR	↑ 189%		↑ 146%		

Table 2. *Cont.*

PGPR	Treatments	GP	PhoPs	AEs	MDA	Ion Content	Pro	Ethylene	Sources
	250 mM NaCl + Unprimed	↓ 35% SL, ↓ 52% RL, ↓ 40% SDW, ↓ 76% RDW, ↓ 61% SpDW, ↓ 41% GY, ↓ 29% GPr		↑ 25% SOD ↑ 80% CAT ↑ 75% GR	↑ 260%		↑ 171%		
	75 mM NaCl + FAB10	↓ 13% SL, ↓ 11% RL, ↓ 13% SDW, ↓ 20% RDW, ↓ 4% SpDW, ↓ 3% GY, ↓ 11% GPr		↑ 5% SOD ↓ 20% CAT ↓ 25% GR	↑ 103%		↑ 77%		
	125 mM NaCl + FAB10	↓ 17% SL, ↓ 22% RL, ↓ 18% SDW, ↓ 43% RDW, ↓ 9% SpDW, ↓ 13% GY, ↓ 16% GPr		↑ 10% SOD ↑ 20% CAT ↑ 0% GR	↑ 180%		↑ 123%		
	250 mM NaCl + FAB10	↓ 25% SL, ↓ 24% RL, ↓ 18% SDW, ↓ 51% RDW, ↓ 61% SpDW, ↓ 25% GY, ↓ 27% GPr		↑ 12% SOD ↑ 20% CAT ↑ 0% GR	↑ 237%		↑ 139%		
	Saline soil Control								
Consortium (*R. leguminosarum* + *A. chroococcum*) and/or Mel. *Vicia faba* seeds were treated with the consortium as peat-based inoculant and/or Mel solution	25 µM Mel + Unprimed	↑ 6% SL, ↑ 37% NL, ↑ 18% SFW, ↑ 24% SDW, ↑ 23% Y	↑ 6% Chl a ↑ 11% Chl b ↑ 9% Car	N/A	N/A	N/A	↑ 17%	N/A	[124]
	50 µM Mel + Unprimed	↑ 24% SL, ↑ 56% NL, ↑ 41% SFW, ↑ 36% SDW, ↑ 41% Y	↑ 30% Chl a ↑ 26% Chl b ↑ 18% Car				↑ 30%		
	100 µM Mel + Unprimed	↑ 41% SL, ↑ 93% NL, ↑ 72% SFW, ↑ 78% SDW, ↑ 58% Y	↑ 44% Chl a ↑ 80% Chl b ↑ 35% Car				↑ 39%		
	0 µM Mel + Primed	↑ 18% SL, ↑ 43% NL, ↑ 25% SFW, ↑ 36% SDW, ↑ 48% Y	↑ 31% Chl a ↑ 35% Chl b ↑ 28% Car				↑ 44%		
	25 µM Mel + Primed	↑ 30% SL, ↑ 79% NL, ↑ 49% SFW, ↑ 56% SDW, ↑ 71% Y	↑ 42% Chl a ↑ 59% Chl b ↑ 43% Car				↑ 57%		
	50 µM Mel + Primed	↑ 70% SL, ↑ 97% NL, ↑ 68% SFW, ↑ 68% SDW, ↑ 82% Y	↑ 56% Chl a ↑ 107% Chl b ↑ 66% Car				↑ 89%		
	100 µM Mel + Primed	↑ 74% SL, ↑ 118% NL, ↑ 98% SFW, ↑ 104% SDW, ↑ 96% Y	↑ 71% Chl a ↑ 118% Chl b ↑ 71% Car				↑ 110%		

Table 2. *Cont.*

PGPR	Treatments	GP	PhoPs	AEs	MDA	Ion Content	Pro	Ethylene	Sources
	Non-saline soil Control								
	25 μM Mel + Unprimed	↑ 6% SL, ↑ 14% NL, ↑ 10% SFW, ↑ 21% SDW, ↑ 18% Y	↑ 9% Chl a ↑ 10% Chl b ↑ 4% Car				↑ 5%		
	50 μM Mel + Unprimed	↑ 12% SL, ↑ 35% NL, ↑ 23% SFW, ↑ 31% SDW, ↑ 29% Y	↑ 11% Chl a ↑ 24% Chl b ↑ 21% Car				↑ 11%		
	100 μM Mel + Unprimed	↑ 16% SL, ↑ 49% NL, ↑ 38% SFW, ↑ 69% SDW, ↑ 32% Y	↑ 16% Chl a ↑ 34% Chl b ↑ 24% Car				↑ 45%		
	0 μM Mel + Primed	↑ 20% SL, ↑ 28% NL, ↑ 14% SFW, ↑ 27% SDW, ↑ 17% Y	↑ 17% Chl a ↑ 24% Chl b ↑ 21% Car				↑ 36%		
	25 μM Mel + Primed	↑ 29% SL, ↑ 49% NL, ↑ 24% SFW, ↑ 46% SDW, ↑ 25% Y	↑ 23% Chl a ↑ 30% Chl b ↑ 30% Car				↑ 94%		
	50 μM Mel + Primed	↑ 45% SL, ↑ 55% NL, ↑ 33% SFW, ↑ 57% SDW, ↑ 38% Y	↑ 26% Chl a ↑ 42% Chl b ↑ 33% Car				↑ 110%		
	100 μM Mel + Primed	↑ 49% SL, ↑ 76% NL, ↑ 52% SFW, ↑ 87% SDW, ↑ 45% Y	↑ 30% Chl a ↑ 29% Chl b ↑ 40% Car				↑ 139%		
A. aneurinilyticus ACC02, *Paenibacillus* ACC06 and Consortium (ACC02+ ACC06). French bean seeds inoculated with ACC02, ACC06 and consortium (ACC02 + ACC06)	Control								
	0 mM NaCl + ACC02	↑ 10% SL, ↑ 50% RL, ↑ 158% SFW, ↑ 10% SDW, ↑ 50% RFW, ↑ 21% RDW	↑ 36% Tchl	N/A	N/A	N/A	N/A	↑ 9%	[88]
	0 mM NaCl + ACC06	↑ 30% SL, ↑ 30% RL, ↑ 216% SFW, ↑ 10% SDW, ↑ 60% RFW, ↑ 14% RDW	↑ 29% Tchl					↓ 9%	
	0 mM NaCl + Consortium	↑ 50% SL, ↑ 70% RL, ↑ 233% SFW, ↑ 80% SDW, ↑ 90% RFW, ↑ 85% RDW	↑ 57% Tchl					↑ 27%	
	25 mM NaCl + Unprimed								

Table 2. *Cont.*

PGPR	Treatments	GP	PhoPs	AEs	MDA	Ion Content	Pro	Ethylene	Sources
	25 mM NaCl + ACC02	↑ 33% SL, ↑ 79% RL, ↑ 120% SFW, ↑ 300% SDW, ↑ 46% RFW, ↑ 182% RDW	↑ 28% Tchl	N/A	N/A	N/A	N/A	↓ 38%	
	25 mM NaCl + ACC06	↑ 47% SL, ↑ 58% RL, ↑ 120% SFW, ↑ 350% SDW, ↑ 36% RFW, ↑ 142% RDW	↑ 35% Tchl					↓ 42%	
	25 mM NaCl + Consortium	↑ 60% SL, ↑ 110% RL, ↑ 255% SFW, ↑ 425% SDW, ↑ 81% RFW, ↑ 220% RDW	↑ 57% Tchl					↓ 61%	
P. fluorescence, B. pumilus, E. aurantiacum and consortium (*P. fluorescence + B. pumilus + E.aurantiacum*) Wheat seeds soaked in bacterial inoculant containing single PGPR strains or the consortium of three bacterial cultures for 2 h. Saline soil EC$_e$ 13.41	Unprimed seeds								
	Seeds primed with *P. fluorescence*	**Galaxy-13:** ↑ 5% SL, ↑ 7% RL, ↑ 3% SFW, ↑ 2% SDW, ↑ 33% 100 GW, ↓ 13% RFW, ↓ 29% RDW **Aas-11:** ↑ 11% SL, ↑ 24% RL, ↑ 48% SFW, ↑ 144% RFW, ↑ 57% SDW, ↑ 75% RDW, ↑ 23% 100 GW	N/A	**Galaxy-13:** ↓ 30% SOD ↓ 0% POD ↑ 27% CAT **Aas-11:** ↓ 57% SOD ↓ 14% POD ↑ 25% CAT	N/A	**Galaxy-13:** **In root:** ↑ 50% Na$^+$, ↑ 40% K$^+$, Na$^+$/K$^+$ ratio ~0.19 **In shoot:** ↑ 28% Na$^+$, ↑ 23% K$^+$, Na$^+$/K$^+$ ratio ~3.9 **Aas-11:** **In root:** ↓ 13% Na$^+$, ↑ 99% K$^+$, Na$^+$/K$^+$ ratio ~0.16. **In shoot:** ↑ 92% Na$^+$, ↑ 16% K$^+$, Na$^+$/K$^+$ ratio ~3.4	**Galaxy-13:** ↓ 20% **Aas-11:** ↑ 33%	N/A	[122]
	Seeds primed with *B. pumilus*	**Galaxy-13:** ↓ 7% SL, ↓ 18% SFW, ↓ 26% SDW, ↓ 8% RFW, ↓ 57% RDW, ↑ 67% RL, 31% 100 GW **Aas-11:** ↑ 13% SL, ↑ 21% RL, ↑ 61% SFW, ↑ 678% RFW, ↑ 66% SDW, ↑ 838% RDW, ↑ 53% 100 GW		**Galaxy-13:** ↓ 35% SOD ↓ 5% POD ↑ 4% CAT **Aas-11:** ↓ 65% SOD ↓ 38% POD ↑ 35% CAT		**Galaxy-13:** **In root:** ↑ 0% Na$^+$, ↑ 34% K$^+$, Na$^+$/K$^+$ ratio ~0.13 **In shoot:** ↓ 8% Na$^+$, ↓ 19% K$^+$, Na$^+$/K$^+$ ratio ~4.3 **Aas-11:** **In root:** ↓ 13% Na$^+$, ↑ 195% K$^+$, Na$^+$/K$^+$ ratio ~0.11. **In shoot:** ↑ 59% Na$^+$, ↓ 12% K$^+$, Na$^+$/K$^+$ ratio ~3.7	**Galaxy-13:** ↑ 287% **Aas-11:** ↑ 150%		

Table 2. *Cont.*

PGPR	Treatments	GP	PhoPs	AEs	MDA	Ion Content	Pro	Ethylene	Sources
	Seeds primed with *E.aurantiacum*	**Galaxy-13:** ↑ 6% SL, ↑ 47% RL, ↑ 3% SFW, ↑ 49% 100 GW, ↓ 17% RFW, ↓ 2% SDW, ↓ 28% RDW **Aas-11:** ↑ 10% SL, ↑ 7% RL, ↑ 52% SFW, ↑ 511% RFW, ↑ 71% SDW, ↑ 713% RDW, ↑ 47% 100 GW		**Galaxy-13:** ↑ 2% SOD ↑ 48% CAT ↓ 43% POD **Aas-11:** ↓ 65% SOD ↓ 57% POD ↓ 5% CAT		**Galaxy-13:** **In root:** ↑ 33% Na$^+$, ↑ 34% K$^+$, Na$^+$/K$^+$ ratio ~0.18 **In shoot:** ↑ 27% Na$^+$, ↑ 0% K$^+$, Na$^+$/K$^+$ ratio ~4.77. **Aas-11:** In root: ↓ 13% Na$^+$, ↑ 286% K$^+$, Na$^+$/K$^+$ ratio ~0.08 **In shoot:** ↑ 40% Na$^+$, ↑ 22% K$^+$, Na$^+$/K$^+$ ratio ~2.36	Galaxy-13: ↑ 227% Aas-11: ↑ 110%		
	Seeds primed with a consortium	**Galaxy-13:** ↓ 1% SL, ↑ 73% RL, ↑ 6% SFW, ↑ 30% RFW, ↑ 7% SDW, ↑ 43% RDW, ↑ 53% 100 GW **Aas-11:** ↑ 13% SL, ↑ 3% RL, ↑ 65% SFW, ↑ 556% RFW, ↑ 77% SDW, ↑ 725% RDW, ↑ 48% 100 GW		**Galaxy-13:** ↑ 37% SOD ↓ 32% POD ↓ 6% CAT **Aas-11:** ↓ 57% SOD ↑ 24% POD ↑ 28% CAT		**Galaxy-13:** **In root:** ↑ 0% Na$^+$, ↑ 114% K$^+$, Na$^+$/K$^+$ ratio ~0.08 **In shoot:** ↑ 17% Na$^+$, ↑ 15% K$^+$, Na$^+$/K$^+$ ratio ~3.8 **Aas-11:** **In root:** ↑ 0% Na$^+$, ↑ 173% K$^+$, Na$^+$/K$^+$ ratio ~0.13 **In shoot:** ↑ 68% Na$^+$, ↑ 30% K$^+$, Na$^+$/K$^+$ ratio ~2.67	Galaxy-13: ↑ 327% Aas-11: ↑ 17%		
Sphingobacterium BHU-AV3. Bacterized tomato seeds with BHU-AV3 for 24 h	Control								
	200 mM NaCl + Unprimed	↓ 52% SL, ↓ 49% RL, ↓ 54% TPDW	↓ 44% Tchl	In shoot: ↑ 90% SOD ↑ 260% POD ↑ 100% PPO In root: ↑ 83% SOD ↑ 100% POD ↑ 53% PPO	N/A	In shoot: ↑ 258% Na$^+$, ↓ 63% K$^+$, Na$^+$/K$^+$ ratio ~3.6 In root: ↑ 190% Na$^+$, ↓ 53% K$^+$, Na$^+$/K$^+$ ratio ~3.5	In shoot: ↑ 153% In root: ↑ 56%	N/A	[47]

Table 2. *Cont.*

PGPR	Treatments	GP	PhoPs	AEs	MDA	Ion Content	Pro	Ethylene	Sources
	0 mM NaCl + BHU–AV3	↓ 11.3% SL, ↑ 16% RL, ↑ 11% TPDW	↑ 5% Tchl	In shoot: ↓ 20% SOD ↓ 0% POD ↓ 25% PPO In root: ↑ 16% SOD ↑ 6% POD ↓ 12% PPO		In shoot: ↑ 9% Na^+, ↑ 9% K^+, Na^+/K^+ ratio ~0.3 In root: ↓ 5% Na^+, ↑ 8% K^+, Na^+/K^+ ratio ~0.5	In shoot: ↓ 7% In root: ↓ 5%		
	200 mM NaCl + BHU-AV3	↓ 30% SL, ↓ 22% RL, ↓ 29% TPDW	↓ 14% Tchl	In shoot: ↑ 10% SOD ↑ 1000% POD ↑ 50% PPO In root: ↑ 117% SOD ↑ 200% POD ↑ 71% PPO		In shoot: ↑ 130% Na^+, ↓ 20% K^+, Na^+/K^+ ratio ~1 In root: ↑ 115% Na^+, ↓ 24% K^+, Na^+/K^+ ratio ~1.6	In shoot: ↑ 84% In root: ↑ 111%		
K. sacchari MSK1. Mung bean seeds primed with MSK1	Control							N/A	[36]
	50 mM NaCl + Unprimed	↓ 8% SL, ↓ 8% RL, ↓ 5% SDW, ↓ 5% RDW, ↓ 15% SY, ↓ 3% GP	↓ 15% Tchl ↓ 4% Car	↑ 5% GR, ↑ 33% CAT ↑ 13% SOD ↑ 23% APX	↑ 32%	In shoor: ↑ 100% Na^+, ↑ 44% K^+, Na^+/K^+ ratio ~0.46, ↓ 4% N, ↓ 19% P	↑ 63%		
	100 mM NaCl + Unprimed	↓ 16% SL, ↓ 20% RL, ↓ 10% SDW, ↓ 15% RDW, ↓ 21% SY, ↓ 6% GP	↓ 35% Tchl ↓ 9% Car	↑ 15% GR ↑ 58% CAT ↑ 39% SOD ↑ 45% APX	↑ 47%	In shoot: ↑ 200% Na^+, ↑ 100% K^+, Na^+/K^+ ratio ~0.5, ↓ 4% N, ↓ 28% P	↑ 88%		
	200 mM NaCl + Unprimed	↓ 24% SL, ↓ 28% RL, ↓ 21% SDW, ↓ 35% RDW, ↓ 26% SY, ↓ 8% GP	↓ 42% Tchl ↓ 19% Car	↑ 35% GR ↑ 108% CAT ↑ 52% SOD ↑ 73% APX	↑ 84%	In shoot: ↑ 450% Na^+, ↑ 222% K^+, Na^+/K^+ ratio ~0.57, ↓ 12% N, ↓ 44% P	↑ 213%		
	400 mM NaCl + Unprimed	↓ 41% SL, ↓ 52% RL, ↓ 34% SDW, ↓ 55% RDW, ↓ 34% SY, ↓ 26% GP	↓ 62% Tchl ↓ 33% Car	↑ 64% GR ↑ 208% CAT ↑ 96% SOD ↑ 102% APX	↑ 153%	In shoot: ↑ 800% Na^+, ↑ 378% K^+, Na^+/K^+ ratio ~0.63, ↓ 21% N, ↓ 59% P	↑ 350%		
	0 mM NaCl + MSK1	↑ 5% SL, ↑ 12% RL, ↑ 7% SDW, ↑ 15% RDW, ↑ 9% SY, ↑ 7% GP	↑ 29% Tchl ↑ 7% Car	↓ 9% GR ↓ 33% CAT ↓ 22% SOD ↓ 9% APX	↓ 37%	In shoot: ↓ 67% Na^+, ↓ 22% K^+, Na^+/K^+ ratio ~0.14, ↑ 9% N, ↑ 15% P	↓ 25%		

Table 2. *Cont.*

PGPR	Treatments	GP	PhoPs	AEs	MDA	Ion Content	Pro	Ethylene	Sources
	50 mM NaCl + MSK1	↓ 3% SL, ↑ 4% RL, ↓ 2% SDW, ↑ 3% RDW, ↑ 10% SY, ↑ 2% GP	↓ 3% Tchl ↓ 0% Car	↑ 2% GR ↑ 8% CAT ↑ 9% SOD ↑ 9% APX	↑ 11%	In shoot: ↑ 33% Na^+, ↑ 22% K^+, Na^+/K^+ ratio ~0.36, ↓ 1% N, ↓ 41% P	↑ 30%		
	100 mM NaCl + MSK1	↓ 8% SL, ↓ 12% RL, ↓ 7% SDW, ↓ 10% RDW, ↓ 19% SY, ↓ 5% GP	↓ 31% Tchl ↓ 8% Car	↑ 11% GR ↑ 50% CAT ↑ 22% SOD ↑ 32% APX	↑ 37%	In shoot: ↑ 183% Na^+, ↑ 89% K^+, Na^+/K^+ ratio ~0.5, ↓ 4% N, ↓ 22% P	↑ 75%		
	200 mM NaCl + MSK1	↓ 22% SL, ↓ 16% RL, ↓ 17% SDW, ↓ 25% RDW, ↓ 24% SY, ↓ 7% GP	↓ 35% Tchl ↓ 13% Car	↑ 33% GR ↑ 100% CAT ↑ 48% SOD ↑ 64% APX	↑ 79%	In shoot: ↑ 433% Na^+, ↑ 211% K^+, Na^+/K^+ ratio ~0.57, ↓ 9% N, ↓ 37% P	↑ 200%		
	400 mM NaCl + MSK1	↓ 35% SL, ↓ 24% RL, ↓ 32% SDW, ↓ 48% RDW, ↓ 32% SY, ↓ 25% GP	↓ 54% Tchl ↓ 27% Car	↑ 60% GR ↑ 192% CAT ↑ 91% SOD ↑ 91% APX	↑ 137%	In shoot: ↑ 783% Na^+, ↑ 367% K^+, Na^+/K^+ ratio ~0.63, ↓ 19% N, ↓ 57% P	↑ 325%		

Note: All calculations in the Table 2 represent the comparisons between the treated plants and the control plants (non-stress conditions and un-priming) except that the control plants in the study of Nawaz et al. (2020) [122] were cultivated in the saline soil ECe ~13.41. The up arrowhead (↑) indicates an increase in a tested parameter as compared to the control. The down arrowhead (↓) displays a reduction in a tested parameter relative to the control. Abbreviation in the Table 2: *A. calcoaceticus, Acinetobacter calcoaceticus*; *A. aneurinilyticus, Aneurinibacillus aneurinilyticus*; *A. chroococcum, Azotobacter chroococcum*; AEs, Antioxidant enzymes; APX, Ascorbate peroxidase; *B. pumilus, Bacillus pumilus*; Car, Carotenoids; CAT, Catalase; Chl, Chlorophyll; *E. aurantiacum, Exiguobacterium aurantiacum*; GP, Germination percentage; GPr, Grain protein; GR, Glutathione reductase; GW, Grain weight; GY, Grain yield; *K. sacchari, Kosakonia sacchari*; MDA, Malondialdehyde; Mel, Melatonin; N/A, Not available; NL, Number of leaves per plant; *P. fluorescence, Pseudomonas fluorescence*; PhoPs, Photosynthetic pigments; POD, Peroxidase; *P. putida, Pseudomonas putida*; PPO, Polyphenol oxidase; RDW, Root dry weight; RFW, Root fresh weight; *R. leguminosarum, Rhizobium leguminosarum*; RL, Root length; SDW, Shoot dry weight; SFW, Shoot fresh weight; SL, Shoot length; *S. maltophilia, Stenotrophomonas maltophilia*; SOD, Superoxide dismutase; SpDW, Spike dry weight; SVI, Seedling vigor index; SY, Seed yield; Tchl, Total chlorophyll; TPDW, Total plant dry weight; Y, Yield.

3.3. The Reduction in AEs and OS in PGPR-Inoculated Plants and PGPR-Primed Seedlings under Salt Stress

The changes in AEs and osmo-regulators have been noticed in both uninoculated- and inoculated plants under normal and salinity conditions. However, the reduction or increase of these enzymes in PGPB-inoculated plants in response to salt conditions remains controversial. The decreased profiles of OS and/or ROS-scavenging enzymes were remarked by Kang et al. (2014 a) [43], Kang et al. (2014 b) [41], Barnawal et al. (2014) [95], Khan et al. (2016) [40], Bhise et al. (2017) [99], Abd_Allah et al. (2018) [35], Sapre et al. (2018) [44], Ansari et al. (2019) [51], Alexander et al. (2020) [48], Misra and Chauhan (2020) [106], and Shahid et al. (2021) [36]. Specifically, in the study of Kang et al. (2014a) [43], the activities of CAT, PPO, and POD enzymes and the PPs contents in the inoculated plants (e.g., *B. cepacia* SE4, *Promicromonospora* sp. SE188 or *A. calcoaceticus* SE370) were lower than those in the uninoculated plants under salt stress (120 mM of NaCl). The reduced profiles of AEs in Kang and his colleagues' findings were in agreement with their another study on soybean using the bacterium *P. putida* H-2-3 [41] and also in line with the study of Sapre et al. [44]. According to Sapre and colleagues' finding, the *Klebsiella*-treated wheat plants increased by 96% SOD and 286% POD, while the SOD and POD in untreated plants were increased by 353% and 540%, respectively. These data were in agreement with those found by Shahid et al. [36], and Sarkar et al. [49] as these investigators found that the highest antioxidant enzyme activities were recorded in the non-inoculated mung bean and rice plants, respectively. In parallel with the report of Sarkar et al. [49], Rojas-Tapias et al. [128] also recorded the highest Pro content was found in the non-inoculated maize seedling leaves under salt stress. The increase of PP contents in bacterized plants was also recorded in [41,43], however, to a lesser extent than those in the untreated plants. Similarly, Pro accumulations in the tissues of the control oat plants and the control rice plants were much higher than those in the *Klebsiella*-inoculated oat plants and *Entorobacter*-inoculated rice plants (230% and 175%, respectively vs. 155% and 75%, respectively) [44,49]. These studies showed similar findings with Manaf and Zayed [158] as the SOD activity and proline content in the cowpea plants treated with mycorrhizae or *P. fluorescence* alone were lower than those in the untreated plants under 3000 ppm NaCl irrigation regime. Manaf and Zayed assumed that the harmful effects of high salinity made the plants lose the ability to control their metabolites [158], whereas Sapre et al. [44] speculated that the treated plants did not sense much stress as the untreated plants did, leading the lower levels of AEs, NEAs, and osmoregulators in their tissues. Misra and Chauhan [106] proposed that the reduced Pro and AEs in *Bacillus*-treated maize plants may be due to the formation of EPS and biofilm on plant root surfaces that prevented plants from over-uptake Na^+, thereby attenuating the detrimental effects of toxic ions on plants. This assumption was corroborated by a study of Mukherjee et al. [110], who found that the amount of EPS-bound Na^+ increased with the increase in NaCl concentration in the solution, thus confirming an efficient role of EPS in NaCl sequestration. In addition, Sarkar et al. [49] explained that the increased antioxidant enzyme activities of *Enterobacter* sp. P23 under saline stress could indirectly quench a significant amount of ROS in rice seedlings, thus delaying the urge to synthesize ROS scavengers by stressed plants.

3.4. Genetic Diversities of Plant and Microbe, Plant–Microbe Interactions and Microbe–Microbe Interactions Are Key Players in Regulating AEs Profiles

In the first case where different plant species inoculated with PGPR species from the same genus *Curtobacterium*, the reduced PPO and POD activities were observed in the *Curtobacterium oceanosedimentum* SAK1-treated soybean plants [159], whereas the increase in POD, CAT, SOD, and APX were recorded in the *Curtobacterium albidum* SRV4-treated paddy plants [154].

In the second situation, soybean plants inoculated with different PGPR species also revealed the contrasting antioxidant enzyme profiles. For instance, the soybean plants cv. Giza 35 treated with *Bacillus firmus* SW5 showed a significant increase in APX, SOD, CAT, and POD activities [153], whereas the SOD and DPPH scavenging activities were relatively decreased in the soybean plants cv. Taekwang inoculated with *Pseudomonas putida* H-2-3 [41]. Moreover, variation in enzyme activities were found in the maize variety cv. Maharaja inoculated with different PGPR species [106]. Specifically, the maize seedlings Maharaja treated with *Bacillus subtilis* (NBNI 28B) had higher GPX and CAT as compared to the control, whereas the seedlings treated with *B. subtilis* (NBRI 33 N) and *B. safensis* (NBRI 12 M) exhibited lower SOD, APX, GPX, CAT, and PPO than those in the control and those in the NBNI 28B-bacterized seedlings [106].

Regarding the effect of consortium treatment on AEs, the POD activity in the salt-sensitive wheat genotype Galaxy-13 inoculated with individual *Pseudomonas fluorescence*, *Bacillus pumilus*, and *Exiguobacterium aurantiacum* was always higher than that in the salt-tolerant wheat genotype Aas-11 [122]. The effect of the consortium (*Pseudomonas fluorescence, Bacillus pumilus*, and *Exiguobacterium aurantiacum*), however, resulted in the lower POD activity in the treated Galaxy-13 in respect of the treated Aas-11. Similarly, CAT activity was higher in the Galaxy-13 treated alone with *P. fluorescence* and *E. aurantiacum*, in comparison with that in Aas-11, but the CAT activity of Galaxy-13 was lower than that of Aas-11 in the consortium treatment. In contrast, Galaxy-13 had lower SOD activity in a single inoculation with *P. fluorescence* and *B. pumilus*, but it exhibited higher SOD activity than Aas-11 in the consortium treatment.

It is worthy to note that antioxidant response to salinity was varied in different cultivars in the same plant species. Kharusi et al. (2019) [160] noticed that the salt-tolerant date palm cultivar Umsila maintained a normal concentration of ROS by accumulating elevated NEAs and by stimulating greater AEs activities with respect to the salt-sensitive date palm cultivar Zabad. The activities of SOD, CAT, APX and the contents of GSH, FLA, PCs, and Pro in Umsila were statistically significantly greater than those in Zabad when exposed to salt stress.

In summary, the findings in these previous studies, taken together, suggest that an increase or a reduction in the activities of AEs and/or OS in PGPR-inoculated plants during salt stress adaptation depends mainly on the specificities of plant species, on PGPR species, interactions between PGPR in consortia, and on plant–microbe interactions. These controversial data indicate not only that the fine-tuning of the ROS quenchers might be critical for plants to tolerate better to salt stress, but also pose questions concerning the exact mechanisms of salt stress tolerance imposed by PGPR. So far, investigators mainly based on their personal assumption, but not on scientific evidence, to elucidate the fluctuation in AEs. Integrated Omics approach would be necessary to gain insight into this interesting issue. The main message of the present review was displayed in the Figure 1.

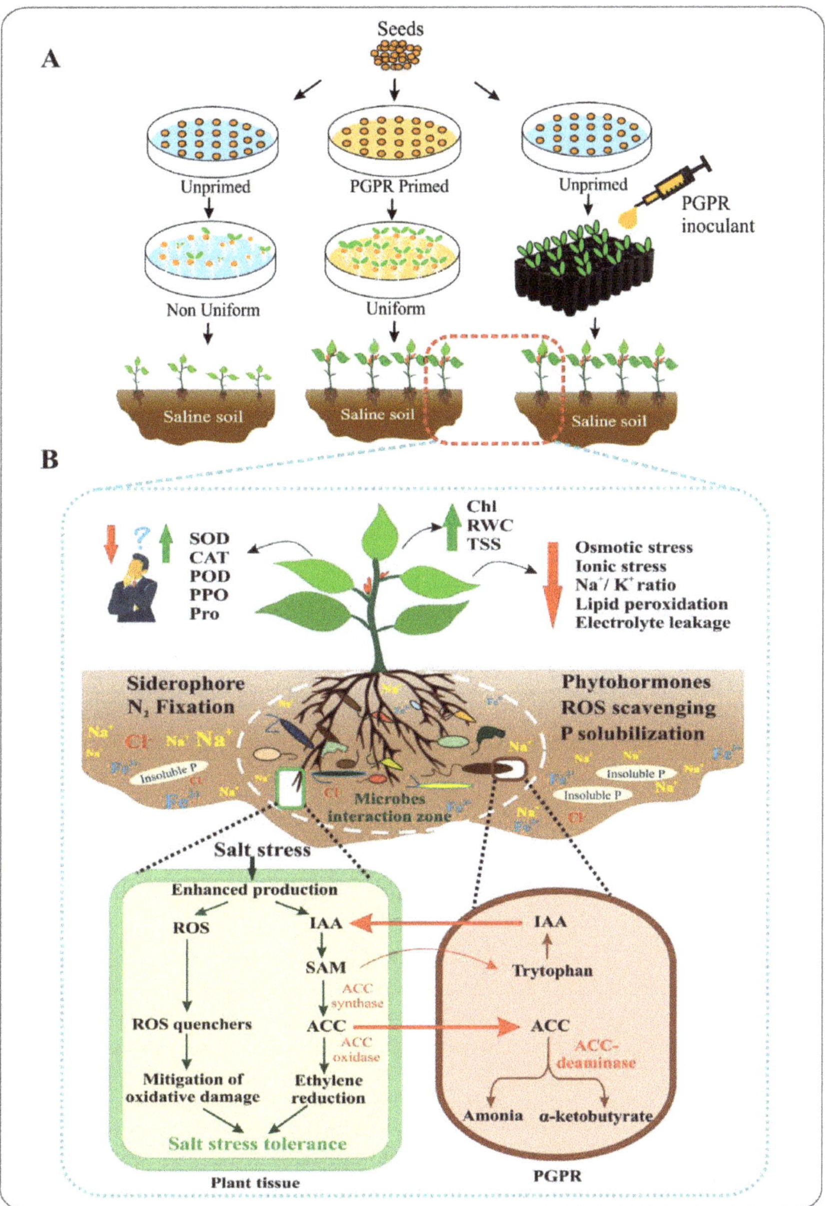

Figure 1. Roles of PGPR in alleviating salinity stress in plants. (**A**) represents the application of PGPR as microbial beneficial tools in seed biopriming technique and as green bioinoculants in seedlings treatment. The primed seeds demonstrate rapid germination and robust, uniform seedlings. (**B**) shows positive effects of PGPR on vegetative parameters and physio-biochemical indexes in PGPR-inoculated plants via various mechanisms e.g., production of OS, AEs to reduce osmotic and ionic stress, and EPS suppress toxic ions uptake and ion exposure. The fluctuation of AEs and OS profiles in PGPR-treated plants is also displayed in the left panel. The middle panel demonstrates key characteristics of PGPR including the production of Sid, phytohormones, EPS, N fixation and P solubilization. The lower panel emphasizes the importance of ACC deaminase-producing PGPR in ameliorating the inhibitory effects of excess ethylene on plant growth.

4. Roles of Multi-Omics Techniques in Deciphering Plant–Microbe Interactions

The modes of action of PGPR on plant salt-stress response mechanism are diverse and complex and remain largely unclear, especially at the molecular level. In the study of Kim et al. (2014) [96], the colonization of *Enterobacter* sp. EJ01 in Arabidopsis root tissues conferred salt stress resistance by inducing salt stress responsive signaling pathways. Specifically, after EJ01 inoculation, the expression levels of *DREB2b*, *RD29A*, and *RAB18* genes related to ABA-dependent and ABA-independent pathways were upregulated, even in the absence of salinity. The expression pattern of *RD29B*, however, was dependent on salt treatment. In addition, *P5CS1*, a Pro biosynthesis-related gene, was upregulated under salinity conditions. The inoculation of EJ01 into Arabidopsis plants induced the host basal innate immunity, as well as the rapid defense responses at systemic level so called induced systemic resistance (ISR). PGPR-elicited ISR was previously observed in Arabidopsis seedlings treated with VOCs from *Bacillus subtilis* GB03 and *Bacillus amyloliquefaciens* IN937a. The use of transgenic and mutant lines of Arabidopsis indicated that the ISR activated by the VOCs from GB03 was based on ethylene-dependent signaling pathway, whereas the ISR was triggered by VOCs from IN937a through an ethylene-independent signaling pathway [161]. In the study of Chen et al. [162], the upregulation of Na^+/H^+ antiporter (*NHX*) and H^+-*PPase* genes in *Bacillus amyloliquefaciens* SQR9-inoculated maize shoots facilitated Na^+ sequestration into vacuoles. The recirculating of Na^+ from shoot to root via an elevated expression of high-affinity K^+ transporter 1 (*HKT1*) gene was observed in the treated maize plants. However, contrary to the results of Chen et al. [162], the lowest expression pattern of *HKT1* gene was found in the *Pseudomonas simiae* + sodium nitroprusside (SNP) treated soybean plants [147]. A stable photosynthetic activity was maintained by the highly expressed *RBCS* (RubisCo small subunit), *RBCL* (Rubisco large subunit) genes [162], similar to the upregulation of the Rubisco-encoding gene *rbcL* in the *Klebsiella*-treated oat seedlings [44]. Moreover, the senescence rate in SQR9-inoculated treated maize was properly controlled by the reduced expression of *NCED*, a key gene in ABA synthesis pathway.

In all treatments, the expression levels of AEs-encoding genes *POD* and *CAT* were the highest in the *P. simiae* + SNP-treated soybean plants [147]. Genes associated with Aux, CK, and GA signaling pathways in the *Paenibacillus polymyxa* YC0136-treated tobacco plants (*Nicotiana tabacum* L.) were found to be upregulated relative to the uninoculated control plants, along with the elevated expression of WRKY and MYB transcription factors (TFs) [163]. The WRKY and MYB TFs are responsible for gene regulations and have a great influence in every aspect of plant growth and development, as well as in plant stress responses [164,165]. Changes in the expression pattern of WRKY TF gene under salinity conditions were reported in previous studies [166–168]. These findings were in line with the transcriptome profiles of the rice roots bacterized with *Azospirillum brasilense*, in which the hormones-encoding genes (e.g., Aux efflux carriers, Aux-responsive genes, Aux response factors, ACC oxidase genes, ethylene insensitive 2, cytokinin-O-glucosyltransferases, and cytokinin dehydrogenase precursors) were significantly upregulated, as well as the major plant TFs families, namely, AP2/ERF family, MYB family, WRKY family, and the GRAS family [169]. The enhanced expressions of MYB and WRKY TFs were also noticed in the *Dietzia natronolimnaea*-inoculated wheat plants under salinity stress [152]. Furthermore, 9 genes in the SA pathway and 6 genes encoding phenylalanine ammonia lyase (PAL), a key enzyme in the phenylpropanoids metabolic pathway, were upregulated with respect to the control plants, resulting the induction of systemic resistance in tobacco host plant [163]. Malviya et al. [170] also found that the infection of *Burkholderia anthina* MYSP113 into the sugarcane plantlets cv. GXB-9 induced the upregulation of phenylpropanoid biosynthesis genes and amino acid biosynthesis pathways, in accordance with the findings from Liu et al. [163]. Likewise, phenylpropanoid biosynthesis was the most enriched pathway in both differentially expressed genes (DEGs) and abundant metabolites (DAMs) among all salt stress responses in barley rootzones [171]. In the *Arthrobacter nitroguajacolicus*-inoculated wheat roots treated with 200 mM NaCl,

8 genes responsible for Fe uptake, and 2 phosphatase-encoding genes were upregulated, as well as the upregulation of several transporter genes which were in charge of ions, sugars, oligopeptide, and amino acids transports [172].

On the one hand, PGPR influence the expression patterns in the host plants. On the other hand, changes in their transcriptome in response to their colonized plants were also recorded [173]. During the interaction with the host plant, 2 genes *ilvB* and *PPYC1_23850* related to Aux biosynthesis, 3 genes belonging to cell motility category (e.g., *fliG, fliH, fliF*), 31 genes related to transport proteins including 16 genes belong to ATP-binding cassette (ABC), and 3 genes associated with a major facilitator superfamily (MFS) in *P. polymyxa* YC0136 were significantly upregulated [163]. It was thought that root exudates from tobacco attracted *P. polymyxa* YC0136 and may play roles as nutrients source for the growth of YC0136 strain. This explained the upregulation of numerous transport and cell motility genes in the bacterium. In response to host rice seedlings, *Bacillus subtilis* OKB105 also altered its transcriptomic patterns, in which 52 genes related to nutrients transport and metabolism were upregulated, suggesting the bacterium used carbohydrates and amino acids secreted by rice seedlings as carbon and energy sources. In contrast to the data in [163], many genes involved in chemotaxis and motility, however, were downregulated [173].

In summary, plants and PGPR influence each other in a mutualistic relationship. Regarding plant resistance to salinity, the microbes regulate the WRKY and MYB TFs which are widely distributed in higher plants. Subsequently, these master regulators will regulate the expression of their key downstream stress responsive genes. The plants, in turn, provide nutrients via root exudates for the growth of the microbes. This interaction benefits plants in non-stress conditions, and also in environmental challenging conditions.

5. Promise, Limitations, and Future Directions

Considerable PGPB-related studies that have been carried out in the last decades help to improve our knowledge concerning advantageous characteristics of PGPB, in both basic and applied aspects. However, most studies focused on estimating the parameters in vegetative growth stage, but rarely on evaluating the parameters that are related to reproductive stage such as GW and FW, numbers of flower, numbers of seed, fruit per plant, and plant yield. We found a scarcity of studies that evaluated beneficial effects of PGPB on attenuating yield loss and on improving nutrient values. In our opinion, this could be one of the main drawbacks of PGPB-related studies thus far if we consider the improvement of crop yields, productivity, and the quality of fruit/grain under high saline conditions to be our main goal in plant agriculture studies. In addition, in some studies, the lack of important measurements regarding ion contents, ROS levels, phytohormone concentrations, and electrolyte leakage in many studies make them difficult to evaluate the overall effects of PGPB on plants. Furthermore, the short exposure of plants to salinity in some studies unlikely reflects the real situation in fields where a variety of biotic and abiotic stresses endures simultaneously and lasts permanently.

The recognition of PGPR as safe, efficient, and appropriate bioinoculants for agricultural practice is widely accorded. However, the primary mechanisms employed by PGPR to promote plant defense against salt stress need to be deeply unraveled, especially changes in both Omics profiles (e.g., proteomics, transcriptomics, and metabolomics) in the treated plants and in the microbes during interaction with their hosts. In addition, the highly genetic variations of plants and PGPR are useful traits in coping with diverse environmental issues. However, this attribute also makes the reproducibility from previous studies' findings challenging. As discussed in the present review, the patterns of ROS quencher in PGPR-treated plants exhibited great differences from case to case mainly due to the genetic diversity. The synergistic and/or antagonistic effects between PGPR in consortia on plant growth and defense system, which occur commonly in terrestrial soil ecosystems, should also be thoroughly deciphered. Consequently, an integration of Omics technologies and systems biology should be considered in future studies to provide broader picture and more detailed information concerning plant–microbe interactions in a more complex scenario.

Author Contributions: Conceptualization, D.M.H.-T. and T.T.M.N.; Visualization, T.T.M.N.; Writing—Original draft preparation, D.M.H.-T., T.T.M.N., E.H., S.-H.H., and C.-C.H.; Writing-Revised manuscript D.M.H.-T.; Supervision, C.-C.H. All authors have read and agreed to the published version of the manuscript.

Funding: This work received no external funding.

Institutional Review Board Statement: Not applicable.

Informed Consent Statement: Not applicable.

Data Availability Statement: Not applicable.

Conflicts of Interest: The authors declare no conflict of interest.

Abbreviations

A. aneurinilyticus	*Aneurinibacillus aneurinilyticus*
ABA	Abscisic acid
ABC	ATP-binding cassette
A. brasilense	*Azospirillum brasilense*
A. calcoaceticus	*Acinetobacter calcoaceticus*
ACC	1-aminocyclopropane-1-carboxylate
ALA	5-aminolevulinic acid
Alg	Alginate
A. macrostachyum	*Arthrocnemum macrostachyum*
A. protophormiae	*Arthrobacter protophormiae*
APX	Ascorbate peroxidase
ASC	Ascorbate
Aux	Auxin
B. aquimaris	*Bacillus aquimaris*
B. gibsonii	*Bacillus gibsonii*
B. iodinum	*Brevibacterium iodinum*
B. megaterium	*Bacillus megaterium*
B. pumilus	*Bacillus pumilus*
BR	Brassinosteroids
B. safensis	*Bacillus safensis*
B. subtilis	*Bacillus subtilis*
CA	Citric acid
Car	Carotenoids
CAT	Catalase
C. gleum	*Chryseobacterium gleum*
Chl	Chlorophyll
CK	Cytokinins
DHAR	dehydroascorbate reductase
DPPH	2,2-diphenyl-1-picryl-hydrazyl-hydrate
E. aurantiacum	*Exiguobacterium aurantiacum*
EL	Electrolyte leakage
EPS	Exopolysaccharide
ETC	Electron transport chains
Fd_{ox}	Oxidized ferredoxin
FD_{red}	Reduced ferredoxin
FD	Fruit diameter
FL	Fruit length
FLA	Flavonoids
FLI	Feather lysate inoculum
FMY	Fruit marketable yield

FW	Fruit weight
FY	Fruit yield
GA	Gibberellins
GLVs	Green leaf volatiles
GP	Germination percentage
G-POD	Guaiacol peroxidase
GPr	Grain protein
GR	Glutathione reductase
GRA	Germination rate
gs	Stomatal conductance
GSH	Glutathione
GST	Glutathione-S-transferase
GW	Grain weight
GY	Grain yield
H_2O_2	Hydrogen peroxide
IAA	Indole-3-acetic acid
ISR	Induced systemic resistance
JA	Jasmonates
K	Potassium
K. sacchari	*Kosakonia sacchari*
L. adecarboxylata	*Leclercia adecarboxylata*
LP	Lipid peroxidation
MA	Malic acid
MDA	Malondialdehyde
MDAR	monodehydroascorbate reductase
Mel	Melatonin
MeSA	Methyl salicylate
MFS	Major facilitator superfamily
M. oleivorans	*Microbacterium oleivorans*
N	Nitrogen
N/A	Not available
NF	Number of fruits per plant
NL	Number of leaves per plant
NT	Number of tillers per plant
$O_2^{\bullet-}$	Superoxide radical
OS	Osmolytes
P	Phosphate
PAs	Polyamines
P. agglomerans	*Pantoea agglomerans*
PAL	Phenylalanine ammonia lyase
P. argentinensis	*Pseudomonas argentinensis*
P. azotoformans	*Pseudomonas azotoformans*
P. putida	*Pseudomonas putida*
PCs	Phenolic compounds
P. fluorescence	*Pseudomonas fluorescence*
PGPR	Plant growth- promoting rhizobacteria
PH	Plant height
PHE	Phenols
PhoPs	Photosynthetic pigments
Pn	Net photosynthetic rate
POD	Peroxidase
PPs	Polyphenols
PPO	Polyphenol oxidase
P. putida	*Pseudomonas putida*
Pro	Proline
PSI	Photosystem I

P. yonginensis	*Paenibacillus yonginensis*
RDW	Root dry weight
RFW	Root fresh weight
RL	Root length
R. massiliae	*Rhizobium massiliae*
ROS	Reactive oxygen species
RWC	Relative water content
SA	Salicylic acid
SAM	S-adenosyl-L-methionine
SDM	Water-soluble dry matter
SDW	Shoot dry weight
SFW	Shoot fresh weight
Sid	Siderophore
SL	Shoot length
S. maltophilia	*Stenotrophomonas maltophilia*
SNP	Sodium nitroprusside
SOD	Superoxide dismutase
SpDW	Spike dry weight
StDW	Stem dry weight
StFW	Stem fresh weight
SVI	Seedling vigor index
SW	Seed weight
SY	Seed yield
Tchl	Total chlorophyll
TCP	Tocopherol
TFs	Transcription factors
TLA	Total leaves area per plant
TPDW	Total plant dry weight
TPFW	Total plant fresh weight
Tr	Transpiration rate
Tre	Trehalose
TSS	Total soluble sugar
X. autotrophicus	*Xanthobacter autotrophicus*
Y	Yield
Z. halotolerans	*Zhihengliuella halotolerans*

References

1. de Lima-Neto, A.; Cavalcante, L.; Mesquita, F.d.O.; Souto, A.d.L.; dos Santos, G.; dos Santos, J.; de Mesquita, E. Papaya seedlings irrigation with saline water in soil with bovine biofertilizer. *Chil. J. Agric. Res.* **2016**, *76*, 236–242. [CrossRef]
2. Reints, J.; Dinar, A.; Crowley, D. Dealing with Water Scarcity and Salinity: Adoption of Water Efficient Technologies and Management Practices by California Avocado Growers. *Sustainability* **2020**, *12*, 3555. [CrossRef]
3. Tri, D.Q.; Tuyet, Q.T.T. Effect of Climate Change on the Salinity Intrusion: Case Study Ca River Basin, Vietnam. *J. Clim. Chang.* **2016**, *2*, 91–101. [CrossRef]
4. Gondek, M.; Weindorf, D.; Thiel, C.; Kleinheinz, G. Soluble Salts in Compost and Their Effects on Soil and Plants: A Review. *Compost Sci. Util.* **2020**, *28*, 59–75. [CrossRef]
5. Rasool, S.; Hameed, A.; Azooz, M.; Muneeb-u-Rehman; Siddiqi, T.; Ahmad, P. Salt Stress: Causes, Types and Responses of Plants. In *Ecophysiology and Responses of Plants under Salt Stress*; Ahmad, P., Azooz, M., Prasad, M., Eds.; Springer: New York, NY, USA, 2013; ISBN 978-1-4614-4747-4.
6. Attia, H.; Alamer, K.; Ouhibi, C.; Oueslati, S.; Lachaâl, M. Interaction Between Salt Stress and Drought Stress on Some Physiological Parameters in Two Pea Cultivars. *Int. J. Bot.* **2020**, *16*. [CrossRef]
7. Maathuis, F.; Amtmann, A. K$^+$ Nutrition and Na$^+$ Toxicity: The Basis of Cellular K$^+$/Na$^+$ Ratios. *Ann. Bot.* **1999**, *84*, 123–133. [CrossRef]
8. Bernstein, N.; Meiri, A. Root Growth of Avocado is More Sensitive to Salinity than Shoot Growth. *J. Am. Soc. Hortic. Sci.* **2004**, *129*, 188–192. [CrossRef]
9. Neves, G.; Marchiosi, R.; Ferrarese, M.; Siqueira-Soares, R.; Ferrarese-Filho, O. Root Growth Inhibition and Lignification Induced by Salt Stress in Soybean. *J. Agron. Crop Sci.* **2010**, *196*, 467–473. [CrossRef]
10. Kafi, M.; Rahimi, Z. Effect of salinity and silicon on root characteristics, growth, water status, proline content and ion accumulation of purslane (*Portulaca oleracea* L.). *Soil Sci. Plant Nutr.* **2011**, *57*. [CrossRef]

11. Fita, A.; Rodríguez-Burruezo, A.; Boscaiu, M.; Prohens, J.; Vicente, O. Breeding and Domesticating Crops Adapted to Drought and Salinity: A New Paradigm for Increasing Food Production. *Front. Plant Sci.* **2015**, *6*. [CrossRef]

12. Wang, W.; Vinocur, B.; Altman, A. Plant responses to drought, salinity and extreme temperatures: Towards genetic engineering for stress tolerance. *Planta* **2003**, *218*, 1–14. [CrossRef]

13. Jha, S. Chapter 14: Transgenic Approaches for Enhancement of Salinity Stress Tolerance in Plants. In *Molecular Approaches in Plant Biology and Environmental Challenges*; Singh, S., Upadhyay, S., Pandey, A., Kumar, S., Eds.; Energy, Environment and Sustainability; Springer Nature Singapore Pte Ltd.: Singapore, 2019; ISBN 978-981-15-0690-1.

14. Kaya, C.; Kirnak, H.; Higgs, D.; Saltali, K. Supplementary calcium enhances plant growth and fruit yield in strawberry cultivars grown at high (NaCl) salinity. *Sci. Hortic.* **2002**, *93*, 65–74. [CrossRef]

15. Alzahrani, Y.; Kuşvuran, A.; Alharby, H.; Kuşvuran, S.; Rady, M. The defensive role of silicon in wheat against stress conditions induced by drought, salinity or cadmium. *Ecotoxicol. Environ. Saf.* **2018**, *154*, 187–196. [CrossRef]

16. Hassanvand, F.; Nejad, A.; Fanourakis, D. Morphological and physiological components mediating the silicon-induced enhancement of geranium essential oil yield under saline conditions. *Ind. Crops Prod.* **2019**, *134*, 19–25. [CrossRef]

17. Tian, S.; Guo, R.; Zou, X.; Zhang, X.; Yu, X.; Zhan, Y.; Ci, D.; Wang, M.; Wang, Y.; Si, T. Priming With the Green Leaf Volatile (Z)-3-Hexeny-1-yl Acetate Enhances Salinity Stress Tolerance in Peanut (*Arachis hypogaea* L.) Seedlings. *Front. Plant Sci.* **2019**, *10*. [CrossRef] [PubMed]

18. Yan, Z.-G.; Wang, C.-Z. Wound-induced green leaf volatiles cause the release of acetylated derivatives and a terpenoid in maize. *Phytochemistry* **2006**, *67*, 34–42. [CrossRef] [PubMed]

19. Wu, Y.; Liao, W.; Dawuda, M.; Hu, L.; Yu, J. 5-Aminolevulinic acid (ALA) biosynthetic and metabolic pathways and its role in higher plants: A review. *Plant Growth Regul.* **2019**, *87*, 357–374. [CrossRef]

20. Wu, Y.; Jin, X.; Liao, W.; Hu, L.; Dawuda, M.; Zhao, X.; Tang, Z.; Gong, T.; Yu, J. 5-Aminolevulinic Acid (ALA) Alleviated Salinity Stress in Cucumber Seedlings by Enhancing Chlorophyll Synthesis Pathway. *Front. Plant Sci.* **2018**, *9*, 635. [CrossRef]

21. Nawaz, K.; Chaudhary, R.; Sarwar, A.; Ahmad, B.; Gul, A.; Hano, C.; Abbasi, B.; Anjum, S. Melatonin as Master Regulator in Plant Growth, Development and Stress Alleviator for Sustainable Agricultural Production: Current Status and Future Perspectives. *Sustainability* **2021**, *13*, 294. [CrossRef]

22. Ke, Q.; Ye, J.; Wang, B.; Ren, J.; Yin, L.; Deng, X.; Wang, S. Melatonin Mitigates Salt Stress in Wheat Seedlings by Modulating Polyamine Metabolism. *Front. Plant Sci.* **2018**, *9*, 914. [CrossRef]

23. Liu, J.; Shabala, S.; Zhang, J.; Ma, G.; Chen, D.; Shabala, L.; Zeng, F.; Chen, Z.-H.; Zhou, M.; Venkataraman, G.; et al. Melatonin improves rice salinity stress tolerance by NADPH oxidase-dependent control of the plasma membrane K+ transporters and K+ homeostasis. *Plant Cell Environ.* **2020**, *43*, 2591–2605. [CrossRef]

24. Abd El-Azeem, S.; Elwan, M.; Sung, J.-K.; Ok, Y. Alleviation of Salt Stress in Eggplant (Solanum melongena L.) by Plant-Growth-Promoting Rhizobacteria. *Commun. Soil Sci. Plant Anal.* **2012**, *43*, 1303–1315. [CrossRef]

25. Hanin, M.; Ebel, C.; Ngom, M.; Laplaze, L.; Masmoudi, K. New Insights on Plant Salt Tolerance Mechanisms and Their Potential Use for Breeding. *Front. Plant Sci.* **2016**, *7*, 1–17. [CrossRef]

26. Santos, A.; da Silveira, J.; Bonifacio, A.; Rodrigues, A.; Figueiredo, M. Antioxidant response of cowpea co-inoculated with plant growth-promoting bacteria under salt stress. *Braz. J. Microbiol.* **2018**, *49*, 513–521. [CrossRef] [PubMed]

27. Gill, S.; Tuteja, N. Reactive oxygen species and antioxidant machinery in abiotic stress tolerance in crop plants. *Plant Physiol. Biochem.* **2010**, *48*, 909–930. [CrossRef] [PubMed]

28. Jogawat, A. Chapter 5: Osmolytes and their Role in Abiotic Stress Tolerance in Plants. In *Molecular Plant Abiotic Stress: Biology and Biotechnology*; Roychoudhury, A., Tripathi, D., Eds.; John Wiley & Sons Ltd.: Hoboken, NJ, USA, 2019; pp. 91–104. ISBN 978-1-119-46366-5.

29. Tester, M.; Davenport, R. Na+ tolerance and Na+ transport in higher plants. *Ann. Bot.* **2003**, *91*, 503–527. [CrossRef] [PubMed]

30. Cheng, Z.; Woody, O.; McConkey, B.; Glick, B. Combined effects of the plant growth-promoting bacterium Pseudomonas putida UW4 and salinity stress on the Brassica napus proteome. *Appl. Soil Ecol.* **2012**, *61*, 255–263. [CrossRef]

31. Shahid, M.; Sarkhosh, A.; Khan, N.; Balal, R.; Ali, S.; Rossi, L.; Gómez, C.; Mattson, N.; Nasim, W.; Garcia-Sanchez, F. Insights into the Physiological and Biochemical Impacts of Salt Stress on Plant Growth and Development. *Agronomy* **2020**, *10*, 938. [CrossRef]

32. Manishankar, P.; Wang, N.; Köster, P.; Alatar, A.; Kudla, J. Calcium signaling during salt stress and in the regulation of ion homeostasis. *J. Exp. Bot.* **2018**, *69*, 4215–4226. [CrossRef] [PubMed]

33. Zhang, M.; Fang, Y.; Ji, Y.; Jiang, Z.; Wang, L. Effects of salt stress on ion content, antioxidant enzymes and protein profile in different tissues of Broussonetia papyrifera. *S. Afr. J. Bot.* **2013**, *85*, 1–9. [CrossRef]

34. Taïbi, K.; Taïbi, F.; Abderrahim, L.; Ennajah, A.; Belkhodja, M.; Mulet, J. Effect of salt stress on growth, chlorophyll content, lipid peroxidation and antioxidant defence systems in *Phaseolus vulgaris* L. *S. Afr. J. Bot.* **2016**, *105*, 306–312. [CrossRef]

35. Abd_Allah, E.; Alqarawi, A.; Hashem, A.; Radhakrishnan, R.; Al-Huqail, A.; Al-Otibi, F.; Malik, J.; Alharbi, R.; Egamberdieva, D. Endophytic bacterium Bacillus subtilis (BERA 71) improves salt tolerance in chickpea plants by regulating the plant defense mechanisms. *J. Plant Interact.* **2018**, *13*, 37–44. [CrossRef]

36. Shahid, M.; Ameen, F.; Maheshwari, H.; Ahmed, B.; AlNadhari, S.; Khan, M. Colonization of Vigna radiata by a halotolerant bacterium Kosakonia sacchari improves the ionic balance, stressor metabolites, antioxidant status and yield under NaCl stress. *Appl. Soil Ecol.* **2021**, *158*, 103809. [CrossRef]

37. Magán, J.; Gallardo, M.; Thompson, R.; Lorenzo, P. Effects of salinity on fruit yield and quality of tomato grown in soil-less culture in greenhouses in Mediterranean climatic conditions. *Agric. Water Manag.* **2008**, *95*, 1041–1055. [CrossRef]

38. Navarro, J.; Garrido, C.; Flores, P.; Martínez, V. The effect of salinity on yield and fruit quality of pepper grown in perlite. *Span. J. Agric. Res.* **2010**, *8*, 142–150. [CrossRef]

39. Wungrampha, S.; Joshi, R.; Singla-Pareek, S.; Pareek, A. Photosynthesis and salinity: Are these mutually exclusive? *Photosynthetica* **2018**, *56*, 366–381. [CrossRef]

40. Khan, A.; Sirajuddin;; Zhao, X.; Javed, M.; Khan, K.; Bano, A.; Shen, R.; Masood, S. Bacillus pumilus enhances tolerance in rice (*Oryza sativa* L.) to combined stresses of NaCl and high boron due to limited uptake of Na^+. *Environ. Exp. Bot.* **2016**, *124*, 120–129. [CrossRef]

41. Kang, S.-M.; Radhakrishnan, R.; Khan, A.; Kim, M.-J.; Park, J.-M.; Kim, B.-R.; Shin, D.-H.; Lee, I.-J. Gibberellin secreting rhizobacterium, Pseudomonas putida H-2-3 modulates the hormonal and stress physiology of soybean to improve the plant growth under saline and drought conditions. *Plant Physiol. Biochem.* **2014**, *84*, 115–124. [CrossRef]

42. Li, H.; Jiang, X. Inoculation with plant growth-promoting bacteria (PGPB) improves salt tolerance of maize seedling. *Russ. J. Plant Physiol.* **2017**, *64*, 235–241. [CrossRef]

43. Kang, S.-M.; Khan, A.; Waqas, M.; You, Y.-H.; Kim, J.-H.; Kim, J.-G.; Hamayun, M.; Lee, I.-J. Plant growth-promoting rhizobacteria reduce adverse effects of salinity and osmotic stress by regulating phytohormones and antioxidants in Cucumis sativus. *J. Plant Interact.* **2014**, *9*, 673–682. [CrossRef]

44. Sapre, S.; Gontia-Mishra, I.; Tiwari, S. Klebsiella sp. confers enhanced tolerance to salinity and plant growth promotion in oat seedlings (Avena sativa). *Microbiol. Res.* **2018**, *206*, 25–32. [CrossRef]

45. Akram, W.; Aslam, H.; Ahmad, S.; Anjum, T.; Yasin, N.; Khan, W.; Ahmad, A.; Guo, J.; Wu, T.; Luo, W.; et al. Bacillus megaterium strain A12 ameliorates salinity stress in tomato plants through multiple mechanisms. *J. Plant Interact.* **2019**, *14*, 506–518. [CrossRef]

46. Orozco-Mosqueda, M.; Duan, J.; DiBernardo, M.; Zetter, E.; Campos-García, J.; Glick, B.; Santoyo, G. The Production of ACC Deaminase and Trehalose by the Plant Growth Promoting Bacterium Pseudomonas sp. UW4 Synergistically Protect Tomato Plants Against Salt Stress. *Front. Microbiol.* **2019**, *10*, 1392. [CrossRef] [PubMed]

47. Vaishnav, A.; Singh, J.; Singh, P.; Rajput, R.; Singh, H.; Sarma, B. Sphingobacterium sp. BHU-AV3 Induces Salt Tolerance in Tomato by Enhancing Antioxidant Activities and Energy Metabolism. *Front. Microbiol.* **2020**, *11*. [CrossRef] [PubMed]

48. Alexander, A.; Singh, V.; Mishra, A. Halotolerant PGPR Stenotrophomonas maltophilia BJ01 Induces Salt Tolerance by Modulating Physiology and Biochemical Activities of Arachis hypogaea. *Front. Microbiol.* **2020**, *11*, 8289. [CrossRef]

49. Sarkar, A.; Ghosh, P.; Pramanik, K.; Mitra, S.; Soren, T.; Pandey, S.; Mondal, M.; Maiti, T. A halotolerant Enterobacter sp. displaying ACC deaminase activity promotes rice seedling growth under salt stress. *Res. Microbiol.* **2018**, *169*, 20–32. [CrossRef]

50. Sukweenadhi, J.; Balusamy, S.; Kim, Y.-J.; Lee, C.; Kim, Y.-J.; Koh, S.; Yang, D. A Growth-Promoting Bacteria, Paenibacillus yonginensis DCY84T Enhanced Salt Stress Tolerance by Activating Defense-Related Systems in Panax ginseng. *Front. Plant Sci.* **2018**, *9*, 813. [CrossRef]

51. Ansari, F.; Ahmad, I.; Pichtel, J. Growth stimulation and alleviation of salinity stress to wheat by the biofilm forming Bacillus pumilus strain FAB10. *Appl. Soil Ecol.* **2019**, *143*, 45–54. [CrossRef]

52. Hamann, T. The Plant Cell Wall Integrity Maintenance Mechanism—Concepts for Organization and Mode of Action. *Plant Cell Physiol.* **2015**, *56*, 215–223. [CrossRef]

53. Yin, H.; Xu, L.; Porter, N. Free Radical Lipid Peroxidation: Mechanisms and Analysis. *Chem. Rev.* **2011**, *111*, 5944–5972. [CrossRef]

54. Corpas, F.; Gupta, D.; Palma, J. Chapter 1: Production Sites of Reactive Oxygen Species (ROS) in Organelles from Plant Cells. In *Reactive Oxygen Species and Oxidative Damage in Plants*; Gupta, D., Palma, J., Corpas, F., Eds.; Springer International Publishing: Cham, Switzerland, 2015; ISBN 978-3-319-20421-5.

55. Gutteridge, J. Lipid Peroxidation and Antioxidants as Biomarkers of Tissue Damage. *Clin. Chem.* **1995**, *41*, 1819–1828. [CrossRef] [PubMed]

56. Awasthi, J.; Saha, B.; Chowardhara, B.; Devi, S.; Borgohain, P.; Panda, S. Qualitative Analysis of Lipid Peroxidation in Plants under Multiple Stress Through Schiff's Reagent: A Histochemical Approach. *Bio-Protocol* **2018**, *8*, e2807. [CrossRef]

57. Sarkar, A.; Pramanik, K.; Mitra, S.; Soren, T.; Maiti, T. Enhancement of growth and salt tolerance of rice seedlings by ACC deaminase-producing Burkholderia sp. MTCC 12259. *J. Plant Physiol.* **2018**, *231*, 434–442. [CrossRef]

58. Khalid, M.; Bilal, M.; Hassani, D.; Iqbal, H.; Wang, H.; Huang, D. Mitigation of salt stress in white clover (Trifolium repens) by Azospirillum brasilense and its inoculation effect. *Bot. Stud.* **2017**, *58*. [CrossRef] [PubMed]

59. Ishibashi, Y.; Kasa, S.; Sakamoto, M.; Aoki, N.; Kai, K.; Yuasa, T.; Hanada, A.; Yamaguchi, S.; Iwaya-Inoue, M. A Role for Reactive Oxygen Species Produced by NADPH Oxidases in the Embryo and Aleurone Cells in Barley Seed Germination. *PLoS ONE* **2015**, *10*, e143173. [CrossRef]

60. Tsukagoshi, H.; Busch, W.; Benfey, P. Transcriptional Regulation of ROS Controls Transition from Proliferation to Differentiation in the Root. *Cell* **2010**, *143*, 606–616. [CrossRef]

61. Tognetti, V.; Bielach, A.; Hrtyan, M. Redox regulation at the site of primary growth: Auxin, cytokinin and ROS crosstalk. *Plant Cell Environ.* **2017**, *40*, 2586–2605. [CrossRef]

62. Zeng, J.; Dong, Z.; Wu, H.; Tian, Z.; Zhao, Z. Redox regulation of plant stem cell fate. *EMBO J.* **2017**, *36*, 2844–2855. [CrossRef]

63. Aarti, P.; Tanaka, R.; Tanaka, A. Effects of oxidative stress on chlorophyll biosynthesis in cucumber (*Cucumis sativus*) cotyledons. *Physiol. Plant.* **2006**, *128*, 186–197. [CrossRef]

64. Huang, H.; Ullah, F.; Zhou, D.-X.; Yi, M.; Zhao, Y. Mechanisms of ROS Regulation of Plant Development and Stress Responses. *Front. Plant Sci.* **2019**, *10*, 800. [CrossRef]

65. Abdelgawad, H.; Zinta, G.; Hegab, M.; Pandey, R.; Asard, H.; Abuelsoud, W. High Salinity Induces Different Oxidative Stress and Antioxidant Responses in Maize Seedlings Organs. *Front. Plant Sci.* **2016**, *7*, 276. [CrossRef]

66. Nimse, S.; Pal, D. Free radicals, natural antioxidants, and their reaction mechanisms. *RSC Adv.* **2015**, *5*, 27986–28006. [CrossRef]

67. Per, T.; Khan, N.; Reddy, P.; Masood, A.; Hasanuzzaman, M.; Khan, M.; Anjum, N. Approaches in modulating proline metabolism in plants for salt and drought stress tolerance: Phytohormones, mineral nutrients and transgenics. *Plant Physiol. Biochem.* **2017**, *115*, 126–140. [CrossRef] [PubMed]

68. Hayat, S.; Hayat, Q.; Alyemeni, M.; Wani, A.; Pichtel, J.; Ahmad, A. Role of proline under changing environments. *Plant Signal. Behav.* **2012**, *7*. [CrossRef] [PubMed]

69. Matysik, J.; Alia; Bhalu, B.; Mohanty, B. Molecular mechanisms of quenching of reactive oxygen species by proline under stress in plants. *Curr. Sci.* **2002**, *82*, 525–532.

70. Hossain, M.; Hoque, M.; Burritt, D.; Fujita, M. Chapter 16: Proline Protects Plants Against Abiotic Oxidative Stress: Biochemical and Molecular Mechanisms. In *Oxidative Damage to Plants-Antioxidant Networks and Signaling*; Ahmad, P., Ed.; Academic Press Inc.: Cambridge, MA, USA, 2014; pp. 477–522. ISBN 978-0-12-799963-0.

71. Iqbal, N.; Khan, N.; Ferrante, A.; Trivellini, A.; Francini, A.; Khan, M. Ethylene Role in Plant Growth, Development and Senescence: Interaction with Other Phytohormones. *Front. Plant Sci.* **2017**, *8*, 475. [CrossRef] [PubMed]

72. Boex-Fontvieille, E.; Rustgi, S.; von Wettstein, D.; Pollmann, S.; Reinbothe, S.; Reinbothe, C. An Ethylene-Protected Achilles' Heel of Etiolated Seedlings for Arthropod Deterrence. *Front. Plant Sci.* **2016**, *7*, 1246. [CrossRef] [PubMed]

73. Zapata, P.; Serrano, M.; García-Legaz, M.; Pretel, M.; Botella, M. Short Term Effect of Salt Shock on Ethylene and Polyamines Depends on Plant Salt Sensitivity. *Front. Plant Sci.* **2017**, *8*, 855. [CrossRef] [PubMed]

74. Arraes, F.; Beneventi, M.; de Sa, M.; Paixao, J.; Albuquerque, E.; Marin, S.; Purgatto, E.; Nepomuceno, A.; Grossi-de-Sa, M. Implications of ethylene biosynthesis and signaling in soybean drought stress tolerance. *BMC Plant Biol.* **2015**, *15*. [CrossRef]

75. Gamalero, E.; Glick, B. Bacterial Modulation of Plant Ethylene Levels. *Plant Physiol.* **2015**, *169*, 13–22. [CrossRef]

76. Tabassum, B.; Khan, A.; Tariq, M.; Ramzan, M.; Khan, M.; Shahid, N.; Aaliya, K. Bottlenecks in commercialisation and future prospects of PGPR. *Appl. Soil Ecol.* **2017**, *121*, 102–117. [CrossRef]

77. Ruiu, L. Plant-Growth-Promoting Bacteria (PGPB) against Insects and Other Agricultural Pests. *Agronomy* **2020**, *10*, 861. [CrossRef]

78. Saravanakumar, D.; Thomas, A.; Banwarie, N. Antagonistic potential of lipopeptide producing Bacillus amyloliquefaciens against major vegetable pathogens. *Eur. J. Plant Pathol.* **2019**, *154*, 319–335. [CrossRef]

79. Oteino, N.; Lally, R.; Kiwanuka, S.; Lloyd, A.; Ryan, D.; Germaine, K.; Dowling, D. Plant growth promotion induced by phosphate solubilizing endophytic Pseudomonas isolates. *Front. Microbiol.* **2015**, *6*, 1–9. [CrossRef] [PubMed]

80. Zahid, M.; Abbasi, M.; Hameed, S.; Rahim, N. Isolation and identification of indigenous plant growth promoting rhizobacteria from Himalayan region of Kashmir and their effect on improving growth and nutrient contents of maize (*Zea mays* L.). *Front. Microbiol.* **2015**, *6*, 207. [CrossRef] [PubMed]

81. Dimkpa, C.; Weinand, T.; Asch, F. Plant–rhizobacteria interactions alleviate abiotic stress conditions. *Plant Cell Environ.* **2009**, *32*, 1682–1694. [CrossRef]

82. Arora, N.; Fatima, T.; Mishra, J.; Mishra, I.; Verma, S.; Verma, R.; Verma, M.; Bhattacharya, A.; Verma, P.; Mishra, P.; et al. Halo-tolerant plant growth promoting rhizobacteria for improving productivity and remediation of saline soils. *J. Adv. Res.* **2020**, *26*, 69–82. [CrossRef]

83. Fouda, A.; Hassan, S.; Eid, A.; Ewais, E. The Interaction Between Plants and Bacterial Endophytes Under Salinity Stress. In *Endophytes and Secondary Metabolites*; Jha, S., Ed.; Reference Series in Phytochemistry; Springer: Cham, Switzerland, 2019; pp. 1–17. ISBN 978-3-319-76900-4.

84. Lata, R.; Chowdhury, S.; Gond, S.; White Jr, J. Induction of abiotic stress tolerance in plants by endophytic microbes. *Lett. Appl. Microbiol.* **2018**, *66*, 268–276. [CrossRef]

85. Ashraf, M.; Hasnain, S.; Berge, O.; Mahmood, T. Inoculating wheat seedlings with exopolysaccharide-producing bacteria restricts sodium uptake and stimulates plant growth under salt stress. *Biol. Fertil. Soils* **2004**, *40*, 157–162. [CrossRef]

86. Upadhyay, S.; Singh, J.; Singh, D. Exopolysaccharide-Producing Plant Growth-Promoting Rhizobacteria Under Salinity Condition. *Pedosphere* **2011**, *21*, 214–222. [CrossRef]

87. Atouei, M.; Pourbabaee, A.; Shorafa, M. Alleviation of Salinity Stress on Some Growth Parameters of Wheat by Exopolysaccharide-Producing Bacteria. *Iran. J. Sci. Technol. Trans. Sci.* **2019**, *43*, 2725–2733. [CrossRef]

88. Gupta, S.; Pandey, S. ACC Deaminase Producing Bacteria With Multifarious Plant Growth Promoting Traits Alleviates Salinity Stress in French Bean (*Phaseolus vulgaris*) Plants. *Front. Microbiol.* **2019**, *10*, 1506. [CrossRef]

89. Glick, B.; Todorovic, B.; Czarny, J.; Cheng, Z.; Duan, J.; McConkey, B. Promotion of Plant Growth by Bacterial ACC Deaminase. *Crit. Rev. Plant Sci.* **2007**, *26*, 227–242. [CrossRef]

90. Pierik, R.; Tholen, D.; Poorter, H.; Visser, E.; Voesenek, L. The Janus face of ethylene: Growth inhibition and stimulation. *Trends Plant Sci.* **2006**, *11*, 176–183. [CrossRef]

91. Abeles, F.; Morgan, P.; Saltveit, M., Jr. *Ethylene in Plant Biology*, 2nd ed.; Academic Press Inc.: Cambridge, MA, USA, 1992; ISBN 978-0-08-091628-6.

92. Stearns, J.; Glick, B. Transgenic plants with altered ethylene biosynthesis or perception. *Biotechnol. Adv.* **2003**, *21*, 193–210. [CrossRef]
93. Glick, B. Bacteria with ACC deaminase can promote plant growth and help to feed the world. *Microbiol. Res.* **2014**, *69*, 30–39. [CrossRef] [PubMed]
94. Fuertes-Mendizábal, T.; Bastías, E.; González-Murua, C.; González-Moro, M. Nitrogen Assimilation in the Highly Salt- and Boron-Tolerant Ecotype *Zea mays* L. Amylacea. *Plants* **2020**, *9*, 322. [CrossRef] [PubMed]
95. Barnawal, D.; Bharti, N.; Maji, D.; Chanotiya, C.; Kalra, A. ACC deaminase-containing Arthrobacter protophormiae induces NaCl stress tolerance through reduced ACC oxidase activity and ethylene production resulting in improved nodulation and mycorrhization in Pisum sativum. *J. Plant Physiol.* **2014**, *171*, 884–894. [CrossRef]
96. Kim, K.; Jang, Y.-J.; Lee, S.-M.; Oh, B.-T.; Chae, J.-C.; Lee, K.-J. Alleviation of Salt Stress by Enterobacter sp. EJ01 in Tomato and Arabidopsis Is Accompanied by Up-Regulation of Conserved Salinity Responsive Factors in Plants. *Mol. Cells* **2014**, *37*, 109–117. [CrossRef]
97. Habib, S.; Kausar, H.; Saud, H. Plant Growth-Promoting Rhizobacteria Enhance Salinity Stress Tolerance in Okra through ROS-Scavenging Enzymes. *BioMed Res. Int.* **2016**, *2016*. [CrossRef]
98. Orhan, F. Alleviation of salt stress by halotolerant and halophilic plant growth-promoting bacteria in wheat (*Triticum aestivum*). *Braz. J. Microbiol.* **2016**, *47*, 621–627. [CrossRef]
99. Bhise, K.; Bhagwat, P.; Dandge, P. Synergistic effect of *Chryseobacterium gleum* sp. SUK with ACC deaminase activity in alleviation of salt stress and plant growth promotion in *Triticum aestivum* L. 3 Biotech **2017**, *7*. [CrossRef] [PubMed]
100. Zerrouk, I.; Benchabane, M.; Khelifi, L.; Yokawa, K.; Ludwig-Müller, J.; Baluska, F. A Pseudomonas strain isolated from date-palm rhizospheres improves root growth and promotes root formation in maize exposed to salt and aluminum stress. *J. Plant Physiol.* **2016**, *191*, 111–119. [CrossRef]
101. Hahm, M.-S.; Son, J.-S.; Hwang, Y.-J.; Kwon, D.-K.; Ghim, S.-Y. Alleviation of Salt Stress in Pepper (*Capsicum annum* L.) Plants by Plant Growth-Promoting Rhizobacteria. *J. Microbiol. Biotechnol.* **2017**, *27*, 1790–1797. [CrossRef] [PubMed]
102. Singh, R.; Jha, P. The PGPR Stenotrophomonas maltophilia SBP-9 Augments Resistance against Biotic and Abiotic Stress in Wheat Plants. *Front. Microbiol.* **2017**, *8*, 1945. [CrossRef] [PubMed]
103. Cherif-Silini, H.; Thissera, B.; Bouket, A.; Saadaoui, N.; Silini, A.; Eshelli, M.; Alenezi, F.; Vallat, A.; Luptakova, L.; Yahiaoui, B.; et al. Durum Wheat Stress Tolerance Induced by Endophyte Pantoea agglomerans with Genes Contributing to Plant Functions and Secondary Metabolite Arsenal. *Int. J. Mol. Sci.* **2019**, *20*, 3989. [CrossRef] [PubMed]
104. Kang, S.-M.; Shahzad, R.; Bilal, S.; Khan, A.; Park, Y.-G.; Lee, K.-E.; Asaf, S.; Khan, M.; Lee, I.-J. Indole-3-acetic-acid and ACC deaminase producing Leclercia adecarboxylata MO1 improves Solanum lycopersicum L. growth and salinity stress tolerance by endogenous secondary metabolites regulation. *BMC Microbiol.* **2019**, *19*. [CrossRef] [PubMed]
105. Phour, M.; Sindhu, S. Amelioration of salinity stress and growth stimulation of mustard (*Brassica juncea* L.) by salt-tolerant Pseudomonas species. *Appl. Soil Ecol.* **2020**, *149*. [CrossRef]
106. Misra, S.; Chauhan, P. ACC deaminase-producing rhizosphere competent Bacillus spp. mitigate salt stress and promote *Zea mays* growth by modulating ethylene metabolism. *3 Biotech* **2020**, *10*. [CrossRef]
107. Zhu, Z.; Zhang, H.; Leng, J.; Niu, H.; Chen, X.; Liu, D.; Chen, Y.; Ying, H. Isolation and characterization of plant growth-promoting rhizobacteria and their effects on the growth of *Medicago sativa* L. under salinity conditions. *Antonie Leeuwenhoek* **2020**, *113*, 1263–1278. [CrossRef]
108. Khan, A.; Waqas, M.; Asaf, S.; Kamran, M.; Shahzad, R.; Bilal, S.; Khan, M.; Kang, S.-M.; Kim, Y.-H.; Yun, B.-W.; et al. Plant growth-promoting endophyte Sphingomonas sp. LK11 alleviates salinity stress in Solanum pimpinellifolium. *Environ. Exp. Bot.* **2017**, *133*, 58–69. [CrossRef]
109. Ilyas, N.; Mazhar, R.; Yasmin, H.; Khan, W.; Iqbal, S.; El Enshasy, H.; Dailin, D. Rhizobacteria Isolated from Saline Soil Induce Systemic Tolerance in Wheat (*Triticum aestivum* L.) against Salinity Stress. *Agronomy* **2020**, *10*, 989. [CrossRef]
110. Mukherjee, P.; Mitra, A.; Roy, M. Halomonas Rhizobacteria of Avicennia marina of Indian Sundarbans Promote Rice Growth Under Saline and Heavy Metal Stresses Through Exopolysaccharide Production. *Front. Microbiol.* **2019**, *10*, 1207. [CrossRef] [PubMed]
111. Nadeem, S.; Zahir, Z.; Naveed, M.; Arshad, M.; Shahzad, S. Variation in growth and ion uptake of maize due to inoculation with plant growth promoting rhizobacteria under salt stress. *Soil Environ.* **2006**, *25*, 78–84.
112. Ahmed, A.; Hasnain, S. Auxins as One of the Factors of Plant Growth Improvement by Plant Growth Promoting Rhizobacteria. *Pol. J. Microbiol.* **2014**, *63*, 261–266. [CrossRef] [PubMed]
113. Cordovez, V.; Schop, S.; Hordijk, K.; de Boulois, H.; Coppens, F.; Hanssen, I.; Raaijmakers, J.; Carrión, V. Priming of Plant Growth Promotion by Volatiles of Root-Associated *Microbacterium* spp. *Appl. Environ. Microbiol.* **2018**, *84*. [CrossRef] [PubMed]
114. Tahir, H.; Wu, H.; Raza, W.; Hanif, A.; Wu, L.; Colman, M.; Gao, X. Plant Growth Promotion by Volatile Organic Compounds Produced by Bacillus subtilis SYST2. *Front. Microbiol.* **2017**, *8*, 171. [CrossRef]
115. Saraf, M.; Jha, C.; Patel, D. The Role of ACC Deaminase Producing PGPR in Sustainable Agriculture. In *Plant Growth and Health Promoting Bacteria*; Maheshwari, D., Ed.; Microbiology Monographs; Springer: Berlin/Heidelberg, Germany, 2010; Volume 18, pp. 365–385. ISSN 1862-5584.

116. López-Bucio, J.; Campos-Cuevas, J.; Hernández-Calderón, E.; Velásquez-Becerra, C.; Farías-Rodríguez, R.; Macías-Rodríguez, L.; Valencia-Cantero, E. Bacillus megaterium Rhizobacteria Promote Growth and Alter Root-System Architecture Through an Auxin- and Ethylene-Independent Signaling Mechanism in Arabidopsis thaliana. *MPMI* **2007**, *20*, 207–217. [CrossRef] [PubMed]

117. Chu, T.N.; Bui, L.V.; Hoang, M.T.T. Pseudomonas PS01 Isolated from Maize Rhizosphere Alters Root System Architecture and Promotes Plant Growth. *Microorganisms* **2020**, *8*, 471. [CrossRef]

118. Liotti, R.; da Silva Figueiredo, M.; da Silva, G.; de Mendonçad, E.; Soares, M. Diversity of cultivable bacterial endophytes in Paullinia cupana and their potential for plant growth promotion and phytopathogen control. *Microbiol. Res.* **2018**, *207*, 8–18. [CrossRef]

119. Khare, E.; Mishra, J.; Arora, N. Multifaceted Interactions Between Endophytes and Plant: Developments and Prospects. *Front. Microbiol.* **2018**, *9*, 2732. [CrossRef] [PubMed]

120. Mena-Violante, H.; Olalde-Portugal, V. Alteration of tomato fruit quality by root inoculation with plant growth-promoting rhizobacteria (PGPR): Bacillus subtilis BEB-13bs. *Sci. Hortic.* **2007**, *113*, 103–106. [CrossRef]

121. Kumar, A.; Maurya, B.; Raghuwanshi, R. Isolation and characterization of PGPR and their effect on growth, yield and nutrient content in wheat (*Triticum aestivum* L.). *Biocatal. Agric. Biotechnol.* **2014**, *3*, 121–128. [CrossRef]

122. Nawaz, A.; Shahbaz, M.; Asadullah, A.I.; Marghoob, M.; Imtiaz, M.; Mubeen, F. Potential of Salt Tolerant PGPR in Growth and Yield Augmentation of Wheat (*Triticum aestivum* L.) Under Saline Conditions. *Front. Microbiol.* **2020**, *11*, 2019. [CrossRef]

123. Awad, N.; Turky, A.; Abdelhamid, M.; Attia, M. Ameliorate of environmental salt stress on the growth of Zea mays L. plants by exopolysaccharides producing bacteria. *J. Appl. Sci. Res.* **2012**, *8*, 2033–2044.

124. Abd El-Ghany, M.; Attia, M. Effect of Exopolysaccharide-Producing Bacteria and Melatonin on Faba Bean Production in Saline and Non-Saline Soil. *Agronomy* **2020**, *10*, 316. [CrossRef]

125. Tsao, R. Chemistry and Biochemistry of Dietary Polyphenols. *Nutrients* **2010**, *2*, 1231–1246. [CrossRef]

126. Hichem, H.; Mounir, D.; Naceur, E. Differential responses of two maize (*Zea mays* L.) varieties to salt stress: Changes on polyphenols composition of foliage and oxidative damages. *Ind. Crops Prod.* **2009**, *30*, 144–151. [CrossRef]

127. Ksouri, R.; Megdiche, W.; Debez, A.; Falleh, H.; Grignon, C.; Abdelly, C. Salinity effects on polyphenol content and antioxidant activities in leaves of the halophyte Cakile maritima. *Plant Physiol. Biochem.* **2007**, *45*, 244–249. [CrossRef] [PubMed]

128. Rojas-Tapias, D.; Moreno-Galván, A.; Pardo-Díaz, S.; Obando, M.; Rivera, D.; Bonilla, R. Effect of inoculation with plant growth-promoting bacteria (PGPB) on amelioration of saline stress in maize (*Zea mays*). *Appl. Soil Ecol.* **2012**, *61*, 264–272. [CrossRef]

129. Kasinath, B.; Senthivel, T.; Ganeshmurthy, A.; Nagegowda, N.; Kumar, M. Effect of Magnesium application on chlorophyll content and yield of tomato. *Plant Arch.* **2014**, *14*, 801–804.

130. Hermans, C.; Verbruggen, N. Physiological characterization of Mg deficiency in Arabidopsis thaliana. *J. Exp. Bot.* **2005**, *56*, 2153–2161. [CrossRef]

131. Gutiérrez-Luna, F.; López-Bucio, J.; Altamirano-Hernández, J.; Valencia-Cantero, E.; de la Cruz, H.; Macías-Rodríguez, L. Plant growth-promoting rhizobacteria modulate root-system architecture in Arabidopsis thaliana through volatile organic compound emission. *Symbiosis* **2010**, *51*, 75–83. [CrossRef]

132. Nassal, D.; Spohn, M.; Eltlbany, N.; Jacquiod, S.; Smalla, K.; Marhan, S.; Kandeler, E. Effects of phosphorus-mobilizing bacteria on tomato growth and soil microbial activity. *Plant Soil* **2018**, *427*, 17–37. [CrossRef]

133. Stephen, J.; Shabanamol, S.; Rishad, K.; Jisha, M. Growth enhancement of rice (*Oryza sativa*) by phosphate solubilizing Gluconacetobacter sp. (MTCC 8368) and Burkholderia sp. (MTCC 8369) under greenhouse conditions. *3 Biotech* **2015**, *5*, 831–837. [CrossRef] [PubMed]

134. Brunner, S.; Goos, R.; Swenson, S.; Foster, S.; Schatz, B.; Lawley, Y.; Prischmann-Voldseth, D. Impact of nitrogen fixing and plant growth-promoting bacteria on a phloem-feeding soybean herbivore. *Appl. Soil Ecol.* **2015**, *86*, 71–81. [CrossRef]

135. Gopalakrishnan, S.; Srinivas, V.; Samineni, S. Nitrogen fixation, plant growth and yield enhancements by diazotrophic growth-promoting bacteria in two cultivars of chickpea (*Cicer arietinum* L.). *Biocatal. Agric. Biotechnol.* **2017**, *11*, 116–123. [CrossRef]

136. Navarro-Torre, S.; Barcia-Piedras, J.; Mateos-Naranjo, E.; Redondo-Gomez, S.; Camacho, M.; Caviedes, M.; Pajuelo, E.; Rodríguez-Llorente, I. Assessing the role of endophytic bacteria in the halophyte Arthrocnemum macrostachyum salt tolerance. *Plant Biol.* **2017**, *19*, 249–256. [CrossRef]

137. Pham Thi Thu, H.; Nguyen Thu, T.; Nguyen Dang Nguyen, T.; Le Minh, K.; Do Tan, K. Evaluate the effects of salt stress on physico-chemical characteristics in the germination of rice (*Oryza sativa* L.) in response to methyl salicylate (MeSA). *Biocatal. Agric. Biotechnol.* **2020**, *23*, 1–6. [CrossRef]

138. Dehnavi, A.; Zahedi, M.; Ludwiczak, A.; Perez, S.; Piernik, A. Effect of Salinity on Seed Germination and Seedling Development of Sorghum (*Sorghum bicolor* (L.) Moench) Genotypes. *Agronomy* **2020**, *10*, 859. [CrossRef]

139. Liu, T.; Li, R.; Jin, X.; Ding, J.; Zhu, X.; Sun, C.; Guo, W. Evaluation of Seed Emergence Uniformity of Mechanically Sown Wheat with UAV RGB Imagery. *Remote Sens.* **2017**, *9*, 1241. [CrossRef]

140. Dellaquila, A.; Spada, P. The effect of salinity stress upon protein synthesis of germinating wheat embryos. *Ann. Bot.* **1993**, *72*, 97–101. [CrossRef]

141. Fercha, A.; Capriotti, A.; Caruso, G.; Cavaliere, C.; Stampachiacchiere, S.; Chiozzi, R.; Laganà, A. Shotgun proteomic analysis of soybean embryonic axes during germination under salt stress. *Proteomics* **2016**, *16*, 1537–1546. [CrossRef] [PubMed]

142. Zhang, C.; Luo, W.; Li, Y.; Zhang, X.; Bai, X.; Niu, Z.; Zhang, X.; Li, Z.; Wan, D. Transcriptomic Analysis of Seed Germination Under Salt Stress in Two Desert Sister Species (*Populus euphratica* and P. pruinosa). *Front. Genet.* **2019**, *10*. [CrossRef] [PubMed]

143. Liu, L.; Xia, W.; Li, H.; Zeng, H.; Wei, B.; Han, S.; Yin, C. Salinity Inhibits Rice Seed Germination by Reducing α-Amylase Activity via Decreased Bioactive Gibberellin Content. *Front. Plant Sci.* **2018**, *9*, 275. [CrossRef]

144. Mahmood, A.; Turgay, O.; Farooq, M.; Hayat, R. Seed biopriming with plant growth promoting rhizobacteria: A review. *FEMS Microbiol. Ecol.* **2016**, *92*. [CrossRef]

145. Zulueta-Rodríguez, R.; Hernández-Montiel, L.; Murillo-Amador, B.; Rueda-Puente, E.; Capistrán, L.; Troyo-Diéguez, E.; Córdoba-Matson, M. Effect of Hydropriming and Biopriming on Seed Germination and Growth of Two Mexican Fir Tree Species in Danger of Extinction. *Forests* **2015**, *6*, 3109–3122. [CrossRef]

146. Prasad, J.; Gupta, S.; Raghuwanshi, R. Screening Multifunctional Plant Growth Promoting Rhizobacteria Strains for Enhancing Seed Germination in Wheat (*Triticum aestivum* L.). *Int. J. Agric. Res.* **2017**. [CrossRef]

147. Vaishnav, A.; Kumari, S.; Jain, S.; Varma, A.; Tuteja, N.; Choudhary, D. PGPR-mediated expression of salt tolerance gene in soybean through volatiles under sodium nitroprusside. *J. Basic Microbiol.* **2016**, *56*, 1274–1288. [CrossRef]

148. Sarropoulou, V.; Dimassi-Theriou, K.; Therios, I.; Koukourikou-Petridou, M. Melatonin enhances root regeneration, photosynthetic pigments, biomass, total carbohydrates and proline content in the cherry rootstock PHL-C (Prunus avium × *Prunus cerasus*). *Plant Physiol. Biochem.* **2012**, *61*, 162–168. [CrossRef]

149. Tal, O.; Haim, A.; Harel, O.; Gerchman, Y. Melatonin as an antioxidant and its semi-lunar rhythm in green macroalga *Ulva* sp. *J. Exp. Bot.* **2011**, *62*, 1903–1910. [CrossRef] [PubMed]

150. Uçarlı, C. *Effects of Salinity on Seed Germination and Early Seedling Stage*; IntechOpen: London, UK, 2020.

151. Halo, B.; Khan, A.; Waqas, M.; Al-Harrasi, A.; Hussain, J.; Ali, L.; Adnan, M.; Lee, I. Endophytic bacteria (*Sphingomonas* sp. LK11) and gibberellin can improve Solanum lycopersicum growth and oxidative stress under salinity. *J. Plant Interact.* **2015**, *10*, 117–125. [CrossRef]

152. Bharti, N.; Pandey, S.; Barnawal, D.; Patel, V.; Kalra, A. Plant growth promoting rhizobacteria Dietzia natronolimnaea modulates the expression of stress responsive genes providing protection of wheat from salinity stress. *Sci. Rep.* **2016**, *6*. [CrossRef]

153. El-Esawi, M.; Alaraidh, I.; Alsahli, A.; Alamri, S.; Ali, H.; Alayafi, A. Bacillus firmus (SW5) augments salt tolerance in soybean (*Glycine max* L.) by modulating root system architecture, antioxidant defense systems and stress-responsive genes expression. *Plant Physiol. Biochem.* **2018**, *132*, 375–384. [CrossRef]

154. Vimal, S.; Patel, V.; Singh, J. Plant growth promoting Curtobacterium albidum strain SRV4: An agriculturally important microbe to alleviate salinity stress in paddy plants. *Ecol. Indic.* **2019**, *105*, 553–562. [CrossRef]

155. El-Nahrawy, S.; Yassin, M. Response of Different Cultivars of Wheat Plants (*Triticum aestivum* L.) to Inoculation by Azotobacter sp. under Salinity Stress Conditions. *J. Adv. Microbiol.* **2020**, *20*, 60–79. [CrossRef]

156. Sun, L.; Lei, P.; Wang, Q.; Ma, J.; Zhan, Y.; Jiang, K.; Xu, Z.; Xu, H. The Endophyte Pantoea alhagi NX-11 Alleviates Salt Stress Damage to Rice Seedlings by Secreting Exopolysaccharides. *Front. Microbiol.* **2020**, *10*, 3112. [CrossRef]

157. Pinedo, I.; Ledger, T.; Greve, M.; Poupin, M. Burkholderia phytofirmans PsJN induces long-term metabolic and transcriptional changes involved in Arabidopsis thaliana salt tolerance. *Front. Plant Sci.* **2015**, *6*, 466. [CrossRef] [PubMed]

158. Manaf, H.; Zayed, M. Productivity of cowpea as affected by salt stress in presence of endomycorrhizae and Pseudomonas fluorescens. *Ann. Agric. Sci.* **2015**, *60*, 219–226. [CrossRef]

159. Khan, M.; Asaf, S.; Khan, A.; Ullah, I.; Ali, S.; Kang, S.; Lee, I. Alleviation of salt stress response in soybean plants with the endophytic bacterial isolate Curtobacterium sp. SAK1. *Ann. Microbiol.* **2019**, *69*, 797–808. [CrossRef]

160. Kharusi, L.; Yahyai, R.; Yaish, M. Antioxidant Response to Salinity in Salt-Tolerant and Salt-Susceptible Cultivars of Date Palm. *Agriculture* **2019**, *9*, 8. [CrossRef]

161. Ryu, C.-M.; Farag, M.; Hu, C.-H.; Reddy, M.; Kloepper, J.; Paré, P. Bacterial Volatiles Induce Systemic Resistance in Arabidopsis. *Plant Physiol.* **2004**, *134*, 1017–1026. [CrossRef] [PubMed]

162. Chen, L.; Liu, Y.; Wu, G.; Njeri, K.; Shen, Q.; Zhang, N.; Zhang, R. Induced maize salt tolerance by rhizosphere inoculation of Bacillus amyloliquefaciens SQR9. *Physiol. Plant.* **2016**, *158*, 34–44. [CrossRef] [PubMed]

163. Liu, H.; Wang, J.; Sun, H.; Han, X.; Peng, Y.; Liu, J.; Liu, K.; Ding, Y.; Wang, C.; Du, B. Transcriptome Profiles Reveal the Growth-Promoting Mechanisms of Paenibacillus polymyxa YC0136 on Tobacco (*Nicotiana tabacum* L.). *Front. Microbiol.* **2020**, *11*, 4174. [CrossRef]

164. Li, J.; Han, G.; Sun, C.; Sui, N. Research advances of MYB transcription factors in plant stress resistance and breeding. *Plan Signal. Behav.* **2019**, *14*, 1613131. [CrossRef]

165. Finatto, T.; Viana, V.; Woyann, L.; Busanello, C.; da Maia, L.; de Oliveira, A. Can WRKY transcription factors help plants to overcome environmental challenges? *Genet. Mol. Biol.* **2018**, *41*, 533–544. [CrossRef] [PubMed]

166. Qin, Y.; Tian, Y.; Liu, X. A wheat salinity-induced WRKY transcription factor TaWRKY93 confers multiple abiotic stress tolerance in Arabidopsis thaliana. *Biochem. Biophys. Res. Commun.* **2015**, *464*, 428–433. [CrossRef]

167. Yu, Y.; Wang, N.; Hu, R.; Xiang, F. Genome-wide identification of soybean WRKY transcription factors in response to salt stress. *SpringerPlus* **2016**, *5*. [CrossRef]

168. Wu, B.; Hu, Y.; Huo, P.; Zhang, Q.; Chen, X.; Zhang, Z. Transcriptome analysis of hexaploid hulless oat in response to salinity stress. *PLoS ONE* **2017**, *12*, e171451. [CrossRef]

169. Thomas, J.; Kim, H.; Rahmatallah, Y.; Wiggins, G.; Yang, Q.; Singh, R.; Glazko, G.; Mukherjee, A. RNA-seq reveals differentially expressed genes in rice (*Oryza sativa*) roots during interactions with plant-growth promoting bacteria, *Azospirillum brasilense*. *PLoS ONE* **2019**, *14*, e217309. [CrossRef]
170. Malviya, M.; Solanki, M.; Singh, R.; Htun, R.; Singh, P.; Verma, K.; Yang, L.-T.; Li, Y.-R. Comparative analysis of sugarcane root transcriptome in response to the plant growth-promoting Burkholderia anthina MYSP113. *PLoS ONE* **2020**, *15*. [CrossRef] [PubMed]
171. Ho, W.; Hill, C.; Doblin, M.; Shelden, M.; van de Meene, A.; Rupasinghe, T.; Bacic, A.; Roessner, U. Integrative Multi-omics Analyses of Barley Rootzones under Salinity Stress Reveal Two Distinctive Salt Tolerance Mechanisms. *Plant Commun.* **2020**, *1*. [CrossRef] [PubMed]
172. Safdarian, M.; Askari, H.; Shariati, J.V.; Nematzadeh, G. Transcriptional responses of wheat roots inoculated with Arthrobacter nitroguajacolicus to salt stress. *Sci. Rep.* **2019**, *9*, 1792. [CrossRef] [PubMed]
173. Xie, S.; Wu, H.; Chen, L.; Zang, H.; Xie, Y.; Gao, X. Transcriptome profiling of Bacillus subtilis OKB105 in response to rice seedlings. *BMC Microbiol.* **2015**, *15*. [CrossRef]

International Journal of
Molecular Sciences

Review

Jasmonates and Plant Salt Stress: Molecular Players, Physiological Effects, and Improving Tolerance by Using Genome-Associated Tools

Celia Delgado [1], Freddy Mora-Poblete [1], Sunny Ahmar [1], Jen-Tsung Chen [2] and Carlos R. Figueroa [1,*]

[1] Institute of Biological Sciences, Campus Talca, Universidad de Talca, Talca 3465548, Chile; celia.delgado@utalca.cl (C.D.); fmora@utalca.cl (F.M.-P.); sunny.ahmar@utalca.cl (S.A.)

[2] Department of Life Sciences, College of Science, National University of Kaohsiung, Kaohsiung 811, Taiwan; jentsung@nuk.edu.tw

* Correspondence: cfigueroa@utalca.cl

Abstract: Soil salinity is one of the most limiting stresses for crop productivity and quality worldwide. In this sense, jasmonates (JAs) have emerged as phytohormones that play essential roles in mediating plant response to abiotic stresses, including salt stress. Here, we reviewed the mechanisms underlying the activation and response of the JA-biosynthesis and JA-signaling pathways under saline conditions in *Arabidopsis* and several crops. In this sense, molecular components of JA-signaling such as MYC2 transcription factor and JASMONATE ZIM-DOMAIN (JAZ) repressors are key players for the JA-associated response. Moreover, we review the antagonist and synergistic effects between JA and other hormones such as abscisic acid (ABA). From an applied point of view, several reports have shown that exogenous JA applications increase the antioxidant response in plants to alleviate salt stress. Finally, we discuss the latest advances in genomic techniques for the improvement of crop tolerance to salt stress with a focus on jasmonates.

Keywords: salt stress; jasmonates; jasmonate signaling pathway; crosstalk; exogenous jasmonate applications; GWAS

Citation: Delgado, C.; Mora-Poblete, F.; Ahmar, S.; Chen, J.-T.; Figueroa, C.R. Jasmonates and Plant Salt Stress: Molecular Players, Physiological Effects, and Improving Tolerance by Using Genome-Associated Tools. *Int. J. Mol. Sci.* **2021**, 22, 3082. https://doi.org/10.3390/ijms22063082

Academic Editor: Charlotte Poschenrieder

Received: 12 February 2021
Accepted: 15 March 2021
Published: 17 March 2021

Publisher's Note: MDPI stays neutral with regard to jurisdictional claims in published maps and institutional affiliations.

1. Introduction

Salinity is a serious hazard for agriculture since most of the crop plants are salt-sensitive [1]. Current data show that global soil salinization increased by more than 100 Mha between 1986 and 2016 and it is expanding on a global scale, approximately at a rate of 2 Mha per year [2]. Thus, the future of food supplies for animals and humankind is threatened [3]. Natural processes such as geological deposits due to parent rock constituents, salinized groundwater, marine transgressions, storm flood events, tsunamis, and recurrent drought events and the general increase in temperature [4,5] cause soil salinization. Unfortunately, human interventions have also promoted the increment of saline lands. Wrong irrigation practices, poor drainage conditions [6], the use of fertilizers [7], mismanagement of treated wastewater [8], industrial [9], and mining operation effluents enriched with a salt contribute to the salinity increment in the soil and water [10].

Under salt stress conditions, physiological and metabolic activities are impaired by osmotic-, ionic-, and oxidative stresses, nutritional imbalance, or a combination of these factors [11]. In fact, plant growth and development are limited by salt stress due to the negative influences through the ionic and osmotic components on various biochemical reactions and physiological processes such as photosynthesis, antioxidant metabolism, mineral nutrient homeostasis, osmolyte accumulation, and hormonal signaling [12]. Most species, including crops, activate tolerance mechanisms only after exposure to salt stress. Activation of the tolerance program drives plants to acclimatize under the saline condition and involves altered physiological responses, redirection of metabolism, reinforcement of defense and repair, and changes in developmental programs to adapt morphological and anatomical characteristics [1,13]. Complex coordination of several signaling pathways is needed to activate the plant responses to salt stress. Phytohormone-mediated signaling, for instance, is crucial in the induction of gene networks related to salt tolerance [14].

Among abiotic stress-related hormones, abscisic acid (ABA) is well known to be the key phytohormone in endogenous signaling that allows plants to survive adverse environmental conditions [15]. The role of ABA for salinity adaptation has long been intensively studied and documented [16–18]. More recently, the biological relevance of jasmonate (JA) and its derivatives in the induction of tolerance to abiotic stresses has been demonstrated [19–21]. JAs are critical signaling molecules in various development and defense processes of plants [22,23] and play essential roles in plant response to salt stress [24–26]. Similar to ABA, accumulation of JAs has been reported in salt-tolerant crops compared to sensitive cultivars [27]. Salt stress has been observed to cause increased levels of JA in leaves and roots and the induction of JA biosynthesis-related genes [28–30]. In addition, exogenous application of JA significantly reduces the Na^+ ion content in salinity-tolerant rice and wheat [27,31] and recovers salt-induced defects in seedling development and photosynthetic activity [32,33]. All this evidence is highly correlated with a positive role of JA in the plant response to salt stress. JA-crosstalk with other phytohormones has aroused the interest of researchers since the interaction among multiple plant hormone signaling integrates environmental and development cues. In this sense, JA has been proposed as a core signal in the phytohormone signaling network [34] because it regulates the balance between plant growth and defense [35,36]. The JA signaling components JASMONATE ZIM-DOMAIN (JAZ) and MYC2 have been identified as the main nodes in the orchestration of JA interplay with other hormone signaling pathways [37,38]. However, the cross-coordination between JA and the other phytohormone is far from being completely understood.

Currently, many strategies have been implemented to obtain tolerant plants to salt stress including classical plant breeding and genetic engineering approaches [39]. However, the complexity of the salt tolerance trait requires more in-depth studies to be understood. Exploring germplasm that possesses genetic variability across a wide spectrum of salt tolerance-related traits could provide valuable information. The study of biparental mapping populations and diversity panels has allowed the discovery of beneficial genetic variants (or alleles) that can be examined for traits that have significant components of salt stress tolerance and their associated quantitative trait loci (QTL) [40]. The analysis of these traits by phenotyping and genotyping techniques could reveal new insights into the biological mechanisms underlying the salt tolerance phenotypes [41]. A widely employed approach to identify the association between each genotyped marker and a phenotype of interest that has been scored across several individuals is the genome-wide association study (GWAS) [42]. It offers some advantages such as more accurate positioning and mapping, simultaneous assessments of multiple alleles at a locus, no requirement for linkage group construction [43,44] and can serve as a basic experiment to identify candidates for mutagenesis and transgenics [45].

With the possibility of performance-specific and predictable genetic modifications through genome (or gene) editing (GE) tools, the information obtained by GWASs becomes more valuable for plant breeding and crop improvement efforts. Among GE methods, the CRISPR-Cas9 system has been used to enhance salt tolerance in some crops like rice and tomato [46,47]. Given that JA regulates processes associated with growth, development, and salt response, the components of JA-biosynthesis, metabolism, and signaling can be targets for phytohormone engineering to produce salt-resilient crops with high yields. Furthermore, increased knowledge of the signals and molecular mechanisms involved in the plant salinity response would pave the path to obtain salt-tolerant crops without productivity penalties.

In this review, we update the information of the role of JA in the salt stress response, mainly in the topics related to JA-crosstalk with other phytohormones in salinity conditions, although limited information about JA-crosstalk during salt stress is available, here we bring together the latest studies on this topic. Finally, we reviewed the use of genome-associated tools to improve salt tolerance by manipulation of JA signaling.

2. Jasmonate Metabolism, Signaling, and Response during Salt Stress

2.1. JA Biosynthesis, Signaling, and Catabolism

The JA hormones have traditionally been studied in several plant species in the regulation of aspects regarding development, metabolism, and adaptation to biotic stress. However, JAs have lastly been associated with plant responses to several abiotic stresses, adding to JAs extra functions in plant adaptation [20,48]. The JA pathway consists mainly of the JA-associated biosynthesis, signaling, catabolism, and response. According to the information obtained in *Arabidopsis*, JAs are oxylipins, which biosynthesis begins in the plastids. The α-linolenic acid fatty acid, produced by the action of A_1-type lipases (PLIP and PLA_1), is transformed into several intermediates by reactions catalyzed by 13-lipoxygenase (13-LOX), allene oxide synthase (AOS), and allene oxide cyclase (AOC), leading to 12-oxo-phytodienoic acid (OPDA) [49,50]. OPDA is translocated to the peroxisome where 12-oxo-phytodienoate reductase 3 (OPR3) and several β-oxidation cycles give rise to jasmonic acid (JA) [51]. The biosynthesis of methyl jasmonate (MeJA) and jasmonoyl-isoleucine (JA-Ile) from JA is catalyzed by jasmonic acid methyltransferase (JMT) [52] and jasmonic acid-amide synthetase 1 (JAR1) [53], respectively. Meanwhile, the reconversion of MeJA and JA-Ile to JA is catalyzed by methyl jasmonate esterase (MJE) [54] and JA-Ile hydrolase 1 (JIH1) [55], respectively.

The JA-Ile molecule is the bioactive JA responsible for the activation of JA responses [56]. The physiological effects mediated by JA-Ile require activation of the JA signaling pathway, which has been well characterized in *Arabidopsis* [57]. The F-box CORONATINE INSENSITIVE1 protein (COI1) is part of the Skp-Cullin-F-box-type E3 ubiquitin ligase complex (SCF^{COI1}) and together with JAZ proteins form the JA-Ile receptor [58–62]. In *Arabidopsis*, when JA-Ile level is low, JAZ proteins repress MYC transcriptional activity by recruiting NOVEL INTERACTOR OF JAZ (NINJA) and TOPLESS corepressors [63,64] and by obstructing the association of the coactivator protein mediator complex subunit 25 (MED25) with the transcription initiation complex [65–67]. However, once the JA-Ile level rises, it mediates the COI1-JAZ interaction leading to JAZ proteins ubiquitination and degradation by the 26S proteasome. Then, MYC2 and additional transcription factors (TFs) induce the expression of early JA-responsive genes such as JAZs, MYCs, and JA biosynthetic ones [58,68]. MYC2 is a key transcription factor in the JA pathway and acts as a regulatory hub for several biotic and abiotic stress-related responses. Its homologs, MYC3, MYC4, and MYC5, also act alternatively in the JA pathway in *Arabidopsis* [57,69,70]. MYC2 contains a conserved basic helix-loop-helix (bHLH) domain [71], which is required to form homo- or heterodimers with MYC3 and MYC4 [69], although the strongest activation of JA-regulated genes is provided by the MYC2 tetramer [72]. The basic region of MYC2 protein is involved in binding to the sequence 5′-CACGTG-3′, known as G-box, which is present in the MYC2 target promoters [73]. The G-box contains many other variants and weaker binding sequences [74,75]. Furthermore, MYC2 N-terminal contains a putative transcriptional activation domain (TAD) [69], and a JAZ interaction domain (JID) through which interacts with the C-terminal Jas domain of JAZ proteins [58,69]. In several JA-dependent functions, MYC2 regulates many secondary TFs by binding to their promoters, which, in turn, activate downstream gene promoters, creating a hierarchical transcriptional network of JA-mediated response [76–78].

Recently, an OPR3-independent pathway for JA synthesis has been described. Chini et al. [79] isolated and characterized the complete knockout mutant of the *opr3-3* allele. Similar to wild-type (WT) plants, the *opr3-3* mutants were resistant to necrotrophic pathogens and insect feeding and activated COI1-dependent JA-mediated gene expression. Through OPDA derivatives analysis the 4,5-didehydro-JA (4,5-ddh-JA) was identified to act as a precursor for JA and JA-Ile biosynthesis in OPR3 absence. The authors demonstrated that in the lack of OPR3, OPDA could enter the peroxisomal β-oxidation pathway to produce 4,5-ddh-JA, which after leaving the peroxisome, is reduced by the cytosolic 12-oxo-phytodienoate reductase 2 (OPR2). This pathway takes place naturally in WT plants and is maximized in the *opr3* mutant [79].

Together with JA biosynthetic and signaling genes, the JA-Ile catabolic genes are strongly coregulated indicating the importance to maintain JA homeostasis [80]. There are two JA-catabolic pathways: one is defined by CYP94B3/B1 and CYP94C1, members of the cytochrome P450 enzymes of the subfamily 94 (CYP94) [81–83], and the other one involves the IAA-alanine resistant 3 (IAR3) and IAA-amino acid hydrolase ILR1-like 6 (ILL6) enzymes, members of the amidohydrolase (AH) family [84]. CYP94B3 and CYP94B1 are JA-Ile ω-hydroxylases that generate 12OH-JA-Ile, and CYP94C1 catalyzes a more complete JA-Ile oxidation to 12COOH-JA-Ile [82]. Meanwhile, IAR3 and ILL6 prompt JA-Ile to deactivate by deconjugation reactions [55]. All these enzymes diminish specifically JA-Ile hormone pools to weaken JA signaling as many loss and gain-of-function experiments have demonstrated [85]. The proper regulation and termination of JA-mediated processes are essential to avoid the harmful metabolic effects of a JA amplified response trigger by biotic and abiotic stress. However, it is necessary to highlight that lately, a role of 12OH-JA-Ile in JA signaling has been described [86,87]. Exogenous application of 12OH-JA-Ile mimicked several JA-Ile effects including JA-marker gene expression, anthocyanin accumulation, and trichome induction in *Arabidopsis* [86]. In silico and in vitro assays showed that 12OH-JA-Ile could interact with some COI1-JAZs coreceptors and function as an active jasmonate signal, but more weakly comparing with JA-Ile. It has been proposed that after a strong immune response mediated by JA-Ile, 12OH-JA-Ile modulates JA-Ile activated processes contributing to wound and defense plant response [87].

In Figure 1, a schematic diagram of the biosynthesis, signaling, and catabolism of JAs is presented.

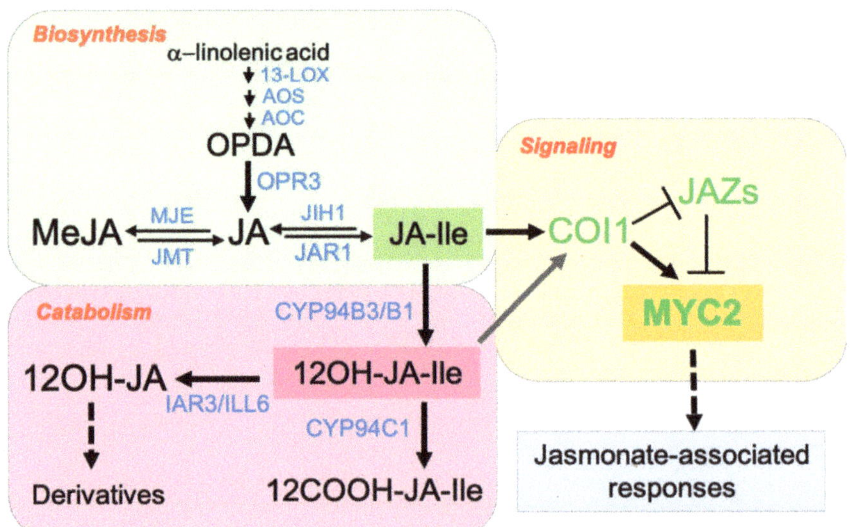

Figure 1. Overview of the jasmonate (JA) pathway including the major molecular players involved in biosynthesis, signaling, and catabolism. Black, blue, and green fonts show the main metabolites, enzymes, and proteins, respectively, for each section of the pathway. Scheme made based on [49,58,80]. For more details, see the text.

2.2. Salt Stress and JA Response

Earlier reports have shown the induction of some JA biosynthetic genes in *Arabidopsis* roots under salt stress conditions [88,89]. In sweet potato, Zhang et al. [90] studied the root transcriptomes of a salt-sensitive variety and a salt-tolerant line revealing a significant upregulation of the genes involved in the JA biosynthesis and signaling pathways under salt stress. The upregulation in the salt-tolerant line was greater than the sensitive line indicating the essential role of JA in the response of sweet potato to salt stress [90]. The *Arabidopsis* lipoxygenase3 (LOX3) was dramatically induced under salt treatment and the *lox3* mutant exhibited salt hypersensitivity. The *lox3* mutant salt sensitivity phenotype was rescued by the MeJA application indicating the association between JA and salt tolerance [91]. The *TaAOC1* gene from bread wheat responds to salinity and its constitutive expression in both bread wheat and *Arabidopsis* enhanced their level of tolerance to salt stress [92]. On the contrary, the impaired function of AOC in the OPDA-deficient rice ALLENE OXIDE CYCLASE mutants (*cpm2* and *hebiba*) conferred salt tolerance [93]. It is not clear what causes this tolerance, the lack of JA or JA-Ile, or the absence of their precursor 12-OPDA [19].

Several studies in *Arabidopsis* have demonstrated the essential role of MYC2 in salt response mediated by JA signaling [94,95]. The salt and ABA inducible gene *responsive to dehydration 22 (RD22)* is regulated by MYC2. The RD22 promoter region contains the MYC and MYB recognition sites [96] and the AtMYC2 and AtMYB2 TFs specifically interact with them, respectively. The *atmyc2* mutant and MYC2 overexpressing (OE) plants treated with ABA showed different *RD22* expression levels, the latter increased the *RD22* expression at low ABA concentration (500 nM) while the former increased *RD22* expression at higher ABA concentration (1 μM) [94]. MYC2 also has an important role in the activation of JA signaling by salt stress on the inhibition of cell elongation in *Arabidopsis* primary roots [95]. Additionally, the salt stress-mediated activation of MYC2 by the MAPK cascade regulates the proline biosynthesis through the delta1-pyrroline-5-carboxylate synthase1 (*P5CS1*) gene, which is a rate-limiting enzyme in the proline biosynthesis pathway [97]. Moreover, Seo et al. [98] demonstrated that the E3 ubiquitin ligase PLANT U-BOX PROTEIN 10 (PUB10), which regulates MYC2 stability [99], positively regulates salt and osmotic stress tolerance during seed germination. It was suggested that PUB10 acts as a negative regulator of ABA signaling through MYC2 and participates in the fine-tuning of ABA signaling and JA crosstalk in the abiotic stress tolerance in plants [98].

In turn, COI1, a core component of the JA-Ile coreceptor [59], is essential for JAZ transcript upregulation in the roots during the response to salt stress [95]. The JAZ upregulation mediated by salt stress in a COI1-dependent manner observed in the roots is likely to follow the canonical JA signaling pathway [100], with proteasome-mediated degradation of JAZ proteins. In this sense, JAZ proteins, the negative regulators of JA signaling, play an important role in the plant response to salt stress. Several JAZ homologous genes were upregulated by NaCl treatment in cotton, *Arabidopsis* roots, tomato, and wheat [95,101–103]. Moreover, the OsJAZ9 overexpression in rice resulted in a higher tolerance to salt stress [104] mainly through regulating the expression of ion transporters for K$^+$ homeostasis [105]. Similarly, enhancing the expression of OsJAZ8 transcripts assured better performance of transgenic rice lines under salt stress [106]. Furthermore, the rice nuclear factor, RICE SALT SENSITIVE3 (RSS3), forms a ternary complex with class-C bHLH TFs and JAZ proteins and regulates root cell elongation during adaptation to salinity [107]. Remarkably, JAZ genes from *Glycine soja* (*GsJAZ2*), *Malus domestica* (*MdJAZ2*), *Triticum durum* (*TdTIFY11*a), and *Pohlia nutans* (*PnJAZ1*) introduced in *Arabidopsis*, granted greater tolerance to salinity [103,108–110]. The introduction of *GaJAZ1* from *Gossypium arboreum* into *Gossypium hirsutum* (upland cotton) significantly increased salt tolerance in upland cotton compared to the WT strain. GaJAZ1-transgenic and WT plants showed many differentially expressed genes involved in JA signaling and biosynthesis, salt stress, and other hormone pathways. In GaJAZ-OE plants, the expression level of *JAZ1/3/6/8* and *JAZ10* were upregulated without NaCl treatment compared to the WT, but under salt conditions they were first downregulated (after 6 h and 12 h of treatment) and then upregulated again (24 h of treatment), indicating a sophisticated regulation of these genes in GaJAZ1-OE plants. Moreover, *MYC2* was significantly upregulated while JAR1 was downregulated in GaJAZ1-OE plants. JA biosynthesis was also affected in GaJAZ1-OE plants since various JA synthesis-related genes (e.g., *LOXs*, *AOS*, *AOC4*, and *JMT*) changed their expression compared to the WT. Additionally, several salt stress-related genes encoding for a vacuolar-associated protein (*VSR1*), an osmotic protein (*OSM34*), and a plasma membrane ion exchanger (*CHX18*) among others showed significant downregulation resulting in the accumulation of osmolytes that protect the plant from salt stress damage. Furthermore, some phytohormone-related genes besides JA-related genes reprogrammed their expression in GaJAZ1-OE plants compared to the WT including those related with ethylene- (*ACS*, *ERF2*, and *ERF4*); ABA- (*ABR1*, *ABA2*, *CBF4*, and *RD26*); and auxin- (*GH3.6*) pathways. In general, these results suggested that ectopic overexpression of *GaJAZ1* affects JA-related genes to increase salt tolerance in *G. hirsutum* plants albeit it also depends on other factors including hormone-crossing signal and salt-inducible genes [111].

Recently, JA-Ile catabolic genes have also been related to salt stress tolerance. Thus, *OsCYP94C2b* overexpression enhanced the viability of the transgenic rice under saline conditions and delayed the salt stress-induced leaf senescence [112]. Similarly, a higher *CYP94C2b* expression has been observed in some salt-tolerant rice varieties indicating that, at least in part, *CYP94C2b* may account for salt tolerance [113]. Hazman et al. [85] analyzed the accumulation of JAs and catabolic compounds in leaves from salt-exposed and control seedlings. OPDA and JA levels were increased by NaCl (100 mM) at most time evaluated points. Correspondingly, the JA-Ile catabolites 12OH-JA-Ile and 12COOH-JA-Ile were enhanced in response to salt exposure. Then, they explored the rice *CYP94* and *AH* gene families and examined the transcriptional response of a gene subset under salt exposure. Among the evaluated *CYP94* genes, only *OsCYP94C2a* was induced by salt stress while the transcripts of the *AH* genes fluctuated marginally. Apparently, *OsCYP94C2a* is the main player of JA-Ile oxidation upon salt stress in rice [85,112,113].

The apparent incongruity in the JA contribution to salt adaptation due to the positive effects of JA exogenous application (see Section 4), the upregulation of JA biosynthesis, and the induction of negative regulators of JA signaling and JA-Ile catabolism, indicates that timing and control of JA are maybe more important than its presence or absence [19,24]. Thus, detailed studies are required to reveal the underlying mechanism of efficient fine-tuning of jasmonate signaling in salt adaptation. Table 1 summarizes the findings regarding the main molecular components of the JA pathway associated with salt tolerance.

It is necessary to highlight that JA does not work independently in the improvement of plant tolerance to salt stress. Instead, its tightened coordination with other phytohormone signaling pathways allows the expression of multiple genes and flux of various metabolic pathways to adjust plant response to stress severity, specifically, in the appropriate time and tissue [18,34].

Table 1. Participation of jasmonate (JA) pathway-associated molecular components in salt stress responses of different plants.

Protein	Function	Salt Stress Response	Species	References
Lipoxygenase3 (LOX3)	JA biosynthesis	Induced under salt stress. Methyl jasmonate (MeJA) rescued the salt sensitivity phenotype of the *lox3* mutant	*Arabidopsis thaliana*	[91]
Allene oxide cyclase (AOC)	JA biosynthesis	Constitutive expression enhances tolerance to salt stress *cpm2* and *hebiba* mutants display salt tolerance	*Triticum aestivum* *A. thaliana* *Oryza sativa*	[92] [93]
MYC2	JA signaling	Transcriptional activator of the salt- and abscisic acid(ABA)-responsive gene *RD22*	*A. thaliana*	[94,96]
		An important role in salt-mediated JA-dependent inhibition of cell elongation in the elongation zone of primary root	*A. thaliana*	[95]
		Its salt stress-mediated activation by MAPK cascade regulates proline biosynthesis	*A. thaliana*	[97]
		Mediates the negative regulation of ABA signaling by PUB10, which acts as a positive regulator for salt and osmotic stress tolerance	*A. thaliana*	[98,99]
Jasmonate ZIM-domain (JAZ)	JA signaling	Induced under salt stress	*A. thaliana* *Gossypium hirsutum* *Solanum lycopersicum*	[95]
		PnJAZ1 inhibited expression of ABA-dependent genes related to seed germination and shoot growth under high salt conditions	*A. thaliana* *Physcomitrella patens*	[110]
		OsJAZ9 and *OsJAZ8* overexpression enhanced salt tolerance	*O. sativa*	[104]
		Heterologous expression of *GsJAZ2* and *MdJAZ2* enhanced tolerance to salinity	*A. thaliana*	[114]
		Overexpression of *TdTIFY11a* variants confer salt tolerance to *Arabidopsis* seedlings	*A. thaliana*	[103]
		GaJAZ1 overexpression significantly increased salt tolerance	*G. hirsutum*	[111]
		GbWRKY1 overexpression negatively affects salt tolerance through an interaction network involving JAZ1 and ABI1	*A. thaliana*	[115]
Cytochrome P450 family (CYP94C2b)	JA catabolism	*OsCYP94C2b* overexpression enhanced viability under salt conditions and delayed the salt stress-induced leaf senescence	*O. sativa*	[112]

99

3. Crosstalk between JA and Other Plant Hormones during Salt Stress

In JA signaling, the proteins MYC, JAZ, and COI1 have been established as the core of the pathway and have been pointed to serve as a link between different hormone signaling [116]. Thus, the JA crosstalk with other phytohormones involves these components.

3.1. JA and ABA

Several studies have reported the combined action of JA and ABA in the plant response to salt stress. Although an earlier study described that ABA and JA antagonistically regulate the expression of transcripts inducible by salt stress in rice (*Oryza sativa*) [28], in more recent studies the synergistic action of both phytohormones has been observed. For instance, the application of ABA together with different concentrations of JA activated the protection mechanism against NaCl-associated stress in strawberry plants (*Fragaria* × *ananassa*) [117]. Likewise, ABA and JAs had a synergistic effect on the inhibition of seed germination under salinity conditions [110]. Moreover, Yang et al. [118] confirmed the cooperation between JA and ABA in the tolerance to salt stress mediated by phytochromes. Albeit the evidence of the concerted action of JA-ABA, the cross-coordination that exists between their signaling pathways is just beginning to be understood [119].

MYC2 seems to be an essential point for JA-ABA crosstalk as demonstrated in several studies. Accordingly, *MYC2-OE Arabidopsis* plants and the *myc2* mutant show boosted and decreased ABA sensitivity, respectively [94,120]. Additionally, the expression of the salt- and ABA-responsive gene *RD22* is promoted by MYC2 [94,96]. MYC2 and ABI5 (a transcription factor activated by the ABA signaling pathway) are modulated at the protein level through MED25, which is a multifunctional subunit of the *Arabidopsis* mediator complex [121]. In turn, the ABA receptor PYL6 directly interacts and alters the transcriptional activity of MYC2 [122]. Meanwhile, PYL4 is involved in the coregulatory effects of ABA and JA on plant growth and metabolism [123]. Furthermore, the induction of *MYC2* by ABA seems to depend on the JA-Ile COI1 receptor according to Lorenzo et al. [120]. Recently, it has been reported that the application of ABA to strawberry plantlets involves the upregulation of the *MYC2* gene (*FabHLH80*) from 0.5 to 12 h post-treatment [124].

Some studies highlight the importance of JAZ proteins as key nodes of JA-ABA crosstalk. The ubiquitin ligase E3 KEEP ON GOING (KEG), a known ABI5 repressor in the ABA signaling pathway, directly interacts with JAZ12 and modulates its stability [125]. Furthermore, JAZ3 interacts with ABI5 in vivo and represses its transcriptional activity [126]. The overexpression of *PnJAZ1* (isolated from the moss *P. nutans*) in *Arabidopsis* plants inhibited the expression of genes of the ABA-dependent pathway related to seed germination and shoot growth under high salt conditions [110]. Furthermore, the transcription factor GbWRKY1 (from *Gossypium barbadense*) negatively regulated ABA signaling through an interaction network involving JAZ1 and ABI1 (the negative regulator of ABA signaling), in the response to salt and drought stress [115]. The evidence demonstrates the role of JAZs in the response to salt stress and JA-ABA crosstalk, however, the mechanisms of action in which these proteins participate in salt tolerance remain to be fully elucidated.

3.2. JA and Other Phytohormones

The nature of jasmonate crosstalk with other phytohormones in salt stress cannot be clearly described due to limited experimental data. At the molecular level, little is known about the convergence points of JA signaling and other phytohormone pathways. Due to the well-established central role of ABA under abiotic stress, more evidence of JA-ABA crosstalk is available [19]. In recent years, the crosstalk among different phytohormone mediating salt stress responses has been described [18]. However, it is necessary to deepen the JA crosstalk with other phytohormones to regulate salt stress tolerance in plants.

3.2.1. JA and Ethylene (ET)

JA–ET crosstalk can be complex and depends on the specific situation. A synergistic effect has been observed in promoting leaf senescence [127,128], but antagonistic interaction was found in controlling apical hook curvature [129]. JA–ET interactions in response to pathogen infections, herbivore attacks, and environmental stress are context-specific. The ET-stabilized transcription factor ETHYLENE-INSENSITIVE3 (EIN3) physically interacts with MYC2 and inhibits its DNA binding activity attenuating JA-regulated plant defense against generalist herbivores [129]. Since the *jaz* decuple mutant showed robust activation of insect and fungal pathogen defenses, the JA–ET crosstalk seems to be mediated via EIN3/ETHYLENE INSENSITIVE 3-like 1 (EIL1) along with JAZs-MYC2 [35]. The two TFs ETHYLENE RESPONSE FACTOR1 (ERF1) and OCTADECANOID-RESPONSIVE ARABIDOPSIS AP2/ERF 59 (ORA59) were previously implicated in integrating JA and ET signaling in *Arabidopsis* [130,131]. Recently, a new convergence-point of JA–ET signaling has been reported. *OsLOX9*, a JA biosynthetic pathway-related gene, is regulated by OsEIL1 in response to piercing-sucking insect attacks [132].

In salt stress, the synergistic activation of the *Arabidopsis* ERF1 by JA and ET is required for inducing tolerance [133]. Moreover, a rice root-specific pathogenesis-related protein (RSOsPR10), induced by high salt and other abiotic stresses, promotes root growth and root mass increasing salt tolerance [134]. JA and ET also induce RSOsPR10, while salicylic acid (SA) almost completely suppresses its induction [135,136]. JA-inducible and OsERF87-dependent expression of *RSOsPR10* were strongly repressed by the SA-inducible OsWRKY76 transcription factor [137]. As OsERF87 and OsWRKY76 bind at the *RSOsPR10* promoter, they antagonistically regulate *RSOsPR10* expression. This mechanism represents a fine-tuning balance between JA/ET and SA signaling in plants under environmental challenges. The description of the molecular components of the synergistic JA–ET crosstalk in the regulation of *RSOsPR10* expression requires investigation. Another gene activated by salt stress and treatment with MeJA or ethephon (an ethylene releasing compound) is *GmCYP82A3*, a gene from the soybean CYP82 family. Transgenic *Nicotiana benthamiana* plants overexpressing *GmCYP82A3* exhibited strong pathogen resistance and enhanced salinity tolerance. Besides, an increased expression of the JA/ET signaling pathway-related genes was observed in the transgenic plants [138]. How GmCYP82A3 is involved in JA–ET crosstalk under salt stress could be an interesting topic to be approached.

3.2.2. JA and SA

The JA–SA antagonism is well known in plant defense pathways and key components of JA–SA crosstalk have been identified. Recently, many of the molecular components in the JA–SA crosstalk that regulate the plant immune network at transcriptional and protein levels are reviewed by Aerts et al. [139]. Among them, MAPKs are involved in the convergence of these phytohormone pathways [140–142]. AtMPK4 negatively regulates the activation of SA- and the repression of JA-mediated defenses under biotic stress [140]. Besides, MPK4 positively regulates the glutaredoxin GRX480 in the SA signaling pathway and negatively regulates MYC2 in the JA signaling pathway, which is necessary for JA responsive genes (*PDF1.2* and *THI2.1*) [100]. SA-induced NON EXPRESSOR OF PR GENE (NPR1) activates GRX480, which can block the JA response gene expression mediated by TGA (a D group of *Arabidopsis* bZIP TFs), confirming SA–JA antagonism [143]. Additionally, SA-induced protein kinase (SIPK) and wound-induced protein kinase (WIPK), which are rapidly activated after perception of herbivory, regulate herbivory-induced JA levels and JA-mediated defense metabolite accumulations. WIPK-signaling is associated with large fitness costs in competing *Nicotiana attenuata* plants, while SIPK acts as an important regulator of plant fitness, possibly modulating SA-JA crosstalk through ethylene signaling [142].

The role of JA and SA in salt stress tolerance has been explored previously [144], but their relationship under salinity is not well known yet. Several reports show that SA and JA can alleviate the hazardous effect of salt stress on plants. In strawberry, improved physiological characters such as increased antioxidant activity and a reduced Na^+/K^+ ratio were observed on MeJA and SA treatments [145]. Likewise, soybean performance under salinity was improved by foliar spraying of JA and SA [146]. Treatment of soybean plants with SA plus JA stimulated H$^+$-ATPase activity of tonoplast, nutrient uptake, and salt tolerance [147]. Additionally, exogenous applications of JA and SA decreased the concentration of Na$^+$ in soybean under different salt stress levels [148]. Moreover, MeJA treatment can protect plants from salt-induced damage by acting role as an antioxidant and cooperating with SA [149]. All the evidence invites a deepening in the JA–SA interplay under salinity conditions for identifying the molecular components of their crosstalk.

3.2.3. JA and Gibberellins (GA)

JA and GA signaling pathway interaction has been described that occurs by the key GA signaling proteins DELLAs and JAZs since they directly interact [150]. The JAZ–DELLA interaction would interfere with the inhibition of MYC2 by JAZ proteins resulting in the activation of MYC2 downstream genes. In presence of GA, DELLA proteins are degraded and JAZs bind MYC2 to inhibit JA signaling [150,151]. On the other hand, the *della* mutant is less sensitive to the plant growth inhibition mediated by JA suggesting that JA delays the degradation of DELLA mediated by GA [152]. Curiously, JA signaling upregulates the transcription of *RGL3*, a DELLA gene, which the promotor is a target of MYC2 [153]. Additionally, RGL3 physically interacts with JAZ1 and JAZ8 [64,153] suggesting the JA-mediated degradation of JAZ1 and consequent release of MYC2 to induce the *RGL3* expression, which in turn binds the non-JA degradable JAZ8 enhancing the MYC2-dependent JA responses [153]. Recently, it was reported that JA and GA synergistically promote fiber cell initiation of cotton (*G. hirsutum*) possibly mediated by the GhJAZ3 and GhSLR1 (a DELLA protein) interaction [154].

The interplay of JA and GA under salt conditions has not been studied in depth. However, there is evidence that in salt-stressed plants of basil (*Ocimum basilicum*) the GA concentration significantly decreased and in non-stressed plants treated with JA. Stressed plants treated with JA also showed a significant decrease in GA concentration [155] showing the antagonistic effect of JA in the GA level under salt stress. The effects of the combination of MeJA and NaCl treatment on the growth regulation and defense response of *Nitraria tangutorum*, a desert halophyte, have been recently investigated [156]. Compared with NaCl treatment alone, MeJA treatment aggravated the growth inhibition of seedlings by antagonizing growth-related hormones like GA. It was demonstrated that the transcript levels of GA-responsive genes *NtPIF3*, *NtGAST1*, and *NtGSAT4* were suppressed by MeJA [156]. More studies are needed to find out the components of JA–GA crosstalk under saline conditions.

3.2.4. JA and Cytokinin (CK)

Interactions between JA and CK could grant some developmental flexibility under stress conditions since they mediate stress response and developmental processes. Although there are scarce reports on JA–CK crosstalk, some data indicate that their interaction could be negative or positive [157,158]. Antagonistic effects of JA and CK have been observed in different processes such as senescence, photosynthesis, RNA and protein synthesis, and vascular formation [157,159–161]. In contrast, a positive interplay between these phytohormones in delaying senescence of the *Iris* flower (*Iris × hollandica*) has been reported [158]. Moreover, CK treatments promoted gene expression of JA-amino synthetase in *Arabidopsis* and tomato plants [162] suggesting an interaction between the JA–CK pathway.

Recently, Avalbaev et al. [163] demonstrated the influence of exogenous MeJA on endogenous CK content in wheat plants under normal and salinity conditions. Low concentrations of MeJA (0.01–1 μM) increased wheat seedling growth while higher concentrations (10 and 100 μM) inhibited it. The hormonal balance of wheat seedlings was shifted by exogenous application of 0.1 μM MeJA. In response to salt stress, MeJA-untreated wheat plants increased the ABA level and gradually decreased indole acetic acid (IAA) and CK contents. Meanwhile, MeJA-pretreated seedlings were characterized by a diminution of ABA accumulation and IAA decreased level induced by salinity. Noticeably, a salinity-induced decline in the CK content was completely preventive by MeJA, which eventually resulted in the maintenance of the wheat seedlings growth rate under salt stress. It was suggested that MeJA application influences CK metabolism since cytokinin oxidase (CKX) gene expression and enzyme activity decreased after 1 h of MeJA treatment [163]. CKX catalyzes the degradation of CKs and controls the CK content in plants [164]. Similar results were obtained on almond rootstocks when MeJA application (0.025–0.05 mM) increased CK concentration in the leaf due to restriction of CKX activity and its gene expression [165]. MeJA-induced protection against salinity was found to be reached by modulating the activity of the antioxidant system and accumulation of osmoprotectants [165,166]. However, the mechanism by which exogenous MeJA mitigates the effect on the growth of salt-stressed plants as a result of the inhibition of CK decline under salt stress is largely unknown.

3.2.5. JA and Auxin (AUX)

The JA–AUX crosstalk was early reported by the identification of the *auxin-resistant1* (*axr1*) mutants with altered jasmonate responsive gene expression [167]. Moreover, a link between JA signaling and AUX homeostasis was evidenced through the JA-mediated modulation of *YUCCA8* and *YUCCA9* gene expression, which are involved in AUX biosynthesis [168]. JA and AUX interaction also involves the ETHYLENE RESPONSE FACTOR109 (ERF109) during JA-induced lateral root formation [169]. Recently, Xu et al. [170] reviewed the novel progresses on the integration of JA and ethylene into AUX signaling in regulating root development of *Arabidopsis thaliana* [170]. Besides, Zhang et al. [171] provided an example of metabolic-level crosstalk between the JA and AUX signaling pathways by demonstrating that wounded leaves JA-inducible amidohydrolases (ILR1, ILL6, and IAR3) contribute to regulate active IAA and JA-Ile levels, promoting AUX signaling while attenuating JA signaling [171].

The transcription factor WRKY57, which is upregulated by IAA, but downregulated by JA, has been described as a convergence node of JA and IAA-mediated signaling. The JAZ4/8 and IAA29 (an AUX/IAA protein) repressors of JA and IAA signaling, respectively, have an opposite function in WRKY57 regulation since both competitively bind WRKY57 during JA-induced leaf senescence in *Arabidopsis* [172]. In earlier work, Jiang et al. [173] found that the *Arabidopsis adt* mutant, which is constitutive to expressing the *WRKY57* gene, exhibited drought, osmotic, and salt tolerance. The enhanced tolerance of the *adt* mutant to these stresses was associated with an increment of ABA content consistently with the upregulation of *RD29A*, *NCED3*, and *ABA3* genes. It was demonstrated that WRKY57 could directly bind the W-box of *RD29A* and *NCED3* promoter sequences, suggesting that WRKY57 could regulate their expression. Since WRKY57 is regulated by JAZ4/8 and IAA29 repressors [172], it could be interesting to evaluate their role in the plant response to salt stress. Perhaps, it would shed light on the JA–AUX interplay under salinity conditions.

The main molecular/physiological effects related to phytohormone crosstalk are summarized in Table 2.

Table 2. Molecular and physiological effects of the jasmonate (JA) crosstalk with abscisic acid (ABA), ethylene (ET), salicylic acid (SA), gibberellins (GA), cytokinin (CK), and auxin (AUX) in different plants under salt stress conditions. For more details, see the text.

Crosstalk	Molecular/Physiological Effects	Species	References
JA-ABA	JA and ABA applications in conjunction activate the antioxidant mechanism against salt stress	*Fragaria* × *ananassa*	[117]
	Synergistic effect on the inhibition of seed germination under salinity conditions	*Arabidopsis thaliana*	[110]
	Synergism in the salt tolerance mediated by phytochrome A and B	*Nicotiana tabacum*	[118]
JA–ET	Synergistic upregulation of *AtERF1* required to induce salt tolerance	*A. thaliana*	[133]
	Synergistic upregulation of *RSOsPR10* which promotes root growth and increases salt tolerance	*Oryza sativa*	[136]
	Synergistic upregulation of *GmCYP82A3* which enhances salinity tolerance	*Glycine max* *Nicotiana benthamiana*	[138]
JA–SA	Methyl jasmonate (MeJA) and SA application increases antioxidant activity and reduced the Na^+/K^+ ratio	*F.* × *ananassa*	[145]
	JA and SA application protects plants from salt-induced damage and improves plant performance under salt conditions	*G. max*	[146]
	JA and SA application stimulates H^+-ATPase activity of tonoplast, nutrient uptake, and salt tolerance	*G. max*	[147]
JA-GA	JA application decreases GA content in salt-stressed plants	*Ocimum basilicum*	[155]
	MeJA application suppresses the transcript levels of the GA-responsive genes *NtPIF3*, *NtGAST1*, and *NtGSAT4*	*Nitraria tangutorum*	[156]
JA-CK	MeJA application prevents the salinity-induced decline of endogenous CK by reducing the cytokinin oxidase enzymatic activity and its related gene expression	*Triticum aestivum* *Prunus dulcis*	[163]
JA-AUX	An opposite function of JAZ4/8 and IAA29 repressors on the regulation of WRKY57. Constitutive activation of WRKY57 in *adt* mutant confers salt tolerance	*A. thaliana*	[172,173]

4. Effects of JA-Exogenous Applications for Improving Salt Stress Tolerance

Exogenous jasmonate applications have effects on different physiological aspects, including protection against biotic and abiotic stresses [174–176]. Methyl jasmonate (MeJA) application induces protection against oxidative stress as has been reported in different species and conditions [176–179].

In *Arabidopsis thaliana*, the total activities of catalase (CAT), peroxidase (POD/POX), superoxide dismutase (SOD), and glutathione reductase (GR) increased considerably in response to MeJA [180]. In crops, such as strawberry, MeJA applications during the preharvest period increase the anthocyanin and ascorbic acid contents, CAT, and ascorbate peroxidase (APX) activities in fruits at postharvest storage [176]. In this sense, increased activity of the antioxidant enzymes, together with higher levels of antioxidant compounds, as a result of MeJA treatment reinforces the antioxidant response of plants to reactive oxidative species (ROS) caused by abiotic stresses, like high salt.

From a physiological point of view, salt stress increases free proline content, photorespiration, and stomatal resistance among others, while decreasing net photosynthetic rates, transpiration, protein, and relative water content (RWC) [181,182]. Pretreatment with 0.1 mM MeJA helps the pea seedlings to counteract the salt stress since RWC and protein content of the treated seedlings were higher in comparison to NaCl-treated seedlings. Moreover, MeJA-treated pea seedlings present a decrease of Na^+ and Cl^- accumulation in the shoot [183]. In another experiment, pretreatment with 0.1 mM MeJA for 3 days before salt treatment diminished the inhibitory effect of NaCl on the rate of $^{14}CO_2$ fixation, and activity and content of ribulose-1,5-bisphosphate carboxylase/oxygenase [184].

Additionally, in rapeseed (*Brassica napus*), exogenously applied MeJA counteracted the inhibitory effects of NaCl by increasing RWC, soluble sugar content, and photosynthesis rate [174]. The application of 0.25 mM MeJA was the most effective treatment to enhance salt tolerance at a concentration of 60 mM NaCl in strawberry (*F.* × *ananassa* 'Camarosa') seedlings [145]. In almond rootstocks, application of MeJA in optimal concentrations of 0.025–0.05 mM alleviated the adverse effect of salt stress by increasing the photosynthetic rate, activity of antioxidant enzymes (APX, SOD, and POX), root and shoot dry mass, and cell membrane integrity [165]. Alleviation of moderate salinity (40 mM NaCl) by foliar application of 5 mM MeJA has also been reported in broccoli [185]. Pretreatment of cowpea seeds with 0.05 mM MeJA improves plant tolerance to salt stress [186].

In other plants, like *Limonium bicolor*, which is a typical recretohalophyte with salt glands in the epidermis, 300 mM NaCl led to a dramatic inhibition of seedling growth that was significantly alleviated by the application of 0.03 mM MeJA, resulting in biomass close to that of plants not subjected to salt stress [187]. Even in high salt concentrations such as 500 mM NaCl, MeJA applied at 0.1 mM has a protective role in the defense response of *Robinia pseudoacacia* especially with a marked increase in the activity of antioxidant enzymes and related gene expression [188]. In another study, foliar applications of 0.5 mM MeJA increased the essential oil content and the antioxidant activities of basil (*O. basilicum* 'Genove') on 30 mM NaCl and have noticeable effects on the main components of the oils [189]. In *Glycyrrhiza uralensis* exposed to 100 mM NaCl, 0.025 or 0.05 mM MeJA increased the root length of salt-stressed *G. uralensis* seedlings but decreased root diameter, stem length, and stem diameter, enhancing peroxidase activity and ascorbate content [190].

In wheat seedlings, NaCl salt stress caused a significant increase in the malondialdehyde (MDA) content and H_2O_2 concentrations and a concomitant decrease in SOD, POD, CAT, and APX activities. Exogenous JA pretreatment (2 mM) combined with NaCl treatment (150 mM) produced a significant decline in MDA and H_2O_2 concentrations and an increase in SOD, POD, CAT, and APX activities. Moreover, a marked upregulation of SOD, POD, CAT, and APX genes was observed in the JA–NaCl combined treatment in comparison to NaCl treatment alone. Additionally, exogenous JA remarkably increased glutathione (GSH) concentration in wheat seedlings treated with NaCl and decreased the deleterious effect of salt stress on the growth of wheat [31]. These results indicate that exogenous JA can effectively scavenge ROS by enhancing the activities of antioxidant enzymes and the concentration of antioxidant compounds in wheat seedlings under salt stress and consequently play an important role in decreasing lipid peroxidation and increase the ability of wheat to resist salt stress [31].

In roselle (*Hibiscus sabdariffa*), the exogenous JA treatment protected roselle seedlings against salt-induced harms through enhancing the activities of both enzymatic and non-enzymatic antioxidants, such as APX, pyrogallol peroxidase (PPX), and PPO, and the accumulation of metabolites non-reducing sugars, total phenols, anthocyanins, flavonoids, and proline. The JA-treated roselle exhibited a significant increase in growth parameters under salt conditions compared to the WT [191]. In forage sorghum (*Sorghum bicolor*), at a high salinity level (200 mM), seeds treated with 10 mM JA showed a positive effect on various growth and physiological parameters such as emergence percentage, emergence rate, shoot length, total fresh weight, salt tolerance index, and total chlorophyll among others [192]. Similarly, seed priming and foliar application with JA enhanced salinity stress tolerance of soybean (*Glycine max*) seedlings. Improved water and osmotic potentials, water use efficiency, net photosynthetic, transpiration rate, stomatal conductance, and total chlorophyll content were observed in JA-primed and treated soybean seedlings compared to the untreated ones. Besides, JA treatment resulted in a reduction of Na^+ concentration and an increment of K^+ concentrations in the leaf and root of the analyzed cultivars despite salinity stress [193].

Although the exogenous application of JA and its derivates constitute a suitable approach to improve the plant response to salt stress, this strategy has some limitations. Due to the JA–SA antagonism during the plant defense response [194], an increment in

JA content by its exogenous application will negatively affect the SA-mediated response to biotrophic pathogens. In addition, JA can inhibit plant growth by reprogramming plant metabolism to produce diverse defense compounds [35], thus the growth-defense trade-off is also an issue. Moreover, the high economic cost of MeJA application in field treatments [176] could prevent its use to mitigate the salt-induced damage in crop plants. It is necessary to keep in mind that the activation of JA-beneficial effects in plant response to salt stress depends on the JA levels and therefore, field experiments will be required to analyze the cost-effective JA doses for exogenous applications.

5. Application of Genome-Associated Tools for Salt Tolerance Mediated by JA

Salt stress tolerance is a complex trait regulated by polygenes [195]. In this context, QTL mapping and GWASs provide a suitable opportunity to identify genes responsible for quantitative trait variation such as salt tolerance. In this regard, genetic factors associated with salt stress have been previously investigated in several crops, such as rice [41,196,197], barley [198,199], wheat [200,201], chickpea [202], sesame [203], cotton [204,205], and soybean [206].

GWAS and QTL mapping have been implemented to identify the genetic factors involved in both osmotic and ionic components of salinity stress. A typical GWAS workflow to identify genes related to salt stress (and genes conferring salt tolerance) in crops is represented in Figure 2. Four main steps can be distinguished: (1) plant genotyping; (2) plant phenotyping based on morphological and physiological traits related to salt stress; (3) identification of marker–trait associations (MTAs); and (4) identification of candidate genes involved in the salt stress response. A diverse panel is genotyped using DNA markers, i.e., single nucleotide polymorphisms (SNPs), diversity arrays technology (DArT), or RNA-sequencing (RNA-seq) for the identification of genetic variants that affect the gene expression level in a specific tissue. Parallelly, this panel is phenotyped for different traits regarding the objectives of the study. In the context of salt stress in plants, the phenotyping is generally carried out considering several morphological and physiological traits, such as leaf area, root and shoot dry weight, seed germination rate, salt stress index (SSI), Na^+/K^+ ratio, chlorophyll content, and MDA, and, remarkably, changes in the phytohormone levels such as JAs. Particularly, these traits are evaluated in plants or seedlings that have been subjected to salt treatment. Subsequently, the genotypic and phenotypic data are combined to associate alleles with particular traits using classical GWAS models, which significantly detect molecular markers associated with the studied traits. Posteriorly, the MTAs can serve as a starting point for the mining of candidate genes. Gene Ontology (GO) annotation of the putative candidate genes can be carried out using BLAST tools.

In a GWAS for salt tolerance (or salt stress), the candidate genes are usually associated with gene annotations, such as abiotic stress-related, anion transport, Ca^+ binding and signaling, phytohormone response elements, and others [207–209]. In this way, many TFs and downstream genes related to phytohormones biosynthesis and signaling have been characterized, including ABA, SA, JA, ET, and others considered as growth promotion hormones, including AUX, GA, and brassinosteroids [41,205,209–213]. Additionally, the candidate genes associated with salt stress or tolerance can be validated by quantitative reverse transcription PCR (RT-qPCR) analysis.

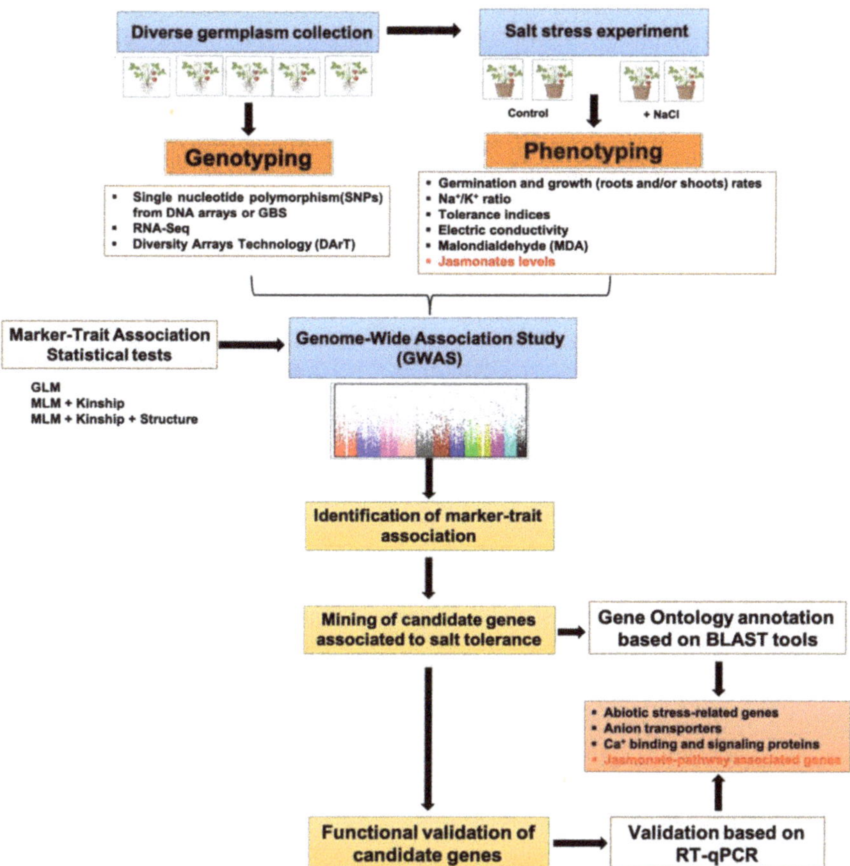

Figure 2. Genome-wide association study (GWAS) pipeline to identify jasmonate pathway-associated genes involved in the salt stress response and salt tolerance in crops. GBS, SSI, GLM, and MLM correspond to genotyping by sequencing, salt stress index, general linear model, and mixed linear model, respectively. For more details, see the text.

To our knowledge, several studies have analyzed the JA response to different stresses, but few have explored JA-dependent genetic mechanisms taking into account natural genetic variation. To et al. [214] performed a GWAS to identify genetic variants associated with exogenous JA treatment responses in rice. They found a high natural variability for the shoot and root growth (in a panel of 150 rice accessions) in response to JA treatment. This GWAS revealed about 230 candidate genes, including several JA-responsive TFs known to play a stress response role. Several GWASs have elucidated the participation of JA in salt tolerance in plants. In this sense, Li et al. [203] reported potential candidate genes related to drought and salt-induced stress in sesame. In this GWAS, the *SiOPR3* gene (detected for drought stress in sesame) is the ortholog of the *Arabidopsis OPR3* gene an essential component of the JA biosynthesis.

Besides, Rohila et al. [41] detected several candidate genes associated with seedling stage salt tolerance by the GWAS approach in a rice core-collection. Particularly, one SNP (on chromosome 3) was located close to the *ALLENE OXIDE CYCLASE 1* (*AOC1*) gene. This gene has been related to salt and other abiotic stresses, and it is a key gene in the JA biosynthetic pathway in *Arabidopsis* [215]. Yuan et al. [205] combined the association mapping and RNA-seq analyzes to explore candidate genes for salt-tolerance in cotton at the germination stage. At least nine genes were associated with signaling or a response to signal factors, such as SA, GA, and JA. In other GWASs, genes related to the JA biosynthesis and signaling pathways have been indirectly identified, in which upstream genes from the JA signaling events and JA-regulated genes during salt stress were detected. For instance, An et al. [216] carried out a GWAS to identify associations conferring salt tolerance in rice. In this study, a significant SNP for seedling length was located on the promoter of a salt stress-related gene (*RSOsPR10*), which has been proposed to be induced by biotic and abiotic stresses, via the JA signaling pathway [135]. Sun et al. [204] reported DNA polymorphisms associated with salt tolerance candidate genes at the cotton seedling stage. Additionally, the expression levels of six genes were reported using salt-tolerant and salt-sensitive varieties. Interestingly, the expression of *Gh_A10G1756*, a homolog of the *Arabidopsis CALCIUM-DEPENDENT PROTEIN KINASE 1* (*AtCPK1*) gene, had higher expression in the salt-sensitive than salt-tolerant types [204]. The *AtCPK1* gene mediates pathogen resistance in *Arabidopsis* and plays a positive role in salt/drought-stress response [217]. According to Coca and San Segundo [218], *AtCPK1*-OE plants showed activation of two components of the chloroplastic pathway for JA biosynthesis (*AOS* and *AOC* genes), which could support the idea that JA is a key element for triggering different responses to salt stress in plants. Patishtan et al. [208] implemented a GWAS using a diversity panel of 306 rice accessions treated with different salt concentrations. The authors characterized more than 30 candidate genes for short, medium, and long-term NaCl treatment. Notably, among these genes, a DNA polymorphism was associated with the *WRKY70* gene, which plays a pivotal role in an antagonistic interaction between SA and JA responses [210,219].

Due to the above-mentioned studies, GWAS and association studies have successfully identified many novel genes associated with traits of interest. These findings will be useful towards the developing of varieties tolerant to salt stress, using new genomic techniques, such GE; an aspect that has not been addressed in depth. In this sense, GE systems provide the ability to modify genes and generate new possibilities for crop improvement precisely. The RNA-guided CRISPR/Cas9 technology is simple to use across different genome editing technologies [220] and has been generally used in major crops such as wheat [221], maize [222], soybean [223], and many others. Even more, the highly efficient multiplex editing toolkit based on an intron-optimized zCas9i gene, which allows assembly of nuclease constructs expressing up to 32 sgRNAs [224], can enable simultaneously targeting of multiple independent loci to generate complex genotypes or to functionally interrogate groups of candidate genes such as those involved in phytohormones signaling.

To improve the salt tolerance in crops employing the CRISPR/Cas9 system with a focus on phytohormones, an example is the targeting of the *OsRR22* gene, which encodes a transcription factor involved in CK signaling, which has effectively enhanced salt tolerance in rice [46]. Recently, Liu et al. [225] enhanced drought tolerance in tomato (*Solanum lycopersicum*) by CRISPR/Cas9 targeted mutagenesis of the *SlLBD40* gene. *SlLBD40* encodes an organ boundaries domain transcription factor, which is highly induced by polyethylene glycol (PEG), salt, and MeJA treatments. The analysis of the *SlLBD40* expression in the *jasmonic acid-insensitive1* (*jai1*) mutant (a mutant in the JA-Ile tomato receptor COI1) and *MYC2*-silenced plants demonstrated that *SlLBD40* depends on JA signaling for its activation and it might be downstream of *SlMYC2* [225]. The previous works offer a good background for targeting key components of the JA pathway to obtain high salinity tolerance in crops using GE tools.

6. Concluding Remarks

Besides the role of development and abiotic stress responses, the jasmonate pathway is certainly involved in plant salt stress responses. Several JA-biosynthetic genes are induced under salt stress, although the lack of jasmonates is also related to this tolerance. Key JA-associated molecular components such as the transcription factor MYC2 and the repressor JAZ seem to be crucial in salt tolerance. Remarkably, overexpression of *JAZ* genes confers salt tolerance in transgenic plants. Regarding phytohormone crosstalk between JA and others in a salt stress context, certainly, more information exists about JA-ABA crosstalk, focused on MYC2 and JAZ interactions with the ABA signaling-components.

In terms of future applied perspectives, on the one hand, several reports demonstrate the positive effects of JA exogenous applications on the physiological status against salt stress. It should be considered as complementary crop management to face not only salinity-derived damages but also all abiotic stresses in a global change framework. On the other hand, genomic tools such as GWAS could help to reveal several JA pathway-associated genes that can serve as a guide in breeding programs and targets in genome editing systems, such as CRISPR/Cas9, to get salt-resistant crops. Surely, the next years will be promising in discoveries and applications related to the role of jasmonates against salt stress in plants.

Author Contributions: Conceptualization, C.R.F. and F.M.-P.; writing—original draft preparation, C.D., C.R.F., F.M.-P. and S.A.; writing—review and editing, C.D., C.R.F., F.M.-P. and J.-T.C.; visualization, C.D., C.R.F., F.M.-P. and S.A.; funding acquisition, C.R.F. All authors have read and agreed to the published version of the manuscript.

Funding: This research and the APC were funded by the National Research and Development Agency (ANID, Chile) grant FONDECYT/Regular 1181310 to C.R.F. Funding approval date: 1 April 2018.

Institutional Review Board Statement: Not applicable.

Informed Consent Statement: Not applicable.

Data Availability Statement: Not applicable.

Acknowledgments: C.D. thanks ANID for a doctoral scholarship (grant "Beca Doctorado Nacional" No. 6647/2019).

Conflicts of Interest: The authors declare no conflict of interest. The funders had no role in the design of the study, in the collection, analyses, or interpretation of data, in the writing of the manuscript, or in the decision to publish the results.

References

1. Zörb, C.; Geilfus, C.M.; Dietz, K.J. Salinity and crop yield. *Plant Biol.* **2019**, *21*, 31–38. [CrossRef] [PubMed]
2. Ivushkin, K.; Bartholomeus, H.; Bregt, A.K.; Pulatov, A.; Kempen, B.; De Sousa, L. Global mapping of soil salinity change. *Remote Sens. Environ.* **2019**, *231*, 111260. [CrossRef]
3. Kumar, K.; Kumar, M.; Kim, S.-R.; Ryu, H.; Cho, Y.-G. Insights into genomics of salt stress response in rice. *Rice* **2013**, *6*, 1–15. [CrossRef] [PubMed]
4. Dasgupta, S.; Hossain, M.M.; Huq, M.; Wheeler, D. Climate change and soil salinity: The case of coastal Bangladesh. *Ambio* **2015**, *44*, 815–826. [CrossRef]
5. Daliakopoulos, I.; Tsanis, I.; Koutroulis, A.; Kourgialas, N.; Varouchakis, A.; Karatzas, G.; Ritsema, C. The threat of soil salinity: A European scale review. *Sci. Total Environ.* **2016**, *573*, 727–739. [CrossRef]
6. Fan, X.; Pedroli, B.; Liu, G.; Liu, Q.; Liu, H.; Shu, L. Soil salinity development in the yellow river delta in relation to groundwater dynamics. *Land Degrad. Dev.* **2012**, *23*, 175–189. [CrossRef]
7. Moreira Barradas, J.; Abdelfattah, A.; Matula, S.; Dolezal, F. Effect of fertigation on soil salinization and aggregate stability. *J. Irrig. Drain. Eng.* **2015**, *141*, 05014010. [CrossRef]
8. Moral, R.; Perez-Murcia, M.; Perez-Espinosa, A.; Moreno-Caselles, J.; Paredes, C.; Rufete, B. Salinity, organic content, micronutrients and heavy metals in pig slurries from South-eastern Spain. *Waste Manag.* **2008**, *28*, 367–371. [CrossRef]
9. Lefebvre, O.; Moletta, R. Treatment of organic pollution in industrial saline wastewater: A literature review. *Water Res.* **2006**, *40*, 3671–3682. [CrossRef]
10. Mateo-Sagasta, J.; Burke, J. Agriculture and water quality interactions: A global overview. *Solaw Backgr. Themat. Rep.* **2011**, *46*.
11. Yang, Y.; Guo, Y. Elucidating the molecular mechanisms mediating plant salt-stress responses. *N. Phytol.* **2018**, *217*, 523–539. [CrossRef]
12. Arif, Y.; Singh, P.; Siddiqui, H.; Bajguz, A.; Hayat, S. Salinity induced physiological and biochemical changes in plants: An omic approach towards salt stress tolerance. *Plant Physiol. Biochem.* **2020**, *156*, 64–77. [CrossRef]
13. Acosta-Motos, J.R.; Ortuño, M.F.; Bernal-Vicente, A.; Diaz-Vivancos, P.; Sanchez-Blanco, M.J.; Hernandez, J.A. Plant responses to salt stress: Adaptive mechanisms. *Agronomy* **2017**, *7*, 18. [CrossRef]
14. Ali, Q.; Shahid, S.; Nazar, N.; Hussain, A.I.; Ali, S.; Chatha, S.A.S.; Perveen, R.; Naseem, J.; Haider, M.Z.; Hussain, B. Use of phytohormones in conferring tolerance to environmental stress. In *Plant Ecophysiology and Adaptation under Climate Change: Mechanisms and Perspectives II*; Springer: Berlin/Heidelberg, Germany, 2020; pp. 245–355. [CrossRef]

15. Raghavendra, A.S.; Gonugunta, V.K.; Christmann, A.; Grill, E. ABA perception and signalling. *Trends Plant Sci.* **2010**, *15*, 395–401. [CrossRef]

16. Suzuki, N.; Bassil, E.; Hamilton, J.S.; Inupakutika, M.A.; Zandalinas, S.I.; Tripathy, D.; Luo, Y.; Dion, E.; Fukui, G.; Kumazaki, A. ABA is required for plant acclimation to a combination of salt and heat stress. *PLoS ONE* **2016**, *11*, e0147625. [CrossRef]

17. Llanes, A.; Andrade, A.; Alemano, S.; Luna, V. Metabolomic approach to understand plant adaptations to water and salt stress. In *Plant Metabolites and Regulation under Environmental Stress*; Elsevier: Amsterdam, The Netherlands, 2018; pp. 133–144. [CrossRef]

18. Yu, Z.; Duan, X.; Luo, L.; Dai, S.; Ding, Z.; Xia, G. How Plant Hormones Mediate Salt Stress Responses. *Trends Plant Sci.* **2020**, *25*, 1117–1130. [CrossRef]

19. Riemann, M.; Dhakarey, R.; Hazman, M.; Miro, B.; Kohli, A.; Nick, P. Exploring jasmonates in the hormonal network of drought and salinity responses. *Front. Plant Sci.* **2015**, *6*, 1077. [CrossRef]

20. Ahmad, P.; Rasool, S.; Gul, A.; Sheikh, S.A.; Akram, N.A.; Ashraf, M.; Kazi, A.; Gucel, S. Jasmonates: Multifunctional roles in stress tolerance. *Front. Plant Sci.* **2016**, *7*, 813. [CrossRef]

21. Wang, J.; Song, L.; Gong, X.; Xu, J.; Li, M. Functions of jasmonic acid in plant regulation and response to abiotic stress. *Int. J. Mol. Sci.* **2020**, *21*, 1446. [CrossRef]

22. Campos, M.L.; Kang, J.-H.; Howe, G.A. Jasmonate-triggered plant immunity. *J. Chem. Ecol.* **2014**, *40*, 657–675. [CrossRef]

23. Wang, J.; Wu, D.; Wang, Y.; Xie, D. Jasmonate action in plant defense against insects. *J. Exp. Bot.* **2019**, *70*, 3391–3400. [CrossRef]

24. Ismail, A.; Seo, M.; Takebayashi, Y.; Kamiya, Y.; Eiche, E.; Nick, P. Salt adaptation requires efficient fine-tuning of jasmonate signalling. *Protoplasma* **2014**, *251*, 881–898. [CrossRef]

25. Abouelsaad, I.; Renault, S. Enhanced oxidative stress in the jasmonic acid-deficient tomato mutant def-1 exposed to NaCl stress. *J. Plant Physiol.* **2018**, *226*, 136–144. [CrossRef]

26. Ahmad, B.; Raina, A.; Naikoo, M.I.; Khan, S. Role of methyl jasmonates in salt stress tolerance in crop plants. In *Plant Signaling Molecules*; Elsevier: Amsterdam, The Netherlands, 2019; pp. 371–384. [CrossRef]

27. Kang, D.J.; Seo, Y.J.; Lee, J.D.; Ishii, R.; Kim, K.; Shin, D.; Park, S.; Jang, S.; Lee, I.J. Jasmonic acid differentially affects growth, ion uptake and abscisic acid concentration in salt-tolerant and salt-sensitive rice cultivars. *J. Agron. Crop. Sci.* **2005**, *191*, 273–282. [CrossRef]

28. Moons, A.; Prinsen, E.; Bauw, G.; Van Montagu, M. Antagonistic effects of abscisic acid and jasmonates on salt stress-inducible transcripts in rice roots. *Plant Cell* **1997**, *9*, 2243–2259. [CrossRef]

29. Tani, T.; Sobajima, H.; Okada, K.; Chujo, T.; Arimura, S.-I.; Tsutsumi, N.; Nishimura, M.; Seto, H.; Nojiri, H.; Yamane, H. Identification of the OsOPR7 gene encoding 12-oxophytodienoate reductase involved in the biosynthesis of jasmonic acid in rice. *Planta* **2008**, *227*, 517. [CrossRef]

30. Du, H.; Liu, H.; Xiong, L. Endogenous auxin and jasmonic acid levels are differentially modulated by abiotic stresses in rice. *Front. Plant Sci.* **2013**, *4*, 397. [CrossRef] [PubMed]

31. Qiu, Z.; Guo, J.; Zhu, A.; Zhang, L.; Zhang, M. Exogenous jasmonic acid can enhance tolerance of wheat seedlings to salt stress. *Ecotoxicol. Environ. Saf.* **2014**, *104*, 202–208. [CrossRef] [PubMed]

32. Yoon, J.Y.; Hamayun, M.; Lee, S.-K.; Lee, I.-J. Methyl jasmonate alleviated salinity stress in soybean. *J. Crop. Sci. Biotechnol.* **2009**, *12*, 63–68. [CrossRef]

33. Javid, M.G.; Sorooshzadeh, A.; Moradi, F.; Modarres Sanavy, S.A.M.; Allahdadi, I. The role of phytohormones in alleviating salt stress in crop plants. *Aust. J. Crop. Sci.* **2011**, *5*, 726.

34. Yang, J.; Duan, G.; Li, C.; Liu, L.; Han, G.; Zhang, Y.; Wang, C. The crosstalks between jasmonic acid and other plant hormone signaling highlight the involvement of jasmonic acid as a core component in plant response to biotic and abiotic stresses. *Front. Plant Sci.* **2019**, *10*, 1349. [CrossRef]

35. Guo, Q.; Yoshida, Y.; Major, I.T.; Wang, K.; Sugimoto, K.; Kapali, G.; Havko, N.E.; Benning, C.; Howe, G.A. JAZ repressors of metabolic defense promote growth and reproductive fitness in Arabidopsis. *Proc. Natl. Acad. Sci. USA* **2018**, *115*, E10768–E10777. [CrossRef]

36. Jang, G.; Yoon, Y.; Choi, Y.D. Crosstalk with jasmonic acid integrates multiple responses in plant development. *Int. J. Mol. Sci.* **2020**, *21*, 305. [CrossRef]

37. Kazan, K.; Manners, J.M. JAZ repressors and the orchestration of phytohormone crosstalk. *Trends Plant Sci.* **2012**, *17*, 22–31. [CrossRef]

38. Zander, M.; Lewsey, M.G.; Clark, N.M.; Yin, L.; Bartlett, A.; Guzmán, J.P.S.; Hann, E.; Langford, A.E.; Jow, B.; Wise, A. Integrated multi-omics framework of the plant response to jasmonic acid. *Nat. Plants* **2020**, *6*, 290–302. [CrossRef]

39. Wani, S.H.; Kumar, V.; Khare, T.; Guddimalli, R.; Parveda, M.; Solymosi, K.; Suprasanna, P.; Kishor, P.K. Engineering salinity tolerance in plants: Progress and prospects. *Planta* **2020**, *251*, 1–29. [CrossRef]

40. Thomson, M.J.; De Ocampo, M.; Egdane, J.; Rahman, M.A.; Sajise, A.G.; Adorada, D.L.; Tumimbang-Raiz, E.; Blumwald, E.; Seraj, Z.I.; Singh, R.K. Characterizing the Saltol quantitative trait locus for salinity tolerance in rice. *Rice* **2010**, *3*, 148–160. [CrossRef]

41. Rohila, J.S.; Edwards, J.D.; Tran, G.D.; Jackson, A.K.; McClung, A.M. Identification of Superior Alleles for Seedling Stage Salt Tolerance in the USDA Rice Mini-Core Collection. *Plants* **2019**, *8*, 472. [CrossRef]

42. Korte, A.; Farlow, A. The advantages and limitations of trait analysis with GWAS: A review. *Plant Methods* **2013**, *9*, 1–9. [CrossRef]

43. Flint-Garcia, S.A.; Thuillet, A.C.; Yu, J.; Pressoir, G.; Romero, S.M.; Mitchell, S.E.; Doebley, J.; Kresovich, S.; Goodman, M.M.; Buckler, E.S. Maize association population: A high-resolution platform for quantitative trait locus dissection. *Plant J.* **2005**, *44*, 1054–1064. [CrossRef]
44. Huang, X.; Han, B. Natural variations and genome-wide association studies in crop plants. *Annu. Rev. Plant Biol.* **2014**, *65*, 531–551. [CrossRef]
45. Zhao, K.; Aranzana, M.J.; Kim, S.; Lister, C.; Shindo, C.; Tang, C.; Toomajian, C.; Zheng, H.; Dean, C.; Marjoram, P. An Arabidopsis example of association mapping in structured samples. *PLoS Genet.* **2007**, *3*, e4. [CrossRef]
46. Zhang, A.; Liu, Y.; Wang, F.; Li, T.; Chen, Z.; Kong, D.; Bi, J.; Zhang, F.; Luo, X.; Wang, J. Enhanced rice salinity tolerance via CRISPR/Cas9-targeted mutagenesis of the OsRR22 gene. *Mol. Breed.* **2019**, *39*, 1–10. [CrossRef]
47. Tran, M.T.; Doan, D.T.H.; Kim, J.; Song, Y.J.; Sung, Y.W.; Das, S.; Kim, E.J.; Son, G.H.; Kim, S.H.; Van Vu, T. CRISPR/Cas9-based precise excision of SlHyPRP1 domain (s) to obtain salt stress-tolerant tomato. *Plant Cell Rep.* **2020**, 1–13. [CrossRef]
48. Kazan, K. Diverse roles of jasmonates and ethylene in abiotic stress tolerance. *Trends Plant Sci.* **2015**, *20*, 219–229. [CrossRef] [PubMed]
49. Wasternack, C.; Strnad, M. Jasmonates: News on occurrence, biosynthesis, metabolism and action of an ancient group of signaling compounds. *Int. J. Mol. Sci.* **2018**, *19*, 2539. [CrossRef] [PubMed]
50. Wasternack, C.; Feussner, I. The oxylipin pathways: Biochemistry and function. *Annu. Rev. Plant Biol.* **2018**, *69*, 363–386. [CrossRef] [PubMed]
51. Wasternack, C.; Kombrink, E. Jasmonates: Structural requirements for lipid-derived signals active in plant stress responses and development. *ACS Chem. Biol.* **2010**, *5*, 63–77. [CrossRef] [PubMed]
52. Seo, H.S.; Song, J.T.; Cheong, J.-J.; Lee, Y.-H.; Lee, Y.-W.; Hwang, I.; Lee, J.S.; Do Choi, Y. Jasmonic acid carboxyl methyltransferase: A key enzyme for jasmonate-regulated plant responses. *Proc. Natl. Acad. Sci. USA* **2001**, *98*, 4788–4793. [CrossRef]
53. Staswick, P.E.; Tiryaki, I. The oxylipin signal jasmonic acid is activated by an enzyme that conjugates it to isoleucine in Arabidopsis. *Plant Cell* **2004**, *16*, 2117–2127. [CrossRef]
54. Koo, Y.J.; Yoon, E.S.; Seo, J.S.; Kim, J.-K.; Do Choi, Y. Characterization of a methyl jasmonate specific esterase in Arabidopsis. *J. Korean Soc. Appl. Biol. Chem.* **2013**, *56*, 27–33. [CrossRef]
55. Woldemariam, M.G.; Onkokesung, N.; Baldwin, I.T.; Galis, I. Jasmonoyl-l-isoleucine hydrolase 1 (JIH1) regulates jasmonoyl-l-isoleucine levels and attenuates plant defenses against herbivores. *Plant J.* **2012**, *72*, 758–767. [CrossRef]
56. Fonseca, S.; Chini, A.; Hamberg, M.; Adie, B.; Porzel, A.; Kramell, R.; Miersch, O.; Wasternack, C.; Solano, R. (+)-7-iso-Jasmonoyl-L-isoleucine is the endogenous bioactive jasmonate. *Nat. Chem. Biol.* **2009**, *5*, 344–350. [CrossRef]
57. Chini, A.; Gimenez-Ibanez, S.; Goossens, A.; Solano, R. Redundancy and specificity in jasmonate signalling. *Curr. Opin. Plant Biol.* **2016**, *33*, 147–156. [CrossRef]
58. Chini, A.; Fonseca, S.; Fernandez, G.; Adie, B.; Chico, J.; Lorenzo, O.; Garcia-Casado, G.; López-Vidriero, I.; Lozano, F.; Ponce, M. The JAZ family of repressors is the missing link in jasmonate signalling. *Nature* **2007**, *448*, 666–671. [CrossRef]
59. Sheard, L.B.; Tan, X.; Mao, H.; Withers, J.; Ben-Nissan, G.; Hinds, T.R.; Kobayashi, Y.; Hsu, F.-F.; Sharon, M.; Browse, J. Jasmonate perception by inositol-phosphate-potentiated COI1–JAZ co-receptor. *Nature* **2010**, *468*, 400–405. [CrossRef]
60. Thines, B.; Katsir, L.; Melotto, M.; Niu, Y.; Mandaokar, A.; Liu, G.; Nomura, K.; He, S.Y.; Howe, G.A.; Browse, J. JAZ repressor proteins are targets of the SCF COI1 complex during jasmonate signalling. *Nature* **2007**, *448*, 661–665. [CrossRef]
61. Garrido-Bigotes, A.; Valenzuela-Riffo, F.; Figueroa, C.R. Evolutionary analysis of JAZ proteins in plants: An approach in search of the ancestral sequence. *Int. J. Mol. Sci.* **2019**, *20*, 5060. [CrossRef]
62. Garrido-Bigotes, A.; Valenzuela-Riffo, F.; Torrejón, M.; Solano, R.; Morales-Quintana, L.; Figueroa, C.R. A new functional JAZ degron sequence in strawberry JAZ1 revealed by structural and interaction studies on the COI1–JA-Ile/COR–JAZs complexes. *Sci. Rep.* **2020**, *10*, 1–17. [CrossRef]
63. Pauwels, L.; Barbero, G.F.; Geerinck, J.; Tilleman, S.; Grunewald, W.; Pérez, A.C.; Chico, J.M.; Bossche, R.V.; Sewell, J.; Gil, E. NINJA connects the co-repressor TOPLESS to jasmonate signalling. *Nature* **2010**, *464*, 788–791. [CrossRef]
64. Shyu, C.; Figueroa, P.; De Pew, C.L.; Cooke, T.F.; Sheard, L.B.; Moreno, J.E.; Katsir, L.; Zheng, N.; Browse, J.; Howe, G.A. JAZ8 lacks a canonical degron and has an EAR motif that mediates transcriptional repression of jasmonate responses in Arabidopsis. *Plant Cell* **2012**, *24*, 536–550. [CrossRef] [PubMed]
65. Zhang, F.; Yao, J.; Ke, J.; Zhang, L.; Lam, V.Q.; Xin, X.-F.; Zhou, X.E.; Chen, J.; Brunzelle, J.; Griffin, P.R. Structural basis of JAZ repression of MYC transcription factors in jasmonate signalling. *Nature* **2015**, *525*, 269–273. [CrossRef] [PubMed]
66. Çevik, V.; Kidd, B.N.; Zhang, P.; Hill, C.; Kiddle, S.; Denby, K.J.; Holub, E.B.; Cahill, D.M.; Manners, J.M.; Schenk, P.M. MEDIATOR25 acts as an integrative hub for the regulation of jasmonate-responsive gene expression in Arabidopsis. *Plant Physiol.* **2012**, *160*, 541–555. [CrossRef] [PubMed]
67. An, C.; Li, L.; Zhai, Q.; You, Y.; Deng, L.; Wu, F.; Chen, R.; Jiang, H.; Wang, H.; Chen, Q. Mediator subunit MED25 links the jasmonate receptor to transcriptionally active chromatin. *Proc. Natl. Acad. Sci. USA* **2017**, *114*, E8930–E8939. [CrossRef] [PubMed]
68. Chung, H.S.; Koo, A.J.; Gao, X.; Jayanty, S.; Thines, B.; Jones, A.D.; Howe, G.A. Regulation and function of Arabidopsis JASMONATE ZIM-domain genes in response to wounding and herbivory. *Plant Physiol.* **2008**, *146*, 952–964. [CrossRef] [PubMed]
69. Fernández-Calvo, P.; Chini, A.; Fernández-Barbero, G.; Chico, J.-M.; Gimenez-Ibanez, S.; Geerinck, J.; Eeckhout, D.; Schweizer, F.; Godoy, M.; Franco-Zorrilla, J.M. The Arabidopsis bHLH transcription factors MYC3 and MYC4 are targets of JAZ repressors and act additively with MYC2 in the activation of jasmonate responses. *Plant Cell* **2011**, *23*, 701–715. [CrossRef]

70. Figueroa, P.; Browse, J. Male sterility in A rabidopsis induced by overexpression of a MYC 5-SRDX chimeric repressor. *Plant J.* **2015**, *81*, 849–860. [CrossRef]

71. Pires, N.; Dolan, L. Origin and diversification of basic-helix-loop-helix proteins in plants. *Mol. Biol. Evol.* **2010**, *27*, 862–874. [CrossRef]

72. Lian, T.-F.; Xu, Y.-P.; Li, L.-F.; Su, X.-D. Crystal structure of tetrameric Arabidopsis MYC2 reveals the mechanism of enhanced interaction with DNA. *Cell Rep.* **2017**, *19*, 1334–1342. [CrossRef]

73. Toledo-Ortiz, G.; Huq, E.; Quail, P.H. The Arabidopsis basic/helix-loop-helix transcription factor family. *Plant Cell* **2003**, *15*, 1749–1770. [CrossRef]

74. Figueroa, P.; Browse, J. The Arabidopsis JAZ2 promoter contains a G-Box and thymidine-rich module that are necessary and sufficient for jasmonate-dependent activation by MYC transcription factors and repression by JAZ proteins. *Plant Cell Physiol.* **2012**, *53*, 330–343. [CrossRef]

75. Godoy, M.; Franco-Zorrilla, J.M.; Pérez-Pérez, J.; Oliveros, J.C.; Lorenzo, Ó.; Solano, R. Improved protein-binding microarrays for the identification of DNA-binding specificities of transcription factors. *Plant J.* **2011**, *66*, 700–711. [CrossRef]

76. Dombrecht, B.; Xue, G.P.; Sprague, S.J.; Kirkegaard, J.A.; Ross, J.J.; Reid, J.B.; Fitt, G.P.; Sewelam, N.; Schenk, P.M.; Manners, J.M. MYC2 differentially modulates diverse jasmonate-dependent functions in Arabidopsis. *Plant Cell* **2007**, *19*, 2225–2245. [CrossRef]

77. Van Moerkercke, A.; Duncan, O.; Zander, M.; Šimura, J.; Broda, M.; Bossche, R.V.; Lewsey, M.G.; Lama, S.; Singh, K.B.; Ljung, K. A MYC2/MYC3/MYC4-dependent transcription factor network regulates water spray-responsive gene expression and jasmonate levels. *Proc. Natl. Acad. Sci. USA* **2019**, *116*, 23345–23356. [CrossRef]

78. Hickman, R.; Van Verk, M.C.; Van Dijken, A.J.; Mendes, M.P.; Vroegop-Vos, I.A.; Caarls, L.; Steenbergen, M.; Van der Nagel, I.; Wesselink, G.J.; Jironkin, A. Architecture and dynamics of the jasmonic acid gene regulatory network. *Plant Cell* **2017**, *29*, 2086–2105. [CrossRef]

79. Chini, A.; Monte, I.; Zamarreño, A.M.; Hamberg, M.; Lassueur, S.; Reymond, P.; Weiss, S.; Stintzi, A.; Schaller, A.; Porzel, A.; et al. An OPR3-independent pathway uses 4,5-didehydrojasmonate for jasmonate synthesis. *Nat. Chem. Biol.* **2018**, *14*, 171–178. [CrossRef]

80. Heitz, T.; Smirnova, E.; Marquis, V.; Poirier, L. Metabolic control within the jasmonate biochemical pathway. *Plant Cell Physiol.* **2019**, *60*, 2621–2628. [CrossRef] [PubMed]

81. Kitaoka, N.; Matsubara, T.; Sato, M.; Takahashi, K.; Wakuta, S.; Kawaide, H.; Matsui, H.; Nabeta, K.; Matsuura, H. Arabidopsis CYP94B3 encodes jasmonyl-L-isoleucine 12-hydroxylase, a key enzyme in the oxidative catabolism of jasmonate. *Plant Cell Physiol.* **2011**, *52*, 1757–1765. [CrossRef]

82. Koo, A.J.; Cooke, T.F.; Howe, G.A. Cytochrome P450 CYP94B3 mediates catabolism and inactivation of the plant hormone jasmonoyl-L-isoleucine. *Proc. Natl. Acad. Sci. USA* **2011**, *108*, 9298–9303. [CrossRef]

83. Heitz, T.; Widemann, E.; Lugan, R.; Miesch, L.; Ullmann, P.; Désaubry, L.; Holder, E.; Grausem, B.; Kandel, S.; Miesch, M. Cytochromes P450 CYP94C1 and CYP94B3 catalyze two successive oxidation steps of plant hormone jasmonoyl-isoleucine for catabolic turnover. *J. Biol. Chem.* **2012**, *287*, 6296–6306. [CrossRef]

84. Widemann, E.; Miesch, L.; Lugan, R.; Holder, E.; Heinrich, C.; Aubert, Y.; Miesch, M.; Pinot, F.; Heitz, T. The amidohydrolases IAR3 and ILL6 contribute to jasmonoyl-isoleucine hormone turnover and generate 12-hydroxyjasmonic acid upon wounding in Arabidopsis leaves. *J. Biol. Chem.* **2013**, *288*, 31701–31714. [CrossRef] [PubMed]

85. Hazman, M.; Sühnel, M.; Schäfer, S.; Zumsteg, J.; Lesot, A.; Beltran, F.; Marquis, V.; Herrgott, L.; Miesch, L.; Riemann, M. Characterization of jasmonoyl-isoleucine (JA-Ile) hormonal catabolic pathways in rice upon wounding and salt stress. *Rice* **2019**, *12*, 1–14. [CrossRef] [PubMed]

86. Poudel, A.N.; Holtsclaw, R.E.; Kimberlin, A.; Sen, S.; Zeng, S.; Joshi, T.; Lei, Z.; Sumner, L.W.; Singh, K.; Matsuura, H. 12-Hydroxy-jasmonoyl-L-isoleucine is an active jasmonate that signals through CORONATINE INSENSITIVE 1 and contributes to the wound response in Arabidopsis. *Plant Cell Physiol.* **2019**, *60*, 2152–2166. [CrossRef]

87. Jimenez-Aleman, G.H.; Almeida-Trapp, M.; Fernández-Barbero, G.; Gimenez-Ibanez, S.; Reichelt, M.; Vadassery, J.; Mithöfer, A.; Caballero, J.; Boland, W.; Solano, R. Omega hydroxylated JA-Ile is an endogenous bioactive jasmonate that signals through the canonical jasmonate signaling pathway. *Biochim. Biophys. Acta Mol. Cell Biol. Lipids* **2019**, *1864*, 158520. [CrossRef]

88. Jiang, Y.; Deyholos, M.K. Comprehensive transcriptional profiling of NaCl-stressed Arabidopsis roots reveals novel classes of responsive genes. *BMC Plant Biol.* **2006**, *6*, 1–20. [CrossRef]

89. Ma, S.; Gong, Q.; Bohnert, H.J. Dissecting salt stress pathways. *J. Exp. Bot.* **2006**, *57*, 1097–1107. [CrossRef]

90. Zhang, H.; Zhang, Q.; Zhai, H.; Li, Y.; Wang, X.; Liu, Q.; He, S. Transcript profile analysis reveals important roles of jasmonic acid signalling pathway in the response of sweet potato to salt stress. *Sci. Rep.* **2017**, *7*, 1–12. [CrossRef]

91. Ding, H.; Lai, J.; Wu, Q.; Zhang, S.; Chen, L.; Dai, Y.-S.; Wang, C.; Du, J.; Xiao, S.; Yang, C. Jasmonate complements the function of Arabidopsis lipoxygenase3 in salinity stress response. *Plant Sci.* **2016**, *244*, 1–7. [CrossRef]

92. Zhao, Y.; Dong, W.; Zhang, N.; Ai, X.; Wang, M.; Huang, Z.; Xiao, L.; Xia, G. A wheat allene oxide cyclase gene enhances salinity tolerance via jasmonate signaling. *Plant Physiol.* **2014**, *164*, 1068–1076. [CrossRef]

93. Hazman, M.; Hause, B.; Eiche, E.; Nick, P.; Riemann, M. Increased tolerance to salt stress in OPDA-deficient rice ALLENE OXIDE CYCLASE mutants is linked to an increased ROS-scavenging activity. *J. Exp. Bot.* **2015**, *66*, 3339–3352. [CrossRef]

94. Abe, H.; Urao, T.; Ito, T.; Seki, M.; Shinozaki, K.; Yamaguchi-Shinozaki, K. Arabidopsis AtMYC2 (bHLH) and AtMYB2 (MYB) function as transcriptional activators in abscisic acid signaling. *Plant Cell* **2003**, *15*, 63–78. [CrossRef]

95. Valenzuela, C.E.; Acevedo-Acevedo, O.; Miranda, G.S.; Vergara-Barros, P.; Holuigue, L.; Figueroa, C.R.; Figueroa, P.M. Salt stress response triggers activation of the jasmonate signaling pathway leading to inhibition of cell elongation in Arabidopsis primary root. *J. Exp. Bot.* **2016**, *67*, 4209–4220. [CrossRef]

96. Iwasaki, T.; Yamaguchi-Shinozaki, K.; Shinozaki, K. Identification of a cis-regulatory region of a gene in Arabidopsis thaliana whose induction by dehydration is mediated by abscisic acid and requires protein synthesis. *Mol. Gen. Genet. MGG* **1995**, *247*, 391–398. [CrossRef]

97. Verma, D.; Jalmi, S.K.; Bhagat, P.K.; Verma, N.; Sinha, A.K. A bHLH transcription factor, MYC2, imparts salt intolerance by regulating proline biosynthesis in Arabidopsis. *FEBS J.* **2020**, *287*, 2560–2576. [CrossRef]

98. Seo, J.S.; Zhao, P.; Jung, C.; Chua, N.-H. Plant U-box protein 10 negatively regulates abscisic acid response in Arabidopsis. *Appl. Biol. Chem.* **2019**, *62*, 39. [CrossRef]

99. Jung, C.; Zhao, P.; Seo, J.S.; Mitsuda, N.; Deng, S.; Chua, N.-H. Plant U-box protein10 regulates MYC2 stability in Arabidopsis. *Plant Cell* **2015**, *27*, 2016–2031. [CrossRef]

100. Wasternack, C.; Hause, B. Jasmonates: Biosynthesis, perception, signal transduction and action in plant stress response, growth and development. An update to the 2007 review in Annals of Botany. *Ann. Bot.* **2013**, *111*, 1021–1058. [CrossRef]

101. Yao, D.; Zhang, X.; Zhao, X.; Liu, C.; Wang, C.; Zhang, Z.; Zhang, C.; Wei, Q.; Wang, Q.; Yan, H. Transcriptome analysis reveals salt-stress-regulated biological processes and key pathways in roots of cotton (Gossypium hirsutum L.). *Genomics* **2011**, *98*, 47–55. [CrossRef] [PubMed]

102. Chini, A.; Ben-Romdhane, W.; Hassairi, A.; Aboul-Soud, M.A. Identification of TIFY/JAZ family genes in Solanum lycopersicum and their regulation in response to abiotic stresses. *PLoS ONE* **2017**, *12*, e0177381. [CrossRef]

103. Ebel, C.; BenFeki, A.; Hanin, M.; Solano, R.; Chini, A. Characterization of wheat (Triticum aestivum) TIFY family and role of Triticum Durum Td TIFY11a in salt stress tolerance. *PLoS ONE* **2018**, *13*, e0200566. [CrossRef]

104. Ye, H.; Du, H.; Tang, N.; Li, X.; Xiong, L. Identification and expression profiling analysis of TIFY family genes involved in stress and phytohormone responses in rice. *Plant Mol. Biol.* **2009**, *71*, 291–305. [CrossRef] [PubMed]

105. Wu, H.; Ye, H.; Yao, R.; Zhang, T.; Xiong, L. OsJAZ9 acts as a transcriptional regulator in jasmonate signaling and modulates salt stress tolerance in rice. *Plant Sci.* **2015**, *232*, 1–12. [CrossRef] [PubMed]

106. Peethambaran, P.K.; Glenz, R.; Höninger, S.; Islam, S.S.; Hummel, S.; Harter, K.; Kolukisaoglu, Ü.; Meynard, D.; Guiderdoni, E.; Nick, P. Salt-inducible expression of OsJAZ8 improves resilience against salt-stress. *BMC Plant Biol.* **2018**, *18*, 1–15. [CrossRef]

107. Toda, Y.; Tanaka, M.; Ogawa, D.; Kurata, K.; Kurotani, K.-I.; Habu, Y.; Ando, T.; Sugimoto, K.; Mitsuda, N.; Katoh, E. Rice salt sensitive 3 forms a ternary complex with JAZ and class-C bHLH factors and regulates jasmonate-induced gene expression and root cell elongation. *Plant Cell* **2013**, *25*, 1709–1725. [CrossRef]

108. Zhu, D.; Cai, H.; Luo, X.; Bai, X.; Deyholos, M.K.; Chen, Q.; Chen, C.; Ji, W.; Zhu, Y. Over-expression of a novel JAZ family gene from Glycine soja, increases salt and alkali stress tolerance. *Biochem. Biophys. Res. Commun.* **2012**, *426*, 273–279. [CrossRef]

109. An, X.-H.; Hao, Y.-J.; Li, E.-M.; Xu, K.; Cheng, C.-G. Functional identification of apple MdJAZ2 in Arabidopsis with reduced JA-sensitivity and increased stress tolerance. *Plant Cell Rep.* **2017**, *36*, 255–265. [CrossRef]

110. Liu, S.; Zhang, P.; Li, C.; Xia, G. The moss jasmonate ZIM-domain protein PnJAZ1 confers salinity tolerance via crosstalk with the abscisic acid signalling pathway. *Plant Sci.* **2019**, *280*, 1–11. [CrossRef]

111. Zhao, G.; Song, Y.; Wang, Q.; Yao, D.; Li, D.; Qin, W.; Ge, X.; Yang, Z.; Xu, W.; Su, Z. Gossypium hirsutum Salt Tolerance is Enhanced by Overexpression of G. arboreum JAZ1. *Front. Bioeng. Biotechnol.* **2020**, *8*, 157. [CrossRef]

112. Kurotani, K.-I.; Hayashi, K.; Hatanaka, S.; Toda, Y.; Ogawa, D.; Ichikawa, H.; Ishimaru, Y.; Tashita, R.; Suzuki, T.; Ueda, M. Elevated levels of CYP94 family gene expression alleviate the jasmonate response and enhance salt tolerance in rice. *Plant Cell Physiol.* **2015**, *56*, 779–789. [CrossRef]

113. Kurotani, K.-I.; Yamanaka, K.; Toda, Y.; Ogawa, D.; Tanaka, M.; Kozawa, H.; Nakamura, H.; Hakata, M.; Ichikawa, H.; Hattori, T. Stress tolerance profiling of a collection of extant salt-tolerant rice varieties and transgenic plants overexpressing abiotic stress tolerance genes. *Plant Cell Physiol.* **2015**, *56*, 1867–1876. [CrossRef]

114. Zhu, D.; Bai, X.; Luo, X.; Chen, Q.; Cai, H.; Ji, W.; Zhu, Y. Identification of wild soybean (Glycine soja) TIFY family genes and their expression profiling analysis under bicarbonate stress. *Plant Cell Rep.* **2013**, *32*, 263–272. [CrossRef] [PubMed]

115. Luo, X.; Li, C.; He, X.; Zhang, X.; Zhu, L. ABA signaling is negatively regulated by GbWRKY1 through JAZ1 and ABI1 to affect salt and drought tolerance. *Plant Cell Rep.* **2020**, *39*, 181–194. [CrossRef] [PubMed]

116. Howe, G.A.; Major, I.T.; Koo, A.J. Modularity in jasmonate signaling for multistress resilience. *Annu. Rev. Plant Biol.* **2018**, *69*, 387–415. [CrossRef] [PubMed]

117. Jamalian, S.; Truemper, C.; Pawelzik, E. Jasmonic and Abscisic Acid Contribute to Metabolism Re-adjustment in Strawberry Leaves under NaCl Stress. *Int. J. Fruit Sci.* **2020**, *20*, 1–22. [CrossRef]

118. Yang, T.; Lv, R.; Li, J.; Lin, H.; Xi, D. Phytochrome A and B negatively regulate salt stress tolerance of Nicotiana tobacum via ABA-jasmonic acid synergistic cross-talk. *Plant Cell Physiol.* **2018**, *59*, 2381–2393. [CrossRef]

119. De Lucas, M.; Brady, S.M. Gene regulatory networks in the Arabidopsis root. *Curr. Opin. Plant Biol.* **2013**, *16*, 50–55. [CrossRef]

120. Lorenzo, O.; Chico, J.M.; Sánchez-Serrano, J.J.; Solano, R. Jasmonate-insensitive 1 encodes a MYC transcription factor essential to discriminate between different jasmonate-regulated defense responses in Arabidopsis. *Plant Cell* **2004**, *16*, 1938–1950. [CrossRef]

121. Chen, R.; Jiang, H.; Li, L.; Zhai, Q.; Qi, L.; Zhou, W.; Liu, X.; Li, H.; Zheng, W.; Sun, J. The Arabidopsis mediator subunit MED25 differentially regulates jasmonate and abscisic acid signaling through interacting with the MYC2 and ABI5 transcription factors. *Plant Cell* **2012**, *24*, 2898–2916. [CrossRef]

122. Aleman, F.; Yazaki, J.; Lee, M.; Takahashi, Y.; Kim, A.Y.; Li, Z.; Kinoshita, T.; Ecker, J.R.; Schroeder, J.I. An ABA-increased interaction of the PYL6 ABA receptor with MYC2 transcription factor: A putative link of ABA and JA signaling. *Sci. Rep.* **2016**, *6*, 1–10. [CrossRef]

123. Lackman, P.; González-Guzmán, M.; Tilleman, S.; Carqueijeiro, I.; Pérez, A.C.; Moses, T.; Seo, M.; Kanno, Y.; Häkkinen, S.T.; Van Montagu, M.C. Jasmonate signaling involves the abscisic acid receptor PYL4 to regulate metabolic reprogramming in Arabidopsis and tobacco. *Proc. Natl. Acad. Sci. USA* **2011**, *108*, 5891–5896. [CrossRef]

124. Zhao, F.; Li, G.; Hu, P.; Zhao, X.; Li, L.; Wei, W.; Feng, J.; Zhou, H. Identification of basic/helix-loop-helix transcription factors reveals candidate genes involved in anthocyanin biosynthesis from the strawberry white-flesh mutant. *Sci. Rep.* **2018**, *8*, 1–15. [CrossRef]

125. Pauwels, L.; Ritter, A.; Goossens, J.; Durand, A.N.; Liu, H.; Gu, Y.; Geerinck, J.; Boter, M.; Bossche, R.V.; De Clercq, R. The ring e3 ligase keep on going modulates jasmonate zim-domain12 stability. *Plant Physiol.* **2015**, *169*, 1405–1417. [CrossRef]

126. Ju, L.; Jing, Y.; Shi, P.; Liu, J.; Chen, J.; Yan, J.; Chu, J.; Chen, K.M.; Sun, J. JAZ proteins modulate seed germination through interaction with ABI 5 in bread wheat and Arabidopsis. *N. Phytol.* **2019**, *223*, 246–260. [CrossRef]

127. Kim, J.H.; Chung, K.M.; Woo, H.R. Three positive regulators of leaf senescence in Arabidopsis, ORE1, ORE3 and ORE9, play roles in crosstalk among multiple hormone-mediated senescence pathways. *Genes Genom.* **2011**, *33*, 373–381. [CrossRef]

128. Lim, C.; Kang, K.; Shim, Y.; Sakuraba, Y.; An, G.; Paek, N.-C. Rice ethylene response factor 101 promotes leaf senescence through Jasmonic acid-mediated regulation of OsNAP and OsMYC2. *Front. Plant Sci.* **2020**, *11*, 1096. [CrossRef] [PubMed]

129. Song, S.; Huang, H.; Gao, H.; Wang, J.; Wu, D.; Liu, X.; Yang, S.; Zhai, Q.; Li, C.; Qi, T. Interaction between MYC2 and ethylene insensitive 3 modulates antagonism between jasmonate and ethylene signaling in Arabidopsis. *Plant Cell* **2014**, *26*, 263–279. [CrossRef]

130. Lorenzo, O.; Piqueras, R.; Sánchez-Serrano, J.J.; Solano, R. Ethylene response factor 1 integrates signals from ethylene and jasmonate pathways in plant defense. *Plant Cell* **2003**, *15*, 165–178. [CrossRef]

131. Pré, M.; Atallah, M.; Champion, A.; De Vos, M.; Pieterse, C.M.; Memelink, J. The AP2/ERF domain transcription factor ORA59 integrates jasmonic acid and ethylene signals in plant defense. *Plant Physiol.* **2008**, *147*, 1347–1357. [CrossRef]

132. Ma, F.; Yang, X.; Shi, Z.; Miao, X. Novel crosstalk between ethylene-and jasmonic acid-pathway responses to a piercing-sucking insect in rice. *N. Phytol.* **2020**, *225*, 474–487. [CrossRef]

133. Cheng, M.-C.; Liao, P.-M.; Kuo, W.-W.; Lin, T.-P. The Arabidopsis ethylene response factor 1 regulates abiotic stress-responsive gene expression by binding to different cis-acting elements in response to different stress signals. *Plant Physiol.* **2013**, *162*, 1566–1582. [CrossRef]

134. Takeuchi, K.; Hasegawa, H.; Gyohda, A.; Komatsu, S.; Okamoto, T.; Okada, K.; Terakawa, T.; Koshiba, T. Overexpression of RSOsPR10, a root-specific rice PR10 gene, confers tolerance against drought stress in rice and drought and salt stresses in bentgrass. *Plant Cell Tissue Organ Cult.* **2016**, *127*, 35–46. [CrossRef]

135. Hashimoto, M.; Kisseleva, L.; Sawa, S.; Furukawa, T.; Komatsu, S.; Koshiba, T. A novel rice PR10 protein, RSOsPR10, specifically induced in roots by biotic and abiotic stresses, possibly via the jasmonic acid signaling pathway. *Plant Cell Physiol.* **2004**, *45*, 550–559. [CrossRef] [PubMed]

136. Takeuchi, K.; Gyohda, A.; Tominaga, M.; Kawakatsu, M.; Hatakeyama, A.; Ishii, N.; Shimaya, K.; Nishimura, T.; Riemann, M.; Nick, P. RSOsPR10 expression in response to environmental stresses is regulated antagonistically by jasmonate/ethylene and salicylic acid signaling pathways in rice roots. *Plant Cell Physiol.* **2011**, *52*, 1686–1696. [CrossRef] [PubMed]

137. Yamamoto, T.; Yoshida, Y.; Nakajima, K.; Tominaga, M.; Gyohda, A.; Suzuki, H.; Okamoto, T.; Nishimura, T.; Yokotani, N.; Minami, E. Expression of RSOsPR10 in rice roots is antagonistically regulated by jasmonate/ethylene and salicylic acid via the activator OsERF87 and the repressor OsWRKY76, respectively. *Plant Direct* **2018**, *2*, e00049. [CrossRef]

138. Yan, Q.; Cui, X.; Lin, S.; Gan, S.; Xing, H.; Dou, D. GmCYP82A3, a soybean cytochrome P450 family gene involved in the jasmonic acid and ethylene signaling pathway, enhances plant resistance to biotic and abiotic stresses. *PLoS ONE* **2016**, *11*, e0162253. [CrossRef]

139. Aerts, N.; Pereira Mendes, M.; Van Wees, S.C. Multiple levels of crosstalk in hormone networks regulating plant defense. *Plant J.* **2021**, *105*, 489–504. [CrossRef]

140. Brodersen, P.; Petersen, M.; Bjørn Nielsen, H.; Zhu, S.; Newman, M.A.; Shokat, K.M.; Rietz, S.; Parker, J.; Mundy, J. Arabidopsis MAP kinase 4 regulates salicylic acid-and jasmonic acid/ethylene-dependent responses via EDS1 and PAD4. *Plant J.* **2006**, *47*, 532–546. [CrossRef]

141. Leon-Reyes, A.; Van der Does, D.; De Lange, E.S.; Delker, C.; Wasternack, C.; Van Wees, S.C.M.; Ritsema, T.; Pieterse, C.M.J. Salicylate-mediated suppression of jasmonate-responsive gene expression in Arabidopsis is targeted downstream of the jasmonate biosynthesis pathway. *Planta* **2010**, *232*, 1423–1432. [CrossRef]

142. Meldau, S.; Ullman-Zeunert, L.; Govind, G.; Bartram, S.; Baldwin, I.T. MAPK-dependent JA and SA signalling in Nicotiana attenuataaffects plant growth and fitness during competition with conspecifics. *BMC Plant Biol.* **2012**, *12*, 213. [CrossRef]

143. Zander, M.; Chen, S.; Imkampe, J.; Thurow, C.; Gatz, C. Repression of the Arabidopsis thaliana jasmonic acid/ethylene-induced defense pathway by TGA-interacting glutaredoxins depends on their C-terminal ALWL motif. *Mol. Plant* **2012**, *5*, 831–840. [CrossRef]

144. Khan, M.I.R.; Syeed, S.; Nazar, R.; Anjum, N.A. An insight into the role of salicylic acid and jasmonic acid in salt stress tolerance. *Phytohorm. Abiotic Stress Toler. Plants* **2012**, 277–300. [CrossRef]

145. Faghih, S.; Ghobadi, C.; Zarei, A. Response of strawberry plant cv'Camarosa'to salicylic acid and methyl jasmonate application under salt stress condition. *J. Plant Growth Regul.* **2017**, *36*, 651–659. [CrossRef]

146. Sheokand, M.; Jakhar, S.; Singh, V.; Sikerwal, V. Effect of salicylic acid, 24-Epibrassinolide and jasmonic acid in modulating morpho-physiological and biochemical constituents in Glycine max L. merill under salt stress. *Int. J. Res. Anal. Rev.* **2018**, *5*, i434–i442.

147. Ghassemi-Golezani, K.; Farhangi-Abriz, S. Foliar sprays of salicylic acid and jasmonic acid stimulate H+-ATPase activity of tonoplast, nutrient uptake and salt tolerance of soybean. *Ecotoxicol. Environ. Saf.* **2018**, *166*, 18–25. [CrossRef]

148. Farhangi-Abriz, S.; Tavasolee, A.; Ghassemi-Golezani, K.; Torabian, S.; Monirifar, H.; Rahmani, H.A. Growth-promoting bacteria and natural regulators mitigate salt toxicity and improve rapeseed plant performance. *Protoplasma* **2020**, *257*, 1035–1047. [CrossRef]

149. Seckin-Dinler, B.; Tasci, E.; Sarisoy, U.; Gul, V. The cooperation between methyl jasmonate and salicylic acid to protect soybean (Glycine max L.) from salinity. *Fresenius Environ. Bull.* **2018**, *27*, 1618–1626.

150. Hou, X.; Lee, L.Y.C.; Xia, K.; Yan, Y.; Yu, H. Dellas modulate jasmonate signaling via competitive binding to JAZs. *Dev. Cell* **2010**, *19*, 884–894. [CrossRef]

151. Boter, M.; Ruíz-Rivero, O.; Abdeen, A.; Prat, S. Conserved MYC transcription factors play a key role in jasmonate signaling both in tomato and Arabidopsis. *Genes Dev.* **2004**, *18*, 1577–1591. [CrossRef]

152. Yang, D.-L.; Yao, J.; Mei, C.-S.; Tong, X.-H.; Zeng, L.-J.; Li, Q.; Xiao, L.-T.; Sun, T.-P.; Li, J.; Deng, X.-W. Plant hormone jasmonate prioritizes defense over growth by interfering with gibberellin signaling cascade. *Proc. Natl. Acad. Sci. USA* **2012**, *109*, E1192–E1200. [CrossRef]

153. Wild, M.; Davière, J.-M.; Cheminant, S.; Regnault, T.; Baumberger, N.; Heintz, D.; Baltz, R.; Genschik, P.; Achard, P. The Arabidopsis DELLA RGA-LIKE3 is a direct target of MYC2 and modulates jasmonate signaling responses. *Plant Cell* **2012**, *24*, 3307–3319. [CrossRef]

154. Xia, X.-C.; Hu, Q.-Q.; Li, W.; Chen, Y.; Han, L.-H.; Tao, M.; Wu, W.-Y.; Li, X.-B.; Huang, G.-Q. Cotton (Gossypium hirsutum) JAZ3 and SLR1 function in jasmonate and gibberellin mediated epidermal cell differentiation and elongation. *Plant Cell Tissue Organ Cult.* **2018**, *133*, 249–262. [CrossRef]

155. Al Timman, W.M.A.; Kamil, T.A. Alleviation of salt stress in Ocimum basilicum plants by jasmonic acid treatment. *Plant Arch.* **2019**, *19*, 1550–1557.

156. Gao, Z.; Gao, S.; Li, P.; Zhang, Y.; Ma, B.; Wang, Y. Exogenous methyl jasmonate promotes salt stress-induced growth inhibition and prioritizes defense response of Nitraria tangutorum Bobr. *Physiol. Plant* **2021**, 1–14. [CrossRef]

157. Stoynova-Bakalova, E.; Petrov, P.; Gigova, L.; Baskin, T. Differential effects of methyl jasmonate on growth and division of etiolated zucchini cotyledons. *Plant Biol.* **2008**, *10*, 476–484. [CrossRef]

158. Van Doorn, W.G.; Çelikel, F.G.; Pak, C.; Harkema, H. Delay of Iris flower senescence by cytokinins and jasmonates. *Physiol. Plant* **2013**, *148*, 105–120. [CrossRef]

159. Mukherjee, I.; Reid, D.M.; Naik, G.R. Influence of cytokinins on the methyl jasmonate-promoted senescence in Helianthus annuus cotyledons. *Plant Growth Regul.* **2002**, *38*, 61–68. [CrossRef]

160. Jang, G.; Chang, S.H.; Um, T.Y.; Lee, S.; Kim, J.-K.; Choi, Y.D. Antagonistic interaction between jasmonic acid and cytokinin in xylem development. *Sci. Rep.* **2017**, *7*, 10212. [CrossRef]

161. Ananieva, K.; Malbeck, J.; Kamínek, M.; Van Staden, J. Methyl jasmonate down-regulates endogenous cytokinin levels in cotyledons of Cucurbita pepo (zucchini) seedlings. *Physiol. Plant* **2004**, *122*, 496–503. [CrossRef]

162. Shi, X.; Gupta, S.; Lindquist, I.E.; Cameron, C.T.; Mudge, J.; Rashotte, A.M. Transcriptome analysis of cytokinin response in tomato leaves. *PLoS ONE* **2013**, *8*, e55090. [CrossRef]

163. Avalbaev, A.; Yuldashev, R.; Fedorova, K.; Somov, K.; Vysotskaya, L.; Allagulova, C.; Shakirova, F. Exogenous methyl jasmonate regulates cytokinin content by modulating cytokinin oxidase activity in wheat seedlings under salinity. *J. Plant Physiol.* **2016**, *191*, 101–110. [CrossRef] [PubMed]

164. Vysotskaya, L.B.; Korobova, A.V.; Veselov, S.Y.; Dodd, I.C.; Kudoyarova, G.R. ABA mediation of shoot cytokinin oxidase activity: Assessing its impacts on cytokinin status and biomass allocation of nutrient-deprived durum wheat. *Funct. Plant Biol.* **2009**, *36*, 66–72. [CrossRef]

165. Tavallali, V.; Karimi, S. Methyl jasmonate enhances salt tolerance of almond rootstocks by regulating endogenous phytohormones, antioxidant activity and gas-exchange. *J. Plant Physiol.* **2019**, *234*, 98–105. [CrossRef]

166. Avalbaev, A.; Allagulova, C.; Maslennikova, D.; Fedorova, K.; Shakirova, F. Methyl jasmonate and cytokinin mitigate the salinity-induced oxidative injury in wheat seedlings. *J. Plant Growth Regul.* **2020**, 1–12. [CrossRef]

167. Tiryaki, I.; Staswick, P.E. An Arabidopsis mutant defective in jasmonate response is allelic to the auxin-signaling mutant axr1. *Plant Physiol.* **2002**, *130*, 887–894. [CrossRef]

168. Hentrich, M.; Böttcher, C.; Düchting, P.; Cheng, Y.; Zhao, Y.; Berkowitz, O.; Masle, J.; Medina, J.; Pollmann, S. The jasmonic acid signaling pathway is linked to auxin homeostasis through the modulation of YUCCA 8 and YUCCA 9 gene expression. *Plant J.* **2013**, *74*, 626–637. [CrossRef]

169. Cai, X.-T.; Xu, P.; Zhao, P.-X.; Liu, R.; Yu, L.-H.; Xiang, C.-B. Arabidopsis ERF109 mediates cross-talk between jasmonic acid and auxin biosynthesis during lateral root formation. *Nat. Commun.* **2014**, *5*, 1–13. [CrossRef] [PubMed]

170. Xu, P.; Zhao, P.-X.; Cai, X.-T.; Mao, J.-L.; Miao, Z.-Q.; Xiang, C.-B. Integration of jasmonic acid and ethylene into auxin signaling in root development. *Front. Plant Sci.* **2020**, *11*, 271. [CrossRef]

171. Zhang, T.; Poudel, A.N.; Jewell, J.B.; Kitaoka, N.; Staswick, P.; Matsuura, H.; Koo, A.J. Hormone crosstalk in wound stress response: Wound-inducible amidohydrolases can simultaneously regulate jasmonate and auxin homeostasis in Arabidopsis thaliana. *J. Exp. Bot.* **2016**, *67*, 2107–2120. [CrossRef]

172. Jiang, Y.; Liang, G.; Yang, S.; Yu, D. Arabidopsis WRKY57 functions as a node of convergence for jasmonic acid-and auxin-mediated signaling in jasmonic acid-induced leaf senescence. *Plant Cell* **2014**, *26*, 230–245. [CrossRef]

173. Jiang, Y.; Liang, G.; Yu, D. Activated expression of WRKY57 confers drought tolerance in Arabidopsis. *Mol. Plant* **2012**, *5*, 1375–1388. [CrossRef]

174. Ahmadi, F.; Karimi, K.; Struik, P. Effect of exogenous application of methyl jasmonate on physiological and biochemical characteristics of Brassica napus L. cv. Talaye under salinity stress. *South. Afr. J. Bot.* **2018**, *115*, 5–11. [CrossRef]

175. Valenzuela-Riffo, F.; Zúñiga, P.E.; Morales-Quintana, L.; Lolas, M.; Cáceres, M.; Figueroa, C.R. Priming of defense systems and upregulation of MYC2 and JAZ1 genes after botrytis cinerea inoculation in methyl Jasmonate-treated strawberry fruits. *Plants* **2020**, *9*, 447. [CrossRef] [PubMed]

176. Zuñiga, P.E.; Castañeda, Y.; Arrey-Salas, O.; Fuentes, L.; Aburto, F.; Figueroa, C.R. Methyl Jasmonate Applications from Flowering to Ripe Fruit Stages of Strawberry (Fragaria × ananassa 'Camarosa') Reinforce the Fruit Antioxidant Response at Post-harvest. *Front. Plant Sci.* **2020**, *11*, 538. [CrossRef] [PubMed]

177. Maksymiec, W.; Krupa, Z. The in vivo and in vitro influence of methyl jasmonate on oxidative processes in Arabidopsis thaliana leaves. *Acta Physiol. Plant* **2002**, *24*, 351–357. [CrossRef]

178. Popova, L.; Ananieva, E.; Hristova, V.; Christov, K.; Georgieva, K.; Alexieva, V.; Stoinova, Z. Salicylic acid-and methyl jasmonate-induced protection on photosynthesis to paraquat oxidative stress. *Bulg J. Plant Physiol.* **2003**, *133*, 152.

179. Singh, I.; Shah, K. Exogenous application of methyl jasmonate lowers the effect of cadmium-induced oxidative injury in rice seedlings. *Phytochemistry* **2014**, *108*, 57–66. [CrossRef]

180. Jung, S. Effect of chlorophyll reduction in Arabidopsis thaliana by methyl jasmonate or norflurazon on antioxidant systems. *Plant Physiol. Biochem.* **2004**, *42*, 225–231. [CrossRef]

181. Turan, M.A.; Elkarim, A.H.A.; Taban, N.; Taban, S. Effect of salt stress on growth, stomatal resistance, proline and chlorophyll concentrations on maize plant. *Afr. J. Agric. Res.* **2009**, *4*, 893–897. [CrossRef]

182. Shahbaz, M.; Ashraf, M.; Akram, N.A.; Hanif, A.; Hameed, S.; Joham, S.; Rehman, R. Salt-induced modulation in growth, photosynthetic capacity, proline content and ion accumulation in sunflower (Helianthus annuus L.). *Acta Physiol. Plant* **2011**, *33*, 1113–1122. [CrossRef]

183. Fedina, I.; Tsonev, T. Effect of pretreatment with methyl jasmonate on the response of Pisum sativum to salt stress. *J. Plant Physiol.* **1997**, *151*, 735–740. [CrossRef]

184. Velitchkova, M.; Fedina, I. Response of photosynthesis of Pisum sativum to salt stress as affected by methyl jasmonate. *Photosynthetica* **1998**, *35*, 89–97. [CrossRef]

185. Del Amor, F.M.; Cuadra-Crespo, P. Alleviation of salinity stress in broccoli using foliar urea or methyl-jasmonate: Analysis of growth, gas exchange, and isotope composition. *Plant Growth Regul.* **2011**, *63*, 55–62. [CrossRef]

186. Sadeghipour, O. Amelioration of salinity tolerance in cowpea plants by seed treatment with methyl jasmonate. *Legume Res. Int. J.* **2017**, *40*. [CrossRef]

187. Yuan, F.; Liang, X.; Li, Y.; Yin, S.; Wang, B. Methyl jasmonate improves tolerance to high salt stress in the recretohalophyte Limonium bicolor. *Funct. Plant Biol.* **2019**, *46*, 82–92. [CrossRef] [PubMed]

188. Jiang, M.; Xu, F.; Peng, M.; Huang, F.; Meng, F. Methyl jasmonate regulated diploid and tetraploid black locust (*Robinia pseudoacacia* L.) tolerance to salt stress. *Acta Physiol. Plant* **2016**, *38*, 106. [CrossRef]

189. Talebi, M.; Moghaddam, M.; Pirbalouti, A.G. Methyl jasmonate effects on volatile oil compounds and antioxidant activity of leaf extract of two basil cultivars under salinity stress. *Acta Physiol. Plant* **2018**, *40*, 1–11. [CrossRef]

190. Yu, X.; Fei, P.; Xie, Z.; Zhang, W.; Zhao, Q.; Zhang, X. Effects of methyl jasmonate on growth, antioxidants, and carbon and nitrogen metabolism of Glycyrrhiza uralensis under salt stress. *Biol. Plant* **2019**, *63*, 89–96. [CrossRef]

191. Sheyhakinia, S.; Bamary, Z.; Einali, A.; Valizadeh, J. The induction of salt stress tolerance by jasmonic acid treatment in roselle (Hibiscus sabdariffa L.) seedlings through enhancing antioxidant enzymes activity and metabolic changes. *Biologia* **2020**, *75*, 681–692. [CrossRef]

192. Ali, A.Y.A.; Ibrahim, M.E.H.; Zhou, G.; Nimir, N.E.A.; Jiao, X.; Zhu, G.; Elsiddig, A.M.I.; Suliman, M.S.E.; Elradi, S.B.M.; Yue, W. Exogenous jasmonic acid and humic acid increased salinity tolerance of sorghum. *Agron. J.* **2020**, *112*, 871–884. [CrossRef]

193. Sheteiwy, M.S.; Shao, H.; Qi, W.; Daly, P.; Sharma, A.; Shaghaleh, H.; Hamoud, Y.A.; El-Esawi, M.A.; Pan, R.; Wan, Q. Seed priming and foliar application with jasmonic acid enhance salinity stress tolerance of soybean (Glycine max L.) seedlings. *J. Sci. Food Agric.* **2021**, *101*, 2027–2041. [CrossRef]

194. Takahashi, H.; Kanayama, Y.; Zheng, M.S.; Kusano, T.; Hase, S.; Ikegami, M.; Shah, J. Antagonistic interactions between the SA and JA signaling pathways in Arabidopsis modulate expression of defense genes and gene-for-gene resistance to cucumber mosaic virus. *Plant Cell Physiol.* **2004**, *45*, 803–809. [CrossRef] [PubMed]

195. Yasir, M.; He, S.; Sun, G.; Geng, X.; Pan, Z.; Gong, W.; Jia, Y.; Du, X. A Genome-Wide Association Study revealed key SNPs/genes associated with salinity stress tolerance in upland cotton. *Genes* **2019**, *10*, 829. [CrossRef] [PubMed]

196. Hossain, H.; Rahman, M.; Alam, M.; Singh, R. Mapping of quantitative trait loci associated with reproductive-stage salt tolerance in rice. *J. Agron. Crop. Sci.* **2015**, *201*, 17–31. [CrossRef]

197. Quan, R.; Wang, J.; Hui, J.; Bai, H.; Lyu, X.; Zhu, Y.; Zhang, H.; Zhang, Z.; Li, S.; Huang, R. Improvement of salt tolerance using wild rice genes. *Front. Plant Sci.* **2018**, *8*, 2269. [CrossRef]

198. Ahmadi-Ochtapeh, H.; Soltanloo, H.; Ramezanpour, S.; Naghavi, M.; Nikkhah, H.; Rad, S.Y. QTL mapping for salt tolerance in barley at seedling growth stage. *Biol. Plant* **2015**, *59*, 283–290. [CrossRef]

199. Fan, Y.; Zhou, G.; Shabala, S.; Chen, Z.-H.; Cai, S.; Li, C.; Zhou, M. Genome-wide association study reveals a new QTL for salinity tolerance in barley (Hordeum vulgare L.). *Front. Plant Sci.* **2016**, *7*, 946. [CrossRef]

200. Díaz De León, J.L.; Escoppinichi, R.; Geraldo, N.; Castellanos, T.; Mujeeb-Kazi, A.; Röder, M.S. Quantitative trait loci associated with salinity tolerance in field grown bread wheat. *Euphytica* **2011**, *181*, 371–383. [CrossRef]

201. Xu, Y.; Li, S.; Li, L.; Zhang, X.; Xu, H.; An, D. Mapping QTL s for salt tolerance with additive, epistatic and QTL × treatment interaction effects at seedling stage in wheat. *Plant Breed.* **2013**, *132*, 276–283. [CrossRef]

202. Vadez, V.; Krishnamurthy, L.; Thudi, M.; Anuradha, C.; Colmer, T.D.; Turner, N.C.; Siddique, K.H.M.; Gaur, P.M.; Varshney, R.K. Assessment of ICCV 2 × JG 62 chickpea progenies shows sensitivity of reproduction to salt stress and reveals QTL for seed yield and yield components. *Mol. Breed.* **2012**, *30*, 9–21. [CrossRef]

203. Li, D.; Dossa, K.; Zhang, Y.; Wei, X.; Wang, L.; Zhang, Y.; Liu, A.; Zhou, R.; Zhang, X. GWAS uncovers differential genetic bases for drought and salt tolerances in sesame at the germination stage. *Genes* **2018**, *9*, 87. [CrossRef]

204. Sun, Z.; Li, H.; Zhang, Y.; Li, Z.; Ke, H.; Wu, L.; Zhang, G.; Wang, X.; Ma, Z. Identification of SNPs and candidate genes associated with salt tolerance at the seedling stage in cotton (Gossypium hirsutum L.). *Front. Plant Sci.* **2018**, *9*, 1011. [CrossRef]

205. Yuan, Y.; Xing, H.; Zeng, W.; Xu, J.; Mao, L.; Wang, L.; Feng, W.; Tao, J.; Wang, H.; Zhang, H.; et al. Genome-wide association and differential expression analysis of salt tolerance in Gossypium hirsutum L at the germination stage. *BMC Plant Biol.* **2019**, *19*, 394. [CrossRef]

206. Do, T.D.; Vuong, T.D.; Dunn, D.; Clubb, M.; Valliyodan, B.; Patil, G.; Chen, P.; Xu, D.; Nguyen, H.T.; Shannon, J.G. Identification of new loci for salt tolerance in soybean by high-resolution genome-wide association mapping. *BMC Genom.* **2019**, *20*, 318. [CrossRef]

207. Hazzouri, K.M.; Khraiwesh, B.; Amiri, K.; Pauli, D.; Blake, T.; Shahid, M.; Mullath, S.K.; Nelson, D.; Mansour, A.L.; Salehi-Ashtiani, K. Mapping of HKT1; 5 gene in barley using GWAS approach and its implication in salt tolerance mechanism. *Front. Plant Sci.* **2018**, *9*, 156. [CrossRef]

208. Patishtan, J.; Hartley, T.N.; Fonseca de Carvalho, R.; Maathuis, F.J. Genome-wide association studies to identify rice salt-tolerance markers. *Plant Cell Environ.* **2018**, *41*, 970–982. [CrossRef]

209. Luo, X.; Wang, B.; Gao, S.; Zhang, F.; Terzaghi, W.; Dai, M. Genome-wide association study dissects the genetic bases of salt tolerance in maize seedlings. *J. Integr. Plant Biol.* **2019**, *61*, 658–674. [CrossRef]

210. Shim, J.S.; Jung, C.; Lee, S.; Min, K.; Lee, Y.W.; Choi, Y.; Lee, J.S.; Song, J.T.; Kim, J.K.; Choi, Y.D. A t MYB 44 regulates WRKY 70 expression and modulates antagonistic interaction between salicylic acid and jasmonic acid signaling. *Plant J.* **2013**, *73*, 483–495. [CrossRef]

211. Kan, G.; Zhang, W.; Yang, W.; Ma, D.; Zhang, D.; Hao, D.; Hu, Z.; Yu, D. Association mapping of soybean seed germination under salt stress. *Mol. Genet. Genom.* **2015**, *290*, 2147–2162. [CrossRef]

212. Pantalião, G.F.; Narciso, M.; Guimarães, C.; Castro, A.; Colombari, J.M.; Breseghello, F.; Rodrigues, L.; Vianello, R.P.; Borba, T.O.; Brondani, C. Genome wide association study (GWAS) for grain yield in rice cultivated under water deficit. *Genetica* **2016**, *144*, 651–664. [CrossRef]

213. Hu, S.; Sanchez, D.L.; Wang, C.; Lipka, A.E.; Yin, Y.; Gardner, C.A.; Lübberstedt, T. Brassinosteroid and gibberellin control of seedling traits in maize (Zea mays L.). *Plant Sci.* **2017**, *263*, 132–141. [CrossRef]

214. To, H.T.M.; Nguyen, H.T.; Dang, N.T.M.; Nguyen, N.H.; Bui, T.X.; Lavarenne, J.; Phung, N.T.P.; Gantet, P.; Lebrun, M.; Bellafiore, S.; et al. Unraveling the Genetic Elements Involved in Shoot and Root Growth Regulation by Jasmonate in Rice Using a Genome-Wide Association Study. *Rice* **2019**, *12*, 69. [CrossRef] [PubMed]

215. Delker, C.; Stenzel, I.; Hause, B.; Miersch, O.; Feussner, I.; Wasternack, C. Jasmonate biosynthesis in Arabidopsis thaliana-enzymes, products, regulation. *Plant Biol.* **2006**, *8*, 297–306. [CrossRef] [PubMed]

216. An, H.; Liu, K.; Wang, B.; Tian, Y.; Ge, Y.; Zhang, Y.; Tang, W.; Chen, G.; Yu, J.; Wu, W. Genome-wide association study identifies QTLs conferring salt tolerance in rice. *Plant Breed.* **2020**, *139*, 73–82. [CrossRef]

217. Huang, K.; Peng, L.; Liu, Y.; Yao, R.; Liu, Z.; Li, X.; Yang, Y.; Wang, J. Arabidopsis calcium-dependent protein kinase AtCPK1 plays a positive role in salt/drought-stress response. *Biochem. Biophys. Res. Commun.* **2018**, *498*, 92–98. [CrossRef] [PubMed]

218. Coca, M.; San Segundo, B. AtCPK1 calcium-dependent protein kinase mediates pathogen resistance in Arabidopsis. *Plant J.* **2010**, *63*, 526–540. [CrossRef] [PubMed]

219. Li, J.; Brader, G.; Palva, E.T. The WRKY70 transcription factor: A node of convergence for jasmonate-mediated and salicylate-mediated signals in plant defense. *Plant Cell* **2004**, *16*, 319–331. [CrossRef]
220. Chen, K.; Wang, Y.; Zhang, R.; Zhang, H.; Gao, C. CRISPR/Cas genome editing and precision plant breeding in agriculture. *Annu. Rev. Plant Biol.* **2019**, *70*, 667–697. [CrossRef]
221. Wang, Y.; Cheng, X.; Shan, Q.; Zhang, Y.; Liu, J.; Gao, C.; Qiu, J.-L. Simultaneous editing of three homoeoalleles in hexaploid bread wheat confers heritable resistance to powdery mildew. *Nat. Biotechnol.* **2014**, *32*, 947–951. [CrossRef]
222. Svitashev, S.; Schwartz, C.; Lenderts, B.; Young, J.K.; Mark Cigan, A. Genome editing in maize directed by CRISPR—Cas9 ribonucleoprotein complexes. *Nat. Commun.* **2016**, *7*, 13274. [CrossRef]
223. Cai, Y.; Chen, L.; Liu, X.; Sun, S.; Wu, C.; Jiang, B.; Han, T.; Hou, W. CRISPR/Cas9-mediated genome editing in soybean hairy roots. *PLoS ONE* **2015**, *10*, e0136064. [CrossRef]
224. Stuttmann, J.; Barthel, K.; Martin, P.; Ordon, J.; Erickson, J.L.; Herr, R.; Ferik, F.; Kretschmer, C.; Berner, T.; Keilwagen, J. Highly efficient multiplex editing: One-shot generation of 8x Nicotiana benthamiana and 12x Arabidopsis mutants. *Plant J.* **2021**. [CrossRef]
225. Liu, L.; Zhang, J.; Xu, J.; Li, Y.; Guo, L.; Wang, Z.; Zhang, X.; Zhao, B.; Guo, Y.-D.; Zhang, N. CRISPR/Cas9 targeted mutagenesis of SlLBD40, a lateral organ boundaries domain transcription factor, enhances drought tolerance in tomato. *Plant Sci.* **2020**, *301*, 110683. [CrossRef]

International Journal of
Molecular Sciences

Review

Advances in Sensing, Response and Regulation Mechanism of Salt Tolerance in Rice

Kimberly S. Ponce [1,2], Longbiao Guo [1,*], Yujia Leng [2,*], Lijun Meng [3] and Guoyou Ye [3,4]

1   State Key Laboratory for Rice Biology, China National Rice Research Institute, Hangzhou 310006, China; kimsuazoponce@gmail.com
2   Jiangsu Key Laboratory of Crop Genetics and Physiology/Key Laboratory of Plant Functional Genomics of the Ministry of Education/Jiangsu Key Laboratory of Crop Genomics and Molecular Breeding/Jiangsu Co-Innovation Center for Modern Production Technology of Grain Crops, Agricultural College of Yangzhou University, Yangzhou 225009, China
3   CAAS-IRRI Joint Laboratory for Genomics-Assisted Germplasm Enhancement, Agricultural Genomics Institute in Shenzhen, Chinese Academy of Agricultural Sciences, Shenzhen 518120, China; menglijun@caas.cn (L.M.); yeguoyou@caas.cn (G.Y.)
4   Strategic Innovation Platform, International Rice Research Institute, DAPO BOX 7777, Metro Manila 1301, Philippines
*   Correspondence: yujialeng@yzu.edu.cn (Y.L.); guolongbiao@caas.cn (L.G.); Tel.: +86-514-8797-4757 (Y.L.); +86-571-6337-0136 (L.G.)

Abstract: Soil salinity is a serious menace in rice production threatening global food security. Rice responses to salt stress involve a series of biological processes, including antioxidation, osmoregulation or osmoprotection, and ion homeostasis, which are regulated by different genes. Understanding these adaptive mechanisms and the key genes involved are crucial in developing highly salt-tolerant cultivars. In this review, we discuss the molecular mechanisms of salt tolerance in rice—from sensing to transcriptional regulation of key genes—based on the current knowledge. Furthermore, we highlight the functionally validated salt-responsive genes in rice.

Keywords: rice; salinity; sensing; signaling; transcription factors; osmoregulation; antioxidation; ion homeostasis

Citation: Ponce, K.S.; Guo, L.; Leng, Y.; Meng, L.; Ye, G. Advances in Sensing, Response and Regulation Mechanism of Salt Tolerance in Rice. *Int. J. Mol. Sci.* **2021**, 22, 2254. https://doi.org/10.3390/ijms 22052254

Academic Editor: Jen-Tsung Chen

Received: 31 January 2021
Accepted: 20 February 2021
Published: 24 February 2021

Publisher's Note: MDPI stays neutral with regard to jurisdictional claims in published maps and institutional affiliations.

1. Background

Soil salinity is one of the most significant abiotic stresses hampering plant growth and development, which ultimately translates to reduced crop yield. Soil salinization is exacerbated by excessive use of chemical fertilizers and soil amendments, improper drainage, and seawater ingress. It is estimated that over 6% of the world's total land area is salt affected, of which over 12 million hectares are irrigated lands posing a serious threat to irrigated agriculture [1].

Rice, being one of the most important staple crops in the world, is crucial for food security in many Asian countries. However, it is the most salt-sensitive cereal crop, with varying responses at different growth stages [2]. It is relatively salt-tolerant at the germination, active tillering, and maturity stages, whereas it is highly sensitive at the early seedling and reproductive stages [1]. Salt sensitivity during the seedling stage often translates to reduced stand density in salt-affected paddies [3]. Meanwhile, sensitivity during the reproductive stage results in yield reduction, as attributed to spikelet sterility [4,5]. Hence, understanding how rice responds to salt stress is crucial in developing rice cultivars that could withstand salt stress.

Salinity imposes two major stresses in rice, (i) osmotic stress, and (ii) ionic stress. Osmotic stress is characterized by hyperosmotic soil solution disrupting cell turgor, similar to drought's effect. In contrast, ionic stress is characterized by altered Na^+ and K^+ concentrations inside the cell, disrupting many biological processes [1]. Both osmotic and ionic stresses are perceived by membrane-bound cytosolic sensors that relay the stress signals to secondary messengers. In turn, the secondary messengers activate the protein phosphorylation cascades required for signal transduction pathways to develop salt-tolerant adaptive traits. In general, osmotic stress triggers the plant for stomatal closure, inhibiting shoot elongation. This ultimately results in reduced overall shoot growth, and, to a lesser extent, reduced root growth [6]. Meanwhile, ionic stress inhibits enzyme activity and therefore disrupts many biological processes, such as nitrogen metabolism [7,8]. Excess uptake of Na^+ ions changes the NH_4^+ assimilation pathway, weakens the glutamate synthase pathway, and elevates the glutamate dehydrogenase pathway, impacting leaf senescence [8]. Thus, plants develop several adaptive mechanisms—namely, Na^+ efflux from the roots to the rhizosphere, Na^+ sequestration into the vacuole, and Na^+ loading and unloading at the xylem—to avert the deleterious effect of Na^+ ions in the cytosol. These mechanisms are mediated by several ion transporters coupled with H^+-pumps.

In the last decades, a large number of salt-responsive genes have been functionally validated in rice (Table 1). However, the overall gene regulatory network of rice responses to salt stress remains elusive. In this review, we aim to discuss the current research progress in gene regulatory networks involved in the development of salt tolerance adaptive mechanisms in rice. We also highlight the key genes involved in salt stress sensing, signaling, transcriptional regulation, and genes encoding downstream functional molecules.

Table 1. List of functionally validated candidate genes involved from sensing to development of salt tolerance adaptive mechanisms in rice.

Gene Name	Gene ID	Functional Annotation	Method of Validation	* Regulation Role	References
		Osmosensing			
SIT1	LOC_Os02g42780	lectin receptor-type protein kinase	Knockdown Overexpression	−	[9]
		Signaling			
OsCam1-1	LOC_Os03g20370	Calmodulin	Overexpression	+	[10]
OsCPK4	LOC_Os02g03410	CAMK_CAMK_like.12—Ca^{2+}/calmodulin-dependent protein kinase(CAMK) includes calcium/calmodulin dependent protein kinases	Knockdown Overexpression	+	[11]
OsCDPK7	LOC_Os04g49510	CAMK_CAMK_like.27—CAMK includes calcium/calmodulin dependent protein kinases	Overexpression	+	[12,13]
OsCPK12	LOC_Os04g47300	CAMK_CAMK_like.26—CAMK includes calcium/calmodulin dependent protein kinases	Overexpression	+	[14]
OsCPK21	LOC_Os08g42750	CAMK_CAMK_like.37—CAMK includes calcium/calmodulin dependent protein kinases	Overexpression	+	[15,16]
OsCIPK15	LOC_Os11g02240	CAMK_Nim1_like.4—CAMK includes calcium/calmodulin dependent protein kinases	Overexpression	+	[17]
OsCIPK31	LOC_Os03g20380	CAMK_Nim1_like.2—CAMK includes calcium/calmodulin dependent protein kinases	Mutant	+	[18]
OsMAPK5	LOC_Os03g17700	CGMC_MAPKCGMC_2_ERK.2—CGMC includes CDA, MAPK, GSK3, and CLKC kinases	Knockdown Overexpression	+	[19]
OsMAPK33	LOC_Os02g05480	CGMC_MAPKCMGC_2_SLT2y_ERK.1—includes cytidine deaminase (CDA), glycogen synthase kinase 3 (GSK3), mitogen-activated protein kinase (MAPK), and CLKC kinases	Knockdown Overexpression	−	[20]
OsMKK1	LOC_Os06g05520	MAPK	Knockdown	+	[21]
OsMKK6	LOC_Os01g32660	STE_MEK_ste7_MAP2K.2—STE kinases	Overexpression	+	[22]
OsMaPKKK63	LOC_Os01g50370	STE_MEKK_ste11_MAP3K.4—STE kinases	Knockdown	−	[23]

Table 1. *Cont.*

Gene Name	Gene ID	Functional Annotation	Method of Validation	* Regulation Role	References
		Transcriptional regulation			
OsDREB1A	LOC_Os09g35030	Dehydration-responsive element (DRE)–binding protein	Overexpression	+	[24]
OsDREB1D	LOC_Os06g06970	DRE–binding protein	Overexpression	+	[25]
OsDREB1F	LOC_Os01g73770	DRE–binding protein	Overexpression	+	[26]
OsDREB2A	LOC_Os01g07120	APETALA2 (AP2) domain containing protein	Overexpression	+	[27,28]
OsDREB2B	LOC_Os05g27930	AP2 domain containing protein	Overexpression	+	[29]
OsAP23	LOC_Os03g05590	AP2 domain containing protein	Overexpression	−	[30]
OsAP37	LOC_Os01g58420	AP2 domain containing protein	Overexpression	+	[31]
OsSTAP1	LOC_Os03g08470	APETALA2/ethylene responsive factor (AP2/ERF)-type transcription factor	Overexpression	+	[32]
OsDREB6	LOC_Os09g20350	ERF transcription factor	Knockdown Overexpression	+	[33]
SERF1	LOC_Os05g34730	ERF020- transcription factor	Knockdown	+	[34]
OsERF922	LOC_Os01g54890	Ethylene-responsive transcription factor 2	Knockdown Overexpression	−	[35]
OsRAV2	LOC_Os01g04800	B3 DNA binding domain containing protein	Mutant	+	[36]
OsNAP	LOC_Os03g21060	No apical meristem (NAM)protein	Overexpression	+	[37]
ONAC022	LOC_Os03g04070	NAM protein	Overexpression	+	[38]
ONAC045	LOC_Os11g03370	NAM protein	Overexpression	+	[39]
ONAC063	LOC_Os08g33910	NAM protein	Overexpression	+	[40]
ONAC106	LOC_Os08g33670	NAM protein	Overexpression	+	[41]
OsNAC2	LOC_Os04g38720	NAM protein	Overexpression	+	[42,43]
OsNAC5	LOC_Os11g08210	NAM protein	Knockdown Overexpression	+	[44,45]
OsNAC6/ SNAC2	LOC_Os01g66120	NAM protein	Overexpression	+	[46,47]
SNAC1	LOC_Os03g60080	NAM, ATAF and CUC (NAC) domain-containing protein 67	Overexpression	+	[48]
OsNAC10	LOC_Os11g03300	NAC domain transcription factor	Overexpression	+	[49]
OsNAC041	-		Knockdown	+	[50]
OsMYB2	LOC_Os03g20090	Myeloblastosis (MYB) family transcription factor	Overexpression	+	[51]
OsMYB3R-2	LOC_Os01g62410	MYB family transcription factor	Overexpression	+	[52]
OsMYB48-1	LOC_Os01g74410	MYB family transcription factor	Overexpression	+	[53]
OsMPS	LOC_Os02g40530	MYB family transcription factor	Overexpression	+	[54]
OsMYB91	LOC_Os12g38400	MYB family transcription factor	Knockdown Overexpression	+	[55]
OsMYBc	LOC_Os09g12770	Adenosine-thymine (AT) hook motif domain containing protein	Mutant	+	[56]
OsABF2	LOC_Os06g10880	Basic leucine-zipper (bZIP) transcription factor	Mutant	+	[57]
OsABI5	LOC_Os01g64000	bZIP transcription factor	Overexpression	−	[58]
OsbZIP23	LOC_Os02g52780	bZIP transcription factor	Overexpression	+	[59]
OsbZIP71	LOC_Os09g13570	CPuORF2—conserved peptide uORF-containing transcript	Knockdown Overexpression	+	[60]
OsHBP1b	LOC_Os01g17260	Transcription factor	Overexpression	+	[61]
DST	LOC_Os03g57240	ZOS3-19—C2H2 zinc finger (ZF) protein	Mutant	−	[62]
OsTZF1	LOC_Os05g10670	ZF CCCH type family protein	Knockdown Overexpression	+	[63]
ZFP179	LOC_Os01g62190	ZOS1-15—C2H2 ZF protein	Overexpression	+	[64]
ZFP182	LOC_Os03g60560	ZOS3-21—C2H2 ZF protein	Overexpression	+	[65]
ZFP185	LOC_Os02g10200	ZF A20 and AN1 domain-containing stress-associated protein	Knockdown Overexpression	−	[66]
ZFP252	LOC_Os12g39400	ZOS12-09—C2H2 ZF protein	Knockdown Overexpression	+	[67]
OsLOL5	LOC_Os01g42710	LSD1-like-type ZF protein	Overexpression	+	[68]
OrbHLH001	LOC_Os01g70310	Inducer of CBF expression 2	Overexpression	+	[69]
OsbHLH035	LOC_Os01g06640	Basic helix-loop-helix (bHLH)	Mutant	+	[70]
Oshox22	LOC_Os04g45810	Homeobox associated leucine zipper	Mutant Overexpression	−	[71]
OsTF1L	LOC_Os08g19590	Homeobox domain containing protein	Knockdown Overexpression	+	[72]
OsMADS25	LOC_Os04g23910	MADS-box family gene with MIKCc type-box	Knockdown Overexpression	+	[73]

Table 1. *Cont.*

Gene Name	Gene ID	Functional Annotation	Method of Validation	* Regulation Role	References
OsWRKY45	LOC_Os05g25770	WRKY45	Knockdown Overexpression	–	[74]
Osmoprotection					
OsBADH1	LOC_Os04g39020	Aldehyde dehydrogenase	Knockdown; Overexpression	+	[75,76]
OsTPP1	LOC_Os02g44230	CPuORF22—conserved peptide uORF-containing transcript	Overexpression	+	[77]
OsTPS1	LOC_Os05g44210	Trehalose-6-phosphate synthase	Overexpression	+	[78]
OsTPS8	LOC_Os08g34580	Trehalose-6-phosphate synthase	Mutant Overexpression	+	[79]
Osmoregulation					
OsPIP1;1	LOC_Os02g44630	Aquaporin protein	Overexpression	+	[80,81]
OsPIP2;2	LOC_Os02g41860	Aquaporin protein	Overexpression	+	[80]
Stomatal Closure					
LP2	LOC_Os02g40240	Receptor kinase	Overexpression	+	[82]
OsSRO1c	LOC_Os03g12820	ATP8	Mutant Overexpression	+	[83]
Antioxidation					
OsCu/Zn-SOD	LOC_Os08g44770	Copper/zinc superoxide dismutase	Overexpression	+	[84]
OsMn-SOD	LOC_Os05g25850	Manganese superoxide dismutase	Overexpression	+	[85]
OsAPx1	LOC_Os03g17690	Cytosolic Ascorbate Peroxidase encoding gene 1-8	Overexpression	+	[86]
OsAPx2	LOC_Os07g49400	Cytosolic Ascorbate Peroxidase encoding gene 4,5,6,8	Knockdown Overexpression	+	[87]
OsGR3	LOC_Os10g28000	Glutathione reductase	Knockdown	+	[88]
OsTRXh1/ OsTrx23	LOC_Os07g08840	Thioredoxin	Knockdown; Overexpression	–	[89]
OsGRX8	LOC_Os02g30850	OsGrx_C8—Glutaredoxin subgroup III	Knockdown; Overexpression	+	[90]
OsGRX20	LOC_Os08g44400	Glutathione S-transferase	Knockdown; Overexpression	+	[91]
Na$^+$ exclusion					
OsHKT1;1	LOC_Os04g51820	Na$^+$ transporter	Natural variation	+	[92]
OsHKT1;4	LOC_Os04g51830	Na+ transporter	Mutant	–	[93]
OsHKT1;5/ SKC1	LOC_Os01g20160	Na$^+$ transporter	Natural variation	+	[94]
OsSOS1	LOC_Os12g44360	Sodium/hydrogen exchanger 7	Mutant	+	[95]
Na$^+$ compartmentation					
OsNHX1	LOC_Os07g47100	transporter, monovalent cation:proton antiporter-2 family	Overexpression	+	[96]
OsVP1	LOC_Os01g68370	B3 DNA binding domain containing protein	Overexpression	+	[96]
K$^+$ uptake					
OsHAK1	LOC_Os04g32920	Potassium transporter	Mutant and overexpression	+	[97]
OsHAK5	LOC_Os01g70490	Potassium transporter	Knockdown overexpression	+	[98]
OsHAK16	LOC_Os03g37840	Potassium transporter	Overexpression	+	[99]
OsHAK21	LOC_Os03g37930	Potassium transporter	Knockdown	+	[100]

* + positive regulation; – negative regulation.

2. Salt Stress Sensing

Stress sensing is the first event in plant response to any abiotic stresses, mounting an effective adaptive strategy. Under salt stress condition, it is presumed that osmotic and ionic stresses are perceived by membrane-bound cytosolic sensors that ultimately trigger early salt-stress signaling routes (Figure 1). However, the current knowledge of how rice sense salt stress is still limited and therefore remains an open question.

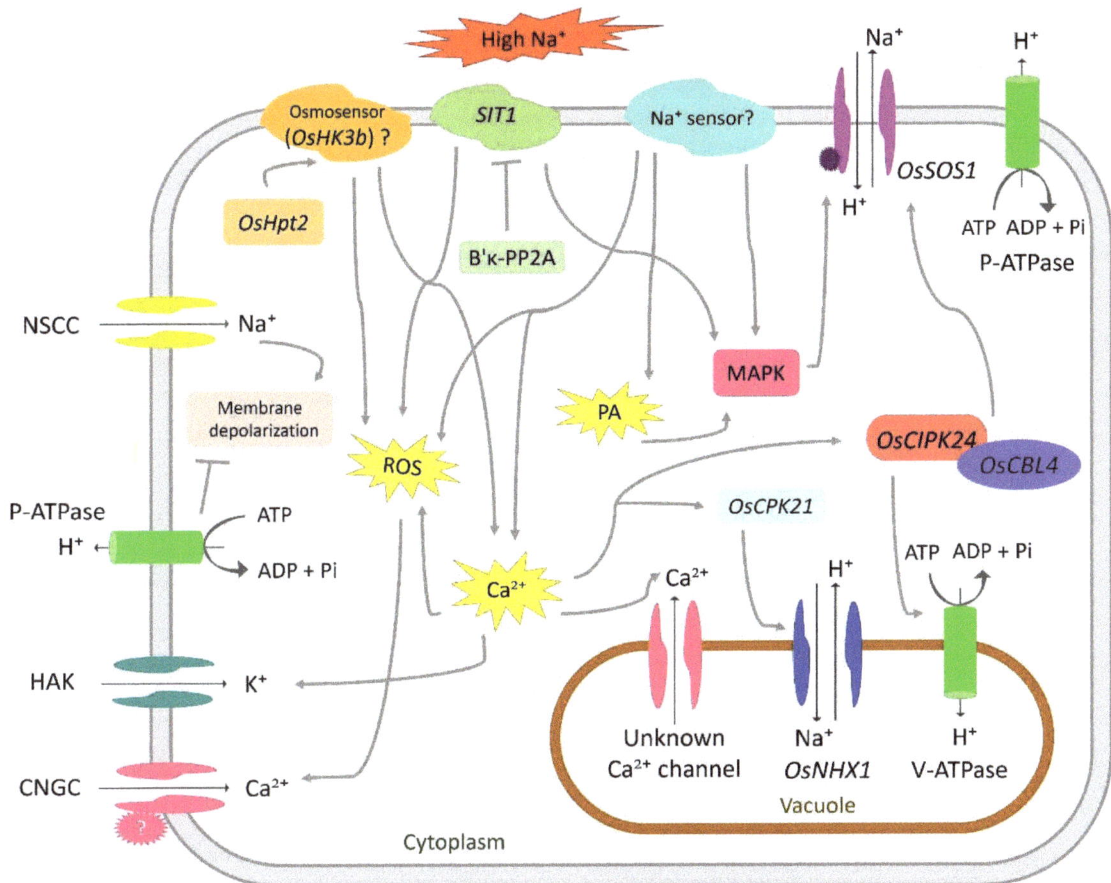

Figure 1. Salt sensing and signaling involved in rice responses to salt stress. Under high salinity, salt-induced osmotic stress begins, which is sensed by putative osmosensor *OsHK3b*, activated by *OsHpt2*. *SIT1* also acts as a sensor via elevated kinase activity and induces reactive oxygen species (ROS) production and mitogen-activated protein kinase (MAPK) signaling. The activity of *SIT1* is deactivated by the B'κ-PP2A subunit. Later, ionic stress occurs and is sensed by an unknown Na⁺ sensor. The Na⁺ enters the mature epidermal cell through nonselective cation channel (NSCC), causing membrane depolarization, and is polarized by P-type ATPases. Excess salt triggers a spike in the concentration of cytosolic secondary messengers, including Ca²⁺, reactive oxygen species (ROS), and phosphatidic acid (PA). ROS triggers Ca²⁺ influx through the cyclic nucleotide-gated ion channel (CNGC), activated by an unknown molecule. Ca²⁺ not only decreases K⁺ efflux but also induces further ROS accumulation; thus, a positive feedback loop exists between Ca²⁺ and ROS. The cytosolic Ca²⁺ also induces vacuolar Ca²⁺ release and activates Ca²⁺-binding proteins, such as *OsCIPK24-OsCBL4* complex. This complex, together with MAPK, activated by phosphatidic acid, upregulates the *OsSOS1* to remove cytosolic Na⁺. The vacuolar *OsNHX1* gene is activated by *OsCPK21*, whereas the V-type ATPase is activated by *OsCIPK24*, establishing a proton gradient and driving the activity of *OsNHX1*.

2.1. Osmosensing

Sensing salt-induced osmotic stress is crucial in early signaling cascades to develop salt tolerance adaptive traits, such as growth retardation, reduction in stomatal conductance, and high abscisic acid (ABA)accumulation. However, little is known about the genetics and physiology of how rice sense hyperosmotic stress.

The transmembrane-protein-receptors, such as histidine kinases and receptor-like kinases (RLKs), function in osmotic stress perception in rice. Histidine kinases perceive osmotic fluctuations and relay the signal to response regulators via phosphotransfer, which is mediated by histidine-containing phosphotransfer protein (HpT) [101]. The first evidence of osmosensing function of histidine kinases was reported in *Arabidopsis*. The *AtHK1*, a histidine kinase encoding gene, interacts with *AtHPt1* and functions as an osmosensor during both drought and salt stress [102,103]. The ortholog of *AtHK1* in rice, *OsHK3b*, interacts with *OsHpt2* and acts as a putative osmosensor [101,104]. However, functional evidence on its osmosensing role in rice is not yet reported.

The RLKs function in drought and salt stress sensing by transmitting signals to downstream signaling pathways [105]. The rice *Salt Intolerance 1* (*SIT1*), a lectin RLK expressed mainly in root epidermal cells, acts as an upstream mediator of salt stress via elevated kinase activity [9]. Recently, Zhao et al. [106] reported that *SIT1* phosphorylates B′κ at Ser_{402}, which in turn promotes the assembly of B′κ-protein phosphatase 2A (B′κ-PP2A) holoenzyme. The B′κ-PP2A subunit positively regulates salt tolerance by deactivating the activity of *SIT1* via dephosphorylation at the $Thr_{515/516}$. *SIT1* kinase activity in turn activates the mitogen-activated protein kinase (MAPK) 3 and MAPK 6 [9]. Thus, it could be pointed out that RLKs are important in MAPK cascade activation during osmotic stress. However, the relationship between the RLKs and MAPKs needs to be further elucidated.

Ca^{2+} permeable stress-gated cation channels (OSCA) also act as hyperosmotic stress sensors. The first evidence of the role of OSCA in osmosensing was reported in *Arabidopsis* with the characterization of *OSCA1*. The *OSCA1* gene forms a hyperosmolality-gated Ca^{2+} permeable channel during osmotic stress, thereby increasing the cytosolic free Ca^{2+} concentration [107]. The rice genome consists of 11 OSCA genes, of which seven (*OsOSCA1.1*, *OsOSCA1.2*, *OsOSCA2.1*, *OsOSCA2.4*, *OsOSCA2.5*, *OsOSCA3.1*, and *OsOSCA4.1*) were upregulated during salt-induced osmotic stress and may function as an osmosensor [108]. However, the Ca^{2+} conducting function of the rice OSCA genes in response to hyperosmotic stress remains an open question.

2.2. Na$^+$ Sensing

The molecular mechanism of Na^+ transport in plants is well understood; however, Na^+ sensing remains elusive. It has been reported that the ion transporters at the plasma membrane are potential Na^+ sensors. For instance, the plasma membrane Na^+/H^+ antiporter *SOS1* (*Salt Overly Sensitive 1*) is thought to be involved in Na^+ sensing [109]. It was later proposed that only the long hydrophilic cytoplasmic tail of *SOS1* could potentially sense Na^+ ions [110]. However, no research experiments have been undertaken to support this hypothesis, and therefore it needs to be clarified. Moreover, it is unlikely that *SOS1* functions as initial Na^+ sensor since the *SOS3/SOS2* complex regulates its activity. Na^+ ions could also be sensed either extracellularly and intracellularly by membrane receptors and unknown cytosolic sensors, respectively [110]. In rice, it was suggested that the intracellular Na^+ ions are sensed by an unknown cytosolic sensor based on the observed elevated levels of free cytosolic Ca^{2+} ions in salt stressed plants. Thus, more research is required to point out the identity of such cytosolic Na^+ sensor [111].

3. Signal Transduction

During salt stress, plants transduce the early stress signals to different cellular machinery called signal transduction. In general, signal transduction starts right after stress sensing, followed by the synthesis of secondary signaling molecules, such as Ca^{2+} and reactive oxygen species (ROS) (Figure 1). The production of secondary signaling molecules modulates the cytosolic Ca^{2+} concentration that binds to different protein kinases, such as calmodulins (CaMs)/CaM-like (CML), calcium-dependent protein kinases (CDPKs), calcineurin B-like interacting protein kinases (CIPKs), and MAPKs. As these protein kinases lack enzymatic activity, they catalyze protein phosphorylation via a Ca^{2+}-dependent manner, resulting in protein conformational change. Thus, protein phosphorylation cascades mainly depend on the cytosolic Ca^{2+} concentration [112,113].

3.1. CaM/CML

CaM/CML proteins are important Ca^{2+} transducers in plant responses to abiotic stress [114,115]. In rice, five CaM-encoding genes—namely, *OsCam1-1*, *OsCam1-2*, *OsCam1-3*, *OsCam2*, and *OsCam3*—were identified [10]. Among these, *OsCam1-1* is highly activated during salt stress. Yuenyong et al. [116] reported that the rice plants overexpressing *OsCam1-1* affected differential expression of genes involved in signaling, hormone-mediated regulation, transcription, lipid metabolism, carbohydrate metabolism, photosynthesis, glycolysis, tricarboxylic acid cycle, and glyoxylate cycle during salt stress. This further suggests that a complex network of downstream cellular processes is involved in the CaM signal transduction pathway. CaM binds with other proteins and interacts with other signaling cascades, such as plant hormone signaling, during stress conditions. For instance, it binds either with MAPK or mitogen-activated protein kinase phosphatase (MKP) to regulate the MAPK cascades [117]. Recently, six novel proteins—namely, *OsLRK5a*, *OsDCNL2*, *OsWD40-139*, *OsGDH1*, *OsCIP*, and *OsERD2*—were identified as targets of *OsCML16* in responses to salt stress through yeast hybridization and bimolecular fluorescence complementation assay. These target genes are involved in plant hormone signaling processes, including auxin and ABA [118]. Interestingly, both *OsCaM1* and *OsCML16* could bind with *OsERD2* and thus could transduce Ca^{2+} via both CaM and CML proteins [118]. Although the functional role of *OsERD2* in response to salt stress is still unknown, it is speculated that it plays a vital role in programmed cell death during innate immunity, similar with *AtERD2* [119].

3.2. CDPK

CDPKs mediate downstream components of the Ca^{2+} signaling cascades by directly binding Ca^{2+} to CaM-like domain. In rice, a total of 29 CDPK genes have been identified [120]. Four rice CDPK genes—namely, *OsCPK4*, *OsCDPK7*, *OsCPK12*, and *OsCPK21*—were functionally validated and act as positive regulators of salt tolerance (Table 1). Overexpression of rice CDPKs upregulate expression of genes involved in lipid metabolism and the active oxygen detoxification system. For instance, overexpression of *OsCPK4* upregulated the genes involved in oxidative stress and redox regulation [11]. Similarly, transgenic rice plants overexpressing *OsCPK12* significantly enhanced the expression of genes encoding reactive oxygen species (ROS) scavenging enzymes, such as *OsAPx2* and *OsAPx8* [14]. *OsCDPK7* positively regulates salt tolerance by regulating salt-stress responsive gene, *rab16A* [12,13]. Meanwhile, *OsCPK21* enhances salt tolerance via regulation of ABA- and salt stress-inducible genes, such as *Rab21*, *OsNAC6*, *OsLEA3*, *OsP5CS*, *OsNHX1*, and *OsSOS1* [15]. Further study revealed that *OsCPK21* regulates salt tolerance by phosphorylating *OsGF14e/Os14-3-3* at the Tyr_{138} [16]. This was the first evidence of 14-3-3 protein-associated phosphorylation of CDPK in rice. Despite intensive work in studying the role of CDPKs in regulation of salt tolerance in rice, their role in different signaling cascades needs to be elucidated.

3.3. Calcineurin B-Like Protein (CBL)/CIPK

CBLs are plant-specific Ca^{2+} sensors that bind with CIPKs to relay perceived Ca^{2+} signal, thereby inducing downstream gene regulation for abiotic stress. The *SOS3–SOS2* complex is the first evidence of CBL–CIPK interaction in plant responses to salt stress [121]. Homologues of *SOS2* and *SOS3* in rice, the *OsCIPK24* and *OsCBL4*, have been cloned, which suggests that the SOS pathway also operates in rice responses to salt stress [122]. Further study revealed that *OsCIPK24/OsSOS2*, *OsCBL4/OsSOS3*, and *OsSOS1* were highly upregulated in salt-tolerant rice cultivars when subjected to salt stress [123]. This suggests that the rice CBL4–CIPK24 complex, together with the Ca^{2+} signal, regulates ion homeostasis similar to *Arabidopsis*. Therefore, the SOS pathway is conserved in both dicots and monocots. Many other CBL and CIPK genes are involved in rice responses to salt stress based on transcriptome analysis [124,125]. However, only *OsCIPK15* and *OsCIPK31* have been functionally validated for their role in salt tolerance. Transgenic rice plants overexpressing *OsCIPK15* showed enhanced salt tolerance with higher free proline and soluble sugar concentration [17]. Similarly, *OsCIPK31* acts as a positive regulator of salt tolerance wherein the loss-of-function mutant *oscipk31:Ds* exhibited hypersensitive phenotype under saline condition [18].

3.4. MAPK

MAPK is considered the last component of the protein phosphorylation cascade in transducing Ca^{2+} ions in response to environmental stimulus. The MAPK signaling pathway activates different transcription factors (TFs) involved in the production and scavenging of ROS [126]. Three classes of MAPKs are found in plants; namely, MAPK kinase kinase (MKKK), MAPK kinase (MKK), and MAPK [127,128]. Rice has 15 MAPKs, 8 MKKs, and 75 MKKKs, of which a few are involved in salt stress response (Table 1) [129–131]. Overexpression and gene silencing validated the role of *OsMAPK5* as a positive regulator of salt tolerance [19]. Further study showed that *OsMAPK5* phosphorylates *SERF1*, a regulator of ROS signaling during initial response to salt stress [34]. Thus, *OsMAPK5* plays an essential role in the ROS signaling pathway. In contrast, *OsMAPK33* acts as a negative regulator and alters the expression of genes involved in Na^+ transport [20]. *OsMAPKKK63* also acts as a negative regulator of salt tolerance and interacts with *OsMKK1* and *OsMKK6* [23]. Both *OsMKK1* and *OsMKK6* are known mediators of rice responses to salt stress. Overexpression of *OsMKK6* enhances salt tolerance by inducing *MAPK* substrate phosphorylation [22]. Similarly, *OsMKK1* acts as a positive regulator with highly upregulated transcripts under saline conditions [21]. Moreover, yeast hybridization and in-vivo/vitro kinase assays revealed that *OsMPK4* is the downstream target of *OsMKK1*. *OsMPK4* is involved in the wounding signaling pathway in rice [132]. However, its functional role in salt tolerance is not well characterized.

4. Transcriptional Regulation

In the past centuries, numerous proteins were reported to play an important role in salt tolerance. Transcriptomic tools have further subdivided these proteins into two major classes, the functional and regulatory proteins. Functional proteins are those that directly function in protecting the plants from stress. These include ion transporters, antioxidant proteins, osmolytes, water channel proteins, heat shock proteins, and late embryogenesis abundant (LEA) proteins. On the contrary, regulatory proteins, such as transcription factors (TFs), are involved in regulating the complex network of signal transduction [133–136].

TFs are key proteins that bind with *cis*-elements in the promoter of target genes, thereby modulating the rate of gene expression in the downstream signaling cascades in response to different environmental cues. A large number of TFs have been identified in rice, with 2025 TFs in *Oryza sativa spp. indica* and 2384 in *spp. japonica* [137]. In recent years, many TFs along with their interacting proteins have been implicated in rice responses to salt stress and regulate a series of signaling pathways (Table 1). Most of these are members of APETALA2/ethylene responsive-factor (AP2/ERF), NAC (NAM, ATAF, and CUC) proteins, myeloblastosis (MYB), basic leucine-zipper (bZIP) type proteins, zinc finger (ZF) and basic helix-loop-helix (bHLH) TFs that regulate many salt stress-responsive genes either through an ABA-dependent or -independent manner (Figure 2). Thus, understanding how TFs, along with their interacting proteins, regulate a network of signaling pathways and their downstream genes is crucial in elucidating the salt tolerance mechanisms of rice.

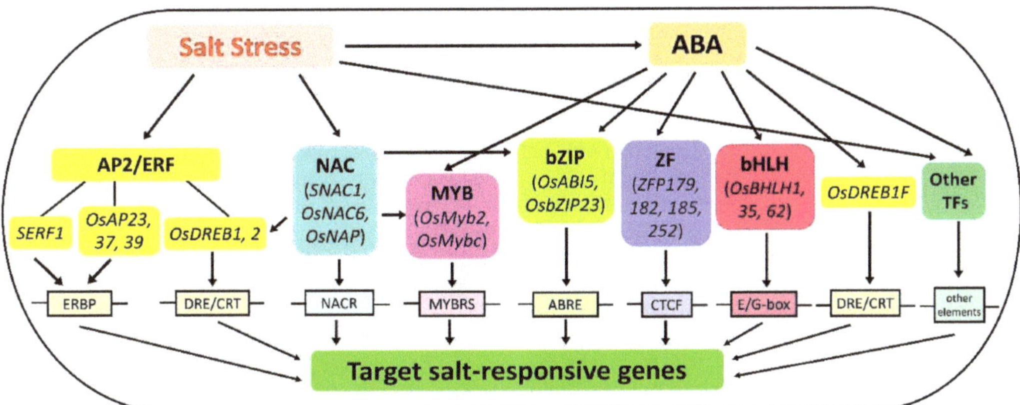

Figure 2. Transcriptional regulation involved in activating salt stress-responsive genes in rice. The transcriptional regulation occurs via abscisic acid (ABA)-dependent and -independent pathway, whereby transcription factors (TFs) bind with their corresponding *cis*-regulatory element. The APETALA2/ethylene responsive factor (AP2/ERF) and NAC (NAM, ATAF, and CUC) TFs operate in an ABA-independent pathway. NAC TFs regulate other TFs, such as dehydration responsive element-binding (DREB), myeloblastosis (MYB), and basic leucine-zipper (bZIP). The MYB, bZIP, zinc finger (ZF), basic-helix-loop-helix (bHLH), DREB, and other TFs are involved in the ABA-dependent pathway.

4.1. APETALA2/Ethylene Responsive Factor (AP2/ERF) Regulation

AP2/ERF-type TFs are characterized by the presence of an AP2 DNA-binding domain of approximately 60 amino acids. In rice, at least 163 AP2/ERF TFs have been identified. This TF family is further subdivided into four subfamilies: the AP2, dehydration responsive element-binding (DREB), ERF, and related to ABI3 and VP1 (RAV) proteins [138]. Among these, DREB is widely involved in rice responses to salt stress, though a few AP2-, ERF-, and RAV-type TFs regulate salt tolerance (Table 1).

DREB binds to the dehydration-responsive element/c-repeat (DRE/CRT) *cis*-elements in the promoter region of stress-responsive genes. DREBs have been isolated in several crops, and their overexpression enhances tolerance to different abiotic stresses, including salinity [139]. Rice DREB1 genes enhance salinity tolerance by regulating osmoprotection, as evident in rice and *Arabidopsis* DREB1 overexpression plants [25,26,140,141]. For instance, *OsDREB1A* targets two dehydrin genes [24]. Dehydrins protect plasma membrane from damage during drought- or salt-induced osmotic stress [142]. Moreover, the level of proline and soluble sugars, which are important for osmotic adjustment, significantly increased in DREB1 overexpression plants [140,143]. DREB genes mainly work in the ABA-independent pathway; however, some also participate in the ABA-dependent pathway, as exemplified by *OsDREB1F*. Transcript profiling in *OsDREB1F* overexpression lines showed expression of ABA-dependent genes, *rd29B* and *RAB18* [26]. DREB2-type genes also act as positive regulators of salt tolerance. Overexpression of *OsDREB2A* and *OsDREB2B* in both rice and *Arabidopsis* improved salt tolerance [24,27–29]. Another DREB gene, *OsDREB6*, classified as an A-6 type of DREB TF positively regulates salt tolerance. Transgenic rice plants overexpressing *OsDREB6* showed high levels of proline, soluble sugars, and catalase. Conversely, the levels of these enzymes were significantly reduced in RNAi plants [33]. This suggests that DREB genes mainly enhance salt tolerance by regulating genes responsible for osmoprotection and antioxidation. Similar to DREB, other TFs in the AP2/ERF family enhance salt tolerance by regulating several downstream genes involved in osmotic stress and antioxidant defense system. For instance, *SERF1* gene regulates ROS-dependent signaling as an initial response to salt stress [34]. Recently, Wang et al. [32] demonstrated that *OsSTAP1*, an AP2/ERF-type TF, positively regulates salt tolerance by activating genes encoding antioxidant enzymes (*OsPOD1*, *OsPOD72*, *GSTT3*) and aquaporin gene (*NIP2-1*).

Unlike most of AP2/ERF-type TFs, *OsERF922* and *OsAP23* act as negative regulators and downregulate the expression of defense-related genes [30,35].

4.2. NAC Regulation

NAC proteins are a plant-specific gene family that regulate both ABA-independent and ABA-dependent inducible genes [144]. Several studies have been carried out to understand the role of rice NAC genes in response to abiotic stimulus, including salinity. Most functionally characterized rice NAC proteins act as positive regulators of salt tolerance (Table 1). *SNAC1*, the first stress-related NAC type TF characterized in rice, enhances both drought and salt tolerance [48]. Transcriptome analysis of transgenic plants overexpressing NAC proteins showed upregulation of many stress-inducible genes. For instance, *OsNAC2*, *OsNAC5*, *ONAC022*, and *ONAC106* target *OsLEA3* [38,41,43,44]; *OsNAP* targets several stress-related genes, including *OsPP2C06/OsABI2*, *OsPP2C09*, *OsPP2C68*, and *OsSalT* [37]; and *OsNAC2* targets genes involved in osmoprotection (*OsP5CS1*), antioxidation (*OsCOX11*), K$^+$-efflux channel genes (*OsGORK* and *OsSKOR*), and ABA-inducible genes (*OsNCED1* and *OsNCED3*) [42,43]. NAC TFs also regulate other stress-related TFs. For instance, *OsNAP* induces the expression of *OsDREB1A* and *OsMYB2* [37]. *ONAC106* binds with the promoter of *OsNAC5*, *OsDREB2A*, and *OsbZIP23* TF genes [41]. Similarly, *ONAC022* targets *OsDREB2a* and *OsbZIP23* (Hong et al. 2016).

4.3. MYB Regulation

MYB proteins are one of the richest TF families in plants, representing at least 155 genes in rice. It is considered as an active player in plant development, secondary metabolism, cell differentiation, organ morphogenesis, and response to both biotic and abiotic stresses [145,146]. These TFs mainly participate in the ABA-dependent pathway, upregulating a number of stress-responsive genes. For example, expression of *OsMPS*, an R2R3 type MYB TF, is significantly induced by ABA and regulates several expansin and glucanase genes [54]. Transcriptome analysis of transgenic rice plants overexpressing *OsMYB48-1* upregulates ABA biosynthesis genes (*OsNCED4* and *OsNCED5*), early signaling genes (*OsPP2C68* and *OSRK1*), and late responsive genes (*RAB21*, *OsLEA3*, *RAB16C*, and *RAB16D*) [53]. Similarly, *OsMYB2* targets *OsLEA3* and *OsRab16A* [51]. MYB TFs also regulate the expression of some transporter genes. For example, *OsMYBc* binds with the AAANATNY motif in the promoter of *OsHKT1;1*, thereby upregulating its expression [56]. Other rice MYB TFs involved in the regulation of salt tolerance are presented in Table 1.

4.4. bZIP Regulation

bZIP TFs are composed of a highly conserved basic region and a leucine zipper domain of about 60 to 80 amino acids in length. Several rice bZIP TFs are involved in transcriptional activation of several stress-responsive genes, most of which participate in the ABA-dependent pathway (Table 1). Overexpression of *OsbZIP71* upregulates several genes that encode ion antiporters (*OsCLC-1*, *OsNHX1*, *OsHKT6* and *OsVHA-B*) and ROS scavenging (*OsCAT*). Interestingly, *OsbZIP71* directly binds to the promoter of *OsNHX1*, an Na$^+$/H$^+$ antiporter gene involved in vacuolar compartmentation of Na$^+$ ions [60]. *OsbZIP23* acts as a key player in salt tolerance by upregulating osmotic stress-inducible genes, such as dehydrins and LEA proteins [59]. *OsHBP1b*, also categorized under the bZIP TF family, could enhance salt tolerance by activating the genes involved in antioxidant defense system [61]. It is worth noting that *OsHBP1b* is localized within the *Saltol* quantitative trait locus (QTL) region, hence an important salt tolerance gene. Moreover, comparative transcript profiling showed that *OsHBP1b* is highly expressed in popular salt-tolerant rice cultivar Pokkali [147]. Meanwhile, *OsABI5* acts as a negative regulator changing the expression of many salt stress-responsive genes. *OsABI5* significantly downregulates the expression of *OsHKT1;5/SKC1* and upregulates *SalT* gene [58]. Transcriptomic analysis showed that many other bZIP TFs play an important role in rice responses to salt stress. However, their regulatory roles have not been functionally studied. Taken together, bZIP TFs mainly regulate salt tolerance via the active oxygen detoxification and ion homeostasis pathways.

4.5. ZF Regulation

ZF proteins are comprised of conserved motifs with cystine (Cys) and histidine (His) residues. These motifs are classified according to the number and order of Cys and His. [148]. Several studies have shown their function in transcriptional activation of several biological processes involved in plant responses to environmental stimulus. Under salt stress conditions, ZF TFs regulate the expression of genes associated with ROS scavenging via ABA-independent and ABA-dependent pathways to reduce oxidative damage. The *ZFP179*, *ZFP182*, and *ZFP252* act as positive regulators of salt tolerance. These ZF TFs transcriptionally activate the *OsDREB1A*, *OsLEA3*, *OsPC5CS*, and *OsProT* genes that are involved in the synthesis of osmolytes, such as proline and soluble sugars [64,65,67]. Conversely, drought and salt tolerance *(DST)* and *ZFP185* act as negative regulators and downregulate several ABA-inducible genes, such as *Prx24* [62,66]. Meanwhile, *OsLOL5*, an LSD1-like-type ZF is involved in transcriptional activation of *OsAPX2*, *OsCAT*, and *OsCu/Zn-SOD* [68]. Thus, ZF TFs play an essential role in the ROS signaling pathway.

4.6. bHLH Regulation

bHLH TFs widely exist in eukaryotic organisms and contain a conserved basic region and a helix-loop-helix (HLH) domain [149]. These TFs play an essential role in several abiotic stress tolerance, wherein several bHLH TF genes have been functionally validated. Concerning salt tolerance, only a few were functionally validated. Three previously reported bHLH TFs enhance salt tolerance in rice by activating ion transporters genes. For instance, *OsbHLH035* enhances salt tolerance by activating Na^+ transporter genes, *OsHKT1;3* and *OsHKT1;5/SKC1*, which are involved in Na^+ loading and unloading [70]. *OrbHLH001* enhances Na^+ efflux and K^+ influx under salt stress by activating *OsAKT1* [69]. Meanwhile, *OsbHLH062* acts as transcriptional activator of *OsHAK21* in response to salt stress [150]. The bHLH TFs therefore regulate salt tolerance via the ion homeostasis pathway. Moreover, these TFs activate gene expression through their interaction with the specific E-box motif in the promoter of the target gene [69,141,151].

4.7. Other TFs Involved in Salt Tolerance

In addition to the TFs previously discussed, many other TF families play an essential role in reprogramming transcriptome during salt stress. The homeodomain-leucine zipper (HD-Zip) TF family is also important for salt tolerance, such as *Oshox22* and *OsTF1L*. *Oshox22* acts a negative regulator of salt tolerance and is upstream to *OsbZIP23* [71]. *OsTF1L* positively regulates salt tolerance mainly by regulating genes involved in stomatal closure and lignin biosynthesis [72].

Apart from *OsbZIP71*, previously discussed, several TFs belonging to different families regulate the expression of the *OsNHX1* transporter gene. The *OsNIN-like4* and *OsPCF2*, a nodule inception (NIN) and teosinte branched 1/cycloidea/proliferating cell (TCP) proteins, respectively, act as transcriptional activators of *OsNHX1*. Conversely, *OsCPP5* and *OsNIN-like2* act as repressors [152]. *OsMADS25*, a MADS-box TF gene, acts as positive regulator by upregulating the expression of genes involved in the ROS detoxification system [73]. Meanwhile, the WRKY-type TF, *OsWRKY45*, negatively regulates salt tolerance [74].

5. Salt Tolerance Adaptive Mechanisms

Several adaptive mechanisms have been observed in plant responses to salt stress. In rice, osmoregulation, stomatal closure, and development of antioxidant enzymes are the immediate responses during salt stress. This is later followed by Na^+ exclusion and sequestration upon uptake of toxic Na^+ ions. The tissue specific localization of genes that regulate salt tolerance adaptive traits in rice is presented in Figure 3.

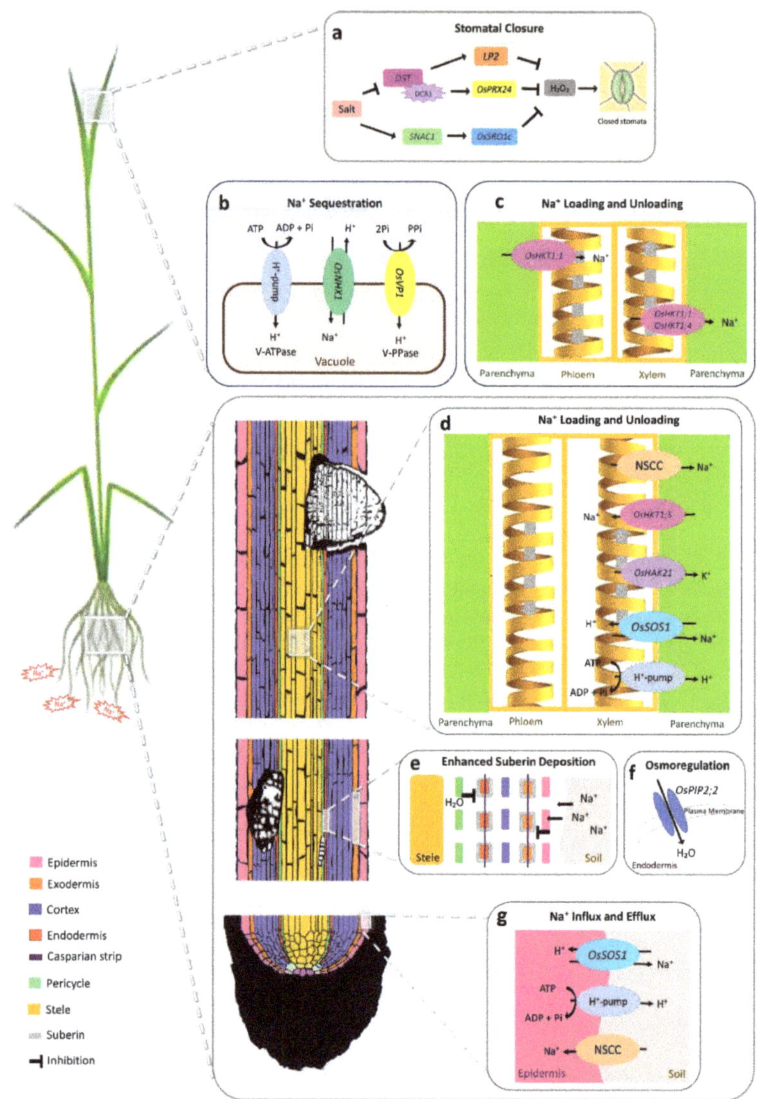

Figure 3. Rice salt tolerance adaptive mechanisms. In the leaf, (**a**) stomatal closure mediated either by *DST* or *SNAC1* is the initial response of rice to salinity. Salt stress downregulates *DST* which interacts with *DCA1* and activates *OsPrx24* and *LP2*. Conversely, *SNAC1* is upregulated, activating the *OsSRO1c*. These downstream genes mediate stomatal closure via H_2O_2 inhibition. (**b,c**) Na^+ content in the leaf cytoplasm is controlled by vacuolar sequestration, xylem unloading, and phloem loading. Excess Na^+ is sequestered into the vacuole via *OsNHX1* coupled with H^+-pump and *OsVP1*, a vacuolar-type H^+-pyrophosphatase encoding gene. Na^+ unloading at the xylem and Na^+ loading at the phloem are mediated by *OsHKT1;4* and *OsHKT1;1*, respectively. In the root, (**d**) Na^+ is loaded at the xylem through nonselective cation channel (NSCC) and *OsSOS1* coupled with H^+-pump. Conversely, *OsHKT1;5* unloads the Na^+ ions from the xylem and shuttles them back to the parenchyma cells. Apart from Na^+, K^+ influx occurs mediated by *OsHAK21*, thereby increasing the K^+/Na^+ ratio. (**e**) Enhanced suberin deposition in the root exodermis and endodermis also inhibits Na^+ influx to the stele. Similarly, it blocks water transport out of the stele. (**f**) The plasma membrane-bound *OsPIP2;2* gene increases hydraulic conductivity in the root endodermis, allowing water uptake. (**g**) Na^+ enters the root epidermis via NSCC and is shuttled back to the external medium via the *OsSOS1* coupled with H^+-pump.

5.1. Osmoprotection and Osmoregulation

Cell dehydration due to low osmotic potential of soil water is the immediate effect of salt stress. Under such a situation, plants (1) synthesize compatible solutes, known as osmolytes, to maintain cell turgor and (2) activate water channel aquaporins that regulate water uptake.

5.1.1. Osmolytes

Several osmolytes, such as trehalose and glycine betaine (GB), have been proven effective in preventing cellular dehydration during salt stress [153]. Thus, exogenous application of osmolytes has been utilized to enhance salt tolerance in rice [154–157]. However, very few studies have been conducted to characterize osmolyte encoding genes for their role in salt tolerance.

The two key enzymes in trehalose biosynthesis, trehalose-6-phosphate phosphatase (TPP) and trehalose-6-phosphate synthase (TPS), are involved in rice responses to salinity. The *OsTPP1*, *OsTPS1*, and *OsTPS8* positively regulate salt tolerance by increasing the accumulation of trehalose and proline in rice overexpression plants [77–79].

GB is also an important osmolyte under salt stress that prevents lipid peroxidation [158]. Additionally, accumulation of high GB enhances photosynthetic activity [159]. The *OsBADH1*, a major gene involved in converting betaine aldehyde to GB, plays an important role in salt tolerance. This gene prevents oxidative damage, protects chlorophyll degradation, and ultimately prevents leaf senescence during salt stress [75]. Moreover, RNAi-directed knockdown of *OsBADH1* enhances the production of ROS, causing lipid peroxidation [76]. Thus, the gene acts as a positive regulator of salt tolerance.

5.1.2. Water Channel Aquaporins

Plant aquaporins also play a significant role in osmoregulation. Aquaporins are membrane-localized channels that are mainly involved in water transport and homeostasis [160,161]. Rice has 33 aquaporins, few of which regulate root hydraulic conductivity under saline condition [162]. Overexpression of *OsPIP1;1* and *OsPIP2;2*, plasma membrane intrinsic proteins (PIPs) family genes in *Arabidopsis*, enhanced salt tolerance by maintaining water homeostasis [80]. Likewise, rice overexpressing *OsPIP1;1* increased root hydraulic conductivity under salt stress [81]. Rice aquaporins might be coordinately orchestrated in maintaining water homeostasis based on their organ-specific transcript expression. Transcript of *OsPIP2* genes were highly expressed in the roots; thus, it could be the predominant gene regulating water uptake in the roots (Figure 3f). Conversely, the *OsPIP1* gene transcript was the highest in the leaves, suggesting its role in leaf water transport [80]. Apart from the PIP genes, several tonoplast intrinsic protein (TIP) genes also play an important role in salt-induced osmotic stress [163].

5.2. Stomatal Closure

Stomatal closure is the initial response of plants under salinity and is controlled by both ABA and ROS signaling [164]. *DST* mainly regulates salt tolerance via stomatal closure under salt-induced osmotic stress. Further study revealed that a leucine-rich repeat (LRR)-RLK gene, *LP2*, required for stomatal closure is downstream to *DST* [82]. Interestingly, *DST* interacts with *DST co-activator 1* (*DCA1*) and regulates the expression of *OsPrx24*, a gene encoding H_2O_2 scavenger [165]. Meanwhile, *OsSRO1c*, expressed in the guard cells and a downstream gene target of *SNAC1* TF, also regulates stomatal closure under both drought- and salt-induced osmotic stress (Figure 3a). Overexpression of *OsSRO1c* in rice plants showed enhanced stomatal closure and maintained H_2O_2 homeostasis under salt stress. Conversely, knockdown mutants showed high sensitivity to osmotic stress [83].

5.3. Antioxidation

ROS synthesis is important in different signaling and physiological processes. However, overproduction of ROS is deleterious to different cellular components, such as proteins, nucleic acids, and membrane lipids. Thus, plants synthesize ROS scavenging enzymes to maintain redox homeostasis [126,166]. In this section, we discuss genes encoding ROS scavenging enzymes that are involved in rice responses to salt stress.

5.3.1. Superoxide Dismutase (SOD)

SODs catalyze the first step in the reactive-oxygen scavenging system by dismutation of the highly toxic O_2^- to H_2O_2. Thus, it is considered the most effective intracellular antioxidant enzyme. Rice has three distinct types of SOD isoforms that are differentiated according to the metals they contain, either Cu/Zn, Mn, or Fe. The activity of these SODs is associated with specific subcellular localization: Mn-SOD is located in both mitochondria and peroxisomes; Fe-SOD is located in the chloroplasts; and Cu/Zn-SOD is located in the chloroplasts, cytosol, and peroxisome [167]. The expression of genes encoding these SOD isoforms is highly influenced by salt stress and is activated by ZF-type TFs, as discussed in Section 4.5. Mishra et al. [168] reported that the increase in SOD activity of salt-tolerant rice cultivar CSR27 exposed to salinity was directly related to the upregulation of Cu/Zn-SOD encoding genes. Similar results were reported by Rossatto et al. [169], who observed upregulation of five Cu/Zn isoforms (*OsCu/Zn-SOD, OsCu/Zn-SOD2, OsCu/Zn-SOD3, OsCu/Zn-SOD4, OsCu/Zn-SODCc1*) under salt stress. Moreover, the rice plants overexpressing chloroplastic *OsCu/Zn-SOD* showed less salt-induced oxidative damage owing to higher ROS detoxification [84]. Upregulation of *OsMn-SOD* was also observed in rice subjected to salt stress. Tanaka et al. [85] reported that overexpression of *OsMn-SOD* in the chloroplasts significantly increased SOD activity and therefore enhanced salt tolerance. Similar results were observed in other plants such as wheat and tall fescue [170,171]. Conversely, salinity downregulates the expression of *OsFe-SOD*, thereby reducing the total SOD activity [172]. This suggests that Cu/Zn-SOD and Mn-SOD isoforms play vital roles in ROS detoxification system during stress condition.

5.3.2. Catalase (CAT)

CATs are strong antioxidant enzymes primarily located in the peroxisome that directly catalyze the conversion of H_2O_2 to water and oxygen [173,174]. Thus, it is indispensable in the ROS detoxification system. Cloning and characterization of the rice CAT genes predicted three isoforms; namely, *OsCatA, OsCatB,* and *OsCatC* [175]. These genes are transcriptionally activated by bZIP- and ZF-type TFs, as described in Sections 4.4 and 4.5. RLK is also involved in transcriptional activation of CAT genes. For instance, the *salt tolerance receptor-like cytoplasmic kinase 1 (STRK1)* activates *OsCatC* via phosphorylation at the Tyr_{120} [176]. Several environmental factors, such as salinity, affect expression of CAT genes. Under saline condition, elevated levels of CAT activity were observed in salt-tolerant rice cultivars [177]. Interestingly, high *OsCatB* and *OsCatC* activity was observed in salt-tolerant plants grown under salt stress [178]. A similar result was reported by Wutipraditkul et al. [179], who observed an inhibitory effect of *OsCatC* in response to salt stress.

5.3.3. Ascorbate Peroxidase (APX)

APXs, which exist in compartment-specific isoforms, have a higher affinity for H_2O_2 than CATs. Thus, they detoxify even at very low H_2O_2 concentrations. Rice has eight APX encoding genes: the cytosolic isoforms *OsAPx1* and *OsAPx2*; the peroxisome isoforms *OsAPx3* and *OsAPx4*; and the chloroplastic isoforms *OsAPx5, OsAPx6, OsAPx7,* and *OsAPx8*. The *OsAPx6* isoform is also localized in the mitochondria [180]. All these APX encoding genes, except *OsAPx3* and *OsAPx5*, were upregulated in rice under salt stress [178,181]. Overexpression of *OsAPx2* showed very high APX activity, thereby enhancing salt tolerance in rice [87]. Likewise, overexpression of either *OsAPx1* or *OsAPx2* exhibited high tolerance to salt stress in *Arabidopsis*; however, *OsAPx2* confers better tolerance than *OsAPx1* [86]. Further study revealed that silencing both *OsAPx1* and *OsAPx2* genes in rice resulted in normal growth and development under salt stress. This is attributed to the upregulation of CAT and APX genes [182,183]. Thus, deficiency of APXs is compensated by other antioxidant enzymes.

5.3.4. Glutathione Reductase (GR)

GRs are flavoprotein oxidoreductases and are important components of the ascorbate (AsA)-glutathione (GSH) cycle [184]. Rice has three GR isoforms: *OsGR1*, located in the cytosol; and *OsGR2* and *OsGR3*, located in both mitochondria and chloroplasts [185]. These rice GRs have been implicated for their role in different abiotic stimuli, including salinity. Salt stress enhances the expression of *OsGR2* and *OsGR3* via the ROS detoxification system [185–187]. Further study demonstrated that *OsGR3*, primarily expressed in the roots, positively regulates salt tolerance [88].

5.3.5. Thioredoxin (TRX) and Glutaredoxin (GRX)

TRXs and glutaredoxin (GRX) are key players in redox regulation, therefore considered as redox-sensing compounds. TRX are reduced by TRX reductase, whereas GRX utilizes glutathione as a cofactor in the ROS scavenging system [188]. The rice genome has 30 and 48 genes encoding TRX and GRX, respectively. However, only a few have been functionally validated for their role in salinity tolerance [189,190]. For instance, *Os-TRXh1*/*OsTRX23* negatively regulates salt tolerance. RNAi-directed knockdown of this gene resulted in salt sensitivity, possibly due to its inhibitory activity on stress-activated MAPKs [89,191]. *OsTRXh1*/*OsTRX23* also inhibits the kinase activity of *OsMPK3* and *OsMPK6* [192]. Meanwhile, *OsGRX8* and *OsGRX20* positively regulate salt tolerance by restraining the accumulation of O_2^- radicals [90,91].

5.4. Na$^+$ Exclusion and Sequestration

Na$^+$ ions are the major toxic element taken up by the plant during salt stress. Maintaining low levels of toxic Na$^+$ ions in the cytosol, either through Na$^+$ exclusion or sequestration, is the most effective strategy to avert the deleterious effects of salinity. Glycophytes, such as rice, exclude Na$^+$ from the shoot either by (i) Na$^+$ efflux from roots to the rhizosphere, (ii) Na$^+$ loading and unloading at the xylem, or (iii) vacuolar Na$^+$ compartmentation.

5.4.1. Na$^+$ Efflux

The efflux of Na$^+$ ions across the root plasma membrane into the external medium is poorly understood. Nevertheless, it is central to the Na$^+$ exclusion mechanisms in plants [1]. To date, only *SOS1*, coupled with H$^+$-ATPases, is the major Na$^+$ efflux transporter that has been genetically characterized in plants [110,193]. The rice *SOS1* ortholog (*OsSOS1*) is expressed in epidermal cells at the root cap and in cells around the xylem similar with *Arabidopsis AtSOS1* [194]. The *OsSOS1* activity, catalyzed by Na$^+$/H$^+$ exchange at the plasma membrane, could suppress Na$^+$ sensitivity of yeast mutant lacking the Na$^+$ efflux system, thus reducing the net cellular Na$^+$ concentration. Similarly, *OsSOS1* complementation in *Arabidopsis* mutant *sos1-1* reduced growth defect in both saline and non-saline conditions [122]. Further study demonstrated that rice *sos1* loss-of-function mutant displayed very high root Na$^+$ uptake and impaired Na$^+$ loading into the xylem [95]. Thus, *OsSOS1* plays a critical role in Na$^+$ efflux from root epidermal cells to the rhizosphere.

5.4.2. Na$^+$ Loading and Unloading

Na$^+$ loading and unloading at the xylem is regulated by high-affinity K$^+$ transporters (HKTs). HKTs are among the most well characterized Na$^+$ and/or K$^+$ plant transporters identified in several plants and play a central role in salt tolerance [195,196]. Two HKTs are highlighted in a proposed two-staged Na$^+$ exclusion mechanism, whereby the (i) *OsHKT1;5*/*SKC1* mediates root-to-shoot Na$^+$ transfer and (ii) *OsHKT1;4* mediates leaf sheath-to-blade Na$^+$ transfer. The Na$^+$ ions entering the root xylem via nonselective cation channel (NSCC) are shuttled back to the parenchyma via *OsHKT1;5*/*SKC1* (Figure 3d). Meanwhile, *OsHKT1;4* not only functions in Na$^+$ unloading to the leaf sheath, but also to the stem during the reproductive stage [197]. Further study revealed that *OsHKT1;4* is involved in leaf Na$^+$ exclusion via Na$^+$ unloading at the xylem (Figure 3c). The mutant line overexpressing *OsHKT1;4* showed salt sensitivity owing to very high root Na$^+$ uptake [93]. Thus, a coordinated balance in root and shoot Na$^+$ exclusion is essential to achieve salt tolerance. Another HKT1 gene, *OsHKT1;1*, transcriptionally activated by *OsMYBc* as previously discussed, is also reported to regulate Na$^+$ exclusion, possibly through both Na$^+$ unloading from the xylem and Na$^+$ loading into the phloem (Figure 3c). The Na$^+$ loaded into the phloem is hypothesized to be recirculated from shoots to roots or from young leaves to old leaves, thereby reducing salt injury in newly emerging leaf [56]. Moreover, it was demonstrated that *OsHKT1;1* is a positive regulator of salt tolerance that mediates Na$^+$ exclusion from the shoot [92]. Recent studies have shown that there are eight and four transcript variations of HKT1 genes with different lengths in *O. sativa spp. indica* and *spp. japonica*, respectively. These eight transcript variations in *O. sativa spp. indica* show different expression levels and transport activities under salt treatment, which suggests the existence of different transport mechanisms [198].

5.4.3. Vacuolar Na$^+$ Sequestration

Few rice cultivars with high Na$^+$ concentrations in the leaves were found to perform well under saline condition. This is mainly due to the active compartmentation of Na$^+$ ions into the vacuole, also known as tissue tolerance, mediated by the tonoplast localized Na$^+$/H$^+$ antiporters (NHX) and energized by a proton motive force (Figure 3b) [193]. This mechanism allows the plant to use Na$^+$ ions in maintaining cell turgor, and hence continuous plant growth under salt [199,200]. Additionally, vacuolar Na$^+$ sequestration maintains cytosolic alkalinity and vacuolar acidity. Maintaining low vacuolar pH is essential since acidity allows the vacuole to isolate and break down misfolded proteins [201]. This phenomenon was only observed in salt-tolerant rice cultivars, such as Pokkali [111].

Four vacuolar NHX genes—namely, *OsNHX1*, *OsNHX2*, *OsNHX3*, and *OsNHX5*— were identified in rice mediating cytosolic Na$^+$ sequestration into the vacuole [202]. Further study revealed that overexpression of *OsNHX1* enhanced tissue tolerance and is regulated by *OsbZIP71* TF [60,96,203]. Very high transcripts of these NHX genes in either flag leaf or panicle has also been observed [202]. This suggests their potential role in enhancing salt tolerance at the reproductive stage.

Functional characterization of vacuolar-type H$^+$-pyrophophatase (H$^+$-PPase) also showed enhanced salt tolerance. H$^+$-PPase is the main driving force for Na$^+$ transport from the cytoplasm to the vacuole (Figure 3b). Overexpression of H$^+$-PPase encoding genes in different plants significantly enhanced salt tolerance [204–206]. In rice, overexpression of *OsVP1*, a H$^+$-PPase encoding gene, resulted in less serious Na$^+$ toxicity under salt stress. Moreover, double overexpression of *OsNHX1* and *OsVP1* conferred better salt tolerance [96]. This is possibly due to the higher electrochemical gradient brought by *OsVP1* overexpression, thereby promoting higher activity of *OsNHX1* (Figure 3b). Interestingly, a similar result has been found in simultaneous expression of *SsNHX1* from *Suaeda salsa* and *AVP1* from *Arabidopsis* in rice [206].

5.5. Suberin Deposition

Suberin deposition is essential in blocking apoplastic leakage of Na$^+$ ions into the stele, resulting in low concentration of Na$^+$ ions that can be transported into the shoot (Figure 3e). In rice, a few studies have reported the role of suberin in salt tolerance. Enhancing suberin in the form of silicon has significantly reduced the root-to-shoot Na$^+$ uptake by preventing apoplastic Na$^+$ transport across the root [207]. Interestingly, the popular salt-tolerant rice, Pokkali, showed higher suberin deposition compared with the salt-sensitive cultivar IR20 [208]. However, the gene regulatory network involved in suberin deposition and salt tolerance in rice is not well understood. The *OsTPS8*, involved in trehalose biosynthesis, was also reported to enhance salt tolerance, mainly by enhancing suberin deposition [79].

5.6. K$^+$ Uptake

Cytosolic K$^+$ concentration has emerged as an important aspect of a plant's adaptability to salt stress, wherein high K$^+$ concentration directly relates to salt tolerance. Four high-affinity K$^+$ transporter (HAK) genes—namely, *OsHAK1*, *OsHAK5*, *OsHAK16*, and *OsHAK21*—play crucial roles in K$^+$ homeostasis under stress conditions [97–100]. Interestingly, differences in spatial expression were observed among these HAK genes. β-glucoronidase (GUS) staining assay showed that *OsHAK1*, *OsHAK5*, and *OsHAK16* were mainly expressed in the root epidermis [97,98,100]. Conversely, *OsHAK21* was mainly expressed in the root xylem parenchyma [99]. Thus, *OsHAK21* is likely the predominant gene mediating K$^+$ influx in the xylem (Figure 3d).

6. Conclusions and Perspectives

Soil salinity, apart from drought and flooding, is a serious menace afflicting global rice production. Being the staple crop of half of the world's population, developing salt-tolerant rice varieties is crucial, requiring a better overview on molecular and physiological responses to salt stress. Rice responds to salinity through different biological processes, starting with salt stress sensing. Sensing is mediated by different sensors. The sensors relay stress signals to secondary messengers that activate protein phosphorylation cascades and finally the transcriptional regulation of stress-responsive genes via abscisic acid (ABA)-independent/ABA-dependent pathways. Rice response to salt stress also involves several signaling components, transcription factors, and functional genes that directly mediate osmoregulation, antioxidation, and ion homeostasis. Despite the characterization of these genes, understanding the molecular mechanism of rice responses to salt stress remains a great challenge.

Over the last few decades, remarkable progress in understanding the genomics-physiology of salinity tolerance in plants has taken place. Several genes have been identified to confer salt tolerance in rice; however, most were achieved through a reverse genetics approach. Thus, a large number of genes need to be identified via forward genetics. The current understanding of the molecular responses of rice to salt stress from sensing and signaling up to the development of adaptive tolerance mechanisms is still obscure and requires further research. In particular, identification of upstream pathways and the molecular mechanisms involved in salt stress sensing is crucial to clearly disentangle the osmotic and Na$^+$ stress responses in rice. To date, only the role of ABA signaling in rice responses to salt stress is widely studied. The crosstalk between signaling pathways and of other hormones, including auxin, gibberellic acid, jasmonic acid, and ethylene, is still not clear and needs further investigation. Studying the epigenetic regulations of salt tolerance in rice is another important field to dissect. Epigenetic mechanisms control the expression of stress-responsive genes in response to internal and environmental cues. Thus, epigenomic variations may provide a useful resource of DNA methylomes that can be used to better understand the complex salt tolerance mechanisms in rice.

Author Contributions: Y.L. and K.S.P. proposed the concept; K.S.P. drafted the manuscript; L.M. and G.Y. participated in the data collection; Y.L. and L.G. revised and finalized the manuscript. All authors have read and agreed to the published version of the manuscript.

Funding: This study was supported by the National Natural Science Foundation of China (Grant No. 31801337), Agricultural Variety Improvement Project of Shandong Province (Grant No. 2019LZGC003), and Dapeng District Industry Development Special Funds (Grant No. KY20180217).

Institutional Review Board Statement: Not applicable.

Informed Consent Statement: Not applicable.

Data Availability Statement: Not applicable.

Acknowledgments: Not applicable.

Conflicts of Interest: The authors declare that they have no competing interests.

Abbreviations

AP2/ERF	APETALA2/ethylene responsive factor
APXAT	Ascorbate peroxidaseAdenosine-thymine
bHLH	Basic-helix-loop-helix
BzipCAMK	Basic leucine-zipperCalcium/calmodulin-dependent protein kinase
CAT	Catalase
CBLCDA	Calcineurin B-like proteinCytidine deaminase
CDPK	Calcium-dependent protein kinase
CIPK	CBL-interacting protein kinase
CaM	Calmodulin
CML	Calmodulin-like protein
CPP	Cysteine-rich poly comb-like protein
DRE/CRT	Dehydration-responsive element/c-repeat
DST	Drought and salt tolerance
DCA1	DST co-activator 1
GB	Glycine betaine
GRGUS	Glutathione reductaseβ-glucoronidase
HAK	High-affinity potassium transporter
HD-Zip	Homeodomain-leucine zipper
HKT	High-affinity K^+ transporter
HpT	Histidine-containing phosphotransfer protein
LEA	Late embryogenesis abundant
LRR-RLK	Leucine-rich repeat-receptor-like kinase
MAPK	Mitogen-activated protein kinase
MKK	MAPK kinase
MKKK	MKK kinase
MKP	Mitogen-activated protein kinase phosphatase
MYB	Myeloblastosis
NAC	NAM, ATAF and CUC
NAM	No apical meristem
NHX	Na^+/H^+ antiporter
NIN	Nodule inception
OSCA	Ca^{2+} permeable stress-gated cation channels
PA	Phosphatidic acid
PIPQTL	Plasma membrane intrinsic proteinQuantitative trait locus
RAV	Related to ABI3 and VP1
RLK	Receptor-like kinase
ROS	Reactive oxygen species
SIT	Salt intolerance
SOD	Superoxide dismutase
SOS	Salt overly sensitive
TCP	Teosinte branched 1/cycloidea/proliferating cell
TF	Transcription factor
TIP	Tonoplast intrinsic
TPS	Trehalose-6-phosphate phosphatase
TPP	Trehalose-6-phosphate synthase
TRX	Thioredoxin
VP	Vacuolar-type H^+-pyrophosphatase
ZF	Zinc finger

References

1. Munns, R.; Tester, M. Mechanisms of salinity tolerance. *Annu. Rev. Plant Biol.* **2008**, *59*, 651–681. [CrossRef]
2. Zeng, L.; Shannon, M.C.; Grieve, C.M. Evaluation of salt tolerance in rice genotypes by multiple agronomic parameters. *Euphytica* **2002**, *127*, 235–245. [CrossRef]
3. Lutts, S.; Kinet, J.M.; Bouharmont, J. Effects of salt stress on growth mineral nutrition and proline accumulation in relation to osmotic adjustment in rice (*Oryza sativa* L.) cultivars differing in salinity resistance. *Plant Growth Regul.* **1996**, *19*, 207–218. [CrossRef]

4. Cui, H.; Takeoka, Y.; Wada, T. Effect of sodium chloride on the panicle and spikelet morphogenesis in rice. *Jpn. J. Crop. Sci.* **1995**, *64*, 593–600. [CrossRef]

5. Khatun, S.; Flowers, T.J. Effects of salinity on seed set in rice. *Plant Cell. Environ.* **1995**, *18*, 61–67. [CrossRef]

6. Munns, R. Genes and salt tolerance: Bringing them together. *New Phytol.* **2005**, *167*, 645–663. [CrossRef]

7. Tester, M.; Davenport, R.J. Na$^+$ tolerance and Na$^+$ transport in higher plants. *Ann. Bot.* **2003**, *91*, 503–527. [CrossRef]

8. Wang, H.; Zhang, M.; Guo, R.; Shi, D.; Liu, B.; Lin, X.; Yang, C. Effects of salt stress on ion balance and nitrogen metabolism of old and young leaves in rice (*Oryza sativa* L.). *BMC Plant Biol.* **2012**, *12*, 194. [CrossRef]

9. Li, C.H.; Wang, G.; Zhao, J.L.; Zhang, L.Q.; Ai, L.F.; Han, Y.F.; Sun, D.Y.; Zhang, S.W.; Sun, Y. The receptor-like kinase SIT1 mediates salt sensitivity by activating MAPK3/6 and regulating ethylene homeostasis in rice. *Plant Cell* **2014**, *26*, 2538–2553. [CrossRef]

10. Boonburapong, B.; Buaboocha, T. Genome-wide identification and analyses of the rice calmodulin and related potential calcium sensor proteins. *BMC Plant Biol.* **2007**, *7*, 4. [CrossRef]

11. Campo, S.; Baldrich, P.; Messeguer, J.; Lalanne, E.; Coca, M.; San Segundo, B. Overexpression of a calcium-dependent protein kinase confers salt and drought tolerance in rice by preventing membrane lipid peroxidation. *Plant Physiol.* **2014**, *165*, 688–704. [CrossRef]

12. Saijo, Y.; Hata, S.; Kyozuka, J.; Shimamoto, S.; Izui, K. Over-expression of a single Ca^{2+}-dependent protein kinase confers both cold and salt/drought tolerance on rice plants. *Plant J.* **2000**, *23*, 319–327. [CrossRef]

13. Saijo, Y.; Kinoshita, N.; Ishiyama, K.; Hata, S.; Kyozuka, J.; Hayakawa, T.; Nakamura, T.; Shimamoto, K.; Yamaya, T.; Izui, K. A Ca^{2+}-dependent protein kinase that endows rice plants with cold- and salt-stress tolerance functions in vascular bundles. *Plant Cell. Physiol.* **2001**, *42*, 1228–1233. [CrossRef]

14. Asano, T.; Hayashi, N.; Kobayashi, M.; Aoki, N.; Miyao, A.; Mitsuhara, I.; Ichikawa, H.; Komatsu, S.; Hirochika HKikuchi, S.; Ohsugi, R. A rice calcium-dependent protein kinase OsCPK12 oppositely modulates salt-stress tolerance and blast disease resistance. *Plant J. Cell. Mol. Biol.* **2011**, *69*, 26–36. [CrossRef] [PubMed]

15. Asano, T.; Hakata, M.; Nakamura, H.; Aoki, N.; Ichikawa, H.; Komatsu, S.; Hirochika, H.; Kikuchi, S.; Ohsugi, R. Functional characterisation of OsCPK21 a calcium-dependent protein kinase that confers salt tolerance in rice. *Plant Mol. Biol.* **2011**, *75*, 179–191. [CrossRef] [PubMed]

16. Chen, Y.X.; Zhou, X.J.; Chang, S.; Chu, Z.L.; Wang, H.M.; Han, S.C.; Wang, Y.D. Calcium-dependent protein kinase 21 phosphorylates 14-3-3 proteins in response to ABA signaling and salt stress in rice. *Biochem. Biophy. Res. Commun.* **2017**, *4*, 1450–1456. [CrossRef]

17. Xiang, Y.; Huang, Y.M.; Xiong, L.Z. Characterization of stress-responsive *CIPK* genes in rice for stress tolerance improvement. *Plant Physiol.* **2007**, *144*, 1416–1428. [CrossRef] [PubMed]

18. Piao, H.L.; Xuan, Y.H.; Park, S.H.; Je, B.I.I.; Park, S.J.; Park, S.H.; Kim, C.H.; Huang, J.; Wang, G.K.; Kim, M.J.; et al. OsCIPK31 a CBL-interacting protein kinase is involved in germination and seedling growth under abiotic stress conditions in rice plants. *Mol. Cells* **2010**, *30*, 19–27. [CrossRef] [PubMed]

19. Xiong, L.Z.; Yang, Y.N. Disease resistance and abiotic stress tolerance in rice are inversely modulated by an abscisic acid–inducible mitogen-activated protein kinase. *Plant Cell* **2003**, *15*, 745–759. [CrossRef] [PubMed]

20. Lee, S.K.; Kim, B.G.; Kwon, T.R.; Jeong, M.J.; Park, S.R.; Lee, J.W.; Byun, M.O.; Kwon, H.B.; Matthews, B.F.; Hong, C.B.; et al. Overexpression of the mitogen-activated protein kinase gene *OsMAPK33* enhances sensitivity to salt stress in rice (*Oryza sativa* L.). *J. Biosci.* **2011**, *36*, 139–151. [CrossRef] [PubMed]

21. Wang, F.Z.; Jing, W.; Zhang, W.H. The mitogen-activated protein kinase cascade MKK1–MPK4 mediates salt signaling in rice. *Plant Sci.* **2014**, *227*, 181–189. [CrossRef]

22. Kumar, K.; Sinha, A.K. Overexpression of constitutively active mitogen activated protein kinase kinase 6 enhances tolerance to salt stress in rice. *Rice* **2013**, *6*, 25. [CrossRef] [PubMed]

23. Na, Y.J.; Choi, H.K.; Park, M.Y.; Choi, S.W.; Vo, K.T.X.; Jeon, J.S.; Kim, S.Y. OsMAPKKK63 is involved in salt stress response and seed dormancy control. *Plant Signal Behav.* **2019**, *14*, 1–6. [CrossRef] [PubMed]

24. Dubouzet, J.G.; Sakuma, Y.; Ito, Y.; Kasuga, M.; Dubouzet, E.G.; Miura, S.; Seki, M.; Shinozaki, K.; Yamaguchi-Shinozaki, K. *OsDREB* genes in rice *Oryza sativa* L., encode transcription activators that function in drought-, high-salt- and cold-responsive gene expression. *Plant J.* **2003**, *33*, 751–763. [CrossRef]

25. Zhang, Y.; Chen, C.; Jin, X.F.; Xiong, A.S.; Peng, R.H.; Hong, Y.H.; Yao, Q.H.; Chen, J.M.; Chen, J.M. Expression of a rice DREB1 gene, *OsDREB1D*, enhances cold and high-salt tolerance in transgenic *Arabidopsis*. *BMB Rep.* **2009**, *42*, 486–492. [CrossRef]

26. Wang, Q.; Guan, Y.; Wu, Y.; Chen, H.; Chen, F.; Chu, C. Overexpression of a rice *OsDREB1F* gene increases salt drought and low temperature tolerance in both *Arabidopsis* and rice. *Plant Mol. Biol.* **2008**, *67*, 589–602. [CrossRef]

27. Cui, M.; Zhang, W.; Zhang, Q.; Xu, Z.; Zhu, Z.; Duan, F.; Wu, R. Induced over-expression of the transcription factor OsDREB2A improves drought tolerance in rice. *Plant Physiol. Biochem.* **2011**, *49*, 1384–1391. [CrossRef]

28. Mallikarjuna, G.; Mallikarjuna, K.; Reddy, M.K.; Kaul, T. Expression of *OsDREB2A* transcription factor confers enhanced dehydration and salt stress tolerance in rice (*Oryza sativa* L.). *Biotechnol. Lett.* **2011**, *33*, 1689–1697. [CrossRef]

29. Matsukura, S.; Mizoi, J.; Yoshida, T.; Todaka, D.; Ito, Y.; Maruyama, K.; Shinozaki, K.; Yamaguchi-Shinozaki, K. Comprehensive analysis of rice *DREB2*-type genes that encode transcription factors involved in the expression of abiotic stress-responsive genes. *Mol. Genet. Genom.* **2010**, *283*, 185–196. [CrossRef] [PubMed]

30. Zhuang, J.; Jiang, H.H.; Wang, F.; Peng, R.H.; Yao, Q.H.; Xiong, A.S. A Rice OsAP23, functioning as an AP2/ERF transcription factor, reduces salt tolerance in transgenic Arabidopsis. *Plant Mol. Biol. Rep.* **2013**, *31*, 1336–1345. [CrossRef]

31. Oh, S.J.; Kim, Y.S.; Kwon, C.W.; Park, H.K.; Jeong, J.S.; Kim, J.K. Overexpression of the transcription factor *AP37* in rice improves grain yield under drought conditions. *Plant Physiol.* **2009**, *150*, 1368–1379. [CrossRef]

32. Wang, Y.X.; Wang, J.; Zhao, X.Q.; Yang, S.; Huang, L.Y.; Du, F.P.; Li, Z.K.; Zhao, X.Q.; Fu, B.Y.; Wang, W. Overexpression of the transcription factor gene *OsSTAP1* increases salt tolerance in rice. *Rice* **2020**, *50*, 1–12. [CrossRef]

33. Ke, Y.G.; Yang, Z.J.; Yu, S.W.; Li, T.F.; Wu, J.H.; Gao, H.; Fu, Y.P.; Luo, L.J. Characterization of *OsDREB6* responsive to osmotic and cold stresses in rice. *J. Plant Biol.* **2014**, *57*, 150–161. [CrossRef]

34. Schmidt, R.; Mieulet, D.; Hubberten, H.M.; Obata, T.; Hoefgen, R.; Fernie, A.R.; Fisahn, J.; San Segundo, B.; Guiderdoni, E.; Schippers, J.H.M.; et al. Salt-responsive ERF1 regulates reactive oxygen species-dependent signaling during the initial response to salt stress in rice. *Plant Cell* **2013**, *25*, 2115–2131. [CrossRef] [PubMed]

35. Liu, D.F.; Chen, X.J.; Liu, J.Q.; Ye, J.C.; Guo, Z.J. The rice ERF transcription factor *OsERF922* negatively regulates resistance to *Magnaporthe oryzae* and salt tolerance. *J. Exp. Bot.* **2012**, *63*, 3899–3911. [CrossRef] [PubMed]

36. Duan, Y.B.; Li, J.; Qin, R.Y.; Xu, R.F.; Li, H.; Yang, Y.C.; Ma, H.; Li, L.; Weng, P.C.; Yang, J.B. Identification of a regulatory element responsible for salt induction of rice *OsRAV2* through ex situ and in situ promoter analysis. *Plant Mol. Biol.* **2016**, *90*, 49–62. [CrossRef] [PubMed]

37. Chen, X.; Wang, Y.F.; Lv, B.; Li, J.; Luo, L.Q.; Lu, S.C.; Zhang, X.; Ma, H.; Ming, F. The NAC family transcription factor OsNAP confers abiotic stress response through the ABA pathway. *Plant Cell. Physiol.* **2016**, *55*, 604–619. [CrossRef] [PubMed]

38. Hong, Y.B.; Zhang, H.J.; Huang, L.; Li, D.Y.; Song, F.M. Overexpression of a stress-responsive NAC transcription factor gene *ONAC022* improves drought and salt tolerance in rice. *Front. Plant Sci.* **2016**, *7*, 4. [CrossRef] [PubMed]

39. Zheng, X.N.; Chen, B.; Lu, G.J.; Han, B. Overexpression of a NAC transcription factor enhances rice drought and salt tolerance. *Biochem. Biophys. Res. Commun.* **2009**, *379*, 985–989. [CrossRef]

40. Yokotani, N.; Ichikawa, T.; Kondou, Y.; Matsui, M.; Hirochika, H.; Iwabuchi, M.; Oda, K. Tolerance to various environmental stresses conferred by the salt-responsive rice gene *ONAC063* in transgenic Arabidopsis. *Planta* **2009**, *229*, 1065–1075. [CrossRef] [PubMed]

41. Sarukaba, Y.; Piao, W.L.; Lim, J.H.; Han, S.H.; Kim, Y.S.; An, G.; Paek, N.C. Rice ONAC106 inhibits leaf senescence and increases salt tolerance and tiller angle. *Plant Cell. Physiol.* **2015**, *56*, 2325–2339. [CrossRef]

42. Mao, C.J.; Ding, J.L.; Zhang, B.; Xi, D.D.; Ming, F. OsNAC2 positively affects salt-induced cell death and binds to the *OsAP37* and *OsCOX11* promoters. *Plant J.* **2018**, *94*, 454–468. [CrossRef] [PubMed]

43. Jiang, D.G.; Zhou, L.Y.; Chen, W.T.; Ye, N.H.; Xia, J.X.; Zhuang, C.X. Overexpression of a microRNA-targeted NAC transcription factor improves drought and salt tolerance in rice via ABA-mediated pathways. *Rice* **2019**, *12*, 76. [CrossRef] [PubMed]

44. Takasaki, H.; Maruyama, K.; Kidokor, S.; Ito, Y.; Fujita, Y.; Shinozaki, K.; Yamaguchi-Shinozaki, K.; Nakashima, K. The abiotic stress-responsive NAC-type transcription factor OsNAC5 regulates stress-inducible genes and stress tolerance in rice. *Mol. Genet. Genom.* **2010**, *284*, 173–183. [CrossRef] [PubMed]

45. Song, S.Y.; Chen, Y.; Chen, J.; Dai, X.Y.; Zhang, W.H. Physiological mechanisms underlying OsNAC5-dependent tolerance of rice plants to abiotic stress. *Planta* **2011**, *234*, 331–345. [CrossRef]

46. Nakashima, K.; Tran, L.S.P.; Nguyen, D.V.; Fujita, M.; Maruyama, K.; Todaka, D.; Ito, Y.; Hayashi, N.; Shinozaki, K.; Yamaguchi-Shinozaki, K. Functional analysis of a NAC-type transcription factor OsNAC6 involved in abiotic and biotic stress-responsive gene expression in rice. *Plant J.* **2007**, *51*, 617–630. [CrossRef]

47. Hu, H.H.; You, J.; Fang, Y.J.; Zhu, X.Y.; Qi, Z.Y.; Xiong, L.Z. Characterization of transcription factor gene *SNAC2* conferring cold and salt tolerance in rice. *Plant Mol. Biol.* **2010**, *67*, 169–181. [CrossRef]

48. Hu, H.H.; Dai, M.Q.; Yao, J.L.; Xiao, B.Z.; Li, X.H.; Zhang, Q.F.; Xiong, L.Z. Overexpressing a NAM ATAF and CUC (NAC) transcription factor enhances drought resistance and salt tolerance in rice. *Proc. Natl. Acad. Sci. USA* **2006**, *103*, 12987–12992. [CrossRef]

49. Jeong, J.S.; Kim, Y.K.; Baek, K.H.; Jung, H.; Ha, S.H.; Choi, Y.D.; Kim, M.; Reuzeau, C.; Kim, J.K. Root-specific expression of *OsNAC10* improves drought tolerance and grain yield in rice under field drought conditions. *Plant Physiol.* **2010**, *153*, 185–197. [CrossRef]

50. Wang, B.; Zhong, Z.H.; Hang, H.H.; Wang, X.; Liu, B.L.; Yang, L.J.; Han, X.Y.; Yu, D.S.; Zheng, X.L.; Wang, C.G.; et al. Targeted mutagenesis of NAC transcription factor gene *OsNAC041* leading to salt sensitivity in rice. *Rice* **2019**, *26*, 98–108. [CrossRef]

51. Yang, A.; Dai, X.Y.; Zhang, W.H. A R2R3-type MYB gene, OsMYB2, is involved in salt, cold, and dehydration tolerance in rice. *J. Exp. Bot.* **2012**, *63*, 2541–2556. [CrossRef]

52. Dai, X.; Xu, Y.; Ma, Q.; Xu, W.; Wang, T.; Xue, Y.; Chong, K. Overexpression of an R1R2R3 MYB gene, *OsMYB3R-2*, increases tolerance to freezing, drought, and salt stress in transgenic Arabidopsis. *Plant Physiol.* **2007**, *143*, 1739–1751. [CrossRef] [PubMed]

53. Xiong, H.Y.; Li, J.J.; Liu, P.L.; Duan, J.Z.; Zhao, Y.; Guo, X.; Li, Y.; Zhang, H.L.; Ali, J.; Li, Z.C. Overexpression of *OsMYB48-1* a novel MYB-related transcription factor enhances drought and salinity tolerance in rice. *PLoS ONE* **2014**, *9*, e92913. [CrossRef] [PubMed]

54. Schmidt, R.; Schippers, J.H.M.; Mieulet, D.; Obata, T.; Fernie, A.R.; Guiderdoni, E.; Mueller-Roeber, B. MULTIPASS, a rice R2R3-type MYB transcription factor, regulates adaptive growth by integrating multiple hormonal pathways. *Plant J.* **2013**, *76*, 258–273. [CrossRef]

55. Zhu, N.; Cheng, S.F.; Liu, X.Y.; Du, H.; Dai, M.Q.; Zhou, D.X.; Yang, W.J.; Zhao, Y. The R2R3-type MYB gene *OsMYB91* has a function in coordinating plant growth and salt stress tolerance in rice. *Plant Sci.* **2015**, *236*, 146–156. [CrossRef] [PubMed]

56. Wang, R.; Jing, W.; Xiao, L.Y.; Jin, Y.K.; Shen, L.K.; Zhang, W.H. The rice High-Affinity Potassium Transporter1;1 is involved in salt tolerance and regulated by an MYB-type transcription factor. *Plant Physiol.* **2015**, *168*, 1076–1090. [CrossRef]

57. Hossain, M.A.; Cho, J.; Han, M.; Ahn, C.H.; Jeon, J.S.; An, G.; Park, P.B. The ABRE-binding bZIP transcription factor OsABF2 is a positive regulator of abiotic stress and ABA signaling in rice. *J. Plant Physiol.* **2010**, *167*, 1512–1520. [CrossRef]

58. Zou, M.J.; Guan, Y.C.; Ren, H.B.; Zhang, F.; Chen, F. A bZIP transcription factor, OsABI5, is involved in rice fertility and stress tolerance. *Plant Mol. Biol.* **2008**, *66*, 675–683. [CrossRef]

59. Xiang, Y.; Tang, N.; Du, H.; Ye, H.Y.; Xiong, L.Z. Characterization of OsbZIP23 as a key player of the basic leucine zipper transcription factor family for conferring abscisic acid sensitivity and salinity and drought tolerance in rice. *Plant Physiol.* **2008**, *148*, 1938–1952. [CrossRef]

60. Liu, C.T.; Mao, B.G.; Ou, S.J.; Wang, W.; Liu, L.C.; Wu, Y.B.; Chu, C.C.; Wang, X.P. OsbZIP71, a bZIP transcription factor, confers salinity and drought tolerance in rice. *Plant Mol. Biol.* **2014**, *89*, 19–36. [CrossRef]

61. Das, P.; Lakra, N.; Nutan, K.K.; Singla-Pareek, S.L.; Pareek, A. A unique bZIP transcription factor imparting multiple stress tolerance in rice. *Rice* **2019**, *12*, 58. [CrossRef]

62. Huang, X.Y.; Chao, D.Y.; Gao, P.J.; Zhu, M.Z.; Shi, M.Z.; Lin, H.X. A previously unknown zinc finger protein, DST, regulates drought and salt tolerance in rice via stomatal aperture control. *Genes Dev.* **2009**, *23*, 1805–1817. [CrossRef]

63. Jan, A.M.K.; Todaka, D.; Kidokoro, S.; Abo, M.; Yoshimura, E.; Shinozaki, K.; Nakashima, K.; Yamaguchi-Shinozaki, K. *OsTZF1*, a CCCH-tandem zinc finger protein, confers delayed senescence and stress tolerance in rice by regulating stress-related genes. *Plant Physiol.* **2013**, *161*, 1202–1216. [CrossRef]

64. Sun, S.J.; Guo, S.Q.; Yang, X.; Bao, Y.M.; Tang, H.J.; Sun, H.; Huang, J.; Zhang, H.S. Functional analysis of a novel Cys2/His2-type zinc finger protein involved in salt tolerance in rice. *J. Exp. Bot.* **2010**, *61*, 2807–2818. [CrossRef]

65. Huang, J.; Sun, S.J.; Xu, D.Q.; Lan, H.X.; Sun, H.; Wang, Z.F.; Bao, Y.M.; Wang, J.F.; Tang, H.J.; Zhang, H.S. A TFIIIA-type zinc finger protein confers multiple abiotic stress tolerances in transgenic rice (*Oryza sativa* L.). *Plant Mol. Biol.* **2012**, *80*, 337–350. [CrossRef] [PubMed]

66. Zhang, Y.; Lan, H.X.; Shao, Q.L.; Wang, R.Q.; Chen, H.; Tang, H.J.; Zhang, H.S.; Huang, J. An A20/AN1-type zinc finger protein modulates gibberellins and abscisic acid contents and increases sensitivity to abiotic stress in rice (*Oryza sativa*). *J. Exp. Bot.* **2016**, *67*, 315–326. [CrossRef] [PubMed]

67. Xu, D.Q.; Huang, J.; Guo, S.Q.; Yang, X.; Bao, Y.M.; Tang, H.J.; Zhang, H.S. Overexpression of a TFIIIA-type zinc finger protein gene *ZFP252* enhances drought and salt tolerance in rice (*Oryza sativa* L.). *FEBS Lett.* **2008**, *582*, 1037–1043. [CrossRef] [PubMed]

68. Guan, Q.J.; Ma, H.Y.; Zhang, Z.J.; Wang, Z.Y.; Bu, Q.Y.; Liu, S.K. A rice LSD1-like-type ZFP gene *OsLOL5* enhances saline-alkaline tolerance in transgenic *Arabidopsis thaliana*, yeast and rice. *BMC Genom.* **2016**, *17*, 142. [CrossRef]

69. Chen, Y.; Li, F.; Ma, Y.; Ching, K.; Xu, Y. Overexpression of OrbHLH001, a putative helix-loop-helix transcription factor, causes increased expression of *AKT1* and maintains ionic balance under salt stress in rice. *J. Plant Physiol.* **2013**, *170*, 93–100. [CrossRef]

70. Chen, H.C.; Cheng, W.H.; Hong, C.Y.; Chang, Y.S.; Chang, M.C. The transcription factor OsbHLH035 mediates seed germination and enables seedling recovery from salt stress through ABA-dependent and ABA-independent pathways respectively. *Rice* **2018**, *11*, 50. [CrossRef] [PubMed]

71. Zhang, S.X.; Haider, I.; Kohlen, W.; Jiang, L.; Bouwmeester, H.; Meijer, A.H.; Schluepmann, H.; Liu, C.M.; Ouwerkerk, P.B.F. Function of the HD-Zip I gene *Oshox22* in ABA-mediated drought and salt tolerances in rice. *Plant Mol. Biol.* **2012**, *80*, 571–585. [CrossRef]

72. Bang, S.W.; Lee, D.K.; Jung, H.; Chung, P.J.; Kim, Y.S.; Choi, Y.D.; Suh, J.W.; Kim, J.K. Overexpression of *OsTF1L* a rice HD-Zip transcription factor promotes lignin biosynthesis and stomatal closure that improves drought tolerance. *Plant Biotechnol. J.* **2019**, *17*, 118–131. [CrossRef]

73. Wu, J.Y.; Yu, C.Y.; Huang, L.L.; Wu, M.J.; Liu, B.H.; Liu, Y.H.; Song, G.; Liu, D.D.; Gan, Y.B. Overexpression of MADS-box transcription factor OsMADS25 enhances salt stress tolerance in Rice and Arabidopsis. *Plant Growth Regul.* **2020**, *90*, 163–171. [CrossRef]

74. Tao, Z.; Kou, Y.J.; Liu, H.B.; Li, X.H.; Xiao, J.H.; Wang, S.P. *OsWRKY45* alleles play different roles in abscisic acid signalling and salt stress tolerance but similar roles in drought and cold tolerance in rice. *J. Exp. Bot.* **2011**, *62*, 4863–4874. [CrossRef]

75. Hasthanssombut, S.; Supaibulwatana, K.; Mii, M.; Nakamura, I. Genetic manipulation of Japonica rice using the *OsBADH1* from Indica rice to improve salinity tolerance. *Plant Cell Tissue Organ. Cult.* **2011**, *104*, 79–89. [CrossRef]

76. Tang, W.; Sun, J.Q.; Liu, J.; Liu, F.F.; Yan, J.; Guo, X.J.; Lu, B.R.; Liu, Y.S. RNAi-directed downregulation of *betaine aldehyde dehydrogenase 1 (OsBADH1)* results in decreased stress tolerance and increased oxidative markers without affecting glycine betaine biosynthesis in rice (*Oryza sativa*). *Plant Mol. Biol.* **2014**, *86*, 443–454. [CrossRef]

77. Ge, L.F.; Chao, D.Y.; Shi, M.; Zhu, M.Z.; Gao, J.P.; Lin, H.X. Overexpression of the trehalose-6-phosphate phosphatase gene *OsTPP1* confers stress tolerance in rice and results in the activation of stress responsive genes. *Planta* **2008**, *228*, 191–201. [CrossRef] [PubMed]

78. Li, H.W.; Zhang, B.S.; Deng, X.W.; Wang, X.P. Overexpression of the trehalose-6-phosphate synthase gene *OsTPS1* enhances abiotic stress tolerance in rice. *Planta* **2011**, *234*, 1007–1018. [CrossRef] [PubMed]

79. Vishal, B.; Krishnamurthy, P.; Ramamoorthy, R.; Kumar, P.P. *OsTPS8* controls yield-related traits and confers salt stress tolerance in rice by enhancing suberin deposition. *New Phytol.* **2018**, *221*, 1369–1386. [CrossRef] [PubMed]

80. Guo, L.; Wang, Z.Y.; Lin, H.; Cui, W.E.; Chen, J.; Liu, M.; Chen, Z.L.; Qu, L.J.; Gu, H. Expression and functional analysis of the rice plasma-membrane intrinsic protein gene family. *Cell Res.* **2006**, *16*, 277–286. [CrossRef] [PubMed]

81. Liu, C.W.; Fukumoto, T.; Matsumoto, T.; Gena, P.; Frascaria, D.; Kaneko, T.; Katsuhara, M.; Zhong, S.H.; Sun, X.L.; Zhu, Y.M.; et al. Aquaporin OsPIP1;1 promotes rice salt resistance and seed germination. *Plant Physiol. Biochem.* **2013**, *63*, 151–158. [CrossRef] [PubMed]

82. Wu, F.Q.; Sheng, P.K.; Tan, J.J.; Chen, X.L.; Lu, G.W.; Ma, W.W.; Heng, Y.Q.; Lin, Q.B.; Zhu, S.S.; Wang, J.L.; et al. Plasma membrane receptor-like kinase leaf panicle 2 acts downstream of the DROUGHT AND SALT TOLERANCE transcription factor to regulate drought sensitivity in rice. *J. Exp. Bot.* **2015**, *66*, 271–281. [CrossRef] [PubMed]

83. You, J.; Zong, W.; Li, X.K.; Ning, J.; Hu, H.H.; Li, X.H.; Xiao, J.H.; Xiong, L.Z. The SNAC1-targeted gene *OsSRO1c* modulates stomatal closure and oxidative stress tolerance by regulating hydrogen peroxide in rice. *J. Exp. Bot.* **2013**, *64*, 569–583. [CrossRef]

84. Guan, Q.; Liao, X.; He, M.; Li, X.; Wang, Z.; Ma, H.; Yu, S.; Liu, S. Tolerance analysis of chloroplast *OsCu/Zn-SOD* overexpressing rice under NaCl and NaHCO$_3$ stress. *PLoS ONE* **2017**, *12*, e0186052. [CrossRef] [PubMed]

85. Tanaka, Y.; Hibino, T.; Tanaka, A.; Kishitani, S.; Takabe, T.; Yokota, S.; Takabe, T. Salt tolerance of transgenic rice overexpressing yeast mitochondrial Mn-SOD in chloroplasts. *Plant Sci.* **1999**, *148*, 131–138. [CrossRef]

86. Lu, Z.Q.; Liu, D.L.; Liu, S.K. Two rice cytosolic ascorbate peroxidases differentially improve salt tolerance in transgenic *Arabidopsis*. *Plant Cell Rep.* **2007**, *26*, 1909–1917. [CrossRef] [PubMed]

87. Zhang, Z.G.; Zhang, Q.A.; Wu, J.X.; Zheng, X.; Zheng, S.; Sun, X.H.; Qiu, Q.S.; Lu, T.G. Gene knockout study reveals that cytosolic ascorbate peroxidase 2 (OsAPX2) plays a critical role in growth and reproduction in rice under drought, salt and cold stresses. *PLoS ONE* **2013**, *8*, e57472. [CrossRef]

88. Wu, T.M.; Lin, W.R.; Kao, Y.T.; Hsu, Y.T.; Yeh, C.H.; Hong, C.Y.; Kao, C.H. Identification and characterization of a novel chloroplast/mitochondria co-localized glutathione reductase 3 involved in salt stress response in rice. *Plant Mol. Biol.* **2013**, *83*, 379–390. [CrossRef] [PubMed]

89. Zhang, C.J.; Guo, Y. OsTRXh1 regulates the redox state of the apoplast and influences stress responses in rice. *Plant Signal Behav.* **2012**, *7*, 440–442. [CrossRef]

90. Sharma, R.; Priya, P.; Jain, M. Modified expression of an auxin-responsive rice CC-type glutaredoxin gene affects multiple abiotic stress responses. *Planta* **2013**, *238*, 871–884. [CrossRef]

91. Ning, X.; Sun, Y.; Wang, C.C.; Zhang, W.L.; Sun, M.H.; Hu, H.T.; Liu, J.Z.; Yang, L. A rice CPYC-type glutaredoxin *OsGRX20* in protection against bacterial blight methyl viologen and salt stresses. *Front. Plant Sci.* **2018**, *9*, 111. [CrossRef] [PubMed]

92. Campbell, M.T.; Bandillo, N.; Razzaq, F.; Shiblawi, A.; Sharma, S.; Liu, K.; Schmitz, A.J.; Zhang, C.; Véry, A.A.; Lorenz, A.J.; et al. Allelic variants of *OsHKT1;1* underlie the divergence between indica and japonica subspecies of rice (*Oryza sativa*) for root sodium content. *PloS Genet.* **2017**, *13*, e1006823. [CrossRef] [PubMed]

93. Oda, Y.; Kobayashi, N.I.; Tanoi, K.; Ma, J.F.; Itou, Y.; Katsuhara, M.; Itou, T.; Horie, T. T-DNA tagging-based gain-of-function of OsHKT1;4 reinforces Na$^+$ exclusion from leaves and stems but triggers Na toxicity in roots of rice under salt stress. *Int. J. Mol. Sci.* **2018**, *19*, 235. [CrossRef] [PubMed]

94. Ren, Z.H.; Gao, J.P.; Li, L.G.; Cai, X.L.; Huang, W.; Chao, D.Y.; Zhu, M.Z.; Wang, Z.Y.; Luan, S.; Lin, H.X. A rice quantitative trait locus for salt tolerance encodes a sodium transporter. *Nat. Genet.* **2005**, *37*, 1141–1146. [CrossRef]

95. El Mahi, H.; Pérez-Hormaeche, J.; De Luca, A.; Villalta, I.; Espartero, J.; Gámez-Arjona, F.; Fernández, J.L.; Bundó, M.; Mendoza, I.; Mieulet, D.; et al. A critical role of sodium flux via the plasma membrane Na$^+$/H$^+$ exchanger SOS1 in the salt tolerance of rice. *Plant Physiol.* **2019**, *180*, 1046–1065. [CrossRef]

96. Liu, S.P.; Zheng, L.Q.; Xue, Y.H.; Zhang, Q.; Wang, L.; Shuo, H.X. Overexpression of *OsVP1* and *OsNHX1* increases tolerance to drought and salinity in rice. *J. Plant Biol.* **2010**, *53*, 444–452. [CrossRef]

97. Chen, G.; Hu, Q.; Luo, L.; Yang, T.; Zhang, S.; Hu, Y.; Xu, G. Rice potassium transporter OsHAK1 is essential for maintaining potassium-mediated growth and functions in salt tolerance over low and high potassium concentration ranges. *Plant Cell. Environ.* **2009**, *38*, 2747–2765. [CrossRef]

98. Yang, T.Y.; Zhang, S.; Hu, Y.B.; Wu, F.C.; Hu, Q.D.; Chen, G.; Cai, J.; Wu, T.; Moran, N.; Yu, L.; et al. The role of a potassium transporter OsHAK5 in potassium acquisition and transport from roots to shoots in rice at low potassium supply levels. *Plant Physiol.* **2014**, *166*, 945–959. [CrossRef] [PubMed]

99. Feng, H.M.; Tang, Q.; Cai, J.; Xu, B.C.; Xu, G.H.; Yu, L. Rice *OsHAK16* functions in potassium uptake and translocation in shoot maintaining potassium homeostasis and salt tolerance. *Planta* **2019**, *250*, 549–561. [CrossRef]

100. Shen, Y.; Shen, L.; Shen, Z.J.W.; Ge, H.; Zhao, J.; Zhang, W. The potassium transporter OsHAK 21 functions in the maintenance of ion homeostasis and tolerance to salt stress in rice. *Plant Cell. Environ.* **2015**, *38*, 2766–2779. [CrossRef] [PubMed]

101. Nongpiur, R.; Soni, P.; Karan, R.; Singla-Pareek, S.L.; Pareek, A. Histidine kinases in plants. *Plant Signal Behav.* **2012**, *7*, 1230–1237. [CrossRef] [PubMed]

102. Tran, L.S.P.; Urao, T.; Qin, F.; Maruyama, K.; Kakimoto, T.; Shonozaki, K.; Yamaguchi-Shinozaki, K. Functional analysis of AHK1/ATHK1 and cytokinin receptor histidine kinases in response to abscisic acid drought and salt stress in *Arabidopsis*. *Proc. Natl. Acad. Sci. USA* **2007**, *104*, 20623–20628. [CrossRef] [PubMed]

103. Kumar, M.N.; Jane, W.N.; Verslues, P.E. Role of the putative osmosensor Arabidopsis *histidine kinase1* in dehydration avoidance and low-water-potential response. *Plant Physiol.* **2013**, *161*, 942–953. [CrossRef]

104. Kushwaha, H.R.; Singla-Pareek, S.L.; Pareek, A. Putative osmosensor–OsHK3b–a histidine kinase protein from rice shows high structural conservation with its ortholog AtHK1 from *Arabidopsis*. *J. Biomol. Struct. Dyn.* **2014**, *32*, 1318–1332. [CrossRef]

105. Ozakabe, Y.; Yamaguchi-Shinozaki, K.; Shinozaki, K.; Tran, L.S.P. Sensing the environment: Key roles of membrane-localized kinases in plant perception and response to abiotic stress. *J. Exp. Bot.* **2013**, *64*, 445–458. [CrossRef]

106. Zhao, J.L.; Zhang, L.Q.; Liu, N.; Xu, S.L.; Yue, Z.L.; Zhang, L.L.; Deng, Z.P.; Burlingame, A.L.; Sun, D.Y.; Wang, Z.Y.; et al. Mutual regulation of receptor-like kinase SIT1 and B′κ-PP2A shapes the early response of rice to salt stress. *Plant Cell* **2019**, *31*, 2131–2151. [CrossRef]

107. Yuan, F.; Yang, H.; Xue, Y.; Kong, D.; Ye, R.; Li, C.; Zhang, J.; Theprungsirikul, L.; Shrift, T.; Krichilsky, B. OSCA1 mediates osmotic-stress-evoked Ca^{2+} increases vital for osmosensing in *Arabidopsis*. *Nature* **2014**, *514*, 367–371. [CrossRef] [PubMed]

108. Li, Y.S.; Yuan, F.; Wen, Z.H.; Li, Y.H.; Wang, F.; Zhu, T.; Zhou, W.Q.; Jin, X.; Wang, Y.D.; Zhao, H.P.; et al. Genome-wide survey and expression analysis of the OSCA gene family in rice. *BMC Plant Biol.* **2015**, *15*, 261. [CrossRef] [PubMed]

109. Shi, H.Z.; Ishitani, M.; Kim, C.S.; Zhu, J.K. The *Arabidopsis thaliana* salt tolerance gene *SOS1* encodes a putative Na^+/H^+ antiporter. *Proc. Natl. Acad. Sci. USA* **2000**, *97*, 6896–6901. [CrossRef]

110. Zhu, J.K. Regulation of ion homeostasis under salt stress. *Curr. Opin. Plant Biol.* **2003**, *6*, 441–445. [CrossRef]

111. Kader, M.A.; Lindberg, S.; Seidel, T.; Golldack, D.; Yemelyanov, V. Sodium sensing induces different changes in free cytosolic calcium concentration and pH in salt-tolerant and -sensitive rice (*Oryza sativa*) cultivars. *Physiol. Plantarum.* **2007**, *130*, 99–111. [CrossRef]

112. Apel, K.; Hirt, H. Reactive oxygen species: Metabolism oxidative stress and signal transduction. *Annu. Rev. Plant Biol.* **2004**, *55*, 373–399. [CrossRef]

113. Batistic, O.; Kudla, J. Integration and channeling of calcium signaling through the CBL calcium sensor/CIPK protein kinase network. *Planta* **2004**, *219*, 915–924. [CrossRef] [PubMed]

114. Bouché, N.; Yellin, A.; Snedded, W.A.; Fromm, H. Plant-specific calmodulin-binding proteins. *Annu. Rev. Plant Sci.* **2005**, *56*, 435–466. [CrossRef] [PubMed]

115. Zeng, H.Q.; Xu, L.Q.; Singh, A.; Wang, H.Z.; Du, L.Q.; Poovaiah, B.W. Involvement of calmodulin and calmodulin-like proteins in plant responses to abiotic stresses. *Front. Plant Sci.* **2015**, *6*, 600. [CrossRef]

116. Yuenyong, W.; Chinpongpanich, A.; Comai, L.; Chadchawan, S.; Buaboocha, T. Downstream components of the calmodulin signaling pathway in the rice salt stress response revealed by transcriptome profiling and target identification. *BMC Plant Biol.* **2018**, *18*, 335. [CrossRef] [PubMed]

117. Katou, S.; Kuroda, K.; Seo, S.; Yanagawa, Y.; Tsuge, T.; Yamazaki, M.; Miyao, A.; Hirochika, H.; Ohashi, Y. A calmodulin-binding mitogen-activated protein kinase phosphatase is induced by wounding and regulates the activities of stress-related mitogen-activated protein kinases in rice. *Plant Cell. Physiol.* **2007**, *48*, 332–344. [CrossRef]

118. Yang, J.; Liu, S.; Ji, L.X.; Tang, X.Y.; Zhu, Y.S.; Xie, G.S. Identification of novel OsCML16 target proteins and differential expression analysis under abiotic stresses in rice. *J. Plant Physiol.* **2020**, *249*, 153–165. [CrossRef] [PubMed]

119. Xu, G.Y.; Li, S.Z.; Xie, K.; Zhang, Q.; Wang, Y.; Tang, Y.; Hong, Y.G.; He, C.Y.; Liu, Y.L. Plant ERD2-like proteins function as endoplasmic reticulum luminal protein receptors and participate in programmed cell death during innate immunity. *Plant J.* **2012**, *72*, 57–69. [CrossRef]

120. Asano, T.; Tanaka, N.; Yang, G.X.; Hayashi, N.; Komatsu, S. Genome-wide Identification of the Rice Calcium-dependent Protein Kinase and its Closely Related Kinase Gene Families: Comprehensive Analysis of the *CDPKs* Gene Family in Rice. *Plant Cell. Physiol.* **2005**, *46*, 356–366. [CrossRef]

121. Huang, S.L.; Jiang, S.F.; Liang, J.S.; Chen, M. Roles of plant CBL-CIPK systems in abiotic stress responses. *Turk. J. Bot.* **2019**, *43*, 271–280. [CrossRef]

122. Martínez-Atienza, J.; Jiang, X.; Garciadeblas, B.; Mendoza, I.; Zhu, J.K.; Pardo, J.M.; Quintero, F.J. Conservation of the salt overly sensitive pathway in rice. *Plant Physiol.* **2007**, *143*, 1001–1012. [CrossRef]

123. Yang, C.W.; Zhang, T.Y.; Wang, H.; Zhao, N.; Liu, B. Heritable alteration in salt tolerance in rice induced by introgression from wild rice (*Zizania latifolia*). *Rice* **2012**, *5*, 36. [CrossRef] [PubMed]

124. Kanwar, P.; Sanyal, S.K.; Tokas, I.; Yadav, A.K.; Pandey, A.; Kapoor, S.; Pandey, G.K. Comprehensive structural interaction and expression analysis of CBL and CIPK complement during abiotic stresses and development in rice. *Cell Calcium* **2014**, *56*, 81–95. [CrossRef]

125. Chen, X.; Gu, Z.; Liu, F.; Ma, B.; Zhang, H. Molecular analysis of rice CIPKs Involved in both biotic and abiotic stress responses. *Rice Sci.* **2016**, *18*, 1–9. [CrossRef]

126. Mittler, R.; Vanderauwera, S.; Gollery, M.; Van Breusegem, F. Reactive oxygen gene network of plants. *Trends Plant Sci.* **2004**, *9*, 490–498. [CrossRef] [PubMed]

127. Rohila, J.S.; Yang, Y. Rice mitogen-activated protein kinase gene family and its role in biotic and abiotic stress response. *J. Integr. Plant Biol.* **2007**, *49*, 751–759. [CrossRef]

128. Rodriguez, M.C.; Petersen, M.; Mundy, J. Mitogen-activated protein kinase signaling in plants. *Annu. Rev. Plant Biol.* **2010**, *61*, 621–649. [CrossRef]

129. MAPK Group. Mitogen-activated protein kinase cascades in plants: A new nomenclature. *Trends Plant Sci.* **2002**, *7*, 301–308. [CrossRef]

130. Sinha, A.K.; Jaggi, M.; Raghuram, B.; Tuteja, N. Mitogen-activated protein kinase signalling in plants under abiotic stress. *Plant Signal Behav.* **2011**, *16*, 196–203. [CrossRef]

131. Kumar, K.; Wankhede, D.P.; Sinha, A.K. Signal convergence through the lenses of MAP kinases: Paradigms of stress and hormone signaling in plants. *Front. Biol.* **2013**, *8*, 109–118. [CrossRef]

132. Yoo, S.J.; Kim, S.H.; Kim, M.J.; Ryu, C.M.; Kim, Y.C.; Cho, B.H.; Yang, K.Y. Involvement of the OsMKK4-OsMPK1 cascade and its downstream transcription factor OsWRKY53 in the wounding response in rice. *Plant Path J.* **2014**, *30*, 168–177. [CrossRef] [PubMed]

133. Bohnert, H.J.; Ayoubi, P.; Borchert, C.; Bressan, R.A.; Burnap, R.L.; Cushman, J.C.; Deyholos MFischer RGalbraith, D.W.; Hasegawa, P.M.; Jenks, M.; Kawasaki, M.; et al. A genomics approach towards salt stress tolerance. *Plant Physiol. Biochem.* **2001**, *39*, 295–311. [CrossRef]

134. Zhu, J.K. Plant salt tolerance. *Trends Plant Sci.* **2001**, *6*, 66–71. [CrossRef]

135. Seki, M.; Narusaka, M.; Abe, H.; Kasuga, M.; Yamaguchi-Shinozaki, K.; Carninci, P.; Hayashizaki, Y.; Shinozaki, K. Monitoring the expression pattern of 1300 Arabidopsis genes under drought and cold stresses by using a full-length cDNA microarray. *Plant Cell* **2001**, *13*, 61–72. [CrossRef]

136. Seki, M.; Ishida, J.; Narusaka, M.; Fujita, M.; Nanjo, T.; Umezawa, T.; Kamiya, A.; Nakajima, M.; Enju, A.; Sakurai, T.; et al. Monitoring the expression pattern of 7000 Arabidopsis genes under ABA treatments by using a full-length cDNA microarray. *Plant J.* **2020**, *2*, 282–291. [CrossRef]

137. Gao, G.; Zhong, Y.F.; Guo, A.; Zhu, Q.; Tang, W.; Zheng, W. DRTF: A database of rice transcription factors. *Bioinformatics* **2006**, *22*, 1286–1287. [CrossRef]

138. Feng, J.X.; Liu, D.; Pan, Y.; Gong, W.; Ma, L.G.; Luo, J.C.; Deng, X.W.; Zhu, Y. An annotation update via cDNA sequence analysis and comprehensive profiling of developmental hormonal or environmental responsiveness of the *Arabidopsis* AP2/EREBP transcription factor gene family. *Plant Mol. Biol.* **2005**, *59*, 853–868. [CrossRef]

139. Agarwal, P.K.; Gupta, K.; Lopato, S.; Agarwal, P. Dehydration responsive element binding transcription factors and their applications for the engineering of stress tolerance. *J. Exp. Bot.* **2017**, *68*, 2135–2148. [CrossRef]

140. Ito, Y.; Katsura, K.; Maruyama, K.; Taji, T.; Kobayashi, M.; Seki, M.; Shinozaki, K.; Yamaguchi-Shinozaki, K. Functional analysis of rice DREB1/CBF-type transcription factors involved in cold-responsive gene expression in transgenic rice. *Plant Cell. Physiol.* **2006**, *47*, 141–153. [CrossRef]

141. Chen, J.Q.; Meng, X.P.; Zhang, Y.; Xia, M.; Wang, X.P. Over-expression of OsDREB genes lead to enhanced drought tolerance in rice. *Biotechnol. Lett.* **2008**, *30*, 2191–2198. [CrossRef]

142. Danyluk, J.; Perron, A.; Houde, M.; Limin, A.; Fowler, B.; Benhamou, N.; Sarhan, F. Accumulation of an acidic dehydrin in the vicinity of the plasma membrane during cold acclimation of wheat. *Plant Cell* **1998**, *10*, 623–638. [CrossRef]

143. Khan, M.S. The role of DREB transcription factors in abiotic stress tolerance of plants. *Agric. Environ. Biotechnol.* **2011**, *25*, 2433–2442. [CrossRef]

144. Puranik, S.; Sahu, P.P.; Srivastava, P.S.; Prasad, M. NAC proteins: Regulation and role in stress tolerance. *Trends Plant Sci.* **2007**, *17*, 369–381. [CrossRef]

145. Li, C.N.; Ng, C.K.Y.; Fan, L.M. MYB transcription factors active players in abiotic stress signaling. *Environ. Exp. Bot.* **2015**, *114*, 80–91. [CrossRef]

146. Li, J.; Han, G.; Sun, C.; Sui, N. Research advances of MYB transcription factors in plant stress resistance and breeding. *Trends Plant Sci.* **2019**, *14*, 1–9. [CrossRef] [PubMed]

147. Lakra, N.; Nutan, K.K.; Das, P.; Anwar KSingla-Pareek, S.L.; Pareek, A. A nuclear-localized histone-gene binding protein from rice (OsHBP1b) functions in salinity and drought stress tolerance by maintaining chlorophyll content and improving the antioxidant machinery. *J. Plant Physiol.* **2015**, *176*, 36–46. [CrossRef] [PubMed]

148. Li, W.T.; He, M.; Wang, J.; Wang, Y.P. Zinc finger protein (ZFP) in plants-A review. *Plant Omics J.* **2013**, *6*, 474–480.

149. Murre, C.; Bain, G.; Dijk, M.A.V.; Engel, I.; Furnari, B.A.; Massari, M.E.; Matthews, J.R.; Quong, M.W.; Rivera, R.R.; Stuiver, M.H. Structure and function of helix–loop–helix proteins. *BBA Biomembr.* **1994**, *1218*, 129–135. [CrossRef]

150. Wu, H.; Ye, H.Y.; Yao, R.F.; Zhang, T.; Xiong, L.Z. OsJAZ9 acts as a transcriptional regulator in jasmonate signaling and modulates salt stress tolerance in rice. *Plant Sci.* **2015**, *232*, 1–12. [CrossRef]

151. Sun, X.; Wang, Y.; Sui, N. Transcriptional regulation of bHLH during plant response to stress. *Biochem. Biophys. Res. Commun.* **2018**, *503*, 397–401. [CrossRef] [PubMed]

152. Almeida, D.M.; Gregorio, G.; Oliveira, M.M.; Saibo, N.J.M. Five novel transcription factors as potential regulators of *OsNHX1* gene expression in a salt tolerant rice genotype. *Plant Mol. Biol.* **2017**, *93*, 61–77. [CrossRef] [PubMed]

153. Rontein, D.; Basset, G.; Hanson, A.D. Metabolic engineering of osmoprotectant accumulation in plants. *Metab. Eng.* **2002**, *4*, 49–56. [CrossRef] [PubMed]

154. Abdallah, M.M.S.; Abdelgawad, Z.A.; El-Bassiuny, H.M.S. Alleviation of the adverse effects of salinity stress using trehalose in two rice varieties. *South Afr. J. Bot.* **2016**, *103*, 275–282. [CrossRef]

155. Bhusan, D.; Das, D.K.; Hossain, M.; Murata, Y.; Hoque, M.A. Improvement of salt tolerance in rice (*Oryza sativa* L.) by increasing antioxidant defense systems using exogenous application of proline. *Aus. J. Crop. Sci.* **2016**, *10*, 50–56.

156. Demiral, T.; Türkan, I. Exogenous glycine betaine affects growth and proline accumulation and retards senescence in two rice cultivars under NaCl stress. *Environ. Exp. Bot.* **2005**, *56*, 72–79. [CrossRef]

157. Harinasut, P.; Tsutsui, K.; Takabe, T.; Nomura, M.; Takabe, T.; Kishitani, S. Exogenous glycinebetaine accumulation and increased salt-tolerance in rice seedlings. *Biosci. Biotech. Bioch.* **1996**, *60*, 366–368. [CrossRef]

158. Wang, Q.; Ding, T.; Zuo, J.; Gao, L.; Fan, L. Amelioration of postharvest chilling injury in sweet pepper by glycine betaine. *Postharvest Biol. Technol.* **2016**, *112*, 114–120. [CrossRef]

159. Chen, T.H.H.; Murata, N. Glycine betaine protects plants against abiotic stress: Mechanisms and biotechnological applications. *Plant Cell. Environ.* **2010**, *34*, 1–20. [CrossRef]

160. Chaumont, F.; Tyerman, S.D. Aquaporins: Highly regulated channels controlling plant water relations. *Plant Physiol.* **2014**, *164*, 1600–1618. [CrossRef]

161. Maurel, C.; Boursiac, Y.; Luu, D.T.; Santoni, V.; Shahzad, Z.; Verdoucq, L. Aquaporins in plants. *Physiol. Rev.* **2015**, *95*, 1321–1358. [CrossRef] [PubMed]

162. Sakurai, J.; Ishikawa, F.; Yamaguchi, T.; Uemura, M.; Maeshima, M. Identification of 33 rice aquaporin genes and analysis of their expression and function. *Plant Cell. Physiol.* **2005**, *46*, 1568–1577. [CrossRef] [PubMed]

163. Kawasaki, S.; Borchert, C.; Deyholos, M.; Wang, H.; Brazille, S.; Kawai, K.; Galbraith, D.; Bohnert, H.J. Gene expression profiles during the initial phase of salt stress in rice. *Plant Cell* **2001**, *13*, 889–905. [CrossRef]

164. Song, Y.; Miao, Y.; Song, C.P. Behind the scenes: The roles of reactive oxygen species in guard cells. *New Phytol.* **2014**, *201*, 1121–1140. [CrossRef] [PubMed]

165. Cui, L.G.; Shan, J.X.; Shi, M.; Gao, J.P.; Lin, H.X. DCA1 acts as a transcriptional co-activator of DST and contributes to drought and salt tolerance in rice. *PLoS Genet.* **2015**, *11*, e1005617. [CrossRef] [PubMed]

166. Gill, S.S.; Tuteja, N. Reactive oxygen species and antioxidant machinery in abiotic stress tolerance in crop plants. *Plant Physiol. Biochem.* **2010**, *48*, 909–930. [CrossRef]

167. Alscher, R.G.; Erturk, N.; Heath, L.S. Role of superoxide dismutases (SODs) in controlling oxidative stress in plants. *J. Exp. Bot.* **2002**, *53*, 1331–1341. [CrossRef]

168. Mishra, P.; Bhoomika, K.; Dubey, R.S. Differential responses of antioxidative defense system to prolonged salinity stress in salt-tolerant and salt-sensitive Indica rice (*Oryza sativa* L.) seedlings. *Protoplasma* **2013**, *250*, 3–19. [CrossRef]

169. Rossatto, T.; do Amaral, M.N.; Benitz, L.C.; Vighi, I.L.; Braga, E.J.B.; Júnior, A.M.D.; Maia, M.A.C.; Pinto, L.D. Gene expression and activity of antioxidant enzymes in rice plants, cv. BRS AG, under saline stress. *Physiol. Mol. Biol. Plants* **2017**, *23*, 865–875. [CrossRef] [PubMed]

170. Lee, S.H.; Ahsan, N.; Lee, K.W.; Kim, D.H.; Lee, D.G.; Kwak, S.S. Simultaneous overexpression of both Cu/Zn superoxide dismutase and ascorbate peroxidase in transgenic tall fescue plants confers increased tolerance to a wide range of abiotic stresses. *J. Plant Physiol.* **2007**, *164*, 1626–1638. [CrossRef]

171. Wang, Y.; Ying, Y.; Chen, J.; Wang, X. Transgenic Arabidopsis, overexpressing Mn-SOD enhanced salt-tolerance. *Plant Sci.* **2004**, *167*, 671–677. [CrossRef]

172. Cheng, Y.W.; Kong, X.W.; Wang, N.; Wang, T.T.; Chen, J.; Shi, Z.Q. Thymol confers tolerance to salt stress by activating anti-oxidative defense and modulating Na$^+$ homeostasis in rice root. *Ecotoxicol. Environ. Saf.* **2020**, *188*, 109894. [CrossRef] [PubMed]

173. Mhamdi, A.; Queval, G.; Chaouch, S.; Vanderauwera, S.; Breusegem, F.V.; Noctor, G. Catalase function in plants: A focus on Arabidopsis mutants as stress-mimic models. *J. Exp. Bot.* **2010**, *61*, 4197–4220. [CrossRef] [PubMed]

174. Yang, T.; Poovaiah, B.W. Hydrogen peroxide homeostasis: Activation of plant catalases by calcium/calmodulin. *Proc. Natl. Acad. Sci. USA* **2002**, *99*, 4097–4102. [CrossRef] [PubMed]

175. Joo, J.; Lee, Y.H.; Song, S.I. Rice CatA, CatB, and CatC are involved in environmental stress response, root growth, and photorespiration, respectively. *J. Plant Biol.* **2002**, *57*, 375–382. [CrossRef]

176. Zhou, Y.B.; Liu, C.; Tang, D.Y.; Yan, L.; Wang, D.; Yang, Y.Z.; Gui, J.S.; Zhao, X.Y.; Li, L.G.; Tang, X.D.; et al. The receptor-like cytoplasmic kinase STRK1 phosphorylates and activates CatC, thereby regulating H$_2$O$_2$ homeostasis and improving salt tolerance in rice. *Plant Cell* **2018**, *30*, 1100–1118. [CrossRef]

177. Kibria, M.G.; Hossain, M.; Murata, Y.; Hoque, M.A. Antioxidant defense mechanisms of salinity tolerance in rice genotypes. *Rice Sci.* **2017**, *24*, 155–162. [CrossRef]

178. Yamane, K.; Mitsuya, S.; Taniguchi, M.; Miyake, H. Transcription profiles of genes encoding catalase and ascorbate peroxidase in the rice leaf tissues under salinity. *Plant Prod. Sci.* **2010**, *13*, 164–168. [CrossRef]

179. Wutipraditkul, N.; Boonkomrat, S.; Buaboocha, T. Cloning and characterization of catalases from rice, Oryza sativa L. *Biosci Biotechnol Biochem* **2011**, *75*, 1900–1906. [CrossRef]

180. Teixeira, F.K.; Menezes-Benzvente, L.; Margis, R.; Margis-Pinheiro, M. Analysis of the molecular evolutionary history of the ascorbate peroxidase gene family: Inferences from the rice genome. *J. Mol. Evol.* **2004**, *59*, 761–770. [CrossRef]

181. Hong, C.Y.; Hsu, Y.T.; Tsai, Y.C.; Kao, C.H. Expression of *Ascorbate Peroxidase 8* in roots of rice (*Oryza sativa* L) seedlings in response to NaCl. *J. Exp. Bot.* **2007**, *58*, 3273–3283. [CrossRef]

182. Bonifacio, A.; Martins, M.O.; Ribeiro, C.V.; Fontenele, A.V.; Carvalho, F.E.; Margis-Pinheiro, M.; Silveira, J.G. Role of peroxidases in the compensation of cytosolic ascorbate peroxidase knockdown in rice plants under abiotic stress. *Plant Cell. Environ.* **2011**, *34*, 1705–1722. [CrossRef]

183. Rosa, S.B.; Caverzan, A.; Teixeira, F.K.; Lazzarotto, F.; Silveira, J.A.G.; Ferreira-Silva, S.L.; Abreu-Neto, J.; Margis, R.; Margis-Pinheiro, M. Cytosolic APx knockdown indicates an ambiguous redox response in rice. *Phytochem* **2010**, *71*, 548–558. [CrossRef]

184. Noctor, G.; Foyer, C.H. Ascorbate and glutathione: Keeping active oxygen under control. *Annu. Rev. Plant Physiol.* **1998**, *49*, 249–279. [CrossRef] [PubMed]

185. Wu, T.M.; Lin, W.R.; Kao, C.H.; Hong, C.Y. Gene knockout of *glutathione reductase 3* results in increased sensitivity to salt stress in rice. *Plant Mol. Biol.* **2015**, *87*, 555–564. [CrossRef] [PubMed]

186. Hong, C.Y.; Chao, Y.Y.; Yang, M.Y.; Cheng, S.Y.; Cho, S.C.; Kao, C.H. NaCl-induced expression of glutathione reductase in roots of rice (*Oryza sativa* L.) seedlings is mediated through hydrogen peroxide but not abscisic acid. *Plant Soil* **2009**, *320*, 103–115. [CrossRef]

187. Tsai, Y.C.; Hong, Y.C.; Kiu, L.F.; Kao, C.H. Expression of ascorbate peroxidase and glutathione reductase in roots of rice seedlings in response to NaCl and H_2O_2. *J. Plant Physiol.* **2005**, *162*, 291–299. [CrossRef] [PubMed]

188. Holmgren, A. Thioredoxin and glutaredoxin systems. *J. Biol. Chem.* **1989**, *264*, 13963–13966. [CrossRef]

189. Garg, R.; Jhanwar, S.; Tyagi, A.K.; Jain, M. Genome-wide survey and expression analysis suggest diverse roles of glutaredoxin gene family members during development and response to various stimuli in rice. *DNA Res.* **2010**, *17*, 353–367. [CrossRef]

190. Nuruzzaman, M.; Gupta, M.; Zhang, C.; Wang, L.; Xie, W.; Xiong, L.; Zhang, Q.; Lian, X. Sequence and expression analysis of the thioredoxin protein gene family in rice. *Mol. Genet. Genom.* **2008**, *280*, 139–151. [CrossRef]

191. Zhang, C.J.; Zhao, B.C.; Ge, W.N.; Zhang, F.Y.; Song, Y.; Sun, D.Y.; Guo, Y. An apoplastic H-type thioredoxin is involved in the stress response through regulation of the apoplastic reactive oxygen species in rice. *Plant Physiol.* **2011**, *157*, 1884–1899. [CrossRef] [PubMed]

192. Xie, G.S.; Kato, H.; Sasaki, K.; Imai, R. A cold-induced thioredoxin h of rice OsTrx23 negatively regulates kinase activities of OsMPK3 and OsMPK6 in vitro. *Febs Lett.* **2009**, *583*, 2734–2738. [CrossRef] [PubMed]

193. Blumwald, E.; Aharon, G.S.; Apse, M.P. Sodium transport in plant cells. *Biochem. Biophys. Acta* **2000**, *1465*, 140–151. [CrossRef]

194. Wu, S.J.; Ding, L.; Zhu, J.K. SOS1, a genetic locus essential for salt tolerance and potassium acquisition. *Plant Cell.* **1996**, *8*, 617–627. [CrossRef] [PubMed]

195. Garciadeblas, B.; Senn, M.E.; Banuelos, M.A.; Rodriguez-Navarro, A. Sodium transport and HKT transporters: The rice model. *Plant J.* **2003**, *34*, 788–801. [CrossRef]

196. Horie, T.; Yoshida, K.; Nakayama, H.; Yamada, K.; Oiki, S.; Shinmyo, A. Two types of HKT transporters with different properties of Na^+ and K^+ transport in *Oryza sativa*. *Plant J.* **2001**, *27*, 129–138. [CrossRef]

197. Suzuki, K.; Yamaji, N.; Costa, A.; Okuma, E.; Kobayashi, N.I.; Kashiwagi, T.; Katsuhara, M.; Wang, C.; Tanoi, K.; Murata, Y.; et al. OsHKT1;4-mediated Na^+ transport in stems contributes to Na^+ exclusion from leaf blades of rice at the reproductive growth stage upon salt stress. *BMC Plant Biol.* **2016**, *16*, 22. [CrossRef]

198. Imran, S.; Horie, T.; Katsuhara, M. Expression and ion transport activity of rice *OsHKT1;1* variants. *Plants* **2020**, *9*, 16. [CrossRef]

199. Flowers, T.J.; Colmer, T.D. Salinity tolerance in halophytes. *New Phytol.* **2008**, *179*, 945–963. [CrossRef]

200. Glenn, E.P.; Brown, J.J.; Blumwald, E. Salt tolerance and crop potential of halophytes. *Crit. Rev. Plant Sci.* **1999**, *18*, 227–255. [CrossRef]

201. Kader, M.A.; Lindberg, S. Cytosolic calcium and pH signaling in plants under salinity stress. *Plant Signal Behav.* **2010**, *5*, 233–238. [CrossRef]

202. Fukuda, A.; Nakamura, A.; Hara, N.; Toki, S.; Tanaka, Y. Molecular and functional analyses of rice NHX-type Na^+/H^+ antiporter genes. *Planta* **2011**, *233*, 175–188. [CrossRef]

203. Amin, U.S.M.; Biswas, S.; Elias, S.M.; Razzaque, S.; Haque, T.; Malo, R.; Seraj, Z.I. Enhanced salt tolerance conferred by the complete 23 kb cDNA of the rice vacuolar Na^+/H^+ antiporter gene compared to 19 kb coding region with 5′ UTR in transgenic lines of rice. *Front. Plant Sci.* **2016**, *7*, 1–14. [CrossRef] [PubMed]

204. Brini, F.; Hanin, M.; Mezghani, I.; Berkowitz, G.A.; Masmou, K. Overexpression of wheat Na^+/H^+ antiporter *TNHX1* and H^+-pyrophosphatase *TVP1* improve salt- and drought-stress tolerance in *Arabidopsis thaliana* plants. *J. Exp. Bot.* **2007**, *58*, 301–308. [CrossRef] [PubMed]

205. Gaxiola, R.A.; Li, J.S.; Undurraga, S.; Dang, L.M.; Allen, G.J.; Alper, S.L.; Fink, G.R. Drought- and salt-tolerant plants result from overexpression of the AVP1 H^+-pump. *Proc. Natl. Acad. Sci. USA* **2005**, *98*, 11444–11449. [CrossRef]

206. Zhao, F.Y.; Zhang, X.J.; Li, P.H.; Zhao, Y.X.; Zhang, H. Co-expression of the *Suaeda salsa SsNHX1* and *Arabidopsis AVP1* confer greater salt tolerance to transgenic rice than the single *SsNHX1*. *Mol. Breed.* **2006**, *17*, 341–353. [CrossRef]

207. Gong, H.J.; Randall, D.P.; Flowers, T.J. Silicon deposition in the root reduces sodium uptake in rice (*Oryza sativa* L.) seedlings by reducing bypass flow. *Plant Cell. Environ.* **2006**, *1273*, 433. [CrossRef]

208. Krishnamurthy, P.; Ranathunge, K.; Franke, R.; Prakash, H.S.; Schreiber, L.; Mathew, M.K. The role of root apoplastic transport barriers in salt tolerance of rice (*Oryza sativa* L.). *Planta* **2009**, *230*, 119–134. [CrossRef] [PubMed]

International Journal of
Molecular Sciences

Article

Overexpression of the *Zygophyllum xanthoxylum* Aquaporin, *ZxPIP1;3*, Promotes Plant Growth and Stress Tolerance

Mengzhan Li, Mingfa Li, Dingding Li, Suo-Min Wang and Hongju Yin *

State Key Laboratory of Grassland Agro-Ecosystems, Key Laboratory of Grassland Livestock Industry Innovation, Ministry of Agriculture and Rural Affairs, College of Pastoral Agriculture Science and Technology, Lanzhou University, Lanzhou 730020, China; limzh2015@lzu.edu.cn (M.L.); limf19@lzu.edu.cn (M.L.); lidd2017@lzu.edu.cn (D.L.); smwang@lzu.edu.cn (S.-M.W.)
* Correspondence: yinhj@lzu.edu.cn

Abstract: Drought and salinity can result in cell dehydration and water unbalance in plants, which seriously diminish plant growth and development. Cellular water homeostasis maintained by aquaporin is one of the important strategies for plants to cope with these two stresses. In this study, a stress-induced aquaporin, ZxPIP1;3, belonging to the PIP1 subgroup, was identified from the succulent xerophyte *Zygophyllum xanthoxylum*. The subcellular localization showed that ZxPIP1;3-GFP was located in the plasma membrane. The overexpression of *ZxPIP1;3* in Arabidopsis prompted plant growth under favorable condition. In addition, it also conferred salt and drought tolerance with better water status as well as less ion toxicity and membrane injury, which led to more efficient photosynthesis and improved growth vigor via inducing stress-related responsive genes. This study reveals the molecular mechanisms of xerophytes' stress tolerance and provides a valuable candidate that could be used in genetic engineering to improve crop growth and stress tolerance.

Keywords: aquaporin; *Zygophyllum xanthoxylum*; plant growth; abiotic stress

Citation: Li, M.; Li, M.; Li, D.; Wang, S.-M.; Yin, H. Overexpression of the *Zygophyllum xanthoxylum* Aquaporin, *ZxPIP1;3*, Promotes Plant Growth and Stress Tolerance. *Int. J. Mol. Sci.* **2021**, 22, 2112. https://doi.org/10.3390/ijms22042112

Academic Editor: Stephan Pollmann

Received: 24 January 2021
Accepted: 16 February 2021
Published: 20 February 2021

1. Introduction

Abiotic stress factors, such as drought and high salinity, are recognized as major environmental threats that may break plant water balance and result in tissue dehydration, thus negatively impacting plant growth and development. As sessile organisms, plants have gradually evolved various strategies to control water flux to cope with environmental constraints [1,2].

Aquaporins (AQPs), a type of major intrinsic protein (MIP) spreading across the plant kingdom, play important roles in maintaining cellular water homeostasis [3,4]. AQPs contain six membrane-spanning domains and two highly conserved Asn-Pro-Ala (NPA) motifs [5]. According to their amino acid sequences and subcellular localization, AQPs can be classified into five subfamilies, including plasma membrane intrinsic proteins (PIPs), tonoplast intrinsic proteins (TIPs), nodulin-26 intrinsic proteins (NIPs), small basic intrinsic proteins (SIPs), and X-intrinsic proteins (XIPs) [6]. PIPs, located at the plasma membrane, are the largest subfamily of plant AQPs and play key roles in transcellular water transport. This subfamily can be subdivided into two groups, PIP1s and PIP2s [7], which are different in the length of N- and C- termini and in water conductivity. PIP2s, commonly possessing a longer C-termini and a shorter N-termini, are more efficient in water movement. However, PIP1s, with shorter C-termini and longer N-termini, have less water conductivity but possess the ability to transport various uncharged small molecule [8,9]. Numerous studies have showed that PIPs are involved in response to salt and drought stresses, and the ectopical expression of some *PIP*s confers abiotic stress tolerance to plants [8]. The overexpression of *OsPIP1-1* and *OsPIP2-2*, two salt and drought-inducible *PIP*s, resulted in a higher salt and drought tolerance of Arabidopsis [10]. In addition, the ectopical expression of *MdPIP1;3* increased the fruit size of tomato and enhanced the drought tolerance of transgenic plants [11]. The overexpression of a salt-inducible *PIP*, *TaAQP8*, increased the salt stress tolerance of transgenic tobacco with promoted root growth [12]. However, a majority of current studies about AQPs mainly focus on glycophytes and rarely concentrate on xerophytes or halophytes containing specific traits generated during their long-term evolution in extremely severe environments [13,14].

Zygophyllum xanthoxylum, a kind of succulent xerophyte belonging to Zygophyllaceae, is widely spread in arid and semiarid land in northwestern China [15]. For its remarkable vitality to survive under adverse drought condition, *Z. xanthoxylum* is often used in sand-fixing as well as water and soil conservation in the desert [16]. Previous studies showed that this species can absorb Na^+ from low salt soil and compartmentalize them into vacuoles as a low-cost osmoregulation substance, which helps *Z. xanthoxylum* maintain lower osmotic potential to absorb water under drought condition. All these studies focused on ion transporters participating in the salt and drought stress tolerance of *Z. xanthoxylum*, while the functions of AQPs in the course of water transportation and cellular water homeostasis maintenance remains unknown [15,17–19]. Previously, in order to understand the mechanisms of *Z. xanthoxylum* to cope with severe environment, 50 mM NaCl-treated and −0.5 MPa-treated transcriptome datasets of *Z. xanthoxylum* roots were analyzed [20,21]. An *AQP*, *ZxPIP1;3*, whose expression was induced under salt and osmotic treatment, was screened.

In this study, *ZxPIP1;3* was cloned, and its expression pattern under salt and osmotic treatment was identified by qRT-PCR. In addition, the transient expression of ZxPIP1;3-GFP fusion protein was used to investigate the subcellular localization in *Nicotiana benthamiana*. Furthermore, ectopical overexpression transgenic Arabidopsis was generated to evaluate the roles of ZxPIP1;3 in plant growth and stress tolerance.

2. Results

2.1. ZxPIP1;3 Is Induced under Osmotic and Salt Treatment

To verify the transcriptome data, 3-week-old *Z. xanthoxylum* seedlings were treated with −0.5 MPa osmotic stress or 50 mM NaCl treatment (Figure 1A). The expression level of *ZxPIP1;3* was significantly increased after 6-h treatment, which confirmed that *ZxPIP1;3* was stress-inducible and suggested that ZxPIP1;3 may participate in plants' salt and drought stress response.

2.2. ZxPIP1;3 Encodes an AQP of PIP1 Subgroup

The full length of the *ZxPIP1;3* open reading frame was 864 bp, encoding 287 amino acids residues (Figure S1A). ZxPIP1;3 contained six putative transmembrane α-helices (Figure S1B). The results of multiple sequence alignment and phylogenetic tree analysis using full-length amino acid sequence indicated that ZxPIP1;3 was highly homologous to DzPIP1;3 (*Durio zibethinus*) and HuPIP1;3 (*Herrania umbratica*) (Figure 1B,C). Through subcellular localization using transient expression driven by the *CaMV35S* promoter in *N. benthamiana* leaves, ZxPIP1;3-GFP was detected at the plasma membrane (Figure 1D).

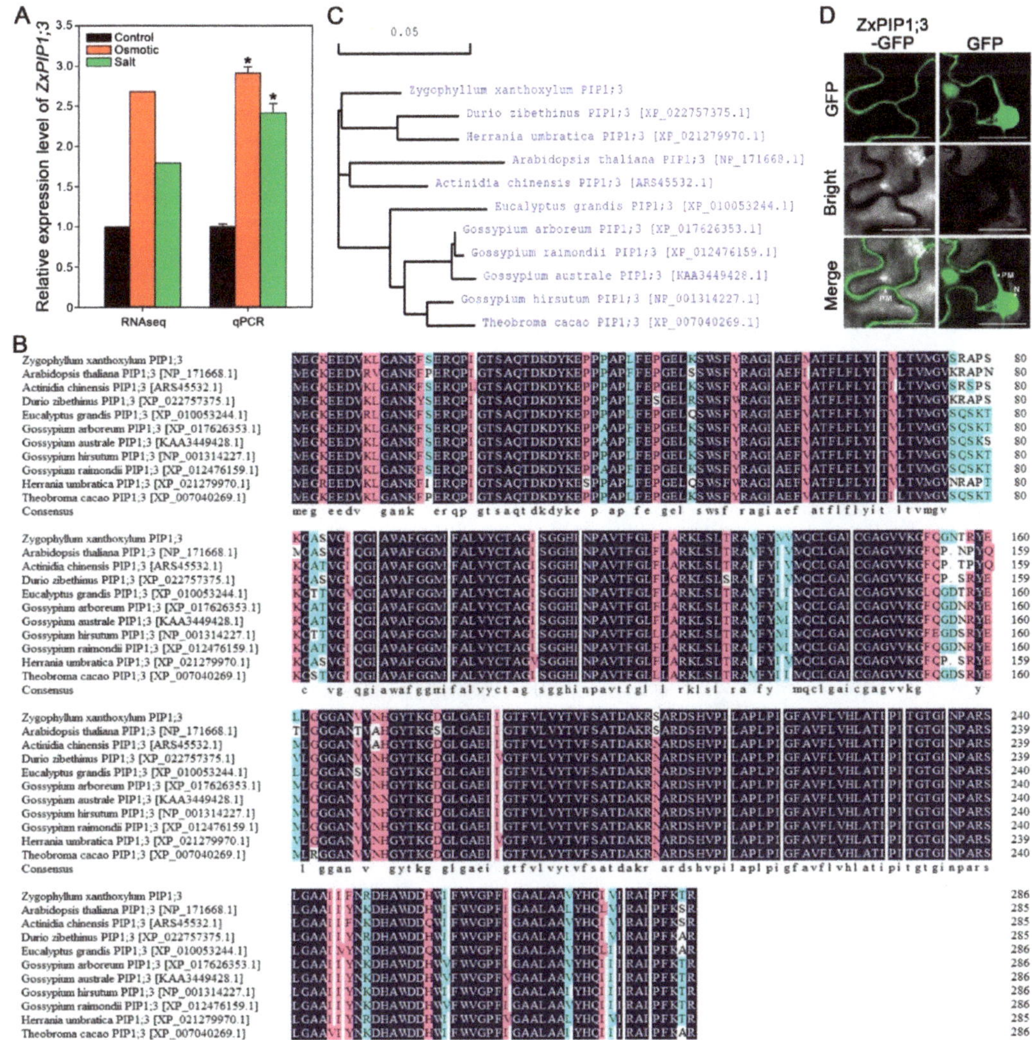

Figure 1. *ZxPIP1;3* encodes a PIP1 protein whose expression level is related with osmotic and salt treatments. (**A**) qRT-PCR validation of RNA sequencing data in *Z. xanthoxylum* roots under osmotic stress or salt treatment for 6 h. (**B,C**), alignment (**B**) and phylogenetic analysis (**C**) of ZxPIP1;3 with other known PIP1 proteins. Dark blue, pink and aqua indicate that the homology levels of these amino acids are 100%, more than 75% and more than 50% respectively. (**D**) Subcellular localization of ZxPIP1;3-GFP in epidermal cells of tobacco leaves. GFP driven by *CaMV35S* promoter served as control. PM, plasma membrane. N, nucleus. Green fluorescence represents GFP. Bar = 10 μm. For (**A**), asterisks indicate significant differences from control condition. Data shown are means of three independent biological replicates (* $p < 0.05$, one way ANOVA).

2.3. Overexpression of ZxPIP1;3 Promotes Plant Growth

To study the performance of ZxPIP1;3 in plant growth and abiotic stress tolerance, *35S::ZxPIP1;3-FLAG* was transformed into Arabidopsis. The expression levels of *ZxPIP1;3* in transgenic lines was detected, and two lines, *OE2* and *OE3*, with different expression level were selected for further analysis (Figure S2).

Col-0 and *OE2*, *OE3* grown vertically on 1/2 Murashige and Skoog (MS) solid medium for 7 days after germination were used to evaluate the effect of ZxPIP1;3 on root growth and development. Comparing with wild type, primary roots of transgenic lines were longer with more emerged lateral root and lateral root primordia (Figure 2A–C), which indicated that *ZxPIP1;3* overexpression promoted root growth under favorable conditions. We further tested the roles of ZxPIP1;3 in shoot growth. The cotyledon of transgenic lines was larger than those of wild-type plants (Figure 2A,D). For 4-week-old seedlings, *OE2* and *OE3* grow better than Col-0 with larger rosette leaves (Figure 2E,F). Additionally, for plants during the reproductive stage, transgenic lines were taller compared with Col-0 (Figure 2G,H). Thus, it was evident that *ZxPIP1;3* overexpression could significantly promote plant growth.

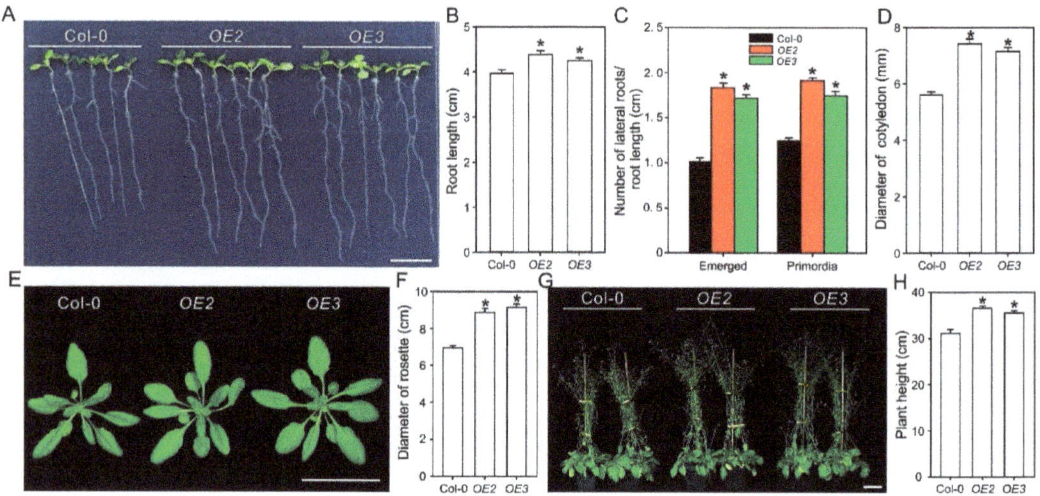

Figure 2. ZxPIP1;3 plays positive roles in plant growth. (**A**) Phenotypes of wild-type (Col-0) and *ZxPIP1;3* overexpression lines (*OE2* and *OE3*) grow vertically on 1/2 Murashige and Skoog (MS) medium for 7 days after germination. Bar = 1 cm. (**B–D**) Primary root length (**B**), number of emerged lateral roots and lateral root primordia (**C**) as well as cotyledon diameter (**D**) of plants treated as described in the legend of (**A**). (**E**) Phenotypes of 4-week-old seedlings grown on soil culture. Bar = 5 cm. (**F**) Rosette leaves diameter of plants treated as described in the legend of (**E**). (**G**) Phenotypes of 7-week-old seedlings on soil culture. Bar = 5 cm. (**H**) Height of plants treated as described in the legend of (**G**). For (**B–D**), (**F,H**), asterisks indicate significant differences from Col-0 (n = 13 per column. * $p < 0.05$, one way ANOVA).

2.4. ZxPIP1;3 Overexpression Improves Salt Tolerance of Transgenic Arabidopsis

To investigate the role of ZxPIP1;3 in response to salt stress, wild-type and transgenic Arabidopsis were grown on 1/2 MS with 150 mM NaCl. Under salt stress treatment, primary roots of *OE2* and *OE3* were longer than those of wild-type plants (Figure S3A). This result indicated that *ZxPIP1;3* transgenic seedlings were less sensitive to salt stress.

To further confirm the function of ZxPIP1;3 in salt stress tolerance, 4-week-old plants were irrigated with 100 mM NaCl for 20 days. Under favorable condition, transgenic plants grow better (Figure 3A). Under salt stress condition, the wild-type plants were etiolated and aborted with sere leaves and inflorescence, while the leaves of transgenic plants were still green, and the shoot apices of them grow well without abortion (Figure 3A,B). In addition, the stems of *OE2* and *OE3* were longer and heavier with more branches than wild type (Figure 3C–E). Physiological parameters including the content of organic osmoregulation substance, relative water content, content of chlorophyll, net photosynthetic rate (Pn), water-use efficiency (WUE) and malondialdehyde (MDA) content as well as K^+/Na^+ ratio were also measured (Figure 4). Salinity has osmotic effects on plants, which lead to water deficiency [22]. To overcome physiological drought, a plant may synthesize organic osmotic substances, such as soluble sugar and proline, to increase water potential. Under salt treatment, transgenic plants contained more soluble sugar and proline than wild type, which resulted in much higher relative water content (Figure 4A–C). The stability of chlorophyll is sensitive to plant water status, and higher relative water content can protect chlorophyll from degradation [2]. Comparing with Col-0, the chlorophyll content of *OE2* and *OE3* was much higher under both normal condition and salt stress treatment, which was consistent with the phenotype (Figures 3A and 4D–F). Consequently, osmotic homeostasis and higher chlorophyll content can also contribute to higher Pn and WUE in *OE2* and *OE3* than Col-0 after salt treatment (Figure 4G,H). As a final product of cell membrane lipid peroxidation, MDA is a good indicator of oxidative damage [23]. MDA content was significantly lower in *OE2* and *OE3* than Col-0 (Figure 4I), which implicated that membrane damage in transgenic lines was not as severe as those in wild type. Salinity can also cause ion toxicity on plants. Moreover, Na^+ at high concentration competes for sites of transporters, which is necessary for K^+ uptake [22]. There was no difference of K^+ and Na^+ content between transgenic and wild-type lines under normal condition. However, after salt treatment, *OE2* and *OE3* contained more K^+ and less Na^+ than Col-0, which resulted in higher K^+/Na^+ ratio (Figure 4J–L). All these results indicated that *ZxPIP1;3* overexpression improved the salt tolerance of transgenic lines.

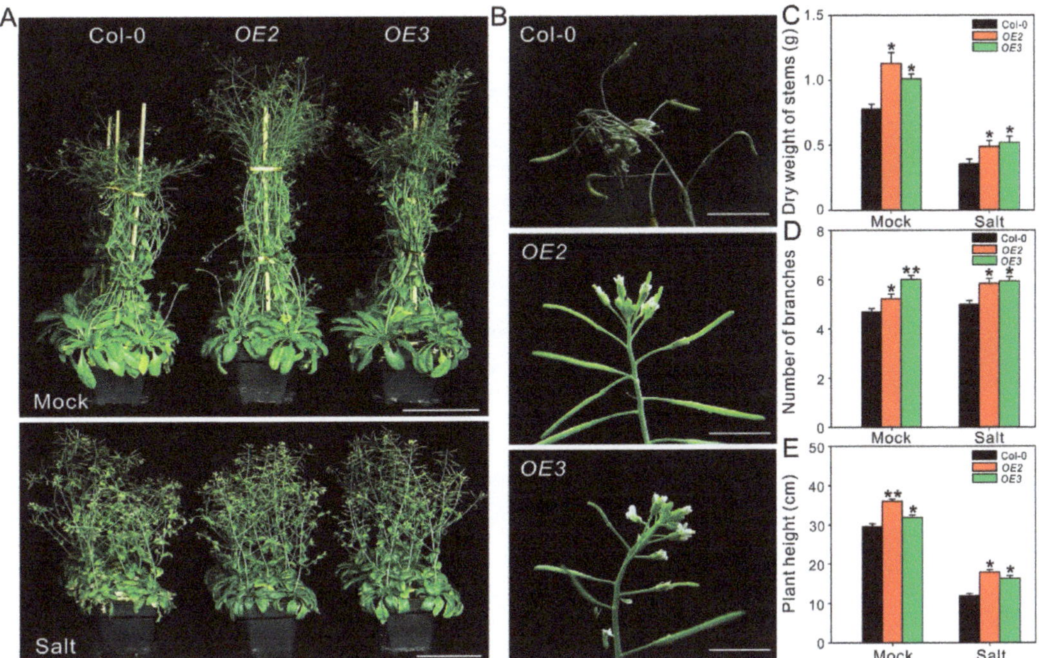

Figure 3. Overexpression of *ZxPIP1;3* improves growth vigor under salt treatment. (**A**) Phenotypes of soil-cultured seedlings treated with (Salt) or without (Mock) 100 mM NaCl for 20 days. Bar = 10 cm. (**B**) Phenotypes of shoot apices of plants under salt treatment as described in (**A**). Bar = 1 cm. (**C–E**) Dry weight of stems (**C**), number of branches (**D**) and plant height (**E**) of plants described in the legend of (**A**). For (**C–E**), asterisks indicate significant differences from Col-0 ($n = 13$ per column. * $p < 0.05$, ** $p < 0.01$, one way ANOVA).

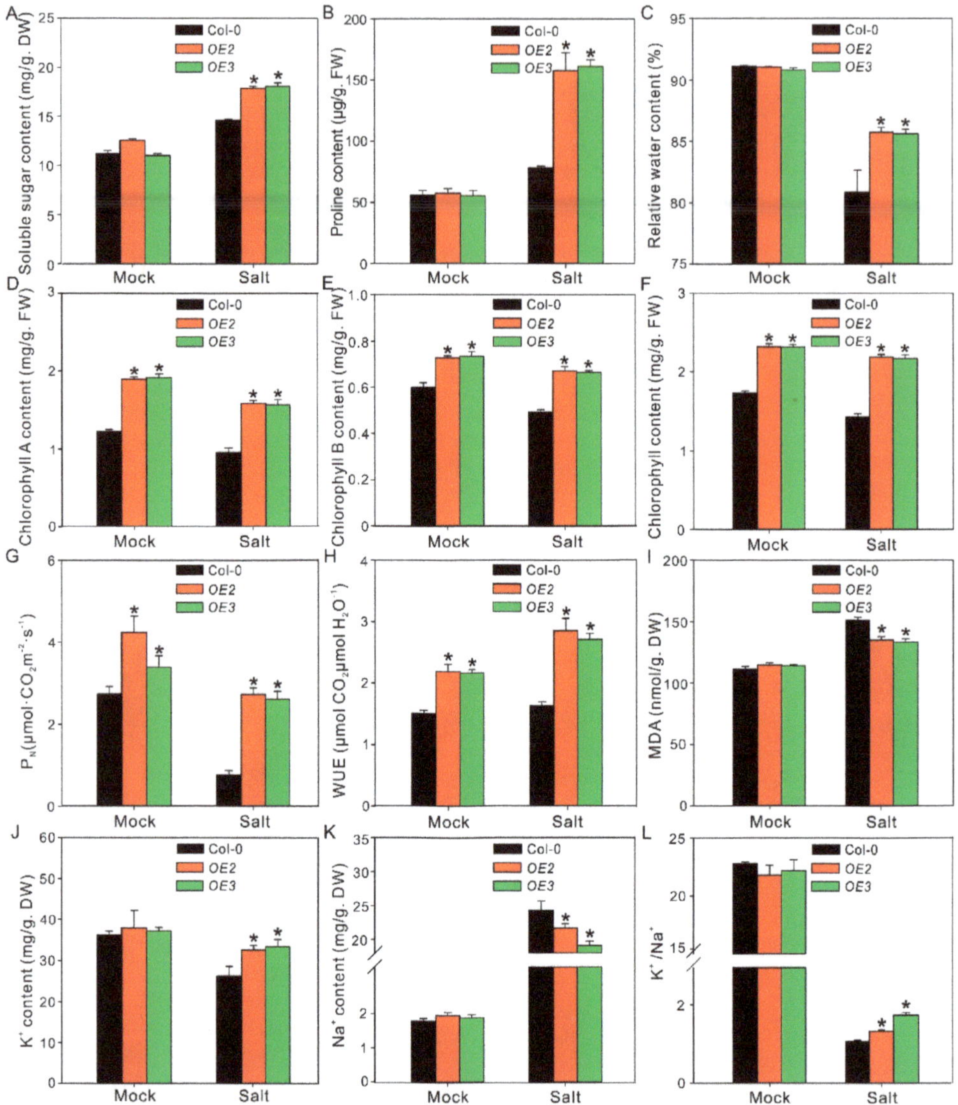

Figure 4. Several physiological parameters of the salt-treated wild-type and transgenic plants. Soluble sugar content (**A**), proline content (**B**), relative water content (**C**), chlorophyll A content (**D**), chlorophyll B content (**E**), chlorophyll content (**F**), net photosynthetic rate (Pn) (**G**), water-use efficiency (WUE) (**H**), content of malondialdehyde (MDA) (**I**), K^+ content (**J**), Na^+ content (**K**), and K^+/Na^+ (**L**) were tested. Asterisks indicate significant differences from Col-0 (Values are mean ± SE of three replicates. * $p < 0.05$, one way ANOVA).

2.5. ZxPIP1;3 Overexpression Confers Drought Tolerance of Transgenic Plants

To verify the functions of ZxPIP1;3 in drought stress response, drought stress treatment was simulated by the cultivation of Col-0 and *OE2*, *OE3* on 1/2 MS solid medium containing 300 mM mannitol. Under osmotic stress treatment, primary roots of *OE2* and *OE3* were longer than those of Col-0 (Figure S3A). This result suggested that *ZxPIP1;3* overexpression decreased plants' sensitivity to simulant drought stress.

Further, the drought tolerance of wild-type and transgenic lines were evaluated in soil culture. Col-0 and *ZxPIP1;3* overexpression plants were grown under well-watered condition for 4 weeks before subjected to drought treatment. After 7 days of water withdrawal, Col-0 began to wilt, while *OE2* and *OE3* grow well with unfolded leaves (Figure S3B). To further evaluate the effect of drought stress on different lines, plants treated with dehydration were watered normally for 7 days for recovery. After that, these plants were subjected to 7 days drought treatment again. Plants were photographed (Figure 5A) and then harvested for analyzing physiological parameters [19]. After period drought treatment, the leaves of Col-0 were etiolated and wilted, whereas *ZxPIP1;3* transgenic lines did not wilt as severely as Col-0 (Figure 5A). In addition, stems' dry weight and the branch numbers of transgenic plants were higher than those of wild type (Figure 5B,C), which indicated that *OE2* and *OE3* showed a stronger growth vigor in comparison to Col-0 under water shortage. Physiological parameters including organic osmoregulation substance content, relative water content, chlorophyll content, and the Pn and WUE as well as MDA content of wild-type and transgenic lines were also measured (Figure 6). After period drought treatment, the contents of organic osmoregulation substance, including soluble sugar and proline, as well as relative water content in transgenic plants were higher than those of wild type (Figure 6A–C). Photosynthesis is influenced by water status and chlorophyll content. In addition, the accumulation of chlorophyll is also related with water status. Compared with wild type, a higher relative water content prevented the chlorophyll of transgenic lines from degradation, which resulted in higher chlorophyll concentration, Pn, and WUE (Figure 6D–H). Water deficiency can also lead to cell membrane destabilization. Under drought treatment, the MDA content of *OE2* and *OE3* was less than those of Col-0 (Figure 6I). All these results indicated that *ZxPIP1;3* overexpression conferred drought tolerance in transgenic plants.

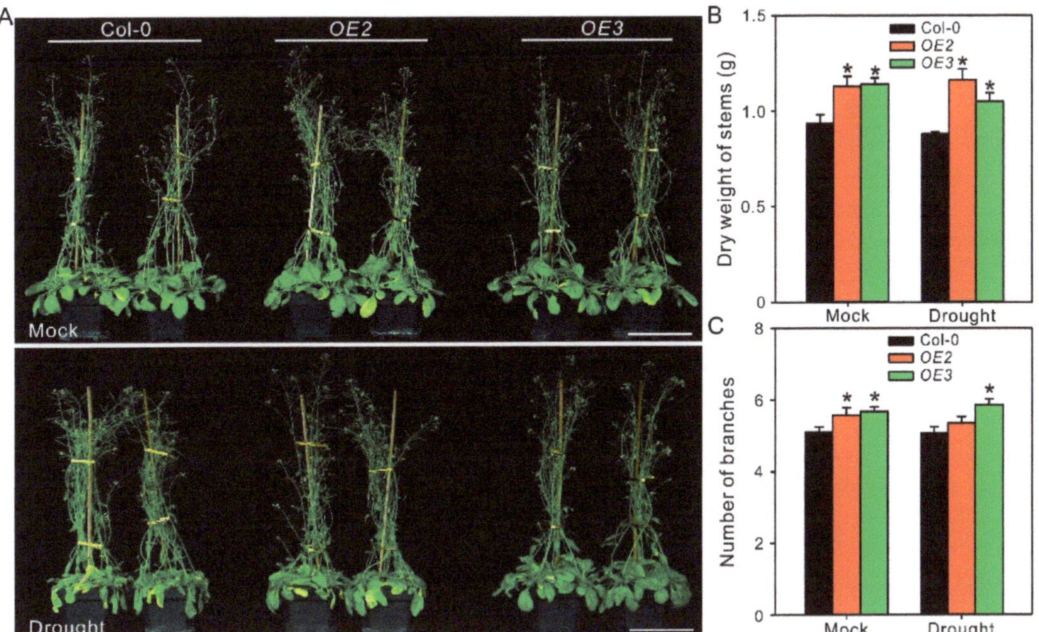

Figure 5. Overexpression of *ZxPIP1;3* improves growth vigor under drought treatment. (**A**) Phenotypes of soil-cultured seedlings treated with (Drought) or without (Mock) period dehydration. Bar = 10 cm. (**B,C**), Dry weight of stems (**B**) and number of branches (**C**) of plants treated as described in the legend of (**A**). For (**B,C**), asterisks indicate significant differences from Col-0 (n = 13 per column. * $p < 0.05$, one way ANOVA).

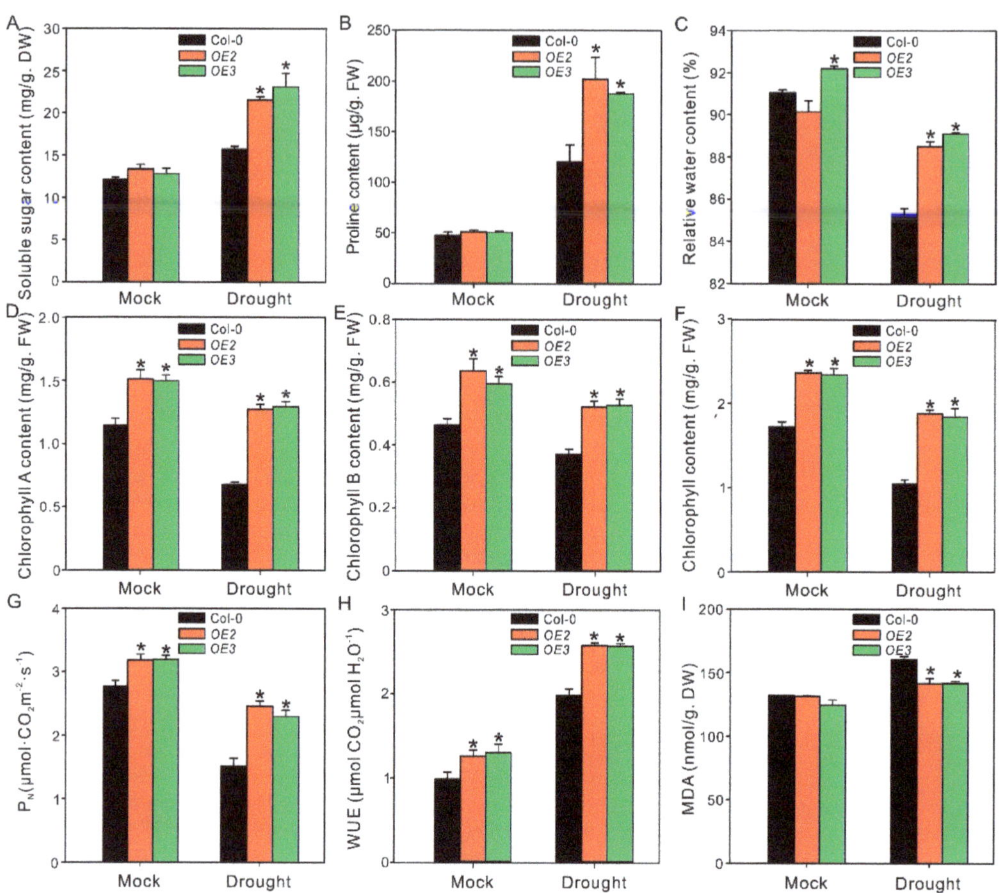

Figure 6. Several physiological parameters of the drought-treated wild-type and transgenic plants. Soluble sugar content (**A**), proline content (**B**), relative water content (**C**), chlorophyll A content (**D**), chlorophyll B content (**E**), chlorophyll content (**F**), Pn (**G**), WUE (**H**), and content of MDA (**I**) were tested. Asterisks indicate significant differences from Col-0 (Values are mean $\pm$ SE of three replicates. * $p < 0.05$, one way ANOVA).

2.6. Expression Level of Stress-Related Genes Is Increased in ZxPIP1;3 Transgenic Plants under Stress Treatment Compared with Wild-Type Plants

To assess the implication of ZxPIP1;3 in the abiotic stress response pathway, the expression of three genes participating in stress response were analyzed via qRT-PCR (Figure 7). Δ1-pyrroline-5-carboxylate synthetase 1 (P5CS1) plays vital roles in proline biosynthesis [24]. The expression level of *P5CS1* in all lines increased under salt and osmotic stresses, and it was higher in transgenic lines compared with wild-type plants (Figure 7A). The expression level of *Response-to-Dehydration 29A* (*RD29A*), an ABA-induced gene related with responsiveness to drought, salt, and cold [25], was significantly higher in transgenic plants comparing with Col-0 under osmotic and salt stress (Figure 7B). *DEHYDRATION-RESPONSIVE ELEMENT-BINDING PROTEIN 1A* (*DREB1A*) is an APETALA2/ethylene-responsive element-binding factor (AP2/ERF)-type transcription factor involved in plant abiotic stress response [26]. The expression of *DREB1A* was remarkably enhanced in *OE2* and *OE3* under the stress-treated condition (Figure 7C).

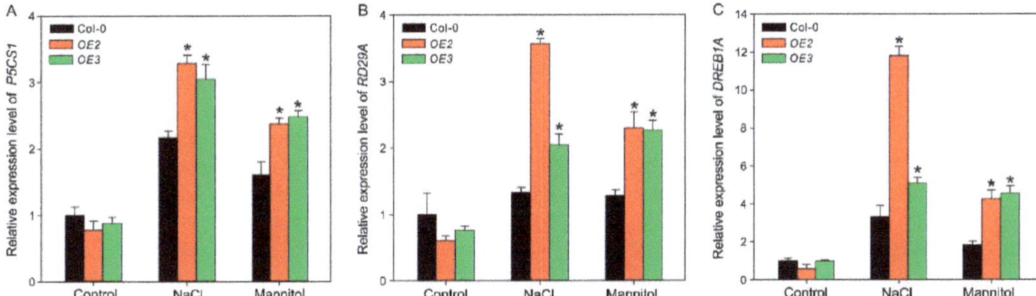

Figure 7. Relative expression levels of stress-related genes in wild-type and transgenic plants. The expression levels of *P5CS1* (**A**), *RD29A* (**B**), and *DREB1A* (**C**) were tested in 7-day-old wild-type (Col-0) and transgenic plants (*OE2, OE3*) under normal (Control), salt (NaCl), and osmotic (Mannitol) treatments. Asterisks indicate significant differences from Col-0. Data shown are means of three independent biological replicates. * $p < 0.05$, one way ANOVA.

3. Discussion

Z. xanthoxylum is widely distributed throughout the desert region of northwestern China, where the mean annual precipitation is usually less than 200 mm [27]. Previous studies have demonstrated that the main strategy for *Z. xanthoxylum* to cope with the extremely arid environment is to absorb Na^+ from the low salt soil and compartmentalize them into vacuoles. Na^+ can be used as a low-cost osmoregulator to decrease the osmotic potential, which helps *Z. xanthoxylum* absorb water under drought stress [17,18,27]. Most studies concerning the stress tolerance of this species were focused on the process of sodium uptake and accumulation [17,18,27]. Various ion transporters and channels were cloned and characterized [15,19]. However, mechanisms of the water influx that result from Na^+-accumulation are obscure. As a majority of transmembrane water flux is dependent on AQPs, the first AQP from *Z. xanthoxylum*, ZxPIP1;3, was isolated, and its functions in plant growth as well as abiotic stress tolerance were evaluated in the present study.

To cope with stressful conditions and the growing demand of food, it is vital to develop cultivars with higher yields and improved tolerance to abiotic stress via genetic engineering breeding [28,29]. However, most identified genes play opposing roles in stress tolerance and plant growth, such as *C repeat/dehydration-responsive element binding factor 1 (CBF1)* and *DWARF AND DELAYED FLOWERING 1 (DDF1)*, whose overexpression conferred stress tolerance at the expense of growth [30–32]. However, in this study, the overexpression of *ZxPIP1;3* can not only promote growth under normal condition (Figure 2) but also decrease the inhibition of salt and drought stress on it (Figures 3 and 5), demonstrating that ZxPIP1;3 plays positive roles in plant growth as well as stress tolerance. Previous studies indicated that AQPs exert an effect on plant growth via impacting water absorption. In addition, the uptake of some nutrients can also be accompanied by water flux through AQPs [33]. Thus, ZxPIP1;3 is an optimal candidate for crop breeding.

Both salinity and water shortage trigger cell dehydration. It is important for plants to retain water from the environment under salt and drought stress. We observed that the relative water content of *ZxPIP1;3* transgenic plants was higher than those of wild type under stress treatment (Figures 4C and 6C), indicating the enhanced ability of transgenic plants to retain water. Similar phenotypes were also reported via studying other *AQPs* transgenic plants, such as Arabidopsis overexpressing *PIP1;1* from banana and potato overexpressing *StPIP1* [34,35]. To investigate the mechanism involved in this process, the content of organic osmotic substances was measured, which were synthesized to adjust osmotic potential. The soluble sugar content of transgenic plants was higher than that of wild type (Figures 4A and 6A), implying that *ZxPIP1;3* overexpression increased the ability of osmotic regulation. This result was consistent with the overexpression of *HvPIP2;5* [36] and *TsPIP1;3* [37], which also play positive roles in osmotic regulation. Plants also synthesize proline to adjust osmotic potential. P5CS participates in proline biosynthesis via reducing glutamate to glutamate semialdehyde [24]. Studies showed that the overexpression of *ZmPIP1;1* and *ScPIP1* can increase the accumulation of proline in transgenic plants via inducing the expression of *P5CS*, which led to enhanced stress tolerance [36,38,39]. In this study, consistent with previous studies, the expression level of *P5CS1* and the accumulation of proline in transgenic plants were also higher than those in wild-type plants (Figures 4B, 6B and 7A). Abiotic stresses induce a rapid accumulation of ROS, which leads to the cell membrane damage [23]. To estimate the membrane injury level, the content of MDA, a product of lipid oxidation, was measured. It was observed that the MDA content of transgenic plants was less than that of wild type (Figures 4I and 6I), indicating that the membrane damage suffered by *ZxPIP1;3* overexpressing plants under dehydration was not as severe as wild type. Our results were consistent with previous studies demonstrating that PIPs participated in reducing membrane damage under different stress [37,40,41]. The water balance mediated by PIPs results in a relatively stable physiological status, which may lead to reduced protein and lipid peroxidation followed by decreased MDA content and membrane damage. High salinity decreases the growth rate of plants via increasing cellular Na^+ concentration. To avert the toxic effects of sodium in cytosol, plants intend to compartmentalize Na^+ into vacuoles. The transport of Na^+ into vacuoles is regulated by Na^+/H^+ antiporters and vacuolar H^+-translocating enzymes, whose activation is related to the stability of the membrane [42]. We surmised that the reduced membrane damage contributed by *ZxPIP1;3* overexpression may help to maintain the functions of transporters localized in the cell membrane and promote the vacuolar Na^+ compartmentation, which reduced the cytotoxic effects of sodium.

Previous studies showed that most *AQPs* transgenic plants with enhanced stress tolerance exhibit a higher expression level of stress-responsive genes under stress-treated condition comparing with wild-type lines [34,38,43,44], which is consistence with our results (Figure 7). Salt and drought stresses are water-related, which can change osmotic gradients. Even though the certain mechanisms remained unclear, studies suggested that AQPs may act as detectors of osmotic gradients and relay information to signaling chains through protein conformation or interaction with downstream signaling elements [45–47]. Thus, there was no significant difference in the expression level of stress-responsive genes between three lines under optimal condition (Figure 7). However, after salt or osmotic treatment, the overexpression of *ZxPIP1;3* may enhance plants' response to these stress signals and result in the higher expression level of stress-related genes indirectly (Figure 7), which elevated plants' stress tolerance.

In conclusion, we identified a stress-induced *AQP*, *ZxPIP1;3*, from *Z. xanthoxylum* and demonstrated that ZxPIP1;3 not only improved growth vigor under favorable condition but also conferred salt and drought tolerance via enhancing the capacity of water retention as well as diminishing membrane injury and ion toxicity. This study reveals the molecular mechanisms of xerophytes' stress tolerance and provides a theoretical basis for environmental protection in the desert area as well as discovers a valuable candidate for crop breeding.

4. Materials and Methods

4.1. Plant Materials and Growth Conditions

Z. xanthoxylum seeds were germinated on wet filter paper at 25 °C in the dark. After germination, seedlings were transformed into a hole plate containing quartz sand, whose grain size was about 0.5–0.8 cm, and them irrigated with modified Hoagland solution as Ma et al. described [21] every 3 days. Seedlings were grown in greenhouse at 28 °C/23 °C (day/night) under a 16-h-light/8-h-dark cycle with the flux density 800 $\mu mol\, m^{-2}\, s^{-1}$. The relative humidity was about 65–70%. Arabidopsis used in this study was in ecotype Columbia-0 (Col-0) background. Arabidopsis seeds were vernalized in sterile water at 4 °C for 3 days before being grown on turfy soil in a greenhouse with relative humidity 65–75% at 22 °C under a 16-h-light/8-h-dark cycle with the flux density of 100–120 $\mu mol\, m^{-2}\, s^{-1}$.

4.2. Expression Pattern Analysis

Three-week-old *Z. xanthoxylum* plants were used for different treatment for 6 h as follows. (i) Control: seedlings were irrigated with modified Hoagland solution; (ii) Salt treatment: seedlings were irrigated with modified Hoagland solution containing 50 mM NaCl; (iii) Osmotic stress: seedlings were irrigated with modified Hoagland solutions supplemented with sorbitol to adjust osmotic potential to -0.5 MPa. Roots of seedlings in each condition were collected and frozen by liquid nitrogen immediately.

Total RNA was extracted by using an RNAprep Pure Plant Plus Kit (Polysaccharides & Polyphenolics-rich) (TIANGEN, Beijing, China), and cDNA was synthesized from DNase-pretreated RNA using a PrimeScript™ RT reagent Kit with gDNA Eraser (TaKaRa Biotechnology, Beijing, China). qRT-PCR was performed in triplicate on three bio-replicates with a StepOne Real-Time PCR Thermocylcer (Applied Biosystems, Foster City, CA, USA) using the Power SYBR™ Green Master Mix (TaKaRa Biotechnology, Beijing, China). *Zx-ACTIN* (GenBank accession no. EU019550) was used as the internal control gene [15,19–21]. Sequences of primers are listed in Table S1. The $2^{-\Delta\Delta Ct}$ method was used to determine the relative expression level [48].

4.3. Cloning of ZxPIP1;3 and Sequence Analysis

Total RNA was extracted from roots of 3-week-old *Z. xanthoxylum* subjected to -0.5 MPa osmotic stress. The cDNA sequence of *ZxPIP1;3* (GenBank accession no. MW590708) was amplified by using a SMART RACE cDNA Amplification Kit (TaKaRa Biotechnology, China). Sequences of primers are listed in Table S1.

Nucleotide and amino acid sequences were analyzed by using DNAMAN (DNAMAN Inc., San Ramon, CA, USA). Transmembrane helices were predicted by TMHMM Server v 2.0 (http://www.cbs.dtu.dk/services/TMHMM/ (accessed on 1 January 2021)). Full-length amino acid sequences of PIP1;3 from Arabidopsis, *Actinidia chinensis, D. zibethinus, Eucalyptus grandis, Gossypium arboretum, Gossypium austral, Gossypium hirsutum, Gossypium raimondii, H. umbratical*, and *Theobroma cacao* were obtained from the NCBI database (https://www.ncbi.nlm.nih.gov/ (accessed on 1 January 2021)). DNAMAN were used for sequence alignment and polygenetic analysis.

4.4. ZxPIP1;3 Expression Vector Construction

ZxPIP1;3 coding sequence was amplified with primers listed in Table S1. The product was cloned into a pDONR-ZERO vector by BP reaction and then inserted into the binary vectors, *pBIB-BASTA-35S-GWR-GFP* and *pBIB-BASTA-35S-GWR-FLAG*, by LR reaction. These binary constructs were introduced into the *Agrobacterium tumefaciens* strain GV3101 separately.

4.5. Subcellular Localization

GV3101 harboring *pBIB-BASTA-35S-ZxPIP1;3-GFP* was used for subcellular localization. GV3101 harboring *pCAMBIA1302* was used as control. After incubating in Luria–Bertani broth containing 10 mM MES (pH 5.7) and 20 mM acetosyringone at 28 °C overnight with shaking, cells were collected and adjusted to an OD600 of 0.6 with MS liquid media containing 10 mM MES (pH 5.7), 10 mM $MgCl_2$, and 150 mM acetosyringone. After incubating at 28 °C for 2 h, the resuspension solution was injected into *N. benthamiana*. Leaves were used for subcellular localization analysis after 48-h infiltration.

4.6. Transgenic Plants Generation

The transformation of Arabidopsis Col-0 plants was performed via the floral dipping method by using GV3101 harboring *pBIB-BASTA-35S-GWR-FLAG*. The expression level of *ZxPIP1;3-FLAG* in transgenic plants with BASTA resistance was investigated via semi-quantitative RT-PCR. In addition, the expression levels of *ZxPIP1;3-FLAG* in wild type and *OE2, OE3* were also evaluated via qRT-PCR. To avoid nonspecific amplification, the reverse primers for semi-quantitative RT-PCR and qRT-PCR were designed by using the

FLAG sequence. *AtACTIN 2* (AT3G18780) was used as the internal control for both semi-quantitative RT-PCR and qRT-PCR. Primers are listed in Table S1. The third generation of homozygous plants were used for further analysis.

4.7. Root and Shoot Growth Analysis

For lateral root growth analysis, seeds of wild-type and transgenic lines were plated on 1/2 MS solid medium and grown for 7 days after germination. Photographs were taken with a digital camera. The number of lateral root and lateral root primordia of 15 seedlings for each line were counted via an Olympus light microscope (magnification $100\times$), and 13 seedlings for each line were used for statistical analysis.

For primary root growth analysis, seeds of wild-type and transgenic plants were plated on 1/2 MS solid medium without or with 150 mM NaCl or 300 mM mannitol and grown for 7 days after germination. Photographs were taken with a digital camera. A primary root length of 15 seedlings for each line under respective treatment was measured by using Digimizer (MedCalc Software Ltd. Ostend, Belgium), and 13 seedlings for each line under respective treatment were used for statistical analysis.

For shoot growth analysis, cotyledon diameter, rosette leaves diameter and plant height, dry weight of stems, and number of branches were measured. For each parameter, 13 out of 15 individuals were used for statistical analysis for different lines.

4.8. Salt Tolerance Analysis of Transgenic Plants

For salt treatment with soil-grown plants, 4-week-old seedlings were irrigated with or without 100 mM NaCl for 20 days every 4 days. After being photographed with a digital camera, plants were harvested for physiological parameters analysis. The content of soluble sugar and proline, relative water content, chlorophyll content, and net photosynthetic rate, K^+/Na^+ ratio as well as malondialdehyde content were determined as described in previous studies, respectively [17,18,36,49,50].

4.9. Drought Tolerance Analysis of Transgenic Plants

For drought tolerance analysis in soil culture, 4-week-old plants were subjected with dehydration for 7 days and photographed with a digital camera (Figure S3B). To further evaluate the effect of drought stress on growth, drought-treated plants were irrigated normally for 7 days for recovery and treated with another 7 days' water withdraw. Plants were photographed via a digital camera (Figure 5A) and then harvested for analyzing physiological parameters. The content of soluble sugar and proline, relative water content, chlorophyll content, and net photosynthetic rate, as well as malondialdehyde content were determined as previous studies, respectively [17,18,36,49,50].

4.10. Expression Pattern Analysis of Stress-Related Genes

Wild-type and transgenic lines grown on 1/2 MS solid medium for 7 days were treated as follows for 6 h: (i) Control: 1/2 MS liquid medium; (ii) Salt stress: 1/2 MS liquid medium containing 150 mM NaCl; (iii) Osmotic stress: 1/2 MS liquid medium containing 300 mM Mannitol. Roots of each treatment were collected, and total RNA were extracted. The transcript level of three stress-related genes *P5CS1* (AT2G39800), *RD29A* (At5G52310), and *DREB1A* (At4G25480) were evaluated, and *AtACTIN 2* (AT3G18780) was used as the internal control. The primer sequences used in this part are listed in Table S1.

4.11. Statistical Analysis

All analyses were performed by using IBM SPSS Statistics, version 22. (SPSS Inc., Chicago, IL, USA). All data shown as mean $\pm$ standard error of means were analyzed using one-way ANOVA, followed by Duncan's test. Statistically significant mean values were denoted as * ($p < 0.05$) or ** ($p < 0.01$).

Supplementary Materials: The following are available online at https://www.mdpi.com/1422-006 7/22/4/2112/s1.

Author Contributions: H.Y. and S.-M.W. designed the experiments; M.L. (Mengzhan Li), M.L. (Mingfa Li) and D.L. performed the experiments; H.Y. and M.L. (Mengzhan Li) discussed the results and wrote the paper. All authors have read and agreed to the published version of the manuscript.

Funding: This work has been jointly supported by the National Natural Science Foundation of China (Grant no.31971621).

Institutional Review Board Statement: Not applicable.

Informed Consent Statement: Not applicable.

Data Availability Statement: Not applicable.

Conflicts of Interest: The authors declare no conflict of interest.

References

1. Aroca, R.; Porcel, R.; Ruiz-Lozano, J.M. Regulation of root water uptake under abiotic stress conditions. *J. Exp. Bot.* **2011**, *63*, 43–57. [CrossRef] [PubMed]
2. Golldack, D.; Li, C.; Mohan, H.; Probst, N. Tolerance to drought and salt stress in plants: Unraveling the signaling networks. *Front. Plant Sci.* **2014**, *5*, 151. [CrossRef] [PubMed]
3. Park, J.H.; Saier, M.H.J. Phylogenetic characterization of the MIP family of transmembrane channel proteins. *J. Membr. Biol.* **1996**, *153*, 171–180. [CrossRef] [PubMed]
4. Kozono, D.; Ding, X.; Iwasaki, I.; Meng, X.; Kamagata, Y.; Agre, P.; Kitagawa, Y. Functional Expression and Characterization of an Archaeal Aquaporin. *J. Biol. Chem.* **2003**, *278*, 10649–10656. [CrossRef]
5. Maurel, C.; Boursiac, Y.; Luu, D.-T.; Santoni, V.; Shahzad, Z.; Verdoucq, L. Aquaporins in Plants. *Physiol. Rev.* **2015**, *95*, 1321–1358. [CrossRef]
6. Kaldenhoff, R.; Fischer, M. Functional aquaporin diversity in plants. *Biochim. Biophys. Acta (BBA) Biomembr.* **2006**, *1758*, 1134–1141. [CrossRef]
7. Benga, G. Water channel proteins (later called aquaporins) and relatives: Past, present, and future. *IUBMB Life* **2009**, *61*, 112–133. [CrossRef]
8. Li, G.; Santoni, V.; Maurel, C. Plant aquaporins: Roles in plant physiology. *Biochim. Biophys. Acta (BBA) Gen. Subj.* **2014**, *1840*, 1574–1582. [CrossRef] [PubMed]
9. Matsui, H.; Hopkinson, B.M.; Nakajima, K.; Matsuda, Y. Plasma-membrane-type aquaporins from marine diatoms function as CO_2/NH_3 channels and provide photoprotection. *Plant Physiol.* **2018**, *178*, 345–357. [CrossRef]
10. Guo, L.; Wang, Z.Y.; Lin, H.; Cui, W.E.; Chen, J.; Liu, M.; Chen, Z.L.; Qu, L.J.; Gu, H. Expression and functional analysis of the rice plasma-membrane intrinsic protein gene family. *Cell Res.* **2006**, *16*, 277–286. [CrossRef] [PubMed]
11. Wang, L.; Li, Q.-T.; Lei, Q.; Feng, C.; Zheng, X.; Zhou, F.; Li, L.; Liu, X.; Wang, Z.; Kong, J. Ectopically expressing MdPIP1;3, an aquaporin gene, increased fruit size and enhanced drought tolerance of transgenic tomatoes. *BMC Plant Biol.* **2017**, *17*, 246. [CrossRef]
12. Hu, W.; Yuan, Q.; Wang, Y.; Cai, R.; Deng, X.; Wang, J.; Zhou, S.; Chen, M.; Chen, L.; Huang, C.; et al. Overexpression of a Wheat Aquaporin Gene, TaAQP8, Enhances Salt Stress Tolerance in Transgenic Tobacco. *Plant Cell Physiol.* **2012**, *53*, 2127–2141. [CrossRef]
13. Shabala, S. Learning from halophytes: Physiological basis and strategies to improve abiotic stress tolerance in crops. *Ann. Bot.* **2013**, *112*, 1209–1221. [CrossRef] [PubMed]
14. Martinez-Ballesta, M.D.C.; Carvajal, M. New challenges in plant aquaporin biotechnology. *Plant Sci.* **2014**, *217–218*, 71–77. [CrossRef] [PubMed]
15. Yuan, H.-J.; Ma, Q.; Wu, G.-Q.; Wang, P.; Hu, J.; Wang, S.-M. ZxNHX controls Na^+ and K^+ homeostasis at the whole-plant level in Zygophyllum xanthoxylum through feedback regulation of the expression of genes involved in their transport. *Ann. Bot.* **2014**, *115*, 495–507. [CrossRef]
16. Zhou, X.; Zhou, Z.; Wu, C. The Research of the breeding characters of *Zygophyllum xanthoxylum*. *Pratacult. Sci.* **2006**, *23*, 38–41.
17. Ma, Q.; Yue, L.-J.; Zhang, J.-L.; Wu, G.-Q.; Bao, A.-K.; Wang, S.-M. Sodium chloride improves photosynthesis and water status in the succulent xerophyte Zygophyllum xanthoxylum. *Tree Physiol.* **2011**, *32*, 4–13. [CrossRef] [PubMed]
18. Yue, L.; Li, S.; Ma, Q.; Zhou, X.; Wu, G.; Bao, A.; Zhang, J.; Wang, S. NaCl stimulates growth and alleviates water stress in the xerophyte Zygophyllum xanthoxylum. *J. Arid. Environ.* **2012**, *87*, 153–160. [CrossRef]
19. Ma, Q.; Hu, J.; Zhou, X.-R.; Yuan, H.-J.; Kumar, T.; Luan, S.; Wang, S.-M. ZxAKT1 is essential for K^+ uptake and K^+/Na^+ ho-meostasis in the succulent xerophyte *Zygophyllum xanthoxylum*. *Plant J.* **2017**, *90*, 48–60. [CrossRef]
20. Yin, H.; Li, M.; Li, D.; Khan, S.-A.; Hepworth, S.R.; Wang, S.-M. Transcriptome analysis reveals regulatory framework for salt and osmotic tolerance in a succulent xerophyte. *BMC Plant Biol.* **2019**, *19*, 88. [CrossRef]

21. Ma, Q.; Bao, A.-K.; Chai, W.-W.; Wang, W.-Y.; Zhang, J.-L.; Li, Y.-X.; Wang, S.-M. Transcriptomic analysis of the succulent xerophyte Zygophyllum xanthoxylum in response to salt treatment and osmotic stress. *Plant Soil* **2016**, *402*, 343–361. [CrossRef]
22. Golldack, D.; Lüking, I.; Yang, O. Plant tolerance to drought and salinity: Stress regulating transcription factors and their functional significance in the cellular transcriptional network. *Plant Cell Rep.* **2011**, *30*, 1383–1391. [CrossRef]
23. Moore, K.; Roberts, L.J. Measurement of Lipid Peroxidation. *Free. Radic. Res.* **1998**, *28*, 659–671. [CrossRef] [PubMed]
24. Szabados, L.; Savouré, A. Proline: A multifunctional amino acid. *Trends Plant Sci.* **2010**, *15*, 89–97. [CrossRef] [PubMed]
25. Yamaguchi-Shinozaki, K.; Shinozak, K. A novel cis-acting element in an Arabidopsis gene 1s involved in responsiveness to drought, low temperature, or high-salt stress. *Plant Cell* **1994**, *6*, 251–264.
26. Liu, Q.; Kasuga, M.; Sakuma, Y.; Abe, H.; Miura, S.; Yamaguchi-Shinozaki, K.; Shinozaki, K. Two transcription factors, DREB1 and DREB2, with an EREBP/AP2 DNA binding domain separate two cellular signal transduction pathways in drought-and low temperature-responsive gene expression, respectively, in Arabidopsis. *Plant Cell* **1998**, *10*, 1391–1406. [CrossRef] [PubMed]
27. Wang, S.; Wan, C.; Wang, Y.; Chen, H.; Zhou, Z.; Fu, H.; Sosebee, R.E. The characteristics of Na$^+$, K+ and free proline distribution in several drought-resistant plants of the Alxa Desert, China. *J. Arid. Environ.* **2004**, *56*, 525–539. [CrossRef]
28. Flowers, T.J. Improving crop salt tolerance. *J. Exp. Bot.* **2004**, *55*, 307–319. [CrossRef]
29. Yang, S.; Vanderbeld, B.; Wan, J.; Huang, Y. Narrowing Down the Targets: Towards Successful Genetic Engineering of Drought-Tolerant Crops. *Mol. Plant* **2010**, *3*, 469–490. [CrossRef] [PubMed]
30. Hsieh, T.H.; Lee, J.T.; Charng, Y.Y.; Chan, M.T. Tomato plants ectopically expressing Arabidopsis CBF1 show enhanced re-sistance to water deficit stress. *Plant Physiol.* **2002**, *130*, 618–626. [CrossRef] [PubMed]
31. Hsieh, T.-H.; Lee, J.-T.; Yang, P.-T.; Chiu, L.-H.; Charng, Y.-Y.; Wang, Y.-C.; Chan, M.-T. Heterology Expression of the Arabidopsis C-Repeat/Dehydration Response Element Binding Factor 1 Gene Confers Elevated Tolerance to Chilling and Oxidative Stresses in Transgenic Tomato. *Plant Physiol.* **2002**, *129*, 1086–1094. [CrossRef]
32. Kang, H.-G.; Kim, J.; Kim, B.; Jeong, H.; Choi, S.H.; Kim, E.K.; Lee, H.-Y.; Lim, P.O. Overexpression of FTL1/DDF1, an AP2 transcription factor, enhances tolerance to cold, drought, and heat stresses in Arabidopsis thaliana. *Plant Sci.* **2011**, *180*, 634–641. [CrossRef]
33. Wang, Y.; Zhao, Z.; Liu, F.; Sun, L.; Hao, F. Versatile Roles of Aquaporins in Plant Growth and Development. *Int. J. Mol. Sci.* **2020**, *21*, 9485. [CrossRef]
34. Xu, Y.; Hu, W.; Liu, J.; Zhang, J.; Jia, C.; Miao, H.; Xu, B.; Jin, Z. A banana aquaporin gene, MaPIP1;1, is involved in tolerance to drought and salt stresses. *BMC Plant Biol.* **2014**, *14*, 59. [CrossRef]
35. Wang, L.; Liu, Y.; Feng, S.; Yang, J.; Li, D.; Zhang, J. Roles of Plasmalemma Aquaporin Gene StPIP1 in Enhancing Drought Tolerance in Potato. *Front. Plant Sci.* **2017**, *8*, 616. [CrossRef] [PubMed]
36. Alavilli, H.; Awasthi, J.P.; Rout, G.R.; Sahoo, L.; Lee, B.-H.; Panda, S.K. Overexpression of a Barley Aquaporin Gene, HvPIP2;5 Confers Salt and Osmotic Stress Tolerance in Yeast and Plants. *Front. Plant Sci.* **2016**, *7*. [CrossRef] [PubMed]
37. Li, W.; Qiang, X.-J.; Han, X.-R.; Jiang, L.-L.; Zhang, S.-H.; Han, J.; He, R.; Cheng, X.-G. Ectopic Expression of a *Thellungiella salsuginea* Aquaporin Gene, TsPIP1;1, Increased the Salt Tolerance of Rice. *Int. J. Mol. Sci.* **2018**, *19*, 2229. [CrossRef] [PubMed]
38. Zhou, L.; Zhou, J.; Xiong, Y.; Liu, C.; Wang, J.; Wang, G.; Cai, Y. Overexpression of a maize plasma membrane intrinsic protein ZmPIP1;1 confers drought and salt tolerance in Arabidopsis. *PLoS ONE* **2018**, *13*, e0198639. [CrossRef]
39. Wang, X.; Gao, F.; Bing, J.; Sun, W.; Feng, X.; Ma, X.; Zhou, Y.; Zhang, G. Overexpression of the Jojoba Aquaporin Gene, ScPIP1, Enhances Drought and Salt Tolerance in Transgenic Arabidopsis. *Int. J. Mol. Sci.* **2019**, *20*, 153. [CrossRef] [PubMed]
40. Zhou, S.; Hu, W.; Deng, X.; Ma, Z.; Chen, L.; Huang, C.; Wang, C.; Wang, J.; He, Y.; Yang, G.; et al. Overexpression of the Wheat Aquaporin Gene, TaAQP7, Enhances Drought Tolerance in Transgenic Tobacco. *PLoS ONE* **2012**, *7*, e52439. [CrossRef] [PubMed]
41. Ayadi, M.; Brini, F.; Masmoudi, K. Overexpression of a Wheat Aquaporin Gene, TdPIP2;1, Enhances Salt and Drought Tolerance in Transgenic Durum Wheat cv. Maali. *Int. J. Mol. Sci.* **2019**, *20*, 2389. [CrossRef] [PubMed]
42. Apse, M.P.; Blumwald, E. Na$^+$ transport in plants. *FEBS Lett.* **2007**, *581*, 2247–2254. [CrossRef] [PubMed]
43. Gao, Z.; He, X.; Zhao, B.; Zhou, C.; Liang, Y.; Ge, R.; Shen, Y.; Huang, Z. Overexpressing a Putative Aquaporin Gene from Wheat, TaNIP, Enhances Salt Tolerance in Transgenic Arabidopsis. *Plant Cell Physiol.* **2010**, *51*, 767–775. [CrossRef] [PubMed]
44. Sun, H.; Li, L.; Lou, Y.; Zhao, H.; Yang, Y.; Wang, S.; Gao, Z. The bamboo aquaporin gene PeTIP4;1−1 confers drought and salinity tolerance in transgenic Arabidopsis. *Plant Cell Rep.* **2017**, *36*, 597–609. [CrossRef]
45. Hill, A.E.; Shachar-Hill, B. What Are Aquaporins For? *J. Membr. Biol.* **2004**, *197*, 1–32. [CrossRef]
46. Macrobbie, E.A.C. Osmotic effects on vacuolar ion release in guard cells. *Proc. Natl. Acad. Sci. USA* **2006**, *103*, 1135–1140. [CrossRef]
47. Ismail, A.; El-Sharkawy, I.; Sherif, S. Salt Stress Signals on Demand: Cellular Events in the Right Context. *Int. J. Mol. Sci.* **2020**, *21*, 3918. [CrossRef] [PubMed]
48. Dang, Z.-H.; Zheng, L.-L.; Wang, J.; Gao, Z.; Wu, S.-B.; Qi, Z.; Wang, Y.-C. Transcriptomic profiling of the salt-stress response in the wild recretohalophyte *Reaumuria trigyna*. *BMC Genom.* **2013**, *14*, 29. [CrossRef]
49. Yamauchi, N.; Watada, A.E. Effectiveness of various phenolic compounds in degradation of chlorophyll by In Vitro peroxi-dase-hydrogen peroxide system. *J. Jpn. Soc. Hortic. Sci.* **1994**, *3*, 439–444. [CrossRef]
50. Chen, Q.; Yang, S.; Kong, X.; Wang, C.; Xiang, N.; Yang, Y.; Yang, Y. Molecular cloning of a plasma membrane aquaporin in *Stipa purpurea*, and exploration of its role in drought stress tolerance. *Gene* **2018**, *665*, 41–48. [CrossRef]

International Journal of
Molecular Sciences

Article

Adaptation Strategies of Halophytic Barley *Hordeum marinum* ssp. *marinum* to High Salinity and Osmotic Stress

Stanislav Isayenkov [1,2,*], Alexander Hilo [1], Paride Rizzo [1], Yudelsy Antonia Tandron Moya [1], Hardy Rolletschek [1], Ljudmilla Borisjuk [1] and Volodymyr Radchuk [1,*]

[1]   Leibniz-Institute of Plant Genetics and Crop Plant Research (IPK), Corrensstrasse 3, 06466 Gatersleben, Germany; hilo@ipk-gatersleben.de (A.H.); rizzo@ipk-gatersleben.de (P.R.); moya@ipk-gatersleben.de (Y.A.T.M.); rollet@ipk-gatersleben.de (H.R.); borisjuk@ipk-gatersleben.de (L.B.)
[2]   Institute of Food Biotechnology and Genomics NAS of Ukraine, Osipovskogo Street, 2a, 04123 Kyiv, Ukraine
*   Correspondence: stan.isayenkov@gmail.com (S.I.); radchukv@ipk-gatersleben.de (V.R.)

Received: 8 October 2020; Accepted: 24 November 2020; Published: 27 November 2020

Abstract: The adaptation strategies of halophytic seaside barley *Hordeum marinum* to high salinity and osmotic stress were investigated by nuclear magnetic resonance imaging, as well as ionomic, metabolomic, and transcriptomic approaches. When compared with cultivated barley, seaside barley exhibited a better plant growth rate, higher relative plant water content, lower osmotic pressure, and sustained photosynthetic activity under high salinity, but not under osmotic stress. As seaside barley is capable of controlling Na^+ and Cl^- concentrations in leaves at high salinity, the roots appear to play the central role in salinity adaptation, ensured by the development of thinner and likely lignified roots, as well as fine-tuning of membrane transport for effective management of restriction of ion entry and sequestration, accumulation of osmolytes, and minimization of energy costs. By contrast, more resources and energy are required to overcome the consequences of osmotic stress, particularly the severity of reactive oxygen species production and nutritional disbalance which affect plant growth. Our results have identified specific mechanisms for adaptation to salinity in seaside barley which differ from those activated in response to osmotic stress. Increased knowledge around salt tolerance in halophytic wild relatives will provide a basis for improved breeding of salt-tolerant crops.

Keywords: halophytic wild barley; salinity; osmotic stress; metabolome; transcriptome; ionome; stress adaptation; *Hordeum marinum*

---

1. Introduction

High salinity is one of the biggest threats to modern agriculture and crop productivity, leading to an annual estimated economic loss of over 10 billion USD [1]. More than 800 million hectares of agricultural land (>6% of the planet's total land area) are considered to be salt-affected [2]. The area of salinized soils is reported to be increasing at a rate of 10% per year, and is an issue in more than 100 countries worldwide [3,4].

High levels of salinity result in impaired plant growth and development through various mechanisms, including osmotic stress (OST) due to loss of cellular water content, cytotoxicity due to excessive uptake of Na^+ and Cl^- ions, oxidative stress due to generation of reactive oxygen species (ROS), and nutritional imbalance [5]. Compared to salt-sensitive plants, or glycophytes, the increased salt tolerance of plants grown in a saline environment, or halophytes, is achieved predominantly by a greater robustness of employed mechanisms rather than qualitative differences [5,6]. These mechanisms involve maintaining the homeostasis of cellular ions, making osmotic adjustments and ROS scavenging. Na^+ and Cl^- ions are themselves the important contributors to the cellular osmotic potential [7]. Because they are toxic if not compartmentalized, these ions have to be sequestrated into vacuoles or endosomal compartments by ion exchangers and the H^+ pumps localized to the tonoplast or endosomal membranes [8]. Organic osmolytic solutes, such as sugars, sugar alcohols, and proline, accumulate in the cytoplasm of halophytic species to balance the osmotic potential of Na^+ and Cl^-, contained in the vacuole, and to maintain the physiological functions of the cell [9]. From an energy-saving aspect, cellular osmotic adjustment is achieved more efficiently by the use of ions than of organic solutes [7].

Plant species have evolved diverse and unique ways to survive in harsh saline environments. Certain dicot halophytic plants, in order to resist or avoid accumulation of toxic ions, have developed special structures and organs, such as epidermal bladder cells, which accumulate excessive Na^+ in their vacuoles, and hydathodes, which actively secrete salt and reduce the concentration of toxic ions in the cells [10,11]. The majority of halophytic monocots do not exhibit such specialized organs, but have developed other ways to survive under saline conditions. Several wild species within the Triticeae tribe, to which the major crops wheat (*Triticum aestivum*) and barley (*Hordeum vulgare*) belong, exhibit exceptional salinity tolerance [10]. The seaside barley (*Hordeum marinum*), a typical Mediterranean halophytic plant of coastal salt marshes, is considered one of the major genetic sources for salinity tolerance [12]. The amphidiploid wheat hybrids with *H. marinum* exhibit improved salt tolerance compared with wheat [13,14]. *H. marinum* possesses a higher water saturation deficit and osmotic potential in comparison with that of cultivated barley due to higher accumulation of proline, glycine betaine, and dehydrins [15,16]. Proteomic analysis also revealed increased levels of proteins involved in energy metabolism [15]. Furthermore, antioxidant enzymes in seaside barley were shown to have significantly higher activity in plants grown at high salinity [17]. Transcriptome studies suggest that the salt-tolerance strategy of *H. marinum* comprises low energy consumption, utilization of inorganic ions as cheap osmotic agents, and changes in the activity of the HmHKT1;5 and HmHKT2;1 transporters [18–20]. However, the molecular mechanisms underlying the biochemical and morphological changes and physiological strategies employed by *H. marinum* during acclimation to salinity remain mostly unexplored.

The aim of the present study was to elucidate the differences in adaptation strategies of *H. marinum* plants to OST and salinity stress (SST) at the molecular, metabolic, morphological, and physiological levels.

2. Results

2.1. Different Physiological Responses of H. marinum and H. vulgare to OST and SST

To evaluate differences in the response of *H. vulgare* and *H. marinum* to SST and to elicit salinity adaptation responses in halophytic seaside barley, the plants were cultivated in hydroponic culture containing 300 mM NaCl, which corresponds to slightly over 500 mOsm osmotic pressure. Preliminary OST experiments showed deleterious effects on plants after treatment with the same osmotic pressure (32% PEG6000), probably due to impermeability of this osmotic agent through cell membranes. Therefore, plants were cultivated in media supplemented by 15% PEG6000, which plants could still tolerate. SST and OST treatments affected the growth of both *H. vulgare* and *H. marinum* plants, albeit to a different extent (Figure 1). The relative growth rate (RGR) of *H. marinum* plants was approximately 2-fold lower under either SST or OST conditions (Figure 1C). A similar decrease in the RGR of *H. vulgare* plants was observed under OST; however, application of SST resulted in a ~95% reduction of growth rate relative to the control (Figure 1F). Under high salinity, the *H. vulgare* plants exhibited leaf chlorosis and wilting as marks of severe salt toxicity, whereas the *H. marinum* plants maintained their strong green color (Figure 1A,B). Under OST, however, they turned a yellow shade. Furthermore, after SST, the osmotic pressure recorded in *H. marinum* plants was slightly lower than that in *H. vulgare* (Figure S1).

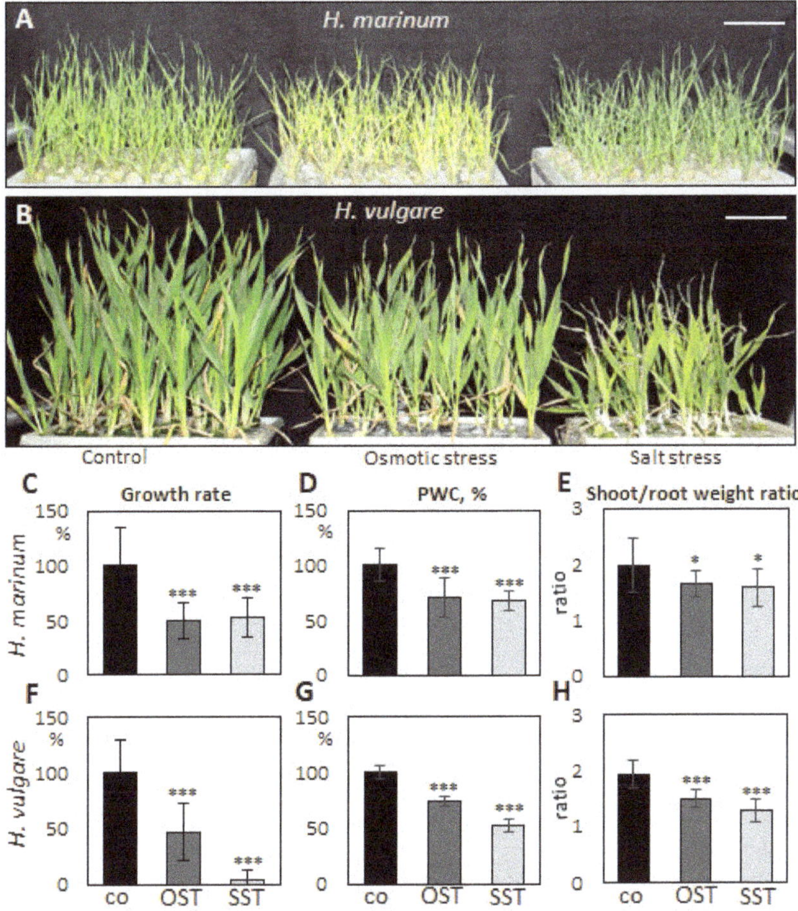

Figure 1. Changes in growth of *Hordeum marinum* and *H. vulgare* plants under osmotic (OST) and salinity (SST) stresses. (**A**,**B**) Morphological characteristics of *H. marinum* (**A**) and *H. vulgare* plants (**B**) under SST and OST after reaching the maximum stress (27 days old); (**C**,**F**) relative growth rate, (**D**,**G**) plant water content (PWC), and (**E**,**H**) shoot/root weight ratio of *H. marinum* (**C**–**E**) and *H. vulgare* (**F**–**H**) plants under control and stressed conditions. Scale bars = 5 cm. Data are mean ± SD; $n = 8$, *t* significant at: *, $p < 0.05$, and ***, $p < 0.001$.

Comparative non-invasive magnetic resonance imaging of *H. marinum* plants demonstrated alterations in the hypocotyl and root structure, compared to the controls, under SST (Figure 2). Numerous root primordia and seminal roots were initiated in the hypocotyl region, resulting in more fibrous roots (Figure 2A,B). NMR models (Figure 2C,D) allowed the calculation of volumes and surface area of roots. While the volumes of individual roots were only marginally decreased under conditions of high salinity compared to the control (0.43 ± 0.20 vs. 0.55 ± 0.10, mm^3), the total surface area of the stressed roots was ~23.8% higher, due to the production of a larger number of thinner roots.

Plant water content (PWC) in *H. marinum* tissues was depleted by ~30% after application of either OST or SST (Figure 1D). In particular, the cortex region of the saline-affected roots contained less water than that of the control roots (Figure 2E,F). A decrease of ~25% in PWC was also observed in the *H. vulgare* plants under OST, and SST resulted in almost 50% less PWC compared to the control plants (Figure 1G). Thus, seaside barley exhibited a greater ability to retain water under conditions of high salinity in comparison to *H. vulgare*. Finally, a greater reduction in shoot:root weight ratio was observed in *H. vulgare* plants under SST than in *H. marinum* (Figure 1E,H). Together, these data indicate that stress treatments, in particular salinity, hinder the growth of *H. vulgare*, while *H. marinum* exhibits stronger resistance to SST, as reflected in enhanced water retention, preserved shoot growth, and, possibly, sustained biosynthetic activity.

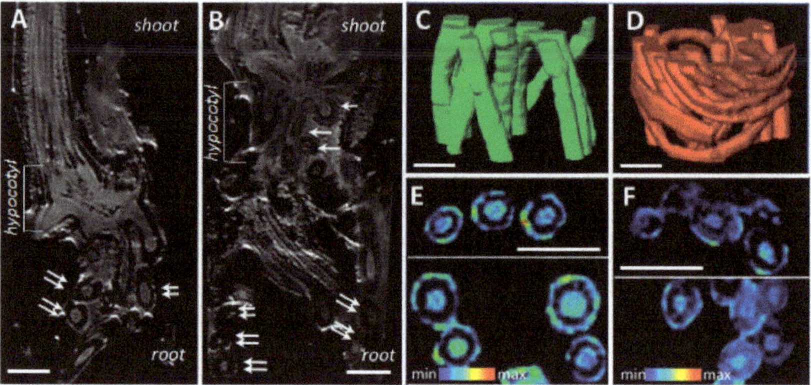

Figure 2. Comparative non-invasive magnetic resonance imaging (MRI) revealed structural changes in SST roots of *H. marinum* when compared to the control. (**A,B**) The representative virtual cross-sections show the internal structure of hypocotyl regions of plants growing under the control condition (**A**) and SST (**B**). Numerous root nodules in the hypocotyl region (white arrows) and root cross-sections (doubled arrows) are visible. (**C,D**) Fragments of the 3D models show spatial arrangement of the fibrous roots in control (**C**, green) and SST (**D**, red). (**E,F**) Relative differences in water distribution across the root tissues are visualized in virtual cross-sections of control (**E**) and SST (**F**) roots. MRI signal in (**E,F**) is at an identical scale and represented using a rainbow-based color scheme. High signal intensities in red (max) indicate high water saturation, while the blue regions (min) indicate lower water saturation. Scale bars = 1 mm.

2.2. Different Photosynthetic Activity and Assimilate Allocation in H. marinum and H. vulgare Plants Under SST

To evaluate the photosynthetic activity and assimilate allocation in *H. marinum* and *H. vulgare* plants under stress conditions, we analyzed the uptake and distribution of assimilates following the treatment of control and stressed shoots with ^{13}C-labeled CO_2 (Figure 3). When compared with domesticated barley, *H. marinum* shoots showed ~2-fold higher efficiency of ^{13}C uptake. In *H. marinum*, the efficiency of ^{13}C assimilation was slightly decreased under OST, and not significantly changed under SST, indicating maintenance of photosynthetic activity rate. On the other hand, the ^{13}C assimilation in *H. vulgare* shoots appeared significantly decreased under OST and was almost negligible under SST (Figure 3A).

The *H. marinum* control plants re-allocated large amounts of ^{13}C-labeled assimilates to the roots (Figure 3B). Application of either OST or SST led to a significant decrease, but not a complete block of the ^{13}C allocation to the roots. The *H. vulgare* roots also accumulated less ^{13}C-labeled assimilates than the equivalent control plants under OST, while ^{13}C accumulation was barely detectable in SST-treated roots due to the inhibited photosynthetic ^{13}C fixation by the shoots.

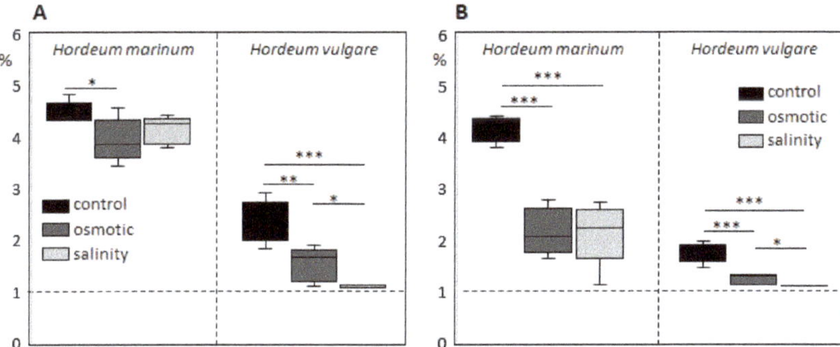

Figure 3. ^{13}C uptake and distribution in shoots (**A**) and roots (**B**) of *Hordeum marinum* and *H. vulgare* plants under control conditions as well as under osmotic and salinity stresses. Dashed lines indicate natural ^{13}C abundance. Data are mean ± SD; $n = 5$, t significant at: *, $p < 0.05$; **, $p < 0.01$; ***, $p < 0.001$.

2.3. Comparative Analysis of Mineral Composition Under SST and OST

We analyzed the mineral content of *H. marinum* roots and shoots under control and stress conditions (Table 1).

Table 1. Element compositions in shoots and roots of *Hordeum marinum* under osmotic and salinity stress compared to control.

Element	Shoots, (µg/g) DW *			Roots (µg/g) DW *		
	Control	Osmotic Stress	Salinity	Control	Osmotic Stress	Salinity
^{11}B	17.9 ± 3.9	13.7 ± 3.1 *	13.3 ± 0.7 **	5.8 ± 1.2	5.7 ± 0.7	5.0 ± 0.8
^{98}Mo	1.8 ± 0.1	2.4 ± 0.5 **	3.4 ± 0.3 ***	1.8 ± 0.9	1.2 ± 0.1	2.8 ± 0.1 **
^{31}P	6510.8 ± 276.4	8006.4 ± 371.0 ***	5696.5 ± 191.9 ***	7152.7 ± 167.5	7850.4 ± 252.0 ***	6463.2 ± 199.8 ***
^{44}Ca	6958.7 ± 1596.1	6364.0 ± 394.0	2604.8 ± 424.6 ***	2806.7 ± 375.5	6850.9 ± 752.6 ***	1388.8 ± 282.9 ***
^{55}Mn	108.1 ± 14.8	205.7 ± 49.2 ***	91.1 ± 37.6	199.4 ± 27.0	379.4 ± 36.6 ***	200.1 ± 23.3
^{60}Ni	4.5 ± 2.3	9.8 ± 3.7 **	6.5 ± 5.8	6.5 ± 3.2	13.8 ± 3.2 ***	16.7 ± 8.2 **
^{63}Cu	20.5 ± 2.1	38.7 ± 48.3	19.8 ± 4.7	297.9 ± 18.0	183.3 ± 13.9 ***	282.8 ± 20.3
^{66}Zn	35.3 ± 2.3	74.1 ± 6.3 ***	56.4 ± 4.8 ***	44.2 ± 5.1	35.7 ± 2.1 ***	111.1 ± 7.0 ***
^{23}Na	539.2 ± 162.3	545.9 ± 68.7	25614.5 ± 2381.9 ***	728.2 ± 81.8	505.7 ± 51.6 ***	36680.6 ± 3394.2 ***
^{26}Mg	3048.6 ± 680.1	2805.7 ± 415.2	1504.1 ± 201.5 ***	1391.7 ± 233.3	1730.4 ± 51.0 **	1075.7 ± 83.3 **
^{34}S	449439 ± 353.8	5845.4 ± 942.7 **	3329.9 ± 249.6 ***	3125.5 ± 191.7	3544.4 ± 162.8 ***	3123.6 ± 155.7
^{39}K	54968.5 ± 3184.1	52948.5 ± 6022.5	42608.6 ± 2273.7 ***	41786.5 ± 1919.9	37967.2 ± 2353.3 **	27505.5 ± 1466.2 ***

* DW, dry weight. Significantly increased contents are highlighted in blue, significantly decreased contents are highlighted in red. Data are means ± SD, n = 8–10, * t significant at $p < 0.05$, ** t significant at $p < 0.01$ and *** t significant at $p < 0.001$.

Following incubation with 300 mM NaCl, a marked elevation of Na content was observed in both tissue types, albeit ~1.4-fold lower in the shoots than in the roots. Contrary to Na, the K and particularly the Ca contents were significantly reduced in the shoots and roots of SST-treated plants. Ca^{2+} is recognized as a crucial second messenger in signaling pathways linking the perception of environmental stimuli to plant adaptive responses [21]. The estimated K/Na ratio was higher in the shoots than in roots (1.66 vs. 0.75) under SST, possibly indicating more efficient K^+ retention in green tissue. OST led to K reduction but Ca elevation in the roots, whereas no change was observed in their levels in the shoots. Zn and Mo contents were also elevated in both tissues under SST and in the shoots under OST. In OST-treated roots, Zn content was reduced, while Mo was not affected. Furthermore, under SST, P and Mg concentrations were decreased in both sample types, and S and B contents were decreased only in the shoots. These results suggest that salinity evokes changes in mineral uptake and allocation in the whole plant to counteract Na and Cl excess and adjust the osmotic pressure. In contrast to SST, OST resulted in a rise in mineral contents (P, Mn, Ni, and S) in the shoots, accompanied by decreases in the Cu and Na contents in the roots, likely reflecting ionic adaptations to high osmotic pressure.

2.4. Alterations of Metabolite Profiles in Response to SST and OST

Changes in metabolome of the roots and shoots of *H. marinum* plants were investigated under OST and SST by untargeted metabolite profiling. In total, 138 and 136 metabolites were identified in the roots and shoots, respectively (Table S1). In the roots, the levels of 59 metabolites were significantly altered by >2-fold (a decrease observed in 44 and an increase in 15 metabolites) following treatment with OST, whereas a change was detected in 61 metabolites (35 decreased and 26 increased) in those under SST conditions. In the shoots of the plants, OST led to a change in the levels of 68 metabolites (45 decreased and 23 increased) and SST resulted in differences for 58 metabolites (40 decreased and 18 increased) (Table S1).

Principal component analysis of the metabolite profiles of *H. marinum* plants revealed differential responses to OST versus SST (Figure S2), with only 14 metabolites being affected under both stresses. The largest increase was detected for the flavonoid 3-methoxy-4-hydroxyhippuric acid under both types of stress (Table S1). The levels of ascorbate and its precursor mannose-6P were increased under OST but decreased under SST in the shoots (Figure 4). A strong increase in gluconolactone, a polyhydroxy acid with metal-chelating and ROS-scavenging activities [22], was specifically detected in SST-treated roots (Table S1). These data suggest differences in ROS production and scavenging in *H. marinum* tissues under OST versus SST.

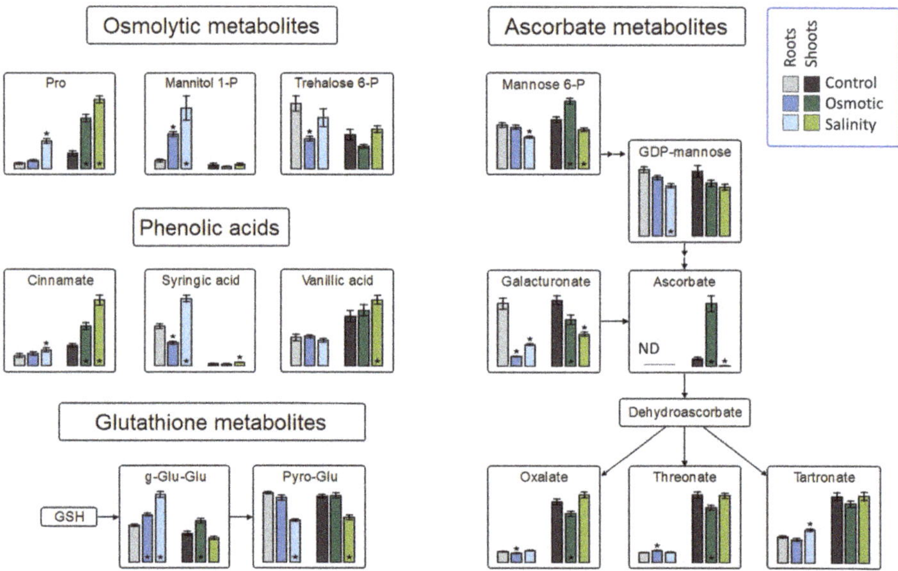

Figure 4. Changes in osmolytic metabolites, and the antioxidant system in roots and shoots of *Hordeum marinum* under osmotic and salinity stresses. Bars represent means of seven independent replicates ± SE. Significant differences to control treatments at specified time points after excision are indicated by asterisks (Wilcoxon, Mann-Whitney U-test; *, $p < 0.05$).

Plants often exhibit an increase in free proline following exposure to hyperosmotic stress or SST [23]. In *H. marinum*, such an increase was larger under SST than OST in both tissue types. The trehalose-6P content was decreased in both tissues under OST, while an increase in mannitol-1P was observed in the roots, but not the shoots, under both stresses (Figure 4). These elevated levels indicate increased mannitol biosynthesis.

The levels of fructose-6P and lactate were decreased in both tissues under OST and SST conditions (Figure 5). An increase in citrate in both tissues under OST and in shoots under SST, but decrease in organic acids associated with malate conversion [2-oxoglutarate, succinate (in both tissues after SST and in roots after OST), fumarate (only in shoots) and malate], were detected in the metabolites associated with the tricarboxylic acid (TCA) cycle (Figure 5). The levels of purine nucleotides generally exhibited trends opposite to those of uric acid, allantoin, and allantoate in roots and shoots, particularly under OST (Figure 6). These results indicate that, in stressed plants, there is a reduction of processes related to nucleotide and energy metabolism and possibly cellular proliferation.

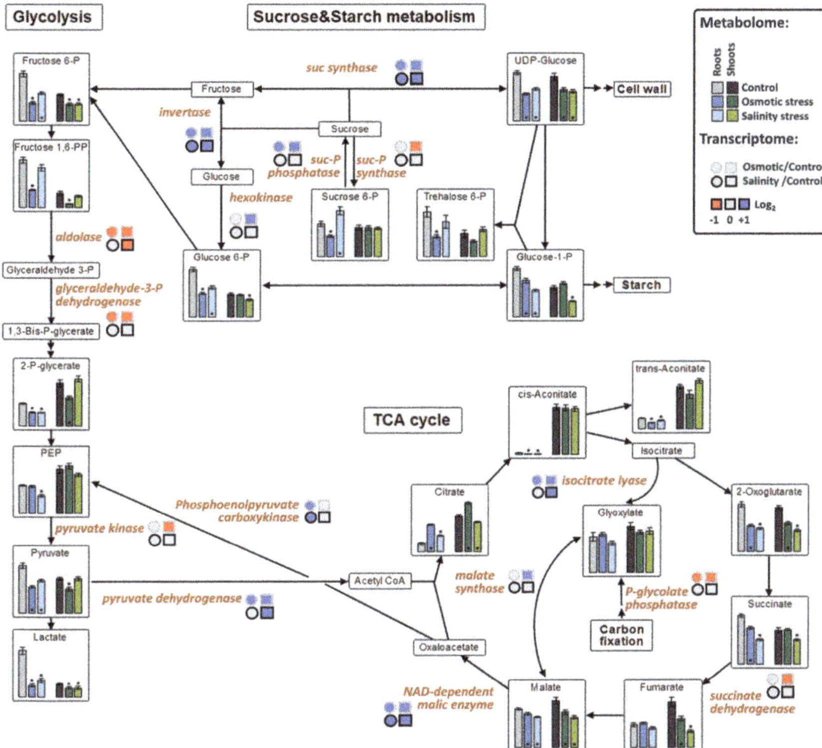

Figure 5. The effect of osmotic and salinity stresses on the sugar and central metabolism and corresponding transcriptomic changes in roots and shoots of *Hordeum marinum*. Metabolic data, presented as bars, are means ± SE; *n* = 7. Significant differences to control treatments at specified time points after excision are indicated by *, *p* < 0.05 (Wilcoxon, Mann-Whitney U-test). Up-regulated genes are labeled by blue, down-regulated by red color.

Homeostasis of the auxin indole-3-acetic acid (IAA) is achieved through amino acid conjugation and catabolism [24]. The levels of the IAA-Ala conjugate (reversible storage compound of IAA) were increased in plants under both type of stress. 2-oxindole-3-acetic acid (oxIAA), a major inactive and irreversible IAA degradation product, was markedly increased in SST-treated roots (Table S1). Thus, adjustment of auxin content may be involved in salinity adaptation.

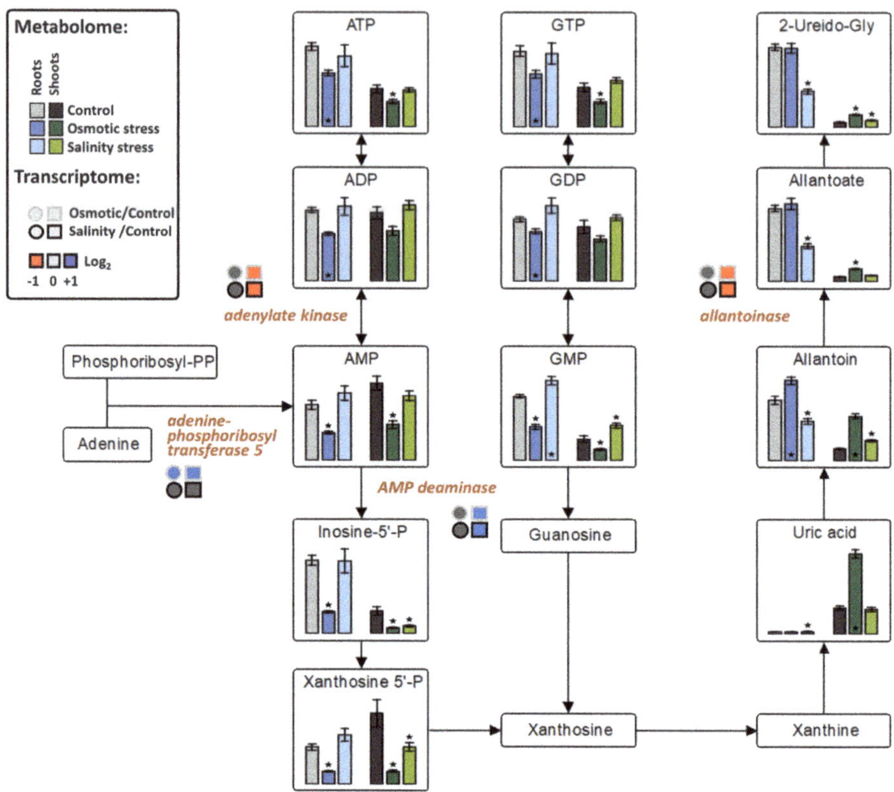

Figure 6. The effect of osmotic and salinity stresses on the purine catabolism and corresponding transcriptomic changes in roots and shoots of *Hordeum marinum*. Metabolic data, presented as bars, are means ± SE; $n = 7$. Significant differences to control treatments at specified time points after excision are indicated by *, $p < 0.05$ (Wilcoxon, Mann-Whitney U-test). Up-regulated genes are labeled by blue, down-regulated by red color, and those with no changes in expression by grey color.

Levels of mevalonate pyrophosphate, involved in the mevalonate pathway of terpenoid backbone biosynthesis [25], were decreased in all tissues, particularly in roots under SST (Table S1). The content of methylerythritol-4P (MEP), part of the MEP pathway of terpenoid synthesis, decreased in shoots under both types of stress (Table S1). Distinct components of the shikimate pathway (e.g., quinic acid and shikimate-3P) were significantly reduced under OST. Tyramine content, a product of tyrosine metabolism and a precursor in alkaloid biosynthesis via the shikimate pathway, was decreased in the roots under both types of stress (Table S1).

2.5. Transcript Profiling in H. marinum Suggests a Stronger Influence by OST than SST

We analyzed changes in RNA transcript abundances in the roots and shoots of *H. marinum* following treatment with OST and SST, using RNA sequencing. In total, 2232 differentially expressed genes (DEGs) with known or predicted function were detected in at least one tissue type as a result of at least one type of stress (fold change in expression ≥ 3, false discovery rate (FDR) < 0.05) (Table S2). The largest changes in transcriptome were observed in the OST-treated shoots with 1210 DEGs detected (821 down- and 389 upregulated), followed by the OST-treated roots with 1063 DEGs (469 down- and 594 upregulated). Notably fewer DEGs were detected under SST in the roots (545 DEGs: 220 down-

and 325 upregulated) and shoots (270 DEGs: 106 down- and 164 upregulated genes). While a difference in DEGs between stressed roots and shoots was expected due to their functional specificity, the overlap between DEGs detected following OST versus SST of the same tissue was also small (Figure 7A,B).

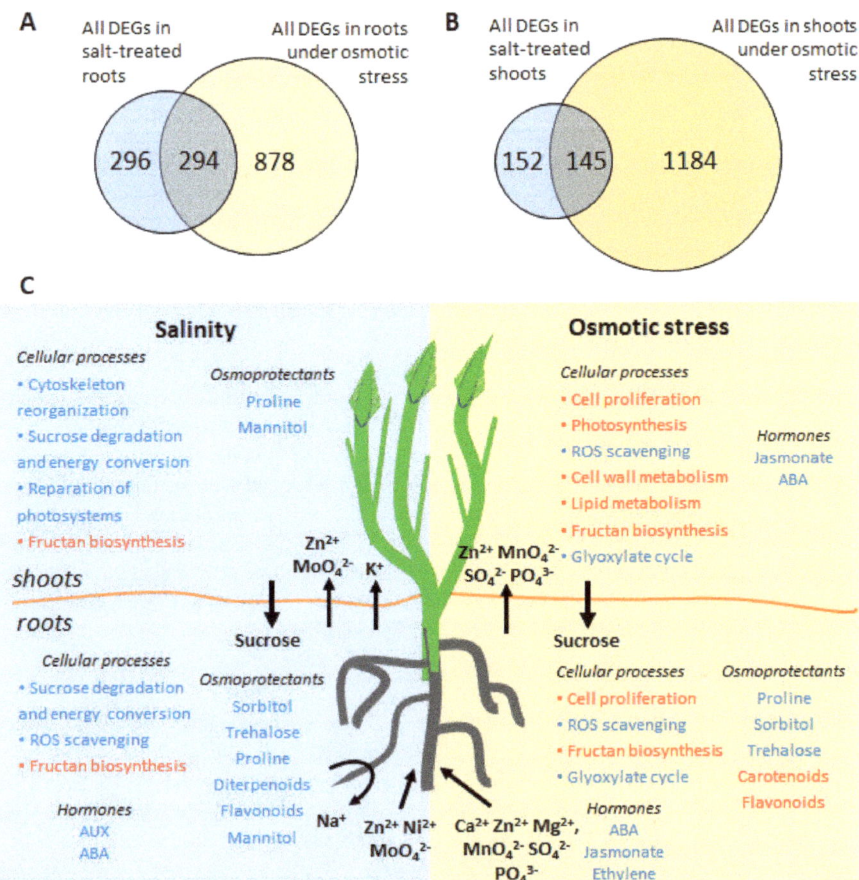

Figure 7. (**A,B**) Venn diagram showing the numbers of common and stress-specific differentially expressed genes (DEG) in roots (**A**) and shoots (**B**) after osmotic (15% PEG6000) and salt (300 mM NaCl) stresses. (**C**) A schematic overview of the main processes occurred during salinity and osmotic stress in *H. marinum*. Activated processes are highlighted in blue, inhibited processes are highlighted in red. Arrows indicate directions of mineral, phytohormone, and sugar redistribution between roots and shoots upon stress treatment.

2.6. OST Differentially Affects Shoot and Root Development

Of the DEGs detected in OST-treated shoots, the majority (117 genes) encoded proteins involved in transcription, translation, and general cellular processes, and included diverse *histones* (43 genes), *ribosomal proteins* (25 genes), *cyclins* (6 genes), *cell division control* (2 genes), and *expansins* (2 genes) (Table S2); all of them were downregulated. The same group of DEGs was also prominent in OST roots (59 genes), in which the genes involved in control of cell division and elongation (*cortical cell-delineating proteins, expansins, mitogen-activated protein kinase, cell cycle control phosphatase*) were also repressed. Under OST conditions, the most highly upregulated gene in both the shoots and roots encoded rRNA N-glycosidase, which is involved in ribosomal degradation [26]. Three *pumilio* genes, whose products may influence mRNA stability [27], were also highly upregulated in the shoots. These data indicate inhibition of cell division and elongation in plants under OST. Furthermore, repression of *actin depolymerizing factor* and *actin* together with activation of *dynein*, two *β-tubulins*, and *flotillin-like protein* genes suggests reorganization of the cytoskeleton and intracellular trafficking under these conditions.

OST resulted in downregulation of genes associated with cell wall metabolism (cellulose synthases, fasciclin-like arabinogalactan proteins, pectinesterases, xyloglucan endotransglucosylases and UDP-glycosyltransferases) in shoots and roots, signifying strong repression of cell wall biosynthesis. In the roots, genes involved in lipid biosynthesis were mostly repressed (including 3-ketoacyl-CoA synthases and fatty acid desaturase), while those responsible for lipid degradation (GDSL esterase/lipases, papatins, and lipoxygenase) exhibited increased transcription. In the shoots, from the repression of GDSL esterase/lipases (9 genes), 3-ketoacyl-CoA synthases (6 genes), bifunctional inhibitor/lipid-transfer protein/seed storage 2S albumin superfamily proteins (6 genes), glycerol-3-phosphate acyltransferases (4 genes), fatty acid desaturases (3 genes), fatty acid hydroxylases (2 genes), lipoxygenases (2 genes), and phospholipases (2 genes), it can be deduced that total lipid metabolism was likely minimized.

In contrast to the response under OST, SST-treated plants revealed only minor transcriptional changes in genes involved in general cellular processes and in the metabolism of lipids and the cell wall (Table S2). Downregulation of *expansins*, GTPase *RsgA*, CRIB domain-containing protein *RIC1*, and *α-tubulin 4*, as well as upregulation of *flotillin* and *β-tubulin 2*, indicate cytoskeleton reorganization in roots under SST. Of 20 DEGs associated with cell wall metabolism, three *pectin lyase* genes, two *fasciclin-like arabinogalactan protein* genes, and one *xyloglucan endotransglucosylase* gene showed decreased expression, while four *UDP-glycosyltransferase* genes and one *xylanase inhibitor* gene revealed increased expression in the roots. Moreover, the transcription of five genes encoding laccase, which is involved in lignin biosynthesis, was increased up to 10-fold in both the shoots and roots, suggesting increased lignification of cell walls under SST conditions.

2.7. OST But Not SST Leads to Strongly Diminished Photosynthetic Processes

OST led to a marked downregulation of genes encoding proteins from the entire photosynthetic machinery (Table 2): *chlorophyll a-b binding proteins* (18 genes), *subunits of reaction center of photosystems I and II* (9), *thylacoid membrane proteins* (7), *RUBISCO* (4), *subunits of cytochrome b6-f complex* (2), *plastocyanin*, *ferrodoxin*, and *ribulose-5P-3-epimerase*. However, this was not observed under SST conditions. *Heme oxygenase*, whose product plays a role in the protection against oxidative damage via ROS scavenging [28], was upregulated in shoots under both OST and SST. The expression of *protein D1*, required for the repair of photosystem II [29], was increased over 44-fold in SST-treated shoots.

2.8. Primary Metabolism and Sugar Conversion Are Altered Under SST and OST

Upregulated expression of genes involved in sucrose cleavage (*sucrose synthases* and *invertases*) as well as starch and glucan degradation (*glucan-1,3-β-glucosidases*, and *β-amylase*), together with the reduction of expression of those responsible for fructan biosynthesis (*sucrose:sucrose 1-fructosyltransferase, sucrose:fructan 6-fructosyltransferase*, and *fructan:fructan 1-fructosyltransferase*), suggest a shift from the production of di- and polysaccharides towards their degradation to hexoses in plants under both types of stress (Figure 5). The transcription of *aldose reductase* and *α-galactosidase*, involved in monosaccharide conversion and sorbitol synthesis, was also increased, as was the expression of *trehalose-P synthase* and *trehalose-6P phosphatase* was also upregulated in the roots (Table S2).

A group of DEGs associated with the TCA cycle and glycolysis showed upregulation in OST-treated plants (Table 2). However, the expression of *pyruvate dehydrogenase* and *NAD-dependent malic enzyme* was also increased in plants under SST (Figure 5). Increased transcription of *isocitrate lyase* and *malate synthase* suggests activation of the glyoxylate bypass [30,31]. The expression of *alanine:glyoxylate aminotransferase* and *NADP-dependent malate dehydrogenase*, whose products are involved in photorespiration, was increased only in plants treated with OST (Table S2).

Table 2. Gene ontology (GO) term enrichment in the differentially expressed genes (DEGs) from the specific tissue under osmotic or salt stresses.

Biological Process	Fold Enrichment in			
	OST Roots	OST Shoots	SST Roots	SST Shoots
Tricarboxylic acid metabolism (GO:0072351 + GO:0072350)	9.56–16.73	7.84–13.55		
Generation of precursor metabolites and energy (GO:0006091)		2.06		
Nicotianamine metabolism (GO:0030418 + GO:0030417)	16.73	13.55		
Amine metabolism (GO:0009309 + GO:0044106)	5.01–5.52			
Cold acclimation and response to cold (GO:0009631 + GO:0009409)	14.6		6.91–29.86	
Nitrate response and transport (GO:0010167 + GO:0015706)	9.61–10.14			
Transition metal ion transport (GO:0000041)	4.27			
Anion transport (GO:0015698 + GO:0006820 + GO:0098656)	2.79–4.25		3.51–5.62	
Ion transport (GO:0006811)	1.97		2.37	
Transmembrane transport (GO:0055085)	1.85		2.06	
Response to inorganic substances (GO:0010035)	3.21			
Response to acid chemical (GO:0001101)	2.85	2.8	3.67	
Oxidation-reduction (GO:0055114 + GO:0072593 + GO:0098869)	1.71	1.84–3.24	2.1	2.13
Tryptophan metabolism (GO:0000162 + GO:0006568)			11.87–15.37	
Indole compound metabolism (GO:0042435 + GO:0042430)			11.87–15.37	
Indolalkylamine metabolism (GO:0046219 + GO:0006586)			11.87–15.37	
Response to abscisic acid (GO:0009737)			5.74	
Response to alcohol (GO:0097305)			5.68	
Response to lipid (GO:0033993)			4.11	
Drug metabolism (GO:0042737+ GO:0017144)		2.54–2.8	3.33	
Photosynthesis (GO:0015979 + GO:0009768 + GO:0009765)		3.88–8.55		
Chromatin organization (GO:0097549 + GO:0045814 + GO:0034401)		5.2–5.41		
Antibiotic metabolism (GO:0016999 + GO:0017001)		3.02–3.14		
Cofactor metabolism (GO:0051187 + GO:0051186)		2.07–3.01		
Cellular detoxification (GO:1990748 + GO:0097237)		2.5		
Small molecule biosynthetic process (GO:0044283)		2.01		

Allantoinase, whose product converts allantoin into allantoate, was repressed in shoots under both stress conditions, while *AMP-deaminase* expression was upregulated, suggesting activation of metabolic conversion of adenine ribonucleotides into allantoin. Similarly, *adenine phosphoribosyltransferase 5*, responsible for *de novo* synthesis of adenosine monophosphate (AMP), was upregulated under OST, whereas *adenylate kinase*, which performs interconversion of adenine nucleotides, was downregulated highlighting a channeling of adenine nucleotides towards catabolism (Figure 6).

2.9. SST and OST Affect the Expression of Distinct Groups of Transporters

H. marinum roots under both OST and SST demonstrated marked changes in expression of genes encoding different membrane transport proteins, including ion and anion transporters (Table 2), the majority of which were upregulated (Table S2). The pattern of salinity-responsive DEGs related to membrane transport, was mostly different to that resulting from OST.

Among anion transporters, three *boron transporter* genes with potential anion efflux activity [32] were specifically and highly upregulated in SST-treated roots, suggesting a role for them in Cl$^-$ removal. Upregulation of S-type anion channels *SLAH2* and *SLAH3* may serve the purpose of enrichment with NO$_3^-$, as a main competitor of Cl$^-$, to minimize Cl$^-$ accumulation [33,34]. In the shoots of SST-treated plants, upregulation of *NRT1/PTR FAMILY 7.3*, a potential anion transporter [35], may be linked to further regulation of root-to-shoot anion transport. In OST-treated roots, however, six other genes from the *NRT1/PTR* family were downregulated. Increased expression of an aluminum-activated *malate transporter* may also be associated with Cl$^-$ efflux [36,37] or malate extrusion into soil to increase phosphate availability [38]. In line with the latter possibility, five *phosphate transporters* were strongly upregulated in the roots under SST.

Among ion transporters, the expression of K$^+$ channel *SKOR* was increased ~11-fold exclusively in SST-treated roots, indicating enhanced re-translocation of K$^+$ as the main Na$^+$ competitor [39,40].

The transcript levels of *cation/H⁺ antiporter 16,* another potential player in the maintenance of Na⁺/K⁺ homeostasis, were also increased. The expression *HKT5* transporters and *NHX* exchangers, shown to be important for adaptation to high salinity [41], was unchanged in *H. marinum* under SST, but decreased under OST (*HKT14;1* and *HKT1;1* in the shoots, *HKT1;5* and *HKT2;1* in the roots). Upregulation of *ammonium transporter 2* in roots under both stress types may indicate increased ammonium transport to foster N demand in stressed plants. Ammonium assimilation is less energy-demanding than nitrate uptake [42]. In line with this, six genes encoding high-affinity nitrate transporters were repressed in roots under OST, further supporting the hypothesis of increased ammonium uptake. OST, but not SST, led to the marked amplification of the expression of seven Zn transporter genes (up to 8-fold), five plant *cadmium resistance protein* (PCRP) genes, two *Zn-facilitator like protein* (ZFLP) genes, and two *YELLOW STRIPE-like proteins* (YSL) genes. ZFLP1 participates in polar auxin transport and drought stress tolerance in *Arabidopsis* [43]; PCRPs are involved in both Zn extrusion and long-distance transport [44]; and YSLs are thought to be implicated in the transport of metals [45]. Two glutamate receptors, *GLR1.3* and *GLR2.8,* both nonselective cation channels [46], were upregulated in OST-treated roots and shoots, respectively. *Mechanosensitive ion channel 10 (MSL10)* was downregulated in roots under both types of stress. Membrane tension during stress may lead to activation of cation conductance via MSL channels [46]. Moreover, two cyclic nucleotide-gated channels, which may also be involved in Na⁺ uptake [47], were downregulated ~10-fold in roots under both stresses.

Different sucrose exporter genes SWEET [48] were upregulated in tissues under particular stress treatments: *SWEET12* and *SWEET14b* in roots under both stresses, *SWEET13a* and *SWEET13b* in SST-treated roots, and *SWEET14a* and *SWEET15b* in OST-treated plants. However, hexose transporter genes *SWEET2b* and *SWEET16* were downregulated in shoots under OST. Two monosaccharide transporters were upregulated under SST. Furthermore, *GDP-mannose transporter 1* was upregulated in both tissue types following both stresses. Of 14 DEGs encoding aquaporins, which mediate water uptake and movement in plants, 12 were repressed in at least one of the two tissues under at least one type of stress.

2.10. Gene Expression Analysis Suggests Differences in Amino Acid and Secondary Metabolite Accumulation Under OST and SST

Regarding DEGs related to amino acid metabolism, the majority were upregulated in SST-treated plants (Table S2). High expression of genes involved in phenylalanine, tyrosine, and tryptophan metabolism (*tyrosine decarboxylase,* 10 genes; *anthranilate synthase,* 2 genes; *tryptophan synthase,* 2 genes; and *anthranilate phosphoribosyltransferase*) indicate a shift towards alkaloid, diterpene, and phenylpropanoid biosynthesis. Transcripts of *proline dehydrogenase 2,* involved in proline degradation, were decreased, but those of *prolyl 4-hydroxylase,* involved in proline synthesis, were increased, in line with proline enrichment in SST-treated plants (Figure 4). Repressed genes were mainly associated with amino acid degradation (*γ-glutamyl P-reductase,* 2 genes; *glutamate decarboxylase; choline dehydrogenase,* 2 genes; *cystathionine β-lyase,* and *phenylalanine ammonia lyase*). An increase in *glutamate synthase* transcripts indicates enhanced synthesis of glutamate, which can regulate ion transport via selective glutamate-gated cation channels [49].

In roots under both types of stress, increased expression of *isovaleryl-CoA-dehydrogenase* and *copalyl-diP synthase,* which use amino acid degradation products to produce diterpenoids, and two *sterol C4-methyl oxidase* genes, further support a shift towards diterpenoid biosynthesis. From the downregulation of two *cinnamoyl-CoA reductase* genes, two *phytoene synthase* genes, one *β-carotene hydroxylase,* and one *β-carotene isomerase,* it appears that the synthesis of isoprenoids and carotenoids may be hampered under OST. The expression of *ascorbate oxidase,* encoding an ascorbate-degrading enzyme, was increased up to 14-fold in roots under OST and SST. Strong upregulation of 12 *nicotianamine synthase* genes was observed specifically in OST-treated shoots and roots.

2.11. Gene Expression Analysis Indicates Changes in Plant Hormone Levels Under Stress

Both types of stress-induced changes in the expression of genes responsive to abscisic acid (ABA), the key phytohormone in adaptation to stress [50]. Strong transcriptional upregulation (up to 12-fold) of genes encoding ABA-induced small hydrophilic proteins [50], together with elevated expression of the ABA-responsive *GRAM domain-containing protein* genes [51], imply an increased ABA level in the roots under both stress conditions. However, downregulation of two *ABA receptor PYR1* genes only in SST-treated roots indicate differences in ABA perception and signaling under the two types of stress.

Strong transcriptional activation in SST-treated roots (up to 12-fold) of two *indole-3-glycerol phosphate synthase* genes, encoding a branch-point enzyme in the tryptophan-independent IAA biosynthetic pathway, points to increased auxin synthesis under conditions of high salinity. Different genes encoding auxin efflux carrier proteins, involved in auxin transport, were upregulated in roots and shoots under both stresses. By contrast, the genes encoding the auxin-responsive proteins IAA23 and IAA4 were downregulated. IAA4 forms part of a module that negatively regulates adventitious root development in *Populus* [52].

Transcripts of *1-aminocyclopropane-1-carboxylate oxidase* and *jasmonate O-methyltransferase*, which encode the key enzymes in ethylene and jasmonate biosynthesis, respectively, were increased in plants under OST, whereas those of diverse *ethylene-responsive transcription factor* genes were decreased in roots under both stresses.

In line with lower Ca^{2+} content in SST-treated plants (Table 1), the transcript levels of three *calmodulin* genes, one Ca^{2+}-*dependent protein kinase*, and one Ca^{2+}-*sensing receptor* were decreased in SST-treated roots.

2.12. Overlap in Expression of Stress-Responsive Genes

Of the 44 stress-related DEGs detected in OST-treated roots, 28 were also detected in roots treated with SST (Table S2). Of these, ten *dehydrin* genes were upregulated under both stresses. Between one (under SST) and eight (under OST) chaperone *DnaJ* genes were also upregulated. While the main function of dehydrins and chaperones is to protect biomolecules, certain dehydrins possess metal-binding capacity and are regarded as ROS scavengers [53]. Regarding gene expression in the shoots, 71 stress-related DEGs were detected under OST, and only 19 under SST, but with an overlap of 13 common DEGs. At least one gene encoding a proline-rich protein was downregulated in a particular tissue, possibly reflecting a redirection of proline into the free pool as a key stress-protecting amino acid. One group of pathogen-related genes (*chitinase*, 9 genes; *germin-like protein*, 3 genes) exhibited increased expression under one or both types of stress, while another (*disease resistance protein*, 4 genes; *defensin*, 2 genes) revealed a decrease. Six *thaumatin* genes were upregulated in OST-treated roots, three of which were also activated under SST. Two genes encoding kiwellin, a protein accumulated to high levels under SST in *H. vulgare* [54], were also upregulated in *H. marinum* under similar conditions. *Rapid alkalinization factor 23*, involved in regulation of salt tolerance in *Arabidopsis* [55], was upregulated in SST-treated shoots and roots. Various repeat domain protein families are also anticipated to be involved in abiotic stress [56]. Expression of a *Kelch repeat-containing protein* was increased in both tissues under the two types of stress, while *WD40* (6 genes), *pentatricopeptide* (7 genes), *tetratricopeptide* (5 genes), and *ankyrin repeat proteins* (2 genes) were additionally upregulated in OST-treated shoots.

In plants, stress generally induces the production of toxic ROS. Alterations in the expression of the *peroxidase* gene superfamily, whose products are involved in ROS scavenging, and lignin production, were observed: nine genes were downregulated while another eight were upregulated in roots under SST and, in part, OST. In the shoots, a total of 25 *peroxidases* were repressed, while only four genes were upregulated under OST, and four genes were downregulated and three were upregulated under SST. Rearrangements were also revealed in the expression of the *thioredoxin* family in plants under OST, but not SST. The increase in expression of two *catalase* genes and *Neighbor of BRCA1 gene 1*, involved in pexophagy [31], was detected only in SST-treated roots. Two genes encoding nonsymbiotic phytoglobin, a NO sensor involved in hypoxia response [57], were strongly upregulated in roots under both stresses. Of genes encoding glutathione S-transferase, which produces a strong nonenzymatic antioxidant glutathione implicated in abiotic stress tolerance, four were upregulated in SST-treated roots and ten in OST.

3. Discussion

While the growth and development of both *H. marinum* and *H. vulgare* plants were affected by OST and SST, domesticated barley plants suffered much more strongly, particularly from SST, as manifested by their leaf chlorosis and wilting, as well as by an almost complete halt in photosynthetic activity and assimilate transport. *H. marinum* plants, however, remained dark-green under SST. They had not even revealed higher photosynthetic efficiency under control conditions but were also able to maintain photosynthetic activity and carbon fixation under high salinity. Genes involved in the defense of the photosystem were transcriptionally boosted in SST-treated *H. marinum* shoots. These observations indicate a more efficient photosynthetic apparatus in *H. marinum* which deserves more detailed investigations in the future. As a result, *H. marinum* plants exhibited better growth capacity, water retention, and shoot development under SST. By contrast, the plants turned yellow under OST, even though exposed to 2.5-fold lower osmotic pressure as compared to SST, likely due to a significant stress response, also supported by observed changes in metabolite and transcript profiles. Genes of the photosynthetic machinery and chlorophyll metabolism were strongly repressed in OST-treated shoots, in line with the yellow color. Transcriptional inhibition of processes related to cell proliferation and differentiation, and activation of lipid and cell wall degradation processes appeared to be triggered predominantly by OST rather than SST. This was in agreement with decreased purine and pyrimidine levels, as well as increased levels of their degradation products under OST. From the fact that a larger number of genes associated with ROS detoxification were upregulated, OST likely caused more severe ROS production than SST. While upregulation of genes involved in the TCA cycle and, in part, glycolysis, coupled with decreased levels of glycolytic intermediates indicated increased energy metabolism during both types of stress, the plant response to OST was possibly more energy demanding than to SST. OST-treated plants likely utilize the glyoxylate cycle for additional energy production. These data are in line with the increased levels of proteins involved in energy metabolism [15]. Overexpression of SWEET proteins is indicative of increased sucrose transport to the roots, and its utilization for energy retrieval. Increased sugar degradation to hexoses, representing the main energy source, is further feasible. A significant increase in glucose content in *H. marinum* plants under salinity has been described recently [18,58].

Seaside barley possesses a higher capacity than domesticated barley to regulate osmotic homeostasis under SST. Differently to OST, where impermeable PEG6000 caused strong stress response, halophytic *H. marinum* might recruit Na^+ and Cl^+ ions under high salinity to regulate osmotic pressure with fewer energy investments for the plant. In addition, the biosynthesis and accumulation of other osmolites may further contribute to osmoregulation. This is in line with the higher proline levels and increased expression of genes encoding hydrophilic dehydrins, germin-like, and other osmolytic proteins detected in SST-treated plants. Additionally, mannitol-1P content was strongly increased in roots, especially under SST. Similarly, halophytic *Prosopis strombulifera* accumulate large amounts of mannitol-1P as an osmoprotective agent [59]. From the upregulation of *trehalose-P synthase* and *trehalose-6P phosphatase*, and decreased levels of trehalose-6P intermediate, SST-treated roots appear to accumulate trehalose, another known osmoprotective agent [60]. Moreover, as follows from the activation of genes of sorbitol synthesis, sorbitol accumulation is also possible. In tomato plants, increased aldose reductase activity and sorbitol synthesis were shown to improve salt tolerance [60]. Enhanced ureide accumulation may be associated with osmoprotection, but also to stabilization of proteins and membranes [61], efficient N utilization due a low C/N ratio of heterocyclic molecules that optimize the transport of organic N under reduced photosynthetic capacity [62], and activation of ABA and jasmonic acid signaling [63].

NMR analysis revealed morphological changes in the *H. marinum* roots, which were associated with adaptation to high salinity. While more roots were maintained in their primordial state under SST, the root surface area was significantly increased, achieving enhanced metabolite uptake and active ion efflux and improved root-rhizosphere interaction. Transcriptional changes also indicate increased deposition of lignin in cell walls, which likely serves to create an apoplastic barrier to prevent water and solute loss and to reduce ionic flow through the apoplastic pathway [64]. High rate of Na^+ accumulation in the tissues of SST treated plants may be caused by Na^+ replacement of Ca^{2+} in cell walls [65–67]. Even though Na^+ was highly accumulated in SST-treated tissues, its concentration in the shoots was 1.4-fold lower than in the roots, implicating active Na^+ recruitment as an additional cheap osmotic agent in the latter, and suggesting prevention of Na^+ transport to photosynthetic tissues. It is worth mentioning that the experimental studies demonstrate much higher level of Na^+ accumulation in shoot tissues of *H. vulgare* [16,18,20]. Accordingly, few changes at the metabolic and transcriptional levels were detected in SST-treated shoots, when compared with the roots (Table 2). Changes in transcription levels of genes encoding transport proteins were particularly extensive in the roots (Table 2) and are considered to correspond to prevention/deceleration of toxic ion accumulation and assurance of nourishment and water uptake. Despite K^+ being generally regarded as a main competitor of Na^+ in uptake and transport [5], K^+ content was decreased in stressed roots, but the K^+/Na^+ ratio was higher in the shoots. The K^+ decrease is likely caused by stress-induced K^+ leakage and competition with Na^+ in root environment [5]. The K^+/Na^+ ratio of 1.4 achieved in our experiments is very similar to that observed in other *H. marinum* ecotypes and differs strongly from the K^+/Na^+ ratio of 4.3 found in *H. vulgare* plants [20]. These results further support the idea that the salt tolerance of *H. marinum* may be based on maintaining Na^+/K^+ balance in its shoots under salinity [16,18,20,59,68]. Despite the proposed roles of SOS1, HKT1;1, HKT1;5, and HKT2;2 in establishing this balance in *H. marinum* under salinity [18], none of the corresponding genes were found to be differentially expressed in our study. A significant decrease of *HmHKT2;1* transcript was observed in one *H. marinum* ecotype but remained unchanged in plants of another ecotype after SST [20]. Instead, other transporters, including K^+ transporter SKOR, an ammonium transporter, a cation/H^+ antiporter, and a Mg^{2+} transporter, were strongly transcriptionally upregulated in SST-treated roots, and may therefore enhance cation uptake and xylem loading to compete with Na^+. Decreased expression of *GLR3.4* and a *cyclic nucleotide-gated channel*, both potentially nonspecific Na^+ channels, may also contribute to minimization of Na^+ uptake by the roots. In SST-treated *H. vulgare* roots, expression of *HvHKT1;5* and *HvSOS1* was also decreased, whereas that of *HvSKOR* was increased [69]. Besides substitution of Na^+ by K^+ in the cytosol, Na^+ compartmentalization into the vacuoles or endosomes may serve as an additional mechanism of salinity tolerance in *H. marinum*. However, this vacuolar or endosomal sequestration could not be explained by the expression of NHX transporters and thus remains unclear. While Ca^{2+} has been described as an early component of salt sensing [5], the Ca^{2+} level under SST was decreased, likely due to its replacement by Na^+ in cell walls and vacuoles, the compartments with the largest Ca^{2+} pools in plants [65–67]. It would be a good option to study functions of Na^+ and K^+ transport proteins, in particular HmSKOR, with further prospective application in domesticated cereals.

The roots of *H. marinum* are also likely to be able to control Cl^- ions under high salinity. The increased expression of S-type anion channels *SLAH2* and *SLAH3* specifically under SST is probably connected with Cl^- retrieval from the xylem and/or efflux from the roots. The repression of *MSL10*, a channel with a moderate Cl^- preference [70], may further reflect the reduction of Cl^- uptake. Upregulation of four boron transporters in SST-treated roots is notable. These transporters belong to the anion exchanger family [71], do not have strict boron selectivity and may transfer other anions including Cl^-, contributing to its removal. Combined salt and boron tolerance have been frequently described [72,73]. Increased expression of several *S* and *P transporter* genes, as well as *NRT1/PTR family protein* genes, may indicate activation of anion uptake to compete with Cl^-. It would be of interest to test the role of these transporters in salinity tolerance and their suitability for biotechnological improvement of other crops.

In OST-treated plants, a different group of specific cation and anion transporters was upregulated, whose activity may result in the increased cation (Mg^{2+}, Ca^{2+}, Zn^{2+}) and anion (MoO_4^{2-}, SO_4^{2-}, PO_4^{3-}) uptake required for osmotic adjustments. Transcriptional depletion of key salinity tolerance genes *HKT1;5* and *HKT1;4*, along with upregulation of *Na*$^+$*/H*$^+$ *exchanger 5* in the shoots, implies endosomal sequestration of Na^+/K^+ for cellular osmotic rearrangements. Marked upregulation of different Zn transporters was specifically detected in OST-treated roots, in line with increased Zn^{2+} accumulation in the shoots. Zn^{2+} has been shown to be involved in ROS scavenging [74] and stomata opening via determination of the K^+ influx rate [75]. Increased Zn content in the shoots under OST is thought to facilitate ROS detoxification and regulation of water management.

Strong transcriptional upregulation of amino acid-degrading enzymes in the roots under both types of stress was coupled with upregulation of genes encoding enzymes involved in utilizing degradation products to produce secondary metabolites. The accumulation of flavonoids occurred specifically in SST-treated roots, whereas their biosynthesis was likely repressed under OST. Accumulation of isoprenoids and carotenoids might be also repressed under OST. While flavonoids may function as ROS scavengers, some (e.g., chalconoids) are able to block voltage-dependent K^+ channels [76]. Ectopic expression of chalcone synthase has been shown to increase salt tolerance [77]. Increased flavonoid accumulation in *H. marinum* is thought to contribute to the inhibition of stress-induced K^+ leakage and ROS balance. Due to transcriptional upregulation of numerous tyrosine decarboxylase genes, involved in the production of little-studied phenylethylamine hordenine [78], the role of alkaloid hordenine in stress tolerance has become obvious and deserves further investigation.

To conclude, the mechanisms of salt tolerance of seaside barley are complex and comprised of adaptations on morphological, physiological, biochemical, and transcriptomic levels (Figure 7C). Seaside barley is likely capable of controlling Na^+ and Cl^- concentrations in its leaves when its roots are subjected to high salinity. However, transporters, shown to achieve salinity tolerance [14,79], were unchanged in the plant line used in the present study, in line with variability of *H. marinum* accessions in salinity tolerance [80,81]. The adaptation of *H. marinum* to SST includes fine-tuning of membrane transport for effective management of both restriction of ion entry and sequestration, as well as accumulation of osmolytes, which help to minimize energy costs. In contrast, markedly more resources and energy are required to overcome the negative consequences of OST, particularly due to the severity of ROS accumulation and nutritional imbalance affecting plant growth under stress. Our results demonstrate that, in order to adapt to salinity, seaside barley has developed specific mechanisms that differ from those which are activated in response to OST.

4. Materials and Methods

4.1. Plant Material and Growth Conditions

Seeds of seaside barley (*Hordeum marinum* ssp. *marinum*), originated from Tuscany region, and cultivated barley (*Hordeum vulgare*) cv. Golden Promise were germinated on moist filter paper in the dark at 20 °C. Seven day-old seedlings were transferred on hydroponic half-strength Hoagland's No. 2 solution (Sigma-Aldrich, St. Louis, MO, USA) and incubated in the growth chamber under irradiance of 350 mmol m^{-2} s^{-1}, 12 h photoperiod, and 20 °C. Due to the halophytic nature of *H. marinum*, 0.2 mM NaCl was added to the control incubation solution [15]. The incubation solution was fully replaced every week by a newly prepared one in order to prevent nutrient depletion. After 14 days of growing, plants were exposed stepwise to the increasing concentrations of 50 mM NaCl or 2.5% PEG6000 per day until 300 mM NaCl (salinity stress) or 15% PEG6000 (osmotic stress) were reached. Control plants were exposed to 0.2 mM NaCl continuously. All plants were sampled after five days (32 days old) of maximum stress and separated into shoot (crown and growing point) and root (2 cm root tips) fractions. To average the genetic background and local environmental influences, a bulk of 10–15 plants, grown in a single hydroponic tank either under the control or stress conditions, were collected for one biological replication. In total, 10 biological replications were harvested, frozen, and used in further analyses.

4.2. Determination of Morphological and Physiological Characteristics

Fresh plant tissues were collected in 1.5 mL Eppendorf microcentrifuge tubes. The tissue sap was excavated by squeezing whole plants with Pellet pestle (Eppendorf, Germany). Osmotic pressure of experimental solutions and sap obtained from squeezed whole plants was measured by Vapor Pressure Osmometer WESCOR 5500 (Thermo Fisher Scientific GmbH, Dreieich, Germany) according to manufacturer's instructions. The relative growth rate (RGR) was calculated from the fresh weight data taken at start of stress application and final harvest using the formula RGR = (ln fresh weight$_2$ − ln fresh weight$_1$)/(t$_2$ − t$_1$), where fresh weight$_1$ = fresh weight (g) at t$_1$; fresh weight$_2$ = fresh weight (g) at t$_2$; and t$_1$ and t$_2$ = time at start and end of experiments in days. The RGRs of individual plants were presented as percentages relative to value in control conditions. Plant water content (PWC) in *H. marinum* and *H. vulgare* plants was determined as follows: PWC = (FW − DW)/DW, where DW was dry weight and FW was fresh weight of an individual plant. The PWC values were converted into percentages relative to the value in the control condition. To estimate shoot/root ratio in tested plants, the FW of shoots and roots of individual plants of *H. marinum* and *H. vulgare* were measured after end of stress application.

4.3. Elemental Analysis

Approximately 10 mg of pulverized and dried (at 65 °C) plant material was weighed into PTFE digestion tubes and 1 mL of concentrated nitric acid (67–69%) was added to each tube. After 4 h incubation, samples were digested under pressure using a high-performance microwave reactor Ultraclave 4 (MLS, Leutkirch, Germany). Samples were then transferred to Greiner centrifuge tubes and diluted with de-ionized water to a final volume of 8 mL. Elemental analysis was carried out using a sector field high resolution mass spectrometer (HR)-ICP-MS ELEMENT 2 (Thermo Fisher Scientific, Dreieich, Germany) with Software version 3.1.2.242. A 10 points external standard calibration curve was set from a certified multiple standards solution (Bernd Kraft, Germany). A least-square regression was applied to best fit the linearity of the curve. Elements Rh and Ge (ICP Standard Certipur®, Merck, Germany) were infused online and used as internal standards for matrix correction.

4.4. Non-Invasive NMR-Imaging and NMR-Spectroscopy of Plant Tissues

The NMR imaging of *H. marinum* tissue was conducted according to [82]. In total, three plants for each type of treatment were analyzed. The internal tissue structure of stressed and control plants was visualized noninvasively with an isotropic resolution of around 40 μm. The NMR analysis was conducted on a Bruker Ascend TD 400 MHz NMR spectrometer (Bruker GmbH, Rheinstetten, Germany). Image processing was performed by application of software MATLAB (The Mathworks, Natick, MA, USA) and AMIRA (Thermo Fisher Scientific GmbH, Dreieich, Germany).

4.5. Measurement of ^{13}C Uptake

400 ppm $^{13}CO_2$ was applied for 24 h to bag-covered hydroponic tanks with control or stressed plants grown in the chamber in the above-described conditions. Afterwards, treated plants were separated into shoot and root fractions, lyophilized, ground, and analyzed on elemental analyzer coupled to stable isotope ratio mass spectrometer (Vario MICRO cube/Isoprime Vision, Elementar Analysensysteme GmbH, Langenselbold, Germany). Five biological repetitions with three technical replicates each were analyzed.

4.6. Untargeted Metabolite Profiling

For the untargeted analysis of central metabolites, the freeze-dried and homogenized samples were incubated for 20 min at 4 °C in 600 µL of extraction buffer consisting of equal volumes of methanol and chloroform (Roth, Karlsruhe, Germany) followed by addition of 300 µL of water and centrifugation at 14,000 rpm at 4 °C for 10 min. Supernatant was transferred into a new tube and stored at −80 °C, prior to the analysis by ion chromatography using Dionex-ICS-5000+HPIC system (Thermo Scientific, Dreieich, Germany) coupled to a Q-Exactive Plus hybrid quadrupol-orbitrap mass spectrometer (Thermo Scientific, Dreieich, Germany). The detailed chromatographic and mass spectrometry (MS) conditions are described in the Table S3. The randomized samples were analyzed in full MS mode. The data-dependent MS-MS analysis for the compound identification was performed in the pooled probe, which also served as a quality control (QC). The batch data was processed using the untargeted metabolomics workflow of the Compound Discoverer 3.0 software (Thermo Fisher Scientific, Dreieich, Germany). The compounds with the maximum relative standard deviation (RSD) below 35% of the QC area were selected for quantification. The compounds were identified using the inhouse library, as well as a public spectral database mzCloud, and the public databases KEGG, NIST and ChEBI via the mass- or formula-based search algorithm. The p-values of the group ratio were calculated by ANOVA and a Tukey-HCD post hoc analysis. Adjusted p-values were calculated using Benjamini-Hochberg correction.

4.7. RNA Extraction, Sequencing and Transcript Analysis

The total RNA was isolated using TRIzol (Thermo Fisher Scientific, Schwerte, Germany) with subsequent DNAse treatment (Thermo Fisher Scientific, Schwerte, Germany) and additional purification using Plant RNA purification kit (Qiagen, Hilden, Germany). cDNA libraries were prepared using Lexogen SENSE RNA-Seq Kit (Lexogen, Vienna, Austria) and sequencing was performed in HiSeq2500 device (Illumina, San Diego, CA, USA). Three repetitions were performed for each data point. The complete data set is deposited at the European Nucleotide Archive with accession number: PRJEB38377.

Adapter trimming was performed using Cutadapt software, version 1.9.1 [83]. Quality trimming was performed using CLC assembly cell software, version 5.0.1 (Qiagen, Hilden, Germany). Read mapping was performed on the *H. vulgare* genome [48] using the software Kallisto, version 0.45.0 [84]. Differential expression was calculated using the R package DESeq2, version 1.18.1 [85]. Differential expression thresholds were set at $\log_2$-fold change > 1.5 and FDR-adjusted p values (according to Benjamini Hochberg) < 0.01. Venn diagrams were created using InteractiVenn [86]. Gene ontology (GO) terms enrichment analysis for biological process carried out using the GO enrichment tool by Panther [87].

4.8. Statistical Analysis

Significance analysis was performed by Student's *t* test using BBBB software (IBM SPSS Statistics, version 16). The difference at $p < 0.05$ was considered as significant. For data presentation, significance was marked as: *, $p < 0.05$; **, $p < 0.01$; ***, $p < 0.001$.

Supplementary Materials: Supplementary materials can be found at http://www.mdpi.com/1422-0067/21/23/9019/s1. Figure S1: Osmotic potential of incubation media and in tissue sap samples derived from squeezed whole *H. marinum* and *H. vulgare* plants under osmotic and salinity stress as compared to control conditions; Figure S2: Principal component analysis of metabolite distribution in roots and shoots of osmotic- and salt-treated plants as compared to control; Table S1: The metabolomic changes in roots and shoots of *Hordeum marinum* L. caused by osmotic and salinity stress; Table S2: List of differentially expressed genes in shoots and roots of *Hordeum marinum* L. under osmotic and salinity stress; Table S3: Chromatographic and mass spectrometry conditions for the untargeted metabolite analysis.

Author Contributions: Conceptualization: S.I., L.B., and V.R.; methodology: S.I., H.R., and L.B.; investigation: S.I, A.H., P.R., Y.A.T.M., H.R., and L.B.; data curation: S.I., H.R., L.B., and V.R.; writing—original draft preparation: S.I. and V.R.; writing—review and editing: L.B. and V.R.; supervision: L.B.; project administration: V.R.; funding acquisition: S.I. and L.B. All authors have read and agreed to the published version of the manuscript.

Funding: This work was supported by German Academic Exchange Service (DAAD, grant number 57314019) for SI, and by Deutsche Forschungsgemeinschaft (DFG, grant numbers BO1917/5-1 and BO1917/5-2).

Acknowledgments: Seeds of *H. marinum* ssp. *marinum* were kindly provided by T. Lombardi (University of Pisa, Italy). We are grateful to S. Wagner and S. Ortleb for their excellent technical assistance, T. Meitzel for ^{13}C measurements, Anne Fiebig for curating the submission to the European Nucleotide Archive, and to Lothar Altschmied for suggestions regarding the RNAseq data analysis.

Conflicts of Interest: The authors declare no conflict of interest.

Abbreviations

ABA	Abscisic acid
DEG	Differentially expressed gene
DW	Dry weight
FW	Fresh weight
IAA	Indole-3-acetic acid
MEP	Methylerythritol-4P
NMR	Nuclear Magnetic Resonance
OST	Osmotic stress
PWC	Plant water content
RGR	Relative growth rate
ROS	Reactive oxygen species
SST	Salinity stress
TCA	Tricarboxylic acid
YSL	YELLOW STRIPE-like protein
ZFLP	Zn-facilitator like protein

References

1. Qadir, M.; Quillerou, E.; Nangia, V.; Murtaza, G.; Singh, M.; Thomas, R.J.; Drechsel, P.; Noble, A.D. Economics of salt-induced land degradation and restoration. *Nat Res. Forum.* **2014**, *38*, 282–295. [CrossRef]
2. Food and Agriculture Organization. *The State of Food and Agriculture: Climate Change, Agriculture and Food Security*; FAO: Roma, Italy, 2016; p. 173.
3. Rengasamy, P. World salinization with emphasis on Australia. *J. Exp. Bot.* **2006**, *7*, 1017–1023. [CrossRef] [PubMed]
4. Shrivastava, P.; Kumar, R. Soil salinity: A serious environmental issue and plant growth promoting bacteria as one of the tools for its alleviation. *Saudi J. Biol. Sci.* **2015**, *22*, 123–131. [CrossRef] [PubMed]
5. Isayenkov, S.V.; Maathuis, F.J.M. Plant salinity stress: Many unanswered questions remain. *Front. Plant Sci.* **2019**, *10*, 80. [CrossRef] [PubMed]
6. Flowers, T.J.; Colmer, T.D. Salinity tolerance in halophytes. *New Phytol.* **2008**, *179*, 945–963. [CrossRef]
7. Flowers, T.J.; Munns, R.; Colmer, T.D. Sodium chloride toxicity and the cellular basis of salt tolerance in halophytes. *Annu. Bot.* **2015**, *115*, 419–431. [CrossRef]
8. Isayenkov, S.V.; Dabravolski, S.A.; Pan, T.; Shabala, S. Phylogenetic diversity and physiological roles of plant monovalent cation/H+ antiporters. *Front. Plant Sci.* **2020**, *11*, 573564. [CrossRef]
9. Slama, I.; Abdelly, C.; Bouchereau, A.; Flowers, T.; Savouré, A. Diversity, distribution and roles of osmoprotective compounds accumulated in halophytes under abiotic stress. *Ann. Bot.* **2015**, *115*, 433–447. [CrossRef]
10. Isayenkov, S.V. Genetic sources for the development of salt tolerance in crops. *Plant Growth Regul.* **2019**, *89*, 1–17. [CrossRef]
11. Yuan, F.; Leng, B.; Wang, B. Progress in studying salt secretion from the salt glands in recretohalophytes: How do plants secrete salt? *Front. Plant Sci.* **2016**, *7*, 977. [CrossRef]

12. Garthwaite, A.J.; Bothmer, R.; Colmer, T.D. Salt tolerance in wild *Hordeum* species is associated with restricted entry of Na$^+$ and Cl$^-$ into the shoots. *J. Exp. Bot.* **2005**, *56*, 2365–2378. [CrossRef] [PubMed]

13. Islam, S.; Malik, A.J.; Islam, A.K.M.R.; Colmer, T.D. Salt tolerance in a *Hordeum marinum-Triticum aestivum* amphiploid, and its parents. *J. Exp. Bot.* **2007**, *58*, 1219–1229. [CrossRef] [PubMed]

14. Munns, R.; James, R.A.; Xu, B.; Athman, A.; Conn, S.J.; Jordans, C.; Byrt, C.S.; Hare, R.A.; Tyerman, S.D.; Tester, M.; et al. Wheat grain yield on saline soils is improved by an ancestral Na$^+$ transporter gene. *Nat. Biotechnol.* **2012**, *30*, 360–364. [CrossRef] [PubMed]

15. Maršálová, L.; Vítámvás, P.; Hynek, R.; Prášil, I.T.; Kosová, K. Proteomic response of *Hordeum vulgare* cv. Tadmor and *Hordeum marinum* to salinity stress: Similarities and differences between a glycophyte and a halophyte. *Front. Plant Sci.* **2016**, *7*, 1154. [CrossRef]

16. Ferchichi, S.; Hessini, K.; Dell'Aversana, E.; D'Amelia, L.; Woodrow, P.; Ciarmiello, L.F.; Fuggi, A.; Carillo, P.C. *Hordeum vulgare* and *Hordeum maritimum* respond to extended salinity stress displaying different temporal accumulation pattern of metabolites. *Funct. Plant Biol.* **2018**, *45*, 1096–1109. [CrossRef]

17. Seckin, B.; Turkan, I.; Sekmen, A.H.; Ozfidan, C. The role of antioxidant defense systems at differential salt tolerance of *Hordeum marinum* Huds. (sea barley grass) and *Hordeum vulgare* L. (cultivated barley). *Environ. Exp. Bot.* **2010**, *69*, 76–85. [CrossRef]

18. Huang, L.; Kuang, L.; Li, X.; Wu, L.; Wu, D.; Zhang, G. Metabolomic and transcriptomic analyses reveal the reasons why *Hordeum marinum* has higher salt tolerance than *Hordeum vulgare*. *Env. Exp. Bot.* **2018**, *156*, 48–61. [CrossRef]

19. Huang, L.; Kuang, L.; Wu, L.; Wu, D.; Zhang, G. Comparisons in functions of HKT1;5 transporters between *Hordeum marinum* and *Hordeum vulgare* in responses to salt stress. *Plant Growth Regul.* **2019**, *89*, 309–319. [CrossRef]

20. Hmidi, D.; Messedi, D.; Corratgï-Faillie, C.; Marhuenda, T.O.; Fizames, C.C.; Zorrig, W.; Abdelly, C.; Sentenac, H.; VïRy, A.N. Investigation of Na$^+$ and K$^+$ transport in halophytes: Functional analysis of the HmHKT2;1 transporter from *Hordeum maritimum* and expression under saline conditions. *Plant Cell Physiol.* **2019**, *60*, 2423–2435. [CrossRef]

21. Ranty, B.; Aldon, D.; Cotelle, V.; Galaud, J.P.; Thuleau, P.; Mazars, C. Calcium sensors as key hubs in plant responses to biotic and abiotic stresses. *Front. Plant Sci.* **2016**, *7*, 327. [CrossRef]

22. Williams, M. *The Merck Index: An Encyclopedia of Chemicals, Drugs, and Biologicals*, 15th ed.; O'Neil, M.J., Ed.; Royal Society of Chemistry: Cambridge, UK, 2013.

23. Meena, M.; Divyanshu, K.; Kumar, S.; Swapnil, P.; Zehra, A.; Shukla, V.; Yadav, M.; Upadhyay, R.S. Regulation of L-proline biosynthesis, signal transduction, transport, accumulation and its vital role in plants during variable environmental conditions. *Heliyon* **2019**, *5*, e02952. [CrossRef] [PubMed]

24. Pencík, A.; Simonovik, B.; Petersson, S.V.; Henyková, E.; Simon, S.; Greenham, K.; Zhang, Y.; Kowalczyk, M.; Estelle, M.; Zazímalová, E.; et al. Regulation of auxin homeostasis and gradients in Arabidopsis roots through the formation of the indole-3-acetic acid catabolite 2-oxindole-3-acetic acid. *Plant Cell* **2013**, *25*, 3858–3870. [CrossRef] [PubMed]

25. Tetali, S.D. Terpenes and isoprenoids: A wealth of compounds for global use. *Planta* **2019**, *249*, 1–8. [CrossRef] [PubMed]

26. Peumans, W.J.; Hao, Q.; Van Damme, E.J. Ribosome-inactivating proteins from plants: More than RNA N-glycosidases? *FASEB J.* **2001**, *15*, 1493–1506. [CrossRef]

27. Wang, M.; Ogé, L.; Perez-Garcia, M.D.; Hamama, L.; Sakr, S. The PUF protein family: Overview on PUF RNA targets, biological functions, and post transcriptional regulation. *Int. J. Mol. Sci.* **2018**, *19*, 410. [CrossRef]

28. Mahawar, L.; Shekhawat, G.S. Haem oxygenase: A functionally diverse enzyme of photosynthetic organisms and its role in phytochrome chromophore biosynthesis, cellular signalling and defence mechanisms. *Plant Cell Environ.* **2018**, *41*, 483–500. [CrossRef]

29. Theis, J.; Schroda, M. Revisiting the photosystem II repair cycle. *Plant Signal. Behav.* **2016**, *11*, e1218587. [CrossRef]

30. Tronconi, M.A.; Fahnenstich, H.; Gerrard Weehler, M.C.; Andreo, C.S.; Flügge, U.I.; Drincovich, M.F.; Maurino, V.G. Arabidopsis NAD-malic enzyme functions as a homodimer and heterodimer and has a major impact on nocturnal metabolism. *Plant Physiol.* **2008**, *146*, 1540–1552. [CrossRef]

31. Pan, R.; Liu, J.; Wang, S.; Hu, J. Peroxisomes: Versatile organelles with diverse roles in plants. *New Phytol.* **2020**, *225*, 1410–1427. [CrossRef]

32. Miwa, K.; Takano, J.; Omori, H.; Seki, M.; Shinozaki, K.; Fujiwara, T. Plants tolerant of high boron levels. *Science* **2007**, *318*, 1417. [CrossRef]

33. Maierhofer, T.; Diekmann, M.; Offenborn, J.N.; Lind, C.; Bauer, H.; Hashimoto, K.S.; Al-Rasheid, K.A.; Luan, S.; Kudla, J.; Geiger, D.; et al. Site- and kinase-specific phosphorylation-mediated activation of SLAC1, a guard cell anion channel stimulated by abscisic acid. *Sci. Signal.* **2014**, *7*, ra86. [CrossRef] [PubMed]

34. Zheng, X.; He, K.; Kleist, T.; Chen, F.; Luan, S. Anion channel SLAH3 functions in nitrate-dependent alleviation of ammonium toxicity in Arabidopsis. *Plant Cell Environ.* **2015**, *38*, 474–486. [CrossRef] [PubMed]

35. Corratgé-Faillie, C.; Lacombe, B. Substrate (un)specificity of Arabidopsis NRT1/PTR FAMILY (NPF) proteins. *J. Exp. Bot.* **2017**, *68*, 3107–3113. [CrossRef] [PubMed]

36. Piñeros, M.A.; Cançado, G.M.A.; Kochian, L.V. Novel properties of the wheat aluminium tolerance organic acid transporter (TaALMT1) revealed by electrophysiological characterization in *Xenopus* Oocytes: Functional and structural implications. *Plant Physiol.* **2008**, *147*, 2131–2146. [CrossRef] [PubMed]

37. Ligaba, A.; Maron, L.; Shaff, J.; Kochian, L.; Piñeros, M. Maize ZmALMT2 is a root anion transporter that mediates constitutive root malate efflux. *Plant Cell Environ.* **2012**, *35*, 1185–1200. [CrossRef] [PubMed]

38. Meyer, S.; Mumm, P.; Imes, D.; Endler, A.; Weder, B.; Al-Rasheid, K.A.S.; Geiger, D.; Marten, I.; Martinoia, E.; Hedrich, R. AtALMT12 represents an R-type anion channel required for stomatal movement in Arabidopsis guard cells. *Plant J.* **2010**, *63*, 1054–1062. [CrossRef]

39. Gaymard, F.; Pilot, G.; Lacombe, B.; Bouchez, D.; Bruneau, D.; Boucherez, J.; Michaux-Ferrière, N.; Thibaud, J.B.; Sentenac, H. Identification and disruption of a plant shaker-like outward channel involved in K+ release into the xylem sap. *Cell* **1998**, *94*, 647–655. [CrossRef]

40. Wegner, L.H.; De Boer, A.H. Activation kinetics of the K^+ outward rectifying conductance (KORC) in xylem parenchyma cells from barley roots. *J. Membr. Biol.* **1999**, *170*, 103–119. [CrossRef]

41. Assaha, D.V.M.; Ueda, A.; Saneoka, H.; Al-Yahyai, R.; Yaish, M.W. The Role of Na^+ and K^+ transporters in salt stress adaptation in glycophytes. *Front. Physiol.* **2017**, *8*, 509. [CrossRef]

42. Bu, Y.; Takano, T.; Liu, S. The role of ammonium transporter (AMT) against salt stress in plants. *Plant Signal. Behav.* **2019**, *14*, e1625696. [CrossRef]

43. Remy, E.; Cabrito, T.R.; Baster, P.; Batista, R.A.; Teixeira, M.C.; Friml, J.; Sá-Correia, I.; Duque, P. A major facilitator superfamily transporter plays a dual role in polar auxin transport and drought stress tolerance in Arabidopsis. *Plant Cell* **2013**, *25*, 901–926. [CrossRef] [PubMed]

44. Song, W.Y.; Choi, K.S.; Kim, D.Y.; Geisler, M.; Park, J.; Vincenzetti, V.; Schellenberg, M.; Kim, S.H.; Lim, Y.P.; Noh, E.; et al. Arabidopsis PCR2 is a zinc exporter involved in both zinc extrusion and long-distance zinc transport. *Plant Cell* **2010**, *22*, 2237–2252. [CrossRef] [PubMed]

45. Zhang, C.; Shinwari, K.I.; Luo, L.; Zheng, L. OsYSL13 is involved in iron distribution in rice. *Int. J. Mol. Sci.* **2018**, *19*, 3537. [CrossRef] [PubMed]

46. Demidchik, V.; Shabala, S.; Isayenkov, S.; Cuin, T.A.; Pottosin, I. Calcium transport across plant membranes: Mechanisms and functions. *New Phytol.* **2018**, *220*, 49–69. [CrossRef]

47. Kugler, A.; Köhler, B.; Palme, K.; Wolff, P.; Dietrich, P. Salt-dependent regulation of a CNG channel subfamily in Arabidopsis. *BMC Plant Biol.* **2009**, *9*, 140. [CrossRef]

48. Mascher, M.; Gundlach, H.; Himmelbach, A.; Beier, S.; Twardziok, S.O.; Wicker, T.; Radchuk, V.; Dockter, C.; Hedley, P.E.; Russell, J.; et al. A chromosome conformation capture ordered sequence of the barley genome. *Nature* **2017**, *544*, 427–433. [CrossRef]

49. Forde, B.G.; Roberts, M.R. Glutamate receptor-like channels in plants: A role as amino acid sensors in plant defence? *F1000 Prime Rep.* **2014**, *6*, 37. [CrossRef]

50. Zan, T.; Li, L.; Xie, T.; Zhang, L.; Li, X. Genome-wide identification and abiotic stress response patterns of abscisic acid stress ripening protein family members in *Triticum aestivum* L. *Genomics* **2020**, *12*, 3794–3802. [CrossRef]

51. Mauri, N.; Fernández-Marcos, M.; Costas, C.; Desvoyes, B.; Pichel, A.; Caro, E.; Gutierrez, C. GEM, a member of the GRAM domain family of proteins, is part of the ABA signaling pathway. *Sci. Rep.* **2016**, *6*, 22660. [CrossRef]

52. Zhang, Y.; Yang, X.; Cao, P.; Xiao, Z.; Zhan, C.; Liu, M.; Nvsvrot, T.; Wang, N. The bZIP53-IAA4 module inhibits adventitious root development in *Populus*. *J. Exp. Bot.* **2020**, *71*, 3485–3498. [CrossRef]

53. Yu, Z.; Wang, X.; Zhang, L. Structural and functional dynamics of dehydrins: A plant protector protein under abiotic stress. *Int. J. Mol. Sci.* **2018**, *19*, 3420. [CrossRef] [PubMed]

54. Riahi, J.; Amri, B.; Chibani, F.; Azri, W.; Mejri, S.; Bennani, L.; Zoghlami, N.; Matros, A.; Mock, H.P.; Ghorbel, A.; et al. Comparative analyses of albumin/globulin grain proteome fraction in differentially salt-tolerant Tunisian barley landraces reveals genotype-specific and defined abundant proteins. *Plant Biol.* **2019**, *21*, 652–661. [CrossRef] [PubMed]

55. Zhao, C.; Zayed, O.; Yu, Z.; Jiang, W.; Zhu, P.; Hsu, C.C.; Zhang, L.; Tao, W.A.; Lozano-Durán, R.; Zhu, J.K. Leucine-rich repeat extensin proteins regulate plant salt tolerance in Arabidopsis. *Proc. Natl. Acad. Sci. USA* **2018**, *115*, 13123–13128. [CrossRef] [PubMed]

56. Sharma, M.; Pandey, G.K. Expansion and function of repeat domain proteins during stress and development in plants. *Front. Plant Sci.* **2016**, *6*, 1218. [CrossRef]

57. Borisjuk, L.; Rolletschek, H. Nitric oxide is a versatile sensor of low oxygen stress in plants. *Plant Signal. Behav.* **2008**, *3*, 391–393. [CrossRef]

58. Medini, W.; Farhat, N.; Al-Rawi, S.; Mahto, H.; Qasim, H.; Ben-Halima, E.; Bessrour, M.; Chibani, F.; Abdelly, C.; Fettke, J.; et al. Do carbohydrate metabolism and partitioning contribute to the higher salt tolerance of *Hordeum marinum* compared to *Hordeum vulgare*? *Acta Physiol. Plant.* **2019**, *41*, 190. [CrossRef]

59. Llanes, A.; Masciarelli, O.; Ordoñez, R.; Islas, M.; Luna, V. Differential growth responses to sodium salts involve different ABA catabolism and transport in the halophyte *Prosopis strombulifera*. *Biol. Plant.* **2014**, *58*, 80–88. [CrossRef]

60. Tari, I.; Kiss, G.; Deér, A.K.; Csiszár, J.; Erdei, L.; Gallé, Á.; Gémes, K.; Horváth, F.; Poór, P.; Szepesi, Á.; et al. Salicylic acid increased aldose reductase activity and sorbitol accumulation in tomato plants under salt stress. *Biol. Plant.* **2010**, *54*, 677–683. [CrossRef]

61. Lescano, C.I.; Martini, C.; González, C.A.; Desimone, M. Allantoin accumulation mediated by allantoinase downregulation and transport by Ureide Permease 5 confers salt stress tolerance to Arabidopsis plants. *Plant Mol. Biol.* **2016**, *91*, 581–595. [CrossRef]

62. Casartelli, A.; Melino, V.J.; Baumann, U.; Riboni, M.; Suchecki, R.; Jayasinghe, N.S.; Mendis, H.; Watanabe, M.; Erban, A.; Zuther, E.; et al. Opposite fates of the purine metabolite allantoin under water and nitrogen limitations in bread wheat. *Plant Mol. Biol.* **2019**, *99*, 477–497. [CrossRef]

63. Takagi, H.; Ishiga, Y.; Watanabe, S.; Konishi, T.; Egusa, M.; Akiyoshi, N.; Matsuura, T.; Mori, I.C.; Hirayama, T.; Kaminaka, H.; et al. Allantoin, a stress-related purine metabolite, can activate jasmonate signaling in a MYC2-regulated and abscisic acid-dependent manner. *J. Exp. Bot.* **2016**, *67*, 2519–2532. [CrossRef] [PubMed]

64. Chen, T.; Cai, X.; Wu, X.; Karahara, I.; Schreiber, L.; Lin, J. Casparian strip development and its potential function in salt tolerance. *Plant Signal. Behav.* **2011**, *6*, 1499–1502. [CrossRef] [PubMed]

65. Proseus, T.E.; Boyer, J.S. Calcium deprivation disrupts enlargement of *Chara corallina* cells: Further evidence for the calcium pectate cycle. *J. Exp. Bot.* **2012**, *63*, 3953–3958. [CrossRef] [PubMed]

66. Schönknecht, G. Calcium signals from the vacuole. *Plants* **2013**, *2*, 589–614. [CrossRef]

67. Byrt, C.S.; Zhao, M.; Kourghi, M.; Bose, J.; Henderson, S.W.; Qiu, J.; Gilliham, M.; Schultz, C.; Schwarz, M.; Ramesh, S.A.; et al. Non-selective cation channel activity of aquaporin AtPIP2;1 regulated by Ca^{2+} and pH. *Plant Cell Environ.* **2017**, *40*, 802–815. [CrossRef]

68. Yousfi, S.; Rabhi, M.; Hessini, K.; Abdelly, C.; Gharsalli, M. Differences in efficient metabolite management and nutrient meta-bolic regulation between wild and cultivated barley grown at high salinity. *Plant Biol.* **2010**, *12*, 650–658.

69. Zhu, M.; Zhou, M.; Shabala, L.; Shabala, S. Physiological and molecular mechanisms mediating xylem Na^+ loading in barley in the context of salinity stress tolerance. *Plant Cell Environ.* **2017**, *40*, 1009–1020. [CrossRef]

70. Maksaev, G.; Haswell, E.S. MscS-Like10 is a stretch-activated ion channel from *Arabidopsis thaliana* with a preference for anions. *Proc. Natl. Acad. Sci. USA* **2012**, *109*, 19015–19020. [CrossRef]

71. Takano, J.; Miwa, K.; Fujiwara, T. Boron transport mechanisms: Collaboration of channels and transporters. *Trends Plant Sci.* **2008**, *13*, 451–457. [CrossRef]

72. Fuertes-Mendizábal, T.; Bastías, E.I.; González-Murua, C.; González-Moro, M.B. Nitrogen assimilation in the highly salt- and boron-tolerant ecotype *Zea mays* L. Amylacea. *Plants* **2020**, *9*, 322. [CrossRef]

73. Pandey, A.; Khan, M.K.; Hakki, E.E.; Gezgin, S.; Hamurcu, M. Combined boron toxicity and salinity stress—An insight into its interaction in plants. *Plants* **2019**, *8*, 364. [CrossRef] [PubMed]

74. Disante, K.B.; Fuentes, D.; Cortina, J. Response to drought of Zn-stressed *Quercus suber* L. seedlings. *Environ. Exp. Bot.* **2010**, *70*, 96–103. [CrossRef]

75. Brennan, R.F. Zinc Application and Its Availability to Plants. Ph.D. Dissertation, School of Environmental Science, Division of Science and Engineering, Murdoch University, Murdoch, WA, Australia, 2005.

76. Yarishkin, O.V.; Ryu, H.W.; Park, J.Y.; Yang, M.S.; Hong, S.G.; Park, K.H. Sulfonate chalcone as new class voltage-dependent K^+ channel blocker. *Bioorg. Med. Chem. Lett.* **2008**, *18*, 137–140. [CrossRef] [PubMed]

77. Lijuan, C.; Huiming, G.; Yi, L.; Hongmei, C. Chalcone synthase EaCHS1 from *Eupatorium adenophorum* functions in salt stress tolerance in tobacco. *Plant Cell Rep.* **2015**, *34*, 885–894. [CrossRef] [PubMed]

78. Schenck, C.A.; Maeda, H.A. Tyrosine biosynthesis, metabolism, and catabolism in plants. *Phytochemistry* **2018**, *149*, 82–102. [CrossRef] [PubMed]

79. Kobayashi, N.I.; Yamaji, N.; Yamamoto, H.; Okubo, K.; Ueno, H.; Costa, A.; Tanoi, K.; Matsumura, H.; Fujii-Kashino, M.; Horiuchi, T.; et al. OsHKT1;5 mediates Na^+ exclusion in the vasculature to protect leaf blades and reproductive tissues from salt toxicity in rice. *Plant J.* **2017**, *91*, 657–670. [CrossRef]

80. Malik, A.I.; English, J.P.; Colmer, T.D. Tolerance of *Hordeum marinum* accessions to O_2 deficiency, salinity and these stresses combined. *Ann. Bot.* **2009**, *103*, 237–248. [CrossRef]

81. Saoudi, W.; Badri, M.; Taamalli, W.; Zribi, O.T.; Gandour, M.; Abdelly, C. Variability in response to salinity stress in natural Tunisian populations of *Hordeum marinum* subsp. *marinum*. *Plant Biol.* **2019**, *21*, 89–100. [CrossRef]

82. Borisjuk, L.; Neuberger, T.; Schwender, J.; Heinzel, N.; Sunderhaus, S.; Fuchs, J.; Hay, J.O.; Tschiersch, H.; Braun, H.P.; Denolf, P.; et al. Seed architecture shapes embryo metabolism in oilseed rape. *Plant Cell* **2013**, *25*, 1625–1640. [CrossRef]

83. Martin, M. Cutadapt removes adapter sequences from high-throughput sequencing reads. *EMBnet. J.* **2011**, *17*, 10–12. [CrossRef]

84. Bray, N.L.; Pimentel, H.; Melsted, P.; Pachter, L. Near-optimal probabilistic RNA-seq quantification. *Nat. Biotechnol.* **2016**, *34*, 525–527. [CrossRef] [PubMed]

85. Love, M.I.; Huber, W.; Anders, S. Moderated estimation of fold change and dispersion for RNA-seq data with DESeq2. *Genome Biol.* **2014**, *15*, 550. [CrossRef] [PubMed]

86. Heberle, H.; Meirelles, V.G.; da Silva, F.R.; Telles, G.P.; Minghim, R. InteractiVenn: A web-based tool for the analysis of sets through Venn diagrams. *BMC Bioinform.* **2015**, *16*, 169. [CrossRef] [PubMed]

87. The Gene Ontology Consortium. The Gene Ontology Resource: 20 years and still GOing strong. *Nucleic Acids Res.* **2019**, *47*, D330–D338. [CrossRef] [PubMed]

International Journal of
Molecular Sciences

Review

Advances and Challenges in the Breeding of Salt-Tolerant Rice

Hua Qin [1,2] **, Yuxiang Li** [1] **and Rongfeng Huang** [1,2,*]

[1]   Biotechnology Research Institute, Chinese Academy of Agricultural Sciences, Beijing 100081, China; qinhua@caas.cn (H.Q.); liyx0929@163.com (Y.L.)
[2]   National Key Facility of Crop Gene Resources and Genetic Improvement, Beijing 100081, China
*   Correspondence: rfhuang@caas.cn

Received: 20 October 2020; Accepted: 7 November 2020; Published: 9 November 2020

Abstract: Soil salinization and a degraded ecological environment are challenging agricultural productivity and food security. Rice (*Oryza sativa*), the staple food of much of the world's population, is categorized as a salt-susceptible crop. Improving the salt tolerance of rice would increase the potential of saline-alkali land and ensure food security. Salt tolerance is a complex quantitative trait. Biotechnological efforts to improve the salt tolerance of rice hinge on a detailed understanding of the molecular mechanisms underlying salt stress tolerance. In this review, we summarize progress in the breeding of salt-tolerant rice and in the mapping and cloning of genes and quantitative trait loci (QTLs) associated with salt tolerance in rice. Furthermore, we describe biotechnological tools that can be used to cultivate salt-tolerant rice, providing a reference for efforts aimed at rapidly and precisely cultivating salt-tolerance rice varieties.

Keywords: biotechnology breeding; high-throughput sequencing; QTLs; rice; salt tolerance

1. Introduction

Plants grow in dynamic environments and frequently experience various abiotic stresses, such as drought, high salinity, cold, and heat [1]. Salt stress is one of the most severe environmental stresses. The effects of salt stress on plants include osmotic stress, ionic toxicity, and nutritional deficiencies and eventually lead to growth inhibition and crop yield losses [2,3]. Soil salinization is mainly caused by poor-quality drainage and irrigation systems, climate change (which leads to sea level rise), and drought [4]. Soil salinity is a global problem that affects more than 20% of cultivated land, including half of all irrigated areas, and this percentage is expected to increase [5]. Therefore, improving the salt tolerance of crops would not only lead to the effective use of saline-alkali land, but also support sustainable agriculture and alleviate the world food crisis.

Rice (*Oryza sativa*) is an important staple food crop worldwide. As the global population continues to rise, rice production also needs to increase. However, global rice production is threatened by climate change [6]. Rice is considered to be a salt-susceptible species [7], and its salt tolerance depends on growth stage, organ type, and genotype [8–10]. Generally, the seedling and reproductive stages are more susceptible to salinity than the vegetative stage, roots are more sensitive than other organs [8], and *japonica* rice is more sensitive than *indica* rice [9]. Salinity stress suppresses photosynthesis and growth, leading to biomass loss, as well as partial sterility, which ultimately results in reductions in rice yield [11,12]. Therefore, the breeding of salt-tolerant rice cultivars is considered to be one of the most economic options to assure food security.

As a semi-aquatic plant, rice lives in a water-saturated environment during most of its life cycle. This environment has led to many distinct adaptations rice—for example, the ethylene response phenotype of rice is different from that of other species [13]. In *Arabidopsis thaliana*, ethylene plays a positive role in regulating salt tolerance, whereas, in rice, ethylene negatively regulates salt tolerance [14]. Since rice is an important food crop, yield and quality are two important criteria for rice producers. However, improving the stress tolerance of rice with less effect on yield and quality has been a challenge for breeders. It is hoped that the breeding of salt-tolerant rice varieties will be accelerated by dissecting the genetics underlying salt tolerance and using biotechnology to generate salt-tolerant plants.

Several reviews have summarized the mechanisms of plant responses to salt stress [15–17], but only a few have focused on rice and the progress toward breeding salt-tolerant rice varieties. In this review, we summarize the current status of salt-tolerant rice breeding, recent advances in the mapping and cloning of salt-tolerant genes/quantitative trait loci (QTLs), and new technologies that can be used for breeding salt-tolerant rice varieties. We also highlight current challenges to breeding salt-tolerant rice, providing a basis for further studies and efforts aimed at breeding salt-tolerant rice varieties.

2. Salt-Tolerant Rice Identification and Evaluation Methods

The development of an efficient and reliable evaluation system is a prerequisite for breeding salt-tolerant rice varieties. The current rice salt tolerance indicators are divided into two aspects: morphological parameters and physiological parameters [18,19]. The morphological parameters evaluation method is to conduct salt treatments at different growth stages of rice and then observe and record the salt damage symptoms of plants, leaves, tillers, and spikelet fertility [20,21]. The standard evaluation score of visual salt injury was proposed by The International Rice Research Institute (IRRI); this method scores the salt tolerance of rice from 1 to 9 based on the tiller number, leaf symptoms, and the growth status of the whole plant under salt stress; the lower score (1) indicates tolerant and higher score (9) denotes sensitive genotypes [22]. However, this method is greatly affected by human qualitativeness, and there are time differences in the rate of leaf death and plant death of different materials. Therefore, this identification and evaluation system cannot completely and accurately reflect the salt tolerance of rice varieties.

Salinity stress induces metabolite changes, and several physiological mechanisms are perceived to contribute to the overall ability of rice plants to cope with excess salts [15,16,23]. Studies have shown that the Na^+/K^+ ratio, proline content, hydrogen peroxide, peroxidase activity, and sugars, etc. are affected under salt stress [21,24]. Therefore, it can be used to screen salt-tolerant rice varieties by comparing the changes of physiological and biochemical indices in rice with or without salt treatment. However, physiological and biochemical parameter methods lack specific evaluation standards, and the measurement of these indices requires corresponding instruments or kits, which are relatively cumbersome to operate.

Rice salt tolerance is a complex genetic and physiological characteristic, and the extent of its sensitivity varies during different growth and developmental stages [8–10]. The salt tolerance during the whole life of rice is a comprehensive reflection of the salt tolerance in each growth and developmental stage, which is closer to production practice and has more practical significance. However, due to soil heterogeneity, climatic factors and other environmental factors may influence the physiological processes; it is difficult to screen salt-tolerant rice varieties at the field level. Hence, screening under laboratory conditions is considered to be advantageous over field screening. Since the salt types in saline-alkali fields are double salts, the salt tolerance identified in the laboratory does not always correlate with that in the field. Therefore, the most reliable way to evaluate the salt tolerance of rice is to compare the changes of morphological parameters and physiological parameters in various growth and developmental stages under salt treatment and normal condition, both in the laboratory and in the field.

3. Salt Stress Affects Rice Growth and Grain Quality

The response and adaptation of rice to salt stress is a complex process. Salt stress causes root growth inhibition, leaf rolling, reduced plant height and tiller number, and spikelet sterility, which, ultimately, leads to a reduced yield [21,25]. In addition to morphological changes, salt stress also causes physiological and biochemical changes, such as the inhibition of photosynthesis; decreased water content; altered metabolism; increased Na transport to the shoot; and decreased K, Zn, and P uptake [12,25,26].

Since rice is a staple food, grain quality is an important driver of marketability. Rice grain is composed mainly of carbohydrates, predominantly starches. The determinants of rice grain quality are grain texture, taste, and visual attributes, which are further determined by the composition and structure of the starch molecules [27]. The effect of salt stress on rice grain starch depends on the salt concentration and the rice genotype. Specifically, salinity treatment leads to a decrease in starch content when the salt concentration is higher than 5-dS/m^2 electrical conductivity, in both salt-tolerant and salt-susceptible cultivars of rice [28,29]. However, grain starch increased in the Nipponbar cultivar when low concentrations of salt (2 or 4-dS/m^2 electrical conductivity) were applied at the anther appearance and seedling stages [30], suggesting that the response of rice to salt stress is complex and highly context-dependent.

Another indicator of rice grain quality is the nutritional value of the grain. Rice grain contains a variety of minerals, such as Ca, Mg, and P, and some trace elements, such as Fe, Cu, Zn, and Mn. Rice also contains the vitamins like thiamine, riboflavin, and niacin [25]. Several studies have shown that the absorption and uptake of micro- and macro-mineral nutrients are altered under salinity stress [31,32]. Saleethong et al. measured the macronutrients and micronutrients in grains of *Pokkali* (a salt-tolerant variety) and *KDML105* (a salt-sensitive variety) under saline conditions and found that salt stress resulted in significant reductions in macronutrient elements but an increase in Ca in brown rice grains of both cultivars. Moreover, the amounts of Mn, Cu, and Zn were higher in *Pokkali* than in *KDML105* [31]. Verma and Neue reported that the contents of Na, Fe, and Zn increased, while those of P and Mn decreased in rice grain with increased salinity, but the contents of N, Mg, Cu, K, and Ca were not affected in the varieties used in this study [32]. In addition to affecting the mineral content, salt stress can increase the rice grain protein content and amino acid levels [25]. Although little is known about the effects of salt stress on vitamins in rice grains, salinity caused a significant reduction in vitamin contents in wheat grains [33]. In summary, these studies show that the effect of salt stress on rice grain quality is multifaceted and depends on the salt concentration and the rice variety.

4. Breeding of Salt-Tolerant Rice Varieties

Developing elite salt-tolerant rice varieties is considered to be the most economically viable and environmentally friendly method to effectively use saline-alkali land. As a food crop, yield is an important indicator to evaluate the merits of rice varieties. Salt-alkali tolerance in rice is defined as the ability to grow on land with a 0.3% saline-alkali concentration, with a yield of more than 4500 kg per hectare. Through years of hard work, breeders have sought out, collected, evaluated, and developed many salt-tolerant rice resources.

Ceylon (modern-day Sri Lanka) was the first country to carry out the screening and cultivation of salt-tolerant rice varieties, introducing the first salt-tolerant rice variety, *Pokkali*, in 1939 [34]. Subsequently, India and the Philippines bred a series of salt-tolerant rice varieties, such as *Kala Rata 1-24*, *Nona Bokra*, *Bhura Rata*, *SR 26B*, *Chin.13*, and *349 Jhona*. Bangladesh bred *BRI*, *BR203-26-2*, *Sail*, and other salt-tolerant rice varieties. Thailand bred the salt-tolerant rice variety *FL530*. Japan bred the salt-tolerant rice varieties *Mantaro rice*, *Kanto 51*, *Hama Minoru*, *Chikushiqing*, and *Lansheng*. The United States bred the salt-tolerant rice variety *American Rice*. South Korea bred the salt-tolerant rice varieties *Dongjinbyeo*, *Ganchukbyeo*, *Gyehwabyeo*, *Ilpumbyeo*, *Seomjimbyeo*, and *Nonganbyeo*. Russia bred 16 the salt-tolerant rice varieties, including *VNIIR8207* and *Fontan* [35–43]. IRRI hosts more than 127,000 rice accessions collected worldwide, providing a rich source of genetic diversity. By evaluating the salt tolerance of these rice varieties, researchers identified approximately 103 varieties that were moderately to highly salt tolerant, including *Nona Bokra*, *Pamodar*, *Jhona349*, *IR4595-4-1-13*, *IR4630-22-2-5-1-3*, *IR9764-45-2-2*, and *IR9884-54-3* [44,45].

China began to study the salt tolerance of rice in the 1950s. In the 1980s, China launched a national collaborative research program on the salt-alkali resistance of rice and wheat. During the Seventh Five-Year Plan period (1986–1990), China began evaluating the salt tolerance of rice germplasm. This large-scale national cooperation resulted in some progress. The Liaoning Saline or Alkaline Land Utilization and Research Institute launched a salt-tolerant rice breeding program in the 1970s and cultivated a series of salt-tolerant *japonica* rice varieties, such as *Liaoyan No. 2, Liaoyan 241*, and *Liaoyan 16*. In 1984, the institute developed highly salt-tolerant *indica* rice varieties *81-210*. Since 1989, salt-tolerant varieties, such as *Salt-resistant No. 100, Yangeng 29, Yanfeng 47*, and *Yangeng 228*, have been cultivated [42]. The Jiangsu Institute of Agricultural Sciences in Coastal Areas began identifying and evaluating salt-tolerant rice germplasm resources in the 1980s. From more than 1300 germplasm resources, this group obtained 61 that were salt-tolerant [43], from which they developed many salt-tolerant materials, such as *Yancheng 156, Yandao No. 10*, and *Yandao No. 12* [43]. In addition, breeding institutes and breeders in China have used existing salt-tolerant germplasms or conventional breeding methods to obtain salt-tolerant rice varieties, such as *Changbai No. 6, Changbai No. 7, Changbai No. 9, Changbai No. 13, Jigeng No. 84*, and *Jinyuan 101* [42].

Sea Rice 86 (SR86) is a new cultivar domesticated from a wild strain of rice that was first found in 1986 in saline-alkaline soil submerged in sea water near the coastal region of the city of Zhanjiang in Southeast China [46]. *SR86* showed a significantly higher ability to cope with high salinity than a highly salt-resistant rice variety, *Yanfen 47*, measured by both germination and salt inhibition rates [46]. After more than 20 years of breeding and selection, *SR86* retains an extraordinary tolerance to salinity and is considered to be a strategic germplasm resource for the development of new rice varieties. *SR86* is being used to investigate the mechanism of salt tolerance and effective breeding strategies.

In recent years, our laboratory has also carried out a breeding program to develop salt-tolerant rice. We collected more than 750 rice accessions, including 500 rice accessions from all over the world and 250 salt-tolerant rice varieties from domestic coastal cities, such as Tianjin, Liaoning, Shandong, and Jiangsu. We investigated the salt tolerance of these varieties in Dongying, Shandong Province (37°31′29″ N 118°33′57″ E), a typical saline-alkali field in China. Briefly, thirty-day-old seedlings were transplanted to a normal field or saline field (0.35% NaCl, pH 8.2) at a spacing of 20 cm × 15 cm, and all agronomic traits were performed at the maturity stage. Through comparing the agronomic traits of the normal field and saline field, we selected a series of varieties with excellent agronomic traits and salt tolerance (Table 1) (unpublished data). Using these varieties, we made more than 300 hybrid combinations, generated more than 100 salt-tolerant genetic populations, and identified more than 1000 high-yielding salt-tolerant recombinant inbred lines.

Table 1. Rice varieties with excellent agronomic traits and salt tolerance results.

Material Code	Plant Height (cm)		Panicle Length (cm)		Tiller Number		Yield (kg/hectare)	
	Control	Salt	Control	Salt	Control	Salt	Control	Salt
DYST1	94.6 ± 1.8	73.1 ± 3.5	21.6 ± 0.7	19.4 ± 1.0	8.0 ± 1.2	5.6 ± 0.5	6438.0	6271.5
DYST2	92.8 ± 3.2	71.8 ± 3.0	18.2 ± 1.8	16.9 ± 1.2	10.6 ± 1.5	9.6 ± 0.9	7159.5	5106.0
DYST3	91.3 ± 4.1	64.9 ± 4.2	19.2 ± 1.6	17.4 ± 1.5	13.0 ± 0.7	7.2 ± 1.1	8325.0	4662.0
DYST4	91.5 ± 3.9	78.8 ± 4.0	19.2 ± 1.1	18.4 ± 1.8	11.4 ± 1.8	8.0 ± 2.0	7270.5	6216.0
DYST5	99.7 ± 2.6	83.3 ± 1.8	22.8 ± 1.3	20.3 ± 1.6	11.8 ± 1.8	8.8 ± 2.2	8325.0	5217.0
DYST6	95.6 ± 1.0	74.6 ± 3.6	21.0 ± 0.6	18.9 ± 0.9	13.2 ± 0.8	11.0 ± 2.9	8103.0	5050.5
DYST7	91.6 ± 2.6	79.4 ± 5.4	21.8 ± 1.3	20.5 ± 0.7	11.8 ± 2.2	10.6 ± 1.7	6993.0	5050.5
DYST8	96.0 ± 3.1	81.5 ± 3.0	21.3 ± 0.7	20.3 ± 1.3	12.0 ± 1.4	10.8 ± 2.3	7992.0	5827.5
DYST9	95.6 ± 2.8	80.6 ± 3.0	21.3 ± 1.2	20.6 ± 1.1	11.6 ± 1.8	10.2 ± 2.9	8268.8	5142.0
DYST10	95.2 ± 1.8	83.0 ± 1.0	20.4 ± 1.6	19.7 ± 1.4	11.2 ± 2.4	10.2 ± 1.6	8880.0	5550.0

"Control" indicates that the variety was grown in a normal field. "Salt" indicates that the variety was grown in a field containing 0.35% NaCl and pH 8.2 throughout its life cycle. "DYST" means Dongying Salt Tolerance.

5. Mapping and Cloning of Salt-Tolerant Genes/QTLs

Salt stress has two main stress effects on rice: osmotic stress and ionic stress [2,3]. Osmotic stress reduces the water uptake by roots and causes internal dehydration. It also leads to the excessive accumulation of reactive oxygen species (ROS), which damages various cellular components and macromolecules and, eventually, leads to plant death [47]. Ionic stress is caused by the excessive accumulation of Na^+ and Cl^- in metabolically active intracellular compartments. High intracellular concentrations of Na^+ inhibit the uptake of other ions, which can disrupt the metabolism and potentially kill the plant [2,3]. Thus, enhancing the ability of rice to minimize and adjust to osmotic and ionic stresses is an effective way to improve the salt tolerance of rice.

Salinity tolerance in rice is a polygenic trait controlled by QTLs [17]. Using mapping populations derived from crosses between salt-sensitive varieties and salt-tolerant varieties, researchers have identified a large number of QTLs (Table 2). Only a few major salt tolerance QTLs or genes have been identified by genomic methods. For example, *qSKC-1* and *qSNC-7* are involved in regulating K^+/Na^+ homeostasis under salt stress and explain 48.5% and 40.1%, respectively, of the total phenotypic variance [48]. Isolation of the *qSKC-1* gene by map-based cloning revealed that it encodes a member of the HKT-type transporter family [49]. The QTL *Saltol*, which acts to maintain shoot Na^+/K^+ homeostasis in the salt-tolerant cultivar *Pokkali*, explained 43% of the variation in the seedling shoot Na^+/K^+ ratio in a recombinant inbred line (RIL) population derived from a cross between *indica* varieties *IR29* and *Pokkali* [50]. *qSE3*, which encodes a K^+ transporter gene, *OsHAK21*, promotes seed germination and seedling establishment under salinity stress in rice [51]. *qST1* and *qST3*, which are located on chromosomes 1 and 3, respectively, conferred salt tolerance at the young seedling stage and explained 36.9% of the total phenotypic variance in the RIL population derived from a cross between *Milyang 23* and *Gihobyeo* [52]. *qST1.1*, which plays a key role in salt tolerance in *SR86*, explained 62.6% of the phenotypic variance [53]. The identification of QTLs is only one step toward determining the genes associated with salinity tolerance; further research is needed to discover the major salt-tolerance genes and investigate their regulatory mechanisms in rice.

Table 2. Quantitative trait loci (QTLs) identified for salt tolerance in rice.

Parents	Number of QTLs	Stage	Reference
Capsule × BRRI dhan29	30	seedling	[54]
DJ15 × Koshihikari	9	seedling	[55]
Hasawi × IR29	20	seedling	[56]
CSR27 × MI48	25	seedling, vegetative and reproductive	[57]
Xiushui09 × IR2061-520-6-9	47	seedling	[58]
At354 × Bg352	6	seedling	[59]
Pokkali × Bengal	50	seedling	[60]
Cheriviruppu8 × Pusa Basmati 1	16	reproductive	[61]
Teqing × Oryza rufipogon Griff	15	seedling	[62]
Milyang 23 × Gihobyeo	2	seedling	[52]
Nona Bokra × Koshihikari	11	seedling	[48]
CSR27 × MI48	8	maturity	[63]
Nona Bokra × Jupiter	33	seedling	[64]
IR75862 × Ce258/Zhongguangxiang1	18	seedling	[65]
Pokkali × IR36	6	maturity	[49]
Tarommahali × Khazar	2	seedling	[66]
Jiucaiqing × IR26	22	seedling	[67]
Jiucaiqing × IR26	16	germination	[68]
9311 × Oryza rufipogon Griff	10	seedling	[69]
Dongnong425 × Changbai10	13	seedling	[70]
Pokkali × IR29	17	seedling	[50]

Recently, great progress has been made toward the identification and cloning of salt-tolerance genes through the molecular genetic analysis of plant responses to salt stress. For example, researchers have been interested in genes involved in ROS metabolism, because maintaining an appropriate level of ROS is essential for the survival of plants under salt stress. The level of ROS in plants depends on two processes: ROS biosynthesis and ROS scavenging. In rice, a salt treatment induces or represses the expression of *respiratory burst oxidase homologs* (*Rbohs*), which catalyze the conversion of O_2 to O_2^- [71], suggesting that *Rbohs* are candidate genes for improving salt tolerance in rice. In addition to the genes involved in ROS biosynthesis, genes related to ROS scavenging also have the potential to improve rice salt tolerance. For example, ascorbate peroxidase (APX) catalyzes the conversion of H_2O_2 to H_2O and O_2 using ascorbate as the specific electron donor [72]. The overexpression of *APX* genes in rice reduces ROS levels and enhances salt tolerance [73–75]. The knockdown of *OsVTC1-1* and *OsVTC1-3*, which are involved in ascorbate synthesis, increases the accumulation of ROS and decreases the salt resistance of rice [76–78].

Genes involved in Na^+/K^+ homeostasis are also candidates for improving salt tolerance in rice. High-salinity stress results in altered K^+/Na^+ ratios, which lead to metabolic changes in the plant [79]. Na^+ causes growth inhibition via Na^+ toxicity, whereas K^+ is essential for plant growth and development. Therefore, restricting the intracellular accumulation of toxic sodium (Na^+) is beneficial for survival under salt stress [2,3]. Members of the high-affinity K^+ transporter (HKT) family in rice function as Na^+ and Na^+/K^+ transporters in controlling Na^+ accumulation, and enhancing or inhibiting the expression of *HKT* genes changes the intracellular Na^+ content and the salt tolerance of rice [80,81]. Thus, modifying ion channels is an option to improve the salt tolerance of rice.

In addition to genes related to ROS and Na^+/K^+ homeostasis, many other genes influence the salt tolerance of rice [16]. For example, *rice expansin 7* (*OsEXPA7*), which encodes a cell wall-loosening protein, positively regulates salt tolerance in rice by coordinating sodium transport, ROS scavenging, and cell-wall loosening [82]. *Response regulator 22* (*OsRR22*) encodes a B-type response regulator protein that acts as a transcription factor regulating genes involved in the osmotic stress response and/or ion transport between parenchyma cells and vascular tissue cells in the root [83]. OsCYP71D8L, a cytochrome P450 monooxygenase in the CYP71 clan in rice, negatively regulates salt tolerance by affecting gibberellin and cytokinin homeostasis [84]. The dominant suppressor of KAR2 (OsDSK2a) is a ubiquitin-binding protein that mediates seedling growth and salt responses in rice through interacting with elongated uppermost internode (EUI) to affect gibberellin metabolism [85]. Although a number of salt-tolerance genes have been identified (Table 3), further information is needed about how to effectively use these genes to improve rice salt tolerance.

Table 3. Genes associated with rice salt tolerance.

Genes Name	Accession Number	Gene Function
OsSOS1	Os12g0641100	Exports Na^+ ions out of cells, positively regulates salt tolerance
OsHKT1;1	Os04g0607500	Mediate Na^+-specific transport, positively regulates salt tolerance
OsHKT1;4	Os04g0607600	Mediate Na^+-specific transport, positively regulates salt tolerance
OsHAK21	Os03g0576200	K^+ transporter, positively regulates salt tolerance
OsEIN2	Os07g0155600	Ethylene signaling component, negatively regulates salt tolerance
OsEIL1	Os03g0324300	Ethylene signaling component, negatively regulates salt tolerance
OsEIL2	Os07g0685700	Ethylene signaling component, negatively regulates salt tolerance
OsAPX2	Os07g0694700	Encoding ascorbate peroxidases, positively regulates salt tolerance
OsCPK12	Os04g0560600	Encoding a calcium-dependent protein kinase, positively regulates salt tolerance
DST	Os03g0786400	Encoding Zinc-finger protein, negatively regulates salt tolerance
SIT1	Os02g0640500	Encoding a lectin receptor-like kinase, positively regulates salt tolerance
OsDOF15	Os03g0764900	Encoding a DOF-binding with one finger transcription factor, negatively regulates salt tolerance
OsTIR1	Os05g0150500	Auxin receptor, positively regulates salt tolerance
OsAFB2	Os04g0395600	Auxin receptor, positively regulates salt tolerance
OsGA2ox5	Os07g0103500	Encoding a gibberellin metabolism enzyme, positively regulates salt tolerance

Table 3. *Cont.*

Genes Name	Accession Number	Gene Function
OsTIR1	Os05g0150500	Auxin receptor, positively regulates salt tolerance
DRO1	Os07g0614400	Associated with root angle modifications, negatively regulates salt tolerance
OsEXPA7	Os03g0822000	Encoding cell wall-loosening proteins, positively regulates salt tolerance
OsPIL14	Os07g0143200	Phytochrome interacting factor like gene, positively regulates salt tolerance
SLR1	Os03g0707600	rice DELLA protein, GA signaling suppressor, positively regulates salt tolerance
OsDSK2a	Os03g0131300	The UBL-UBA protein, negatively regulates salt tolerance
IDS1	Os03g0818800	Encoding an apetala2/ethylene response factor transcription factor, negatively regulates salt tolerance
AGO2	Os04g0615700	Encoding a ARGONAUTE family protein, positively regulates salt tolerance
BG3	Os01g0680200	A putative cytokinin transporter, positively regulates salt tolerance
OsRR2	Os06g0183100	Encoding a B-type response regulator, negatively regulates salt tolerance
OsRR9	Os11g0143300	Negative regulators of cytokinin signaling, negatively regulates salt tolerance
OsRR10	Os12g0139400	Negative regulators of cytokinin signaling, negatively regulates salt tolerance
OsC2DP	Os09g0571200	Encoding a novel C2 domain-containing protein, positively regulates salt tolerance
OsOTS1	Os06g0487900	The ubiquitin-like protease class of SUMO protease, positively regulates salt tolerance
OsZFP179	Os01g0839100	Encoding C2H2-type zinc-finger protein, positively regulates salt tolerance
OsZFP182	Os03g0820300	Encoding TFIIIA-type zinc-finger protein, positively regulates salt tolerance
OsZFP185	Os02g0195600	Encoding A20/AN1-type zinc-finger protein, negatively regulates salt tolerance
OsZFP213	Os12g0617000	Encoding C2H2-type zinc-finger protein, positively regulates salt tolerance
OsZFP245	Os07g0587400	Encoding TFIIIA-type zinc-finger protein, positively regulates salt tolerance
OsZFP252	Os12g0583700	Encoding TFIIIA-type zinc-finger protein, positively regulates salt tolerance
OsMGT1	Os01g0869200	A rice Mg2+ transporter, positively regulates salt tolerance
CYP71D8L	Os02g0184900	Encoding a cytochrome P450 monooxygenases, positively regulates salt tolerance
OsCYP94C2b	Os12g0150200	Encoding a JA-catabolizing enzyme, positively regulates salt tolerance
OsVTC1-1	Os01g0847200	GDP-D-mannose pyrophosphorylase (GMPase) catalyzes the synthesis of GDP-D-mannose, positively regulates salt tolerance
OsVTC1-3	Os03g0268400	GDP-D-mannose pyrophosphorylase (GMPase) catalyzes the synthesis of GDP-D-mannose, positively regulates salt tolerance

6. Biotechnology Promises to Accelerate Breeding of Salt-Tolerant Rice Cultivars

Publication of the rice reference genome and the development of next-generation sequencing (NGS) techniques provide an opportunity to explore genome-wide genetic variations and carry out genotyping in a highly efficient way [86]. The relatively low cost of sequencing enables the use of genome and transcriptome sequencing to discover a large number of sequence polymorphisms and to map some agronomic traits [87]. Genome-wide association studies (GWAS) are a powerful approach for identifying valuable natural variations in trait-associated loci, as well as allelic variations in candidate genes underlying quantitative and complex traits, including those related to growth, salt tolerance, and nutritional quality [88,89]. Compared to a traditional QTL linkage analysis, GWAS is based on high-density variations in natural populations and can detect multiple alleles at the same site [90]. Several loci associated with salt tolerance have been identified in rice based on GWAS. Kumar et al. conducted a GWAS of 12 different salt tolerance-related traits at the reproductive stage and identified 20 single-nucleotide polymorphisms (SNPs) significantly associated with the Na$^+$/K$^+$ ratio and 44 SNPs associated with other traits observed under salt stress conditions [91]. Liu et al. used 708 rice accessions to perform a GWAS to identify the genes associated with rice salt tolerance and identified seven accessions carrying favorable haplotypes of four genes significantly associated with grain yield under salt stress. These promising candidates will provide valuable resources for salt-tolerant rice breeding [89]. Lekklar et al. conducted a GWAS for salt tolerance during rice reproduction, and more than 73% of the identified loci overlapped with the previously reported salt QTLs [92]. Yuan et al. performed a GWAS using 664 cultivated rice accessions from the 3000 Rice genomes, and twenty-one salt-tolerant QTLs and two candidate genes were identified [93]. These studies indicate that GWAS is a powerful strategy for mapping QTLs of salt tolerance in rice.

Effective phenotyping data (phenomics data) is a prerequisite for the discovery of genes/QTLs, association mapping, and GWAS. Despite recent advances in genomics, the lack of appropriate phenomics data limits the progress in genomics-assisted crop improvement programs. Therefore, the acquisition of high-throughput, effective, and comprehensive trait data in rice has become an acute need [94,95]. High-throughput phenotyping offers the opportunity capture phenotypically complex variations underpinning adaptation in traditional phenotypic selection or statistics-based breeding programs [94,95]. With the development of plant phenotyping platforms, we can obtain the phenomics data more effectively and cost-efficiently, and the integration of high-throughput trait phenotyping with genomics will greatly promote the genetic dissection of salt tolerance-related traits.

Genomic and transcriptomic analysis, proteomics, and metabolomics are powerful tools for identifying genes related to salt tolerance in rice. Salt stress causes a series of changes in rice, including changes in gene expression, protein content, and metabolite levels [96–98]. By comparing the transcriptome, proteome, and metabolome of rice under salt stress versus normal conditions, or in salt-tolerant versus salt-sensitive varieties, we can obtain many potential genes related to salt tolerance. For example, a study comparing transcriptome profiles of *FL478* and *IR29* found more than one thousand genes that were differentially expressed under salt stress compared to normal conditions [96]. Sun et al. analyzed the transcriptome data of a salt-tolerant rice landrace called *Changmaogu* and detected a large number of genes that were differentially expressed at the germination and seedling stages under salt stress. A further analysis revealed that most of the differentially expressed genes were clustered in the pathways of ABA signal transduction and carotenoid biosynthesis [97]. Peng et al. used an Isobaric Tags for Relative and Absolute Quantitation (iTRAQ)-based proteomic technique to detect proteins that become more or less abundant under salt stress and identified 332 differentially abundant proteins in seedlings of *salinity-tolerant and dwarf 58* (*sd58*) and *Kitaake* [98]. Although these studies provide many candidate genes for salinity tolerance, whether these genes can be used for breeding salt-tolerant rice remains to be verified. The transgenic approach and genome editing approach are powerful tools for verifying whether genes can be used for salt-tolerant rice breeding. Using the transgenic approach and genome editing approach, ideal materials for target genes can rapidly be obtained and used to assess the salt tolerance, which would accelerate efforts to improve the salinity tolerance in rice.

Based on high-density genome-wide SNP markers detected by next-generation sequencing, the SNP marker-assisted selection method is used to accelerate the process of breeding salt-tolerant rice. Using this method, Rana et al. precisely introgressed the *hitomebore salt tolerant 1* (*hst1*) gene from the salt-tolerant cultivar *Kaijin* into the high-yielding cultivar *Yukinko-mai*. Their offspring, *YNU31-2-4*, had agronomic traits similar to *Yukinko-mai* under normal growth conditions. Under salt stress, the yield of *YNU31-2-4* was significantly higher than that of *Yukinko-mai* [99]; Bimpong et al. used marker-assisted selection to develop salt-tolerant rice cultivars through introgressing *Saltol* into the lowland cultivar, *Rassi*, and obtained 16 introgression lines for further African-wide testing prior to release in six West African countries [100]. Pauyawaew et al. introgressed *Saltol* into *KDML105* by two rounds of marker-assisted backcrossing, and introgression lines with positive *Saltol* alleles are being tested for salinity tolerance in the salt-affected areas in the northeast of Thailand [101]. All these results suggest that SNP marker-assisted selection breeding is a fast and effective method to improve rice salt tolerance.

The most current methods of genomic selection use single SNP markers to predict the genetic merits of individuals, but haplotypes may have several advantages over single markers for genomic selection. A haplotype is the combination of a series of genetic mutations that coexist on a single chromosome; it contains multiple SNPs and can better capture the linkage disequilibrium and genomic similarity in different lines [102,103]. The use of haplotypes can improve the accuracy of genomic prediction. Thus, identifying key haplotypes related to the salt tolerance of rice will provide useful genetic resources and parent materials for breeding salt-tolerant rice.

7. Challenges and Perspectives

Salt stress is one of the main environmental factors affecting rice growth and rice yield worldwide. Improving the salt tolerance of rice is the most direct and effective method to solve this problem. After years of collection and screening, scientists have obtained several salt-tolerant rice germplasms (Table 2), which have laid the foundation for studying the mechanism of salt tolerance in rice and developing salt-tolerant rice cultivars. This review summarizes the progress to date on the breeding of salt-tolerant rice and proposes that genomics and molecular tools for precision breeding will accelerate the development of salt-tolerant cultivars.

Rice salt tolerance is controlled by multiple genes and is a complex physiological characteristic [17]. The evaluation of rice salt tolerance is also complex. Phenotypic traits are typically used to evaluate the salt tolerance of rice varieties [18], an approach that lacks comprehensiveness and accuracy and may result in an assessment of salt tolerance that is not consistent with actual performances in the field. Therefore, a standardized system to evaluate rice salt tolerance is urgently needed. Moreover, salt tolerance is genotype-dependent. Although numerous QTLs controlling salt tolerance traits have been identified in different mapping populations (Table 2), only a few major salt tolerance genes have been isolated from QTLs [48,49]. The inconsistency and variability of QTLs in different genetic backgrounds and environments have limited their applications in breeding programs.

Using traditional breeding methods, scientists have bred a series of salt-tolerant rice cultivars, but the mechanism of salt tolerance in rice is largely unclear. How plants perceive salt stress, how these signals are translated into adaptive responses, and how multiple salt tolerance genes coordinately regulate salt tolerance in rice are all questions that need further investigation. Moreover, traditional breeding methods are time-consuming and inefficient. SNP marker-assisted selection and genetic engineering technology will greatly improve the molecular breeding process [99,104]. Therefore, efforts should be made to capture useful salt tolerance genes as possible genetic markers to introgress into elite rice varieties. In addition, it is difficult to obtain salt-tolerant varieties that can be used in field production by the introduction of a single gene or several genes. To cultivate valuable salt-tolerant rice varieties, multiple salt tolerance genes or haplotypes need to be gathered into elite rice varieties; how to quickly and effectively introduce these genes or haplotypes simultaneously is another problem that needs to be solved. With the innovation of breeding technology, many new breeding technologies, such as MutMap, knock-out/in, and genome editing, will accelerate the process of salt-tolerant rice cultivation [83,105]. Further studies should focus on cloning salt tolerance genes and elucidating their regulatory mechanisms and on investigating how multiple genes or haplotypes can be transferred at the same time with stable inheritance by their offspring.

Due to variations in saline-alkali soils and environmental conditions in different areas, the cultivation of salt-tolerant rice varieties is subject to strong geographical constraints; varieties cultivated in one place are unsuitable for planting in another. Therefore, research efforts should focus on identifying molecular markers and haplotypes associated with salt tolerance and on developing salt-tolerant rice. The main rice varieties in various regions should be transformed with genes associated with improved salt tolerance by using a combination of molecular breeding and traditional breeding approaches. The resulting core rice germplasm with a high salt tolerance and excellent agronomic traits would provide a valuable germplasm resource for breeding programs aimed at developing salt-tolerant rice.

Author Contributions: All the authors H.Q., Y.L., and R.H. discussed and created the review's outline, H.Q. wrote the manuscript, and R.H. edited the manuscript. All authors have read and agreed to the published version of the manuscript.

Funding: This work was funded by the National Natural Science Foundation of China, grant numbers 32030079, 31801445, and 31871551, and the Agricultural Science and Technology Innovation Program of the Chinese Academy of Agricultural Sciences.

Conflicts of Interest: The authors declare no conflict of interest.

References

1. Zhu, J.K. Abiotic stress signaling and responses in plants. *Cell* **2016**, *167*, 313–324. [CrossRef] [PubMed]
2. Yang, Y.Q.; Guo, Y. Elucidating the molecular mechanisms mediating plant salt-stress responses. *New Phytol.* **2018**, *217*, 523–539. [CrossRef] [PubMed]
3. Yang, Y.; Guo, Y. Unraveling salt stress signaling in plants. *J. Integr. Plant Biol.* **2018**, *60*, 796–804. [CrossRef] [PubMed]
4. Rengasamy, P. World salinization with emphasis on Australia. *J. Exp. Bot.* **2006**, *57*, 1017–1023. [CrossRef] [PubMed]
5. Munns, R. Genes and salt tolerance: Bringing them together. *New Phytol.* **2005**, *167*, 645–663. [CrossRef] [PubMed]
6. Lobell, D.B.; Gourdji, S.M. The influence of climate change on global crop productivity. *Plant Physiol.* **2012**, *160*, 1686–1697. [CrossRef]
7. Chinnusamy, V.; Jagendorf, A.; Zhu, J.K. Understanding and improving salt tolerance in plants. *Crop. Sci.* **2005**, *45*, 437–448. [CrossRef]
8. Nam, M.H.; Bang, E.; Kwon, T.Y.; Kim, Y.; Kim, E.H.; Cho, K.; Park, W.J.; Kim, B.G.; Yoon, I.S. Metabolite profiling of diverse rice germplasm and identification of conserved metabolic markers of rice roots in response to long-term mild salinity stress. *Int. J. Mol. Sci.* **2015**, *16*, 21959–21974. [CrossRef]
9. Lee, K.S.; Choi, W.Y.; Ko, J.C.; Kim, T.S.; Gregorio, G.B. Salinity tolerance of *japonica* and *indica* rice (*Oryza sativa* L.) at the seedling stage. *Planta* **2003**, *216*, 1043–1046. [CrossRef] [PubMed]
10. Kurotani, K.; Yamanaka, K.; Toda, Y.; Ogawa, D.; Tanaka, M.; Kozawa, H.; Nakamura, H.; Hakata, M.; Ichikawa, H.; Hattori, T.; et al. Stress tolerance profiling of a collection of extant salt-tolerant rice varieties and transgenic plants overexpressing abiotic stress tolerance genes. *Plant Cell Physiol.* **2015**, *56*, 1867–1876. [CrossRef]
11. Radanielson, A.M.; Angeles, O.; Li, T.; Ismail, A.M.; Gaydon, D.S. Describing the physiological responses of different rice genotypes to salt stress using sigmoid and piecewise linear functions. *Field Crop. Res.* **2018**, *220*, 46–56. [CrossRef]
12. Tsai, Y.C.; Chen, K.C.; Cheng, T.S.; Lee, C.; Lin, S.H.; Tung, C.W. Chlorophyll fluorescence analysis in diverse rice varieties reveals the positive correlation between the seedlings salt tolerance and photosynthetic efficiency. *BMC Plant Biol.* **2019**, *19*, 403. [CrossRef]
13. Yang, C.; Lu, X.; Ma, B.; Chen, S.Y.; Zhang, J.S. Ethylene signaling in rice and *Arabidopsis*: Conserved and diverged aspects. *Mol. Plant* **2015**, *8*, 495–505. [CrossRef] [PubMed]
14. Yang, C.; Ma, B.; He, S.J.; Xiong, Q.; Duan, K.X.; Yin, C.C.; Chen, H.; Lu, X.; Chen, S.Y.; Zhang, J.S. Maohuzi6/ethylene insensitive3-like1 and ethylene insensitive3-like2 regulate ethylene response of roots and coleoptiles and negatively affect salt tolerance in rice. *Plant Physiol.* **2015**, *169*, 148–165. [CrossRef]
15. Van Zelm, E.; Zhang, Y.; Testerink, C. Salt tolerance mechanisms of plants. *Annu. Rev. Plant Biol.* **2020**, *71*, 1–31. [CrossRef]
16. Ganie, S.A.; Molla, K.A.; Henry, R.J.; Bhat, K.V.; Mondal, T.K. Advances in understanding salt tolerance in rice. *Theor. Appl. Genet.* **2019**, *132*, 851–870. [CrossRef]
17. Ismail, A.M.; Horie, T. Genomics, physiology, and molecular breeding approaches for improving salt tolerance. *Annu. Rev. Plant Biol.* **2017**, *68*, 405–434. [CrossRef] [PubMed]

18. Jaiswal, S.; Gautam, R.K.; Singh, R.K.; Krishnamurthy, S.L.; Ali, S.; Sakthivel, K.; Iquebal, M.A.; Rai, A.; Kumar, D. Harmonizing technological advances in phenomics and genomics for enhanced salt tolerance in rice from a practical perspective. *Rice* **2019**, *12*, 89. [CrossRef]

19. Krishnamurthy, S.L.; Sharma, P.C.; Sharma, S.K.; Batra, V.; Kumar, V.; Rao, L.V.S. Effect of salinity and use of stress indices of morphological and physiological traits at the seedling stage in rice. *Indian J. Exp. Biol.* **2016**, *54*, 843–850.

20. Ali, M.N.; Yeasmin, L.; Gantait, S.; Goswami, R.; Chakraborty, S. Screening of rice landraces for salinity tolerance at seedling stage through morphological and molecular markers. *Physiol. Mol. Biol. Plants* **2014**, *20*, 411–423. [CrossRef]

21. Chang, J.; Cheong, B.E.; Natera, S.; Roessner, U. Morphological and metabolic responses to salt stress of rice (*Oryza sativa* L.) cultivars which differ in salinity tolerance. *Plant Physiol. Biochem.* **2019**, *144*, 427–435. [CrossRef]

22. Gregorio, G.B.; Dharmawansa, S.; Mendoza, R.D. *Screening Rice for Salinity Tolerance*; IRRI Discussion Paper Series NO.22; International Rice Research Institute: Manila, Philippines, 1997; pp. 2–23.

23. Ismail, A.M.; Heuer, S.; Thomson, M.J.; Wissuwa, M. Genetic and genomic approaches to develop rice germplasm for problem soils. *Plant Mol. Biol.* **2007**, *65*, 547–570. [CrossRef]

24. Saini, S.; Kaur, N.; Pati, P.K. Reactive oxygen species dynamics in roots of salt sensitive and salt tolerant cultivars of rice. *Anal. Biochem.* **2018**, *550*, 99–108. [CrossRef]

25. Razzaq, A.; Ali, A.; Bin Safdar, L.; Zafar, M.M.; Rui, Y.; Shakeel, A.; Shaukat, A.; Ashraf, M.; Gong, W.K.; Yuan, Y.L. Salt stress induces physiochemical alterations in rice grain composition and quality. *J. Food Sci.* **2020**, *85*, 14–20. [CrossRef]

26. Lekklar, C.; Suriya-arunroj, D.; Pongpanich, M.; Comai, L.; Kositsup, B.; Chadchawan, S.; Buaboocha, T. Comparative genomic analysis of rice with contrasting photosynthesis and grain production under salt stress. *Genes* **2019**, *10*, 562. [CrossRef] [PubMed]

27. Calingacion, M.; Laborte, A.; Nelson, A.; Resurreccion, A.; Concepcion, J.C.; Daygon, V.D.; Mumm, R.; Reinke, R.; Dipti, S.; Bassinello, P.Z.; et al. Diversity of global rice markets and the science required for consumer-targeted rice breeding. *PLoS ONE* **2014**, *9*, e85106. [CrossRef]

28. Rao, P.S.; Mishra, B.; Gupta, S. Effects of soil salinity and alkalinity on grain quality of tolerant, semi-tolerant and sensitive rice genotypes. *Rice Sci.* **2013**, *20*, 284–291. [CrossRef]

29. Siscar-Lee, J.J.; Juliano, B.O.; Qureshi, R.H.; Akbar, M. Effect of saline soil on grain quality of rices differing in salinity tolerance. *Plant Foods Hum. Nutr.* **1990**, *40*, 31–36. [CrossRef] [PubMed]

30. Thitisaksakul, M.; Tananuwong, K.; Shoemaker, C.F.; Chun, A.; Tanadul, O.U.; Labavitch, J.M.; Beckles, D.M. Effects of timing and severity of salinity stress on rice (*Oryza sativa* L.) yield, grain composition, and starch functionality. *J. Agric. Food Chem.* **2015**, *63*, 2296–2304. [CrossRef]

31. Saleethong, P.; Sanitchon, J.; Kong-Ngern, K.; Theerakulpisut, P. Effects of exogenous spermidine (spd) on yield, yield-related parameters and mineral composition of rice (*Oryza sativa* L. ssp. 'indica') grains under salt stress. *Aust. J. Crop Sci.* **2013**, *7*, 1293–1301.

32. Verma, T.; Neue, H. Effect of soil salinity level and zinc application on growth, yield, and nutrient composition of rice. *Plant Soil* **1984**, *82*, 3–14. [CrossRef]

33. Ashraf, M. Stress-induced changes in wheat grain composition and quality. *Crit. Rev. Food Sci. Nutr.* **2014**, *54*, 1576–1583. [CrossRef]

34. Fernando, L.H. The performance of salt resistant paddy, Pokkali in Ceylon. *Trop. Agric.* **1949**, *105*, 124–126.

35. Ghose, R.L.M.; Butany, W.T. Botanical improvement of varieties-general characters of Indian varieties and the application of genetics to rice improvement. *Proc. Indian Acad. Sci. Sect. B* **1959**, *49*, 287–302.

36. Yeo, A.R.; Flowers, T.J. Salinity resistance in rice (*Oryza sativa* L.) and a pyramiding approach to breeding varieties for saline soils. *Aust. J. Plant Physiol.* **1986**, *13*, 161–173. [CrossRef]

37. Fageria, N.K. Salt tolerance of rice cultivars. *Plant Soil* **1985**, *88*, 237–243. [CrossRef]

38. Akbar, M.; Yabuno, Y.; Nakao, S. Breeding for saline resistant varieties of rice. I. Variability for salt-tolerance among some rice varieties. *Jpn. J. Breed.* **1972**, *22*, 277–284. [CrossRef]

39. Fageria, N.K.; Barbosa, M.P.; Gheyi, H.R. Tolerance of rice cultivars to salinty tolerance. *Pesqui. Agropecu. Bras.* **1981**, *16*, 677–681.

40. Bernstein, A.L.; Hayward, H.E. Physiology of salt tolerance. *Annu. Rev. Plant Physiol.* **1958**, *9*, 25–46. [CrossRef]

41. Heenan, D.P.; Lewin, L.G.; Mccaffery, D.W. Salinity tolerance in rice varieties at different growth stages. *Aust. J. Exp. Agric.* **1988**, *28*, 343–349. [CrossRef]

42. Sun, M.F.; Yan, G.H.; Wang, A.M.; Zhu, G.Y.; Tang, H.S.; He, C.X.; Ren, Z.L.; Liu, K.; Zhang, G.Y.; Shi, W.; et al. Research progress on the breeding of salt-tolerant rice varieties. *Barley Cereals Sci.* **2017**, *34*, 1–9.

43. Wang, C.L.; Zhang, Y.D.; Zhao, L.; Lu, K.; Zhu, Z.; Chen, T.; Zhao, Q.Y.; Yao, S.; Zhou, L.H.; Zhao, C.F.; et al. Research status, problems and suggestions on salt-alkali tolerant rice. *China Rice* **2019**, *25*, 1–6.

44. Platten, J.D.; Egdane, J.A.; Ismail, A.M. Salinity tolerance, Na$^+$ exclusion and allele mining of HKT1;5 in *Oryza sativa* and O-glaberrima: Many sources, many genes, one mechanism? *BMC Plant Biol.* **2013**, *13*, 32. [CrossRef] [PubMed]

45. Rahman, M.A.; Thomson, M.J.; Shah-E-Alam, M.; De Ocampo, M.; Egdane, J.; Ismail, A.M. Exploring novel genetic sources of salinity tolerance in rice through molecular and physiological characterization. *Ann. Bot. Lond.* **2016**, *117*, 1083–1097. [CrossRef]

46. Chen, R.S.; Cheng, Y.F.; Han, S.Y.; Van Handel, B.; Dong, L.; Li, X.M.; Xie, X.Q. Whole genome sequencing and comparative transcriptome analysis of a novel seawater adapted, salt-resistant rice cultivar—sea rice 86. *BMC Genom.* **2017**, *18*, 655. [CrossRef]

47. You, J.; Chan, Z.L. ROS regulation during abiotic stress responses in crop plants. *Front. Plant Sci.* **2015**, *6*, 1092. [CrossRef]

48. Lin, H.X.; Zhu, M.Z.; Yano, M.; Gao, J.P.; Liang, Z.W.; Su, W.A.; Hu, X.H.; Ren, Z.H.; Chao, D.Y. QTLs for Na$^+$ and K$^+$ uptake of the shoots and roots controlling rice salt tolerance. *Theor. Appl. Genet.* **2004**, *108*, 253–260. [CrossRef]

49. Ren, Z.H.; Gao, J.P.; Li, L.G.; Cai, X.L.; Huang, W.; Chao, D.Y.; Zhu, M.Z.; Wang, Z.Y.; Luan, S.; Lin, H.X. A rice quantitative trait locus for salt tolerance encodes a sodium transporter. *Nat. Genet.* **2005**, *37*, 1141–1146. [CrossRef]

50. Thomson, M.J.; de Ocampo, M.; Egdane, J.; Rahman, M.A.; Sajise, A.G.; Adorada, D.L.; Tumimbang-Raiz, E.; Blumwald, E.; Seraj, Z.I.; Singh, R.K.; et al. Characterizing the saltol quantitative trait locus for salinity tolerance in Rice. *Rice* **2010**, *3*, 148–160. [CrossRef]

51. He, Y.; Yang, B.; He, Y.; Zhan, C.; Cheng, Y.; Zhang, J.; Zhang, H.; Cheng, J.; Wang, Z. A quantitative trait locus, qSE3, promotes seed germination and seedling establishment under salinity stress in rice. *Plant J.* **2019**, *97*, 1089–1104. [CrossRef] [PubMed]

52. Lee, S.Y.; Ahn, J.H.; Cha, Y.S.; Yun, D.W.; Lee, M.C.; Ko, J.C.; Lee, K.S.; Eun, M.Y. Mapping of quantitative trait loci for salt tolerance at the seedling stage in rice. *Mol. Cells* **2006**, *21*, 192–196.

53. Wu, F.L.; Yang, J.; Yu, D.Q.; Xu, P. Identification and validation a major QTL from "Sea Rice 86" seedlings conferred salt tolerance. *Agronomy* **2020**, *10*, 410. [CrossRef]

54. Rahman, M.A.; Thomson, M.J.; De Ocampo, M.; Egdane, J.A.; Salam, M.A.; Shah-E-Alam, M.; Ismail, A.M. Assessing trait contribution and mapping novel QTL for salinity tolerance using the Bangladeshi rice landrace Capsule. *Rice* **2019**, *12*, 63. [CrossRef]

55. Quan, R.D.; Wang, J.; Hui, J.; Bai, H.B.; Lyu, X.L.; Zhu, Y.X.; Zhang, H.W.; Zhang, Z.J.; Li, S.H.; Huang, R.F. Improvement of salt tolerance using wild rice genes. *Front. Plant Sci.* **2018**, *8*, 2269. [CrossRef]

56. Bizimana, J.B.; Luzi-Kihupi, A.; Murori, R.W.; Singh, R.K. Identification of quantitative trait loci for salinity tolerance in rice (*Oryza sativa* L.) using IR29/Hasawi mapping population. *J. Genet.* **2017**, *96*, 571–582. [CrossRef]

57. Ammar, M.H.M.; Pandit, A.; Singh, R.K.; Sameena, S.; Chauhan, M.S.; Singh, A.K.; Sharma, P.C.; Gaikwad, K.; Sharma, T.R.; Mohapatra, T.; et al. Mapping of QTLs controlling Na$^+$, K$^+$ and Cl$^-$ ion concentrations in salt tolerant *Indica* rice variety CSR27. *J. Plant Biochem. Biotechnol.* **2009**, *18*, 139–150. [CrossRef]

58. Cheng, L.R.; Wang, Y.; Meng, L.J.; Hu, X.; Cui, Y.R.; Sun, Y.; Zhu, L.H.; Ali, J.; Xu, J.L.; Li, Z.K. Identification of salt-tolerant QTLs with strong genetic background effect using two sets of reciprocal introgression lines in rice. *Genome* **2012**, *55*, 45–55. [CrossRef]

59. Dahanayaka, B.A.; Gimhani, D.; Kottearachchi, N.; Samarasighe, W.L.G. QTL mapping for salinity tolerance using an elite rice (*Oryza sativa*) breeding population. *SABRAO J. Breed. Genet.* **2017**, *49*, 123–134.

60. De Leon, T.B.; Linscombe, S.; Subudhi, P.K. Identification and validation of QTLs for seedling salinity tolerance in introgression lines of a salt tolerant rice landrace 'Pokkali'. *PLoS ONE* **2017**, *12*, e0175361. [CrossRef]

61. Hossain, H.; Rahman, M.A.; Alam, M.S.; Singh, R.K. Mapping of quantitative trait loci associated with reproductive-stage salt tolerance in rice. *J. Agron. Crop Sci.* **2015**, *201*, 17–31. [CrossRef]
62. Tian, L.; Tan, L.B.; Liu, F.X.; Cai, H.W.; Sun, C.Q. Identification of quantitative trait loci associated with salt tolerance at seedling stage from *Oryza rufipogon*. *J. Genet. Genom.* **2011**, *38*, 593–601. [CrossRef]
63. Pandit, A.; Rai, V.; Bal, S.; Sinha, S.; Kumar, V.; Chauhan, M.; Gautam, R.K.; Singh, R.; Sharma, P.C.; Singh, A.K.; et al. Combining QTL mapping and transcriptome profiling of bulked RILs for identification of functional polymorphism for salt tolerance genes in rice (*Oryza sativa* L.). *Mol. Genet. Genom.* **2010**, *284*, 121–136. [CrossRef]
64. Puram, V.R.R.; Ontoy, J.; Linscombe, S.; Subudhi, P.K. Genetic dissection of seedling stage salinity tolerance in rice using introgression lines of a salt tolerant landrace Nona Bokra. *J. Hered.* **2017**, *108*, 658–670. [CrossRef] [PubMed]
65. Qiu, X.J.; Yuan, Z.H.; Liu, H.; Xiang, X.J.; Yang, L.W.; He, W.J.; Du, B.; Ye, G.Y.; Xu, J.L.; Xing, D.Y. Identification of salt tolerance-improving quantitative trait loci alleles from a salt-susceptible rice breeding line by introgression breeding. *Plant Breed.* **2015**, *134*, 653–660. [CrossRef]
66. Sabouri, H.R.; Moumeni, A.; Kavousi, A.; Katouzi, M.; Sabouri, A. QTLs mapping of physiologicaltraits related to salt tolerance in young rice seedlings. *Biol. Plant.* **2009**, *53*, 657–662. [CrossRef]
67. Wang, Z.F.; Cheng, J.P.; Chen, Z.W.; Huang, J.; Bao, Y.M.; Wang, J.F.; Zhang, H.S. Identification of QTLs with main, epistatic and QTL x environment interaction effects for salt tolerance in rice seedlings under different salinity conditions. *Theor. Appl. Genet.* **2012**, *125*, 807–815. [CrossRef]
68. Wang, Z.F.; Wang, J.F.; Bao, Y.M.; Wu, Y.Y.; Zhang, H.S. Quantitative trait loci controlling rice seed germination under salt stress. *Euphytica* **2011**, *178*, 297–307. [CrossRef]
69. Wang, S.S.; Cao, M.; Ma, X.; Chen, W.K.; Zhao, J.; Sun, C.Q.; Tan, L.B.; Liu, F.X. Integrated RNA sequencing and QTL mapping to identify candidate genes from *Oryza rufipogon* associated with salt tolerance at the seedling stage. *Front. Plant Sci.* **2017**, *8*, 1427. [CrossRef]
70. Zheng, H.L.; Zhao, H.W.; Liu, H.L.; Wang, J.G.; Zou, D.T. QTL analysis of Na^+ and K^+ concentrations in shoots and roots under NaCl stress based on linkage and association analysis in japonica rice. *Euphytica* **2015**, *201*, 109–121. [CrossRef]
71. Wang, G.F.; Li, W.Q.; Li, W.Y.; Wu, G.L.; Zhou, C.Y.; Chen, K.M. Characterization of rice NADPH oxidase genes and their expression under various environmental conditions. *Int. J. Mol. Sci.* **2013**, *14*, 9440–9458. [CrossRef]
72. Asada, K. The water-water cycle in chloroplasts: Scavenging of active oxygens and dissipation of excess photons. *Annu. Rev. Plant Phys.* **1999**, *50*, 601–639. [CrossRef]
73. Zhang, Z.G.; Zhang, Q.; Wu, J.X.; Zheng, X.; Zheng, S.; Sun, X.H.; Qiu, Q.S.; Lu, T.G. Gene knockout study reveals that cytosolic ascorbate peroxidase 2 (OsAPX2) plays a critical role in growth and reproduction in rice under drought, salt and cold stresses. *PLoS ONE* **2013**, *8*, e57472. [CrossRef]
74. Hong, C.Y.; Hsu, Y.T.; Tsai, Y.C.; Kao, C.H. Expression of ASCORBATE PEROXIDASE 8 in roots of rice (*Oryza sativa* L.) seedlings in response to NaCl. *J. Exp. Bot.* **2007**, *58*, 3273–3283. [CrossRef] [PubMed]
75. Teixeira, F.K.; Menezes-Benavente, L.; Galvao, V.C.; Margis, R.; Margis-Pinheiro, M. Rice ascorbate peroxidase gene family encodes functionally diverse isoforms localized in different subcellular compartments. *Planta* **2006**, *224*, 300–314. [CrossRef] [PubMed]
76. Qin, H.; Deng, Z.; Zhang, C.; Wang, Y.; Wang, J.; Liu, H.; Zhang, Z.; Huang, R.; Zhang, Z. Rice GDP-mannose pyrophosphorylase OsVTC1-1 and OsVTC1-3 play different roles in ascorbic acid synthesis. *Plant Mol. Biol.* **2016**, *90*, 317–327. [CrossRef]
77. Qin, H.; Wang, Y.; Wang, J.; Liu, H.; Zhao, H.; Deng, Z.; Zhang, Z.; Huang, R.; Zhang, Z. Knocking down the expression of GMPase gene OsVTC1-1 decreases salt tolerance of rice at seedling and reproductive stages. *PLoS ONE* **2016**, *11*, e0168650. [CrossRef]
78. Wang, Y.; Zhao, H.; Qin, H.; Li, Z.; Liu, H.; Wang, J.; Zhang, H.; Quan, R.; Huang, R.; Zhang, Z. The synthesis of ascorbic acid in rice roots plays an important role in the salt tolerance of rice by scavenging ROS. *Int. J. Mol. Sci.* **2018**, *19*, 3347. [CrossRef]
79. Blumwald, E. Sodium transport and salt tolerance in plants. *Curr. Opin. Cell Biol.* **2000**, *12*, 431–434. [CrossRef]

80. Campbell, M.T.; Bandillo, N.; Al Shiblawi, F.R.A.; Sharma, S.; Liu, K.; Du, Q.; Schmitz, A.J.; Zhang, C.; Very, A.A.; Lorenz, A.J.; et al. Allelic variants of OsHKT1;1 underlie the divergence between indica and japonica subspecies of rice (*Oryza sativa*) for root sodium content. *PLoS Genet.* **2017**, *13*, e1006823. [CrossRef] [PubMed]

81. Suzuki, K.; Yamaji, N.; Costa, A.; Okuma, E.; Kobayashi, N.I.; Kashiwagi, T.; Katsuhara, M.; Wang, C.; Tanoi, K.; Murata, Y.; et al. OsHKT1;4-mediated Na$^+$ transport in stems contributes to Na$^+$ exclusion from leaf blades of rice at the reproductive growth stage upon salt stress. *BMC Plant Biol.* **2016**, *16*, 22. [CrossRef] [PubMed]

82. Jadamba, C.; Kang, K.; Paek, N.C.; Lee, S.I.; Yoo, S.C. Overexpression of rice expansin7 (Osexpa7) confers enhanced tolerance to salt stress in rice. *Int. J. Mol. Sci.* **2020**, *21*, 454. [CrossRef]

83. Takagi, H.; Tamiru, M.; Abe, A.; Yoshida, K.; Uemura, A.; Yaegashi, H.; Obara, T.; Oikawa, K.; Utsushi, H.; Kanzaki, E.; et al. MutMap accelerates breeding of a salt-tolerant rice cultivar. *Nat. Biotechnol.* **2015**, *33*, 445–449. [CrossRef]

84. Zhou, J.H.; Li, Z.Y.; Xiao, G.Q.; Zhai, M.J.; Pan, X.W.; Huang, R.F.; Zhang, H.W. CYP71D8L is a key regulator involved in growth and stress responses by mediating gibberellin homeostasis in rice. *J. Exp. Bot.* **2020**, *71*, 1160–1170.

85. Wang, J.; Qin, H.; Zhou, S.R.; Wei, P.C.; Zhang, H.W.; Zhou, Y.; Miao, Y.C.; Huang, R.F. The ubiquitin-binding protein OsDSK2a mediates seedling growth and salt responses by regulating gibberellin metabolism in rice. *Plant Cell* **2020**, *32*, 414–428. [CrossRef]

86. Huang, X.H.; Lu, T.T.; Han, B. Resequencing rice genomes: An emerging new era of rice genomics. *Trends Genet.* **2013**, *29*, 225–232. [CrossRef] [PubMed]

87. Wang, W.S.; Mauleon, R.; Hu, Z.Q.; Chebotarov, D.; Tai, S.S.; Wu, Z.C.; Li, M.; Zheng, T.Q.; Fuentes, R.R.; Zhang, F.; et al. Genomic variation in 3,010 diverse accessions of Asian cultivated rice. *Nature* **2018**, *557*, 43–49. [CrossRef]

88. Li, N.; Zheng, H.L.; Cui, J.N.; Wang, J.G.; Liu, H.L.; Sun, J.; Liu, T.T.; Zhao, H.W.; Lai, Y.C.; Zou, D.T. Genome-wide association study and candidate gene analysis of alkalinity tolerance in japonica rice germplasm at the seedling stage. *Rice* **2019**, *12*, 24. [CrossRef]

89. Liu, C.; Chen, K.; Zhao, X.Q.; Wang, X.Q.; Shen, C.C.; Zhu, Y.J.; Dai, M.L.; Qiu, X.J.; Yang, R.W.; Xing, D.Y.; et al. Identification of genes for salt tolerance and yield-related traits in rice plants grown hydroponically and under saline field conditions by genome-wide association study. *Rice* **2019**, *12*, 88. [CrossRef]

90. Flint-Garcia, S.A.; Thornsberry, J.M.; Buckler, E.S. Structure of linkage disequilibrium in plants. *Annu. Rev. Plant Biol.* **2003**, *54*, 357–374. [CrossRef]

91. Kumar, V.; Singh, A.; Mithra, S.V.A.; Krishnamurthy, S.L.; Parida, S.K.; Jain, S.; Tiwari, K.K.; Kumar, P.; Rao, A.R.; Sharma, S.K.; et al. Genome-wide association mapping of salinity tolerance in rice (*Oryza sativa*). *DNA Res.* **2015**, *22*, 133–145. [CrossRef]

92. Lekklar, C.; Pongpanich, M.; Suriya-arunroj, D.; Chinpongpanich, A.; Tsai, H.; Comai, L.; Chadchawan, S.; Buaboocha, T. Genome-wide association study for salinity tolerance at the flowering stage in a panel of rice accessions from Thailand. *BMC Genom.* **2019**, *20*, 76. [CrossRef] [PubMed]

93. Yuan, J.; Wang, X.; Zhao, Y.; Khan, N.U.; Zhao, Z.; Zhang, Y.; Wen, X.; Tang, F.; Wang, F.; Li, Z. Genetic basis and identification of candidate genes for salt tolerance in rice by GWAS. *Sci. Rep.* **2020**, *10*, 9958. [CrossRef]

94. Mir, R.R.; Reynolds, M.; Pinto, F.; Khan, M.A.; Bhat, M.A. High-throughput phenotyping for crop improvement in the genomics era. *Plant Sci.* **2019**, *282*, 60–72. [CrossRef]

95. Rebetzke, G.J.; Jimenez-Berni, J.; Fischer, R.A.; Deery, D.M.; Smith, D.J. Review: High-throughput phenotyping to enhance the use of crop genetic resources. *Plant Sci.* **2019**, *282*, 40–48. [CrossRef]

96. Mansuri, R.M.; Shobbar, Z.S.; Jelodar, N.B.; Ghaffari, M.R.; Nematzadeh, G.A.; Asari, S. Dissecting molecular mechanisms underlying salt tolerance in rice: A comparative transcriptional profiling of the contrasting genotypes. *Rice* **2019**, *12*, 13. [CrossRef]

97. Sun, B.R.; Fu, C.Y.; Fan, Z.L.; Chen, Y.; Chen, W.F.; Zhang, J.; Jiang, L.Q.; Lv, S.; Pan, D.J.; Li, C. Genomic and transcriptomic analysis reveal molecular basis of salinity tolerance in a novel strong salt-tolerant rice landrace Changmaogu. *Rice* **2019**, *12*, 99. [CrossRef]

98. Peng, P.; Gao, Y.D.; Li, Z.; Yu, Y.W.; Qin, H.; Guo, Y.; Huang, R.F.; Wang, J. Proteomic analysis of a rice mutant *sd58* possessing a novel *d1* allele of heterotrimeric G protein alpha subunit (RGA1) in salt stress with a focus on ROS scavenging. *Int. J. Mol. Sci.* **2019**, *20*, 167. [CrossRef]

99. Rana, M.M.; Takamatsu, T.; Baslam, M.; Kaneko, K.; Itoh, K.; Harada, N.; Sugiyama, T.; Ohnishi, T.; Kinoshita, T.; Takagi, H.; et al. Salt tolerance improvement in rice through efficient SNP marker-assisted selection coupled with speed-breeding. *Int. J. Mol. Sci.* **2019**, *20*, 2585. [CrossRef] [PubMed]

100. Bimpong, I.K.; Manneh, B.; Sock, M.; Diaw, F.; Amoah, N.K.A.; Ismail, A.M.; Gregorio, G.; Singh, R.K.; Wopereis, M. Improving salt tolerance of lowland rice cultivar 'Rassi' through marker-aided backcross breeding in West Africa. *Plant Sci.* **2016**, *242*, 288–299. [CrossRef] [PubMed]

101. Punyawaew, K.; Suriya-Arunroj, D.; Siangliw, M.; Thida, M.; Lanceras-Siangliw, J.; Fukai, S.; Toojinda, T. Thai jasmine rice cultivar KDML105 carrying Saltol QTL exhibiting salinity tolerance at seedling stage. *Mol. Breed.* **2016**, *36*, 150. [CrossRef]

102. Lu, X.K.; Fu, X.Q.; Wang, D.L.; Wang, J.Y.; Chen, X.G.; Hao, M.R.; Wang, J.J.; Gervers, K.A.; Guo, L.X.; Wang, S.; et al. Resequencing of cv CRI-12 family reveals haplotype block inheritance and recombination of agronomically important genes in artificial selection. *Plant Biotechnol. J.* **2019**, *17*, 945–955. [CrossRef] [PubMed]

103. Maldonado, C.; Mora, F.; Scapim, C.A.; Coan, M. Genome-wide haplotype-based association analysis of key traits of plant lodging and architecture of maize identifies major determinants for leaf angle: HapLA4. *PLoS ONE* **2019**, *14*, e0212925. [CrossRef]

104. Gimhani, D.R.; Gregorio, G.B.; Kottearachchi, N.S.; Samarasinghe, W.L. SNP-based discovery of salinity-tolerant QTLs in a bi-parental population of rice (*Oryza sativa*). *Mol. Genet. Genom.* **2016**, *291*, 2081–2099. [CrossRef] [PubMed]

105. Chen, K.; Wang, Y.; Zhang, R.; Zhang, H.; Gao, C. CRISPR/Cas genome editing and precision plant breeding in agriculture. *Annu. Rev. Plant Biol.* **2019**, *70*, 667–697. [CrossRef]

International Journal of
Molecular Sciences

Article

The bZIP Transcription Factor GmbZIP15 Negatively Regulates Salt- and Drought-Stress Responses in Soybean

Man Zhang [1], Yanhui Liu [1], Hanyang Cai [1], Mingliang Guo [1], Mengnan Chai [1], Zeyuan She [2], Li Ye [1], Yan Cheng [1], Bingrui Wang [3,*] and Yuan Qin [1,2,*]

[1] Key Lab of Genetics, Breeding and Multiple Utilization of Crops, Ministry of Education, State Key Laboratory of Ecological Pest Control for Fujian and Taiwan Crops, Fujian Provincial Key Laboratory of Haixia Applied Plant Systems Biology, Center for Genomics and Biotechnology, College of Plant Protection, College of Life Sciences, College of Crop Science, Fujian Agriculture and Forestry University, Fuzhou 350002, China; zhangman3043@163.com (M.Z.); yanhuiliu520@gmail.com (Y.L.); caihanyang123@163.com (H.C.); gml604@163.com (M.G.); chaimengnan1@163.com (M.C.); 1180204044@fafu.edu.cn (L.Y.); chengyan1220@hotmail.com (Y.C.)

[2] State Key Laboratory for Conservation and Utilization of Subtropical Agro-Bioresources, Guangxi Key Lab of Sugarcane Biology, College of Agriculture, Guangxi University, Nanning 530004, China; szy0809@yeah.net

[3] College of Plant Science & Technology, Huazhong Agricultural University, Wuhan 430070, China

* Correspondence: brwang@mail.hzau.edu.cn (B.W.); yuanqin@fafu.edu.cn (Y.Q.)

Received: 22 September 2020; Accepted: 18 October 2020; Published: 21 October 2020

Abstract: Soybean (*Glycine max*), as an important oilseed crop, is constantly threatened by abiotic stress, including that caused by salinity and drought. bZIP transcription factors (TFs) are one of the largest TF families and have been shown to be associated with various environmental-stress tolerances among species; however, their function in abiotic-stress response in soybean remains poorly understood. Here, we characterized the roles of soybean transcription factor GmbZIP15 in response to abiotic stresses. The transcript level of *GmbZIP15* was suppressed under salt- and drought-stress conditions. Overexpression of *GmbZIP15* in soybean resulted in hypersensitivity to abiotic stress compared with wild-type (WT) plants, which was associated with lower transcript levels of stress-responsive genes involved in both abscisic acid (ABA)-dependent and ABA-independent pathways, defective stomatal aperture regulation, and reduced antioxidant enzyme activities. Furthermore, plants expressing a functional repressor form of *GmbZIP15* exhibited drought-stress resistance similar to WT. RNA-seq and qRT-PCR analyses revealed that *GmbZIP15* positively regulates *GmSAHH1* expression and negatively regulates *GmWRKY12* and *GmABF1* expression in response to abiotic stress. Overall, these data indicate that GmbZIP15 functions as a negative regulator in response to salt and drought stresses.

Keywords: GmbZIP15; transcription factor; salt stress; drought stress; RNA-seq; soybean

---

1. Introduction

As a result of their sessile nature, plants are subject to variable biotic and abiotic-stress conditions. As the most pertinent abiotic-stress conditions, drought and salinity threaten the growth and productivity of crops. Plants respond and adapt to these stress conditions by activating stress-related pathways, which comprise signal perception and transduction, regulation of gene expression, and biochemical and physiological responses [1]. Signaling pathways, including those involving various phytohormones, multiple secondary metabolism processes, and reactive oxygen species (ROS), are crucial for plant survival under environmental-stress conditions [2–4].

Abiotic stress usually leads to the generation of ROS such as hydrogen peroxide (H_2O_2) and superoxide (O^{2-}); however, ROS overaccumulation is cytotoxic [5,6]. To control the level of ROS accumulation under stress conditions, plants have evolved a wide range of antioxidants to scavenge ROS and reinstate cellular redox homeostasis. Previous studies have found that ROS signaling is linked to abscisic acid (ABA), Ca^{2+} fluxes, and sugar sensing, and is likely to be involved in ABA-dependent signaling pathways activated under abiotic stress [5,7]. Overexpression of *GmSIN1* in soybean (*Glycine max*) promotes root growth and salt tolerance by enhancing cellular ABA and ROS contents [7]. In rice (*Oryza sativa*), OsCPK12 promotes salt-stress resistance, likely through repression of ROS production and/or the participation of the ABA signaling pathway [8].

ABA is one of the most important stress-related phytohormones and plays a pivotal role in signal transduction during abiotic-stress responses [8,9]. Cellular ABA levels in plants increase in response to abiotic stress, leading to the expression of stress-responsive genes, the regulation of metabolic processes, and the quenching of ROS, thereby maintaining plant cell homeostasis under stress conditions [10]. Recent studies have found that plants respond to abiotic-stress conditions mainly through ABA-dependent and ABA-independent signaling pathways, which are regulated by AREB/ABFs and DREB2A transcription factors (TFs), respectively [11]. In *Arabidopsis* under abiotic-stress conditions, ABA enacts responses primarily through four bZIP TFs, namely, ABF1, AREB1/ABF2, AREB2/ABF4, and ABF3, the expression of which is activated by SnRK2s [10–14]. Abiotic stress and ABA has also been shown to induce the expression of *AREB1/ABF2*, *AREB2/ABF4*, and *ABF3* [15], and an increased abundance of these AREB/ABFs in turn increases ABA sensitivity and abiotic-stress resistance [13,16]. *ABF1* has also been shown to function in drought-stress responses, despite its lower expression level compared to other abiotic-stress-induced AREB/ABFs, and thus *areb1areb2abf3abf1* plants display decreased drought resistance compared to *areb1areb2abf3* plants regarding primary root growth [10]. DREB2 proteins are members of the AP2/ERF family of plant-specific TFs and function in an ABA-independent manner [11]. Among the DREB2 genes in *Arabidopsis*, DREB2A is largely induced by drought, salinity, and cold stress [17]. Three DREB homologues, namely, *GmDREBa*, *GmDREBb*, and *GmDREBc*, have been identified in the soybean genome and the transcript levels of *GmDREBa* and *GmDREBb* in the leaves of soybean seedlings were shown to increase following salt-, drought-, and cold-stress treatment [18].

Transcription factors are considered to be the most important regulators of gene expression. Several groups of TFs, such as DREB, NAC, MYB, WRKY, and bZIP, are responsible for abiotic-stress responses [19–23]. The bZIP TFs, which represent one of the largest plant TF families, can be divided into different subfamilies depending on the bZIP domain [24]. Plant bZIP TFs play crucial regulatory roles in multiple abiotic-stress resistances [12,25,26]. In *Arabidopsis*, ABI5 expression is regulated by *ABF3*, which may contribute to salt-stress tolerance [14]. In rice, OsABF1 improves drought tolerance by activating the transcription of *COR413-TM1* [12]. Overexpression of the sweet potato (*Ipomoea batatas*) TF *IbABF4* in *Arabidopsis* resulted in increased ABA sensitivity as well as enhanced drought and salt-stress tolerance [27].

Soybean is one of the most important crops and is widely cultivated worldwide because of its nutritive value. However, soybean growth is threatened by many abiotic-stress factors such as salinity, drought, and extreme temperature. In a previous study, 160 bZIP family members were identified from the soybean genome and were divided into 12 subgroups [24]. Among these, many family members have been characterized to play roles in abiotic-stress responses, including *GmbZIP132*, *GmbZIP110*, *GmbZIP44*, *GmbZIP62*, and *GmbZIP78* [28–30]. However, the function of GmbZIP15, the only member of subfamily K, in response to abiotic stress remains poorly understood. In this study, biochemical and physiological analyses were performed to reveal the regulatory roles of GmbZIP15 in abiotic-stress responses.

2. Results

2.1. GmbZIP15 Expression Pattern in Response to Abiotic-Stress Conditions

We previously identified 160 bZIP genes from soybean and characterized their expression in response to abiotic stress [24]. Among these genes, the expression of *GmbZIP15* was suppressed by drought and flooding stress and it was therefore selected for further investigation. To validate the response of *GmbZIP15* to abiotic-stress conditions, we generated soybean plants overexpressing GmbZIP15 (*OX-GmbZIP15*), and two lines (OE-8, OE-16) with higher expression level were selected for further research (Figure S1). The qRT-PCR analysis was used to determine *GmbZIP15* expression patterns in 2-week-old wild-type (WT) and two overexpression lines under salt and drought treatments. Under normal conditions, *GmbZIP15* expression level in *OX-GmbZIP15* plants was clearly higher than that in WT; however, *GmbZIP15* expression in *OX-GmbZIP15* plants treated with NaCl and mannitol sharply decreased, but was higher than WT plants by 12 h and was nearly undetectable by 24 h post-treatment (Figure S2A,B). In addition, through the GUS staining of 1-week-old *pGmbZIP15:GUS* transgenic *Arabidopsis* seedlings grown on media supplemented with 100 mM NaCl or 200 mM mannitol, *GmbZIP15* promoter activity was observed to be significantly decreased in cotyledons and true leaves under NaCl and mannitol conditions compared to normal conditions (Figure S2C), which was consistent with the results of qRT-PCR (Figure S1A,B). These results suggest that the expression of *GmbZIP15* is suppressed by abiotic stress.

2.2. GmbZIP15 Negatively Regulates Salt and Drought Tolerance in Soybean

To investigate the role of GmbZIP15 in plant response to salt stress, transgenic soybean plants carrying functional repression of *GmbZIP15* (*35S:GmbZIP15-SRDX*) were obtained and two lines (SRDX-15, SRDX-21) with higher expression levels were selected for further research (Figure S1). WT, *OX-GmbZIP15* and *35S:GmbZIP15-SRDX* seedlings were treated with 200 mM NaCl. Following 2 weeks of salt treatment, WT and *35S:GmbZIP15-SRDX* soybean seedlings exhibited a comparable degree of leaf shedding, whereas *OX-GmbZIP15* plants displayed a severe, almost lethal wilt phenotype (Figure 1A). These results suggest that overexpression of *GmbZIP15* in soybean causes sensitivity to salt stress.

To test the function of GmbZIP15 in plant drought responses, WT, *OX-GmbZIP15* and *35S:GmbZIP15-SRDX* seedlings were withheld water for 2 weeks. Compared to WT and *35S:GmbZIP15-SRDX* seedlings, *OX-GmbZIP15* seedlings were severely wilted and almost all leaves displayed a considerable dehydration phenotype (Figure 2A). To test whether the dehydration phenotype could be rescued, we rewatered the drought-treated plants for 3 days. Although there was slight shedding of the old leaves of WT and *35S:GmbZIP15-SRDX* seedlings after rewatering, there was vigorous new leaf growth, suggesting that WT and *35S:GmbZIP15-SRDX* plants can recover from such a dehydration phenotype. In contrast, *OX-GmbZIP15* seedlings did not display growth recovery after rewatering. Combined, these results further support that GmbZIP15 acts as a potential negative regulator of abiotic-stress response.

In addition, the transcription levels of drought-/salt-tolerance marker genes, including *GmDREBb*, *GmMYB118*, and *GmWRKY28*, were evaluated by qRT-PCR in WT and *OX-GmbZIP15* plants. Under mock conditions, the expression levels of each marker were considerably higher in *OX-GmbZIP15* plants compared to WT plants (Figure S3A,B). However, under salt stress, the expression levels of *GmDREBb*, *GmMYB118*, and *GmWRKY28* dramatically decreased in *OX-GmbZIP15* plants compared with that in WT plants (Figure S3A). Similarly, under drought stress, the expression levels of *GmDREBb*, *GmMYB118*, and *GmWRKY28* genes in *OX-GmbZIP15* plants also decreased significantly compared with that in WT plants (Figure S3B). These results indicate that *GmbZIP15*-overexpressing soybean plants are hypersensitive to abiotic stress.

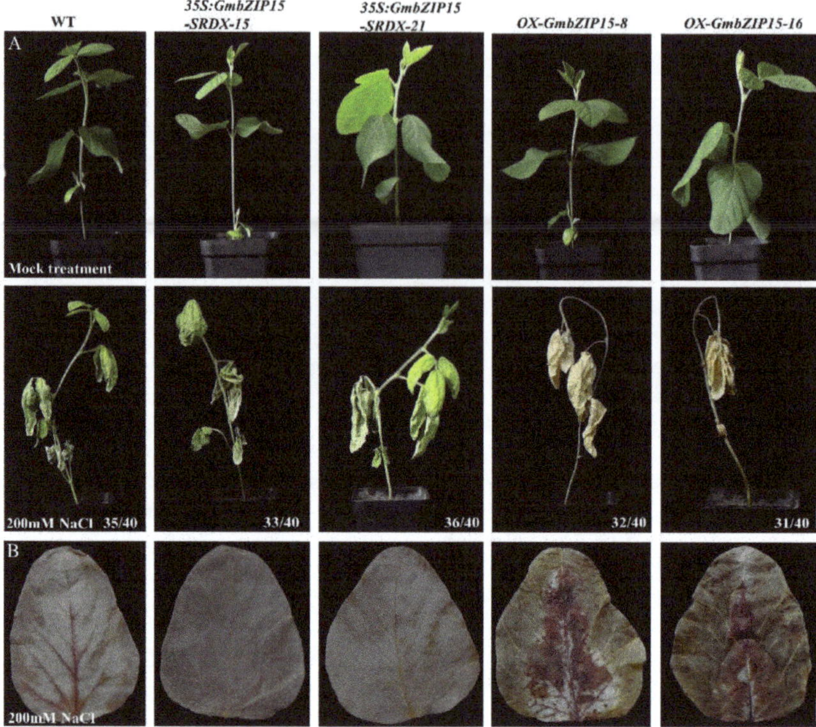

Figure 1. *GmbZIP15* negatively regulates salt-stress resistance in soybean. (**A**) Phenotype observation of transgenic soybean seedlings in response to salt stress. The pictures were obtained before or after 200 mM NaCl treatment for 2 weeks. Numbers in the panels denote the frequencies of the phenotypes shown. (**B**) Diaminobenzidine (DAB) staining of the soybean leaves. All the plants were treated with 200 mM NaCl for 4 days and then the leaves were harvested. The depth of color shows the H_2O_2 content in leaves. Bar = 1 cm.

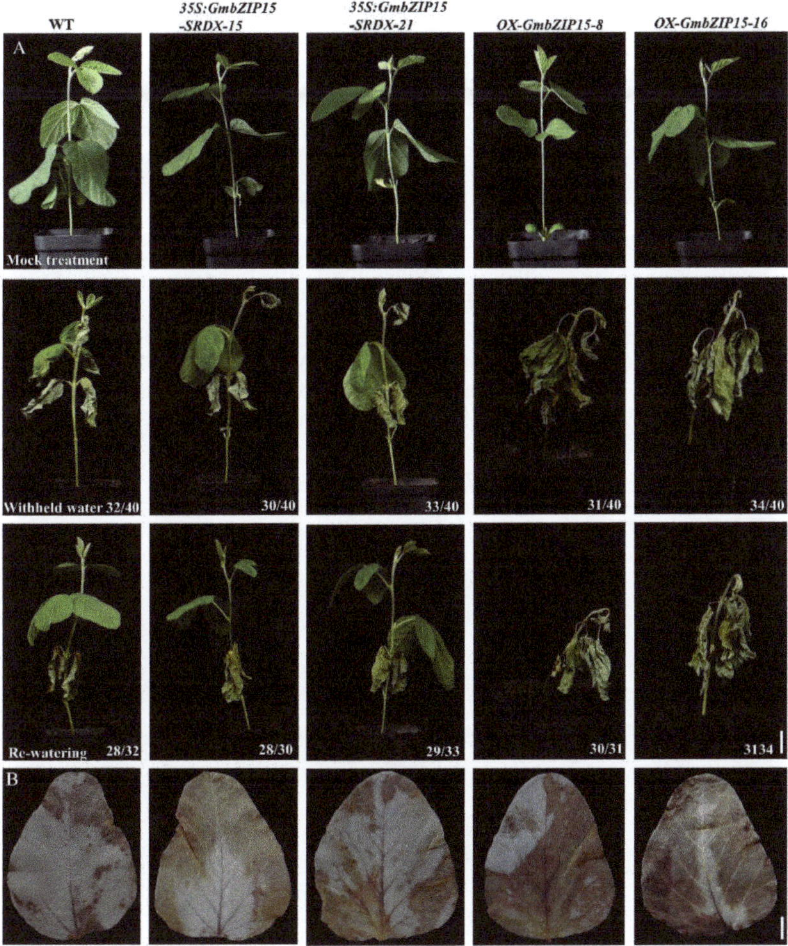

Figure 2. *GmbZIP15* negatively regulates drought-stress resistance in soybean. (**A**) Phenotype observation of transgenic soybean seedlings in response to drought stress. The pictures were obtained under normal conditions; thereafter, the plants were not watered for 2 weeks, then rewatered for 3 days. Numbers in the panels denote the frequencies of the phenotypes shown. (**B**) DAB staining of the soybean leaves. All the plants were not watered for 4 days and then the leaves were harvested. The depth of color shows the H_2O_2 content in the leaves. Bar = 1 cm.

2.3. GmbZIP15 Depresses the ROS Scavenging Ability of Soybean

Abiotic stress can lead to damage to plant cells via oxidative stress involving the generation of ROS [31]. Diaminobenzidine (DAB) staining showed that H_2O_2 levels were largely increased in the leaves of *OX-GmbZIP15* soybean plants compared with WT plants as indicated by the larger amount of reddish-brown precipitate observed following the treatment with 200 mM NaCl and 300 mM mannitol (Figures 1B and 2B). By contrast, the H_2O_2 contents in *35S:GmbZIP15-SRDX* soybean plants were comparable with that in WT plants under salt- or drought-stress conditions (Figures 1B and 2B). We further investigated whether altered H_2O_2 contents reflected altered ROS-scavenging capability in these plants. For this, the activities of the two main antioxidant enzymes involved in ROS scavenging, namely, peroxidase (POD) and catalase (CAT), were determined in WT, *OX-GmbZIP15*, and *35S:GmbZIP15-SRDX* soybean seedlings before and after 24-h treatment with 200 mM NaCl and 300 mM mannitol. Before stress treatment, *OX-GmbZIP15* soybean plant showed higher activities of POD and CAT than WT plants (Figure S4A,B), whereas following salt- and drought-stress treatments, there was a marked decrease in POD and CAT activities in *OX-GmbZIP15* soybean plant compared to WT plants (Figure S4A,B). These results indicate that ROS scavenging was depressed in *GmbZIP15*-overexpressing soybean plants upon abiotic stress.

2.4. Changes of Stomatal Aperture in GmbZIP15 Transgenic Soybean Plants during Abiotic-Stress Conditions

Abiotic stress usually leads to a reduction in plant water loss through the regulation of the stomata aperture [32]. Thus, we analyzed stomatal regulation and its possible association with the stress-sensitive phenotype of *GmbZIP15* transgenic soybean plants. The stomatal apertures (width/length, W/L) of WT, *OX-GmbZIP15*, and *35S:GmbZIP15-SRDX* soybean plants were measured under control and abiotic-stress conditions. As shown in Figure 3, WT and transgenic soybean plants displayed similar stomatal apertures under control conditions. Moreover, following 200 mM NaCl and 300 mM mannitol treatments, there were no obvious differences in stomatal apertures between *35S:GmbZIP15-SRDX* and WT plants; however, a greater stomatal aperture was observed in *OX-GmbZIP15* plants compared to WT and *35S:GmbZIP15-SRDX* plants (Figure 3A,B). These data show that there is defective stomatal closure in *OX-GmbZIP15* plants upon abiotic stress.

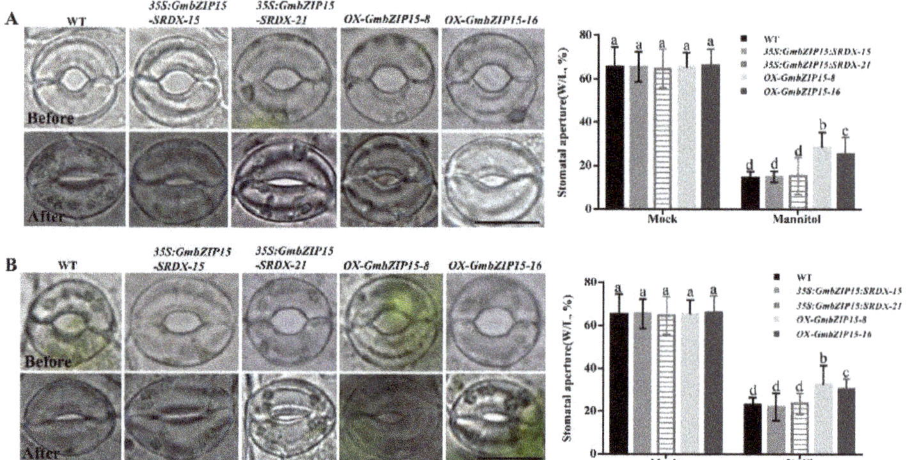

Figure 3. Changes in the stomatal aperture in *GmbZIP15* transgenic soybean plants under salt- and drought-stress conditions. (**A,B**) Comparison of stomatal aperture with width over length before or after 200 mM NaCl treatment for 1 h (**A**) or before or after 300 mM mannitol treatment for 1 h (**B**). Data were calculated from 100 stomata of the leaves of three different soybean plants. The experiments were performed three times with similar results. Bar = 10 μm. Errors bars indicate ± SD of three biological replicates. Significant differences between samples labeled a, b, and c were determined by one-way ANOVA, $p < 0.05$.

2.5. Conservation of GmbZIP15-Mediated Abiotic-Stress Responses in Soybean and Arabidopsis

To further investigate the function of GmbZIP15 in response to abiotic stress, *OX-GmbZIP15* and *35S:GmbZIP15-SRDX* transgenic *Arabidopsis* plants were generated and two lines of each with higher expression levels were selected for further research (Figure 4B). Five-week-old soil-grown WT, *OX-GmbZIP15* and *35S:GmbZIP15-SRDX* transgenic *Arabidopsis* plant were watered with 150 mM NaCl. After 2 weeks, approximately 80% of WT (n = 50) and *35S:GmbZIP15-SRDX* plants (n = 50) remained viable, while nearly 90% of *OX-GmbZIP15* plants (n = 50) died (Figure S5A). These results indicate that, similar to *OX-GmbZIP15* soybean plants, *OX-GmbZIP15 Arabidopsis* plants were also sensitive to salt stress. Similarly, following a 2-week dehydration treatment of WT, *OX-GmbZIP15*, and *35S:GmbZIP15-SRDX* transgenic *Arabidopsis* plants, approximately 85% of WT (n = 50) and *35S:GmbZIP15-SRDX* (n = 50) plants showed some degree of wilting phenotype but remained alive, whereas approximately 90% of the *OX-GmbZIP15* plants (n = 50) displayed a severe, near lethal wilting phenotype (Figure S5B). These results suggest that *OX-GmbZIP15* overexpression causes similar drought- and salt-stress sensitivity in both *Arabidopsis* and soybean.

In addition, seed germination efficiency of WT, *OX-GmbZIP15*, and *35S:GmbZIP15-SRDX Arabidopsis* lines were evaluated under control and drought- and salt-stress conditions. For this, seeds of each line were germinated on 1/2 Murashige and Skoog Medium (MS) with or without 150 mM NaCl or 300 mM mannitol; the growth of two *OX-GmbZIP15* lines were inhibited severely (Figure 4A). In addition, the cotyledon greening rate was much lower in *OX-GmbZIP15* compared to WT and *35S:GmbZIP15-SRDX* transgenic *Arabidopsis* plants (Figure 4C). Taken together, these results indicate that GmbZIP15 plays a conserved role in drought- and salt-stress responses in both soybean and *Arabidopsis*.

To further understand the causal factors behind the drought- and salt-stress hypersensitivity of GmbZIP15-overexpressing *Arabidopsis* plants, we assayed expression levels of several known abiotic-stress-responsive genes in WT and *OX-GmbZIP15* transgenic *Arabidopsis* plants under control and drought- and salt-stress conditions. Transcript levels of each of the analyzed genes, including *AtWRKY33*, *AtCOR6-6*, *AtDREB2A*, and *AtRD29A*, were suppressed in *OX-GmbZIP15* plants compared with WT plants under normal conditions (Figure 4D,E). Under salt stress, the expression levels of *AtCOR6-6*, *AtDREB2A*, and *AtRD29A* were increased in both WT and *OX-GmbZIP15* plants following salt-stress treatment, although the magnitude of expression induction was much lower in *OX-GmbZIP15* plants (Figure 4D). Similar patterns of repressed expression of abiotic stress-responsive genes were detected in *OX-GmbZIP15* plants following drought-stress treatment. The expression levels of *AtWRKY33*, *AtDREB2A*, and *AtRD29A* were induced in both WT and *OX-GmbZIP15* plants by drought stress, but to a smaller extent in *OX-GmbZIP15* plants (Figure 4E). These results indicate that the drought- and salt-stress hypersensitivity caused by GmbZIP15 overexpression in *Arabidopsis* may be due to the repressed expression of drought- and salt-responsive genes.

AtbZIP60 is the homologue of *GmbZIP15* in *Arabidopsis*. To investigate the role of AtbZIP60 in response to drought and salt stress, we analyzed the growth of the *bzip60* mutant (SALK_050203C) under control and drought- and salt-stress conditions. The results showed that the growth of the *bzip60* mutant was significantly repressed as compared to that in WT plants after 150 mM NaCl or 300 mM mannitol treatment (Figure 4A), which was accompanied by a lower cotyledon greening rate (Figure 4C), suggesting that these mutants are sensitive to drought and salt stress. These results agree with the previous findings that overexpression of *AtbZIP60* enhances salt, drought, and cold tolerance in rice [33]. Moreover, these data suggest that the roles of *GmbZIP15* and *AtbZIP60* in response to abiotic stress have diversified.

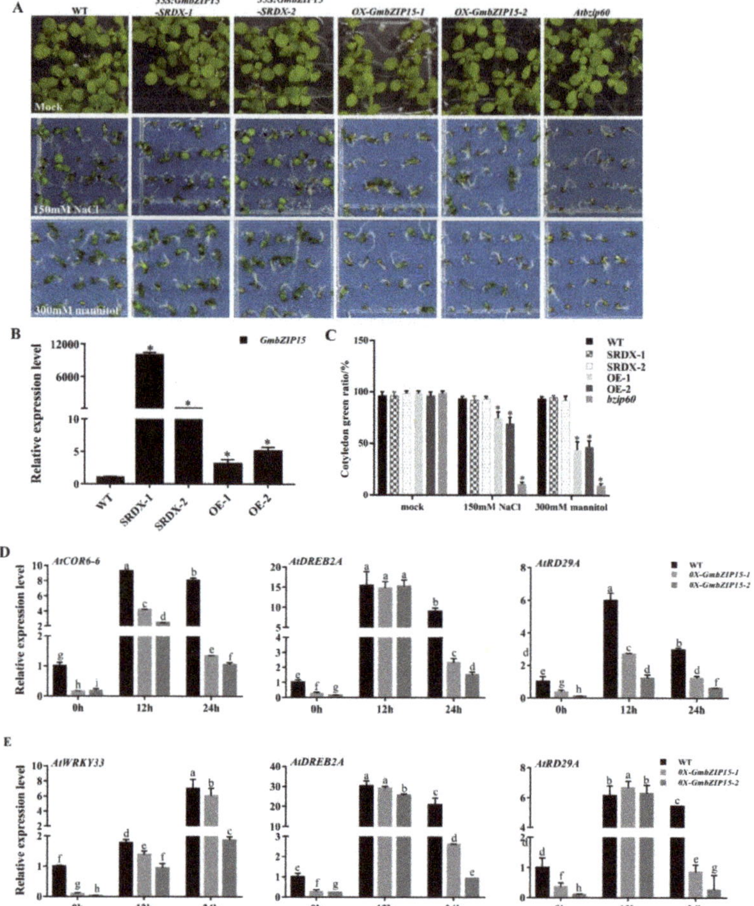

Figure 4. *GmbZIP15*-overexpressed *Arabidopsis* is hypersensitive to salt and drought stresses. (**A**) Phenotype observation of wild-type (WT) and *GmbZIP15* transgenic *Arabidopsis* plants under normal and stress conditions. All the seeds were germinated on the 1/2 Murashige and Skoog Medium (MS) medium under normal conditions or supplemented with 150 mM NaCl or 300 mM mannitol for 1 week. (**B**) Transcript level detection of *GmbZIP15* in transgenic *Arabidopsis* plants. (**C**) Quantification of the cotyledon green rate. (**D**,**E**) *GmbZIP15* regulates stress-responsive gene expression in WT and *GmbZIP15* transgenic *Arabidopsis* plants. Gene expression levels of *AtCOR6-6*, *AtDREB2A,* and *AtRD29A* were quantified by qRT-PCR assays after 150 mM NaCl treatment for 0.12, and 24 h (**D**). Gene expression levels of *AtWRKY33*, *AtDREB2A*, and *AtRD29A* were quantified by qRT-PCR assays after 300 mM mannitol treatment for 0.12, and 24 h (**E**). Errors bars indicate ± SD of three biological replicates. Significant differences between samples labeled a, b, and c were determined by one-way ANOVA, $p < 0.05$.

2.6. Transcriptomic Analysis of OX-GmbZIP15 Transgenic Soybean Plants in Response to Salt and Drought Stress

To further reveal the molecular mechanism behind the abiotic-stress sensitivity caused by *GmbZIP15* overexpression, we conducted RNA sequencing (RNA-seq) analysis using *OX-GmbZIP15-16* (OE) and WT soybean plants grown under either control conditions (mock treated) or treated with NaCl or mannitol. Three biological replicates were collected for each sample. In OE and

WT plants, 2229 and 1693 differentially expressed genes (DEGs) were detected respectively, under salt-stress conditions compared to control conditions (fold change: ≥ 2 and $p \geq 0.05$) (Figure 5A). Moreover, 8917 and 4811 DEGs were detected in OE and WT plants, respectively, under drought-stress conditions compared to control conditions (Figure 5A, Table S1). Thus, there were more DEGs induced by both salt and drought stress in OE plants than in WT plants, indicating that *GmbZIP15* is responsible for gene expression changes upon salt and drought stress. In addition, we detected 1361 common DEGs (695 upregulated and 546 downregulated) in OE plants upon salt and drought stress (Figure 5B). In order to characterize the DEGs downregulated in the OE line upon salt and drought stresses, we studied their gene annotation (GO) term enrichment compared to that for untreated OE plants. As shown in Figure 5, a number of metabolic processes were enriched in both salt-stress and drought-stress downregulated gene sets in OE plants, such as response to stimuli, oxidation-reduction reactions, photosynthesis, hydrolase activity, phenylalanine biosynthesis, and some secondary metabolism processes (Figure 5C–F). These results imply that the above metabolic pathways are repressed in GmbZIP15-overexpressing soybean plants under abiotic stress.

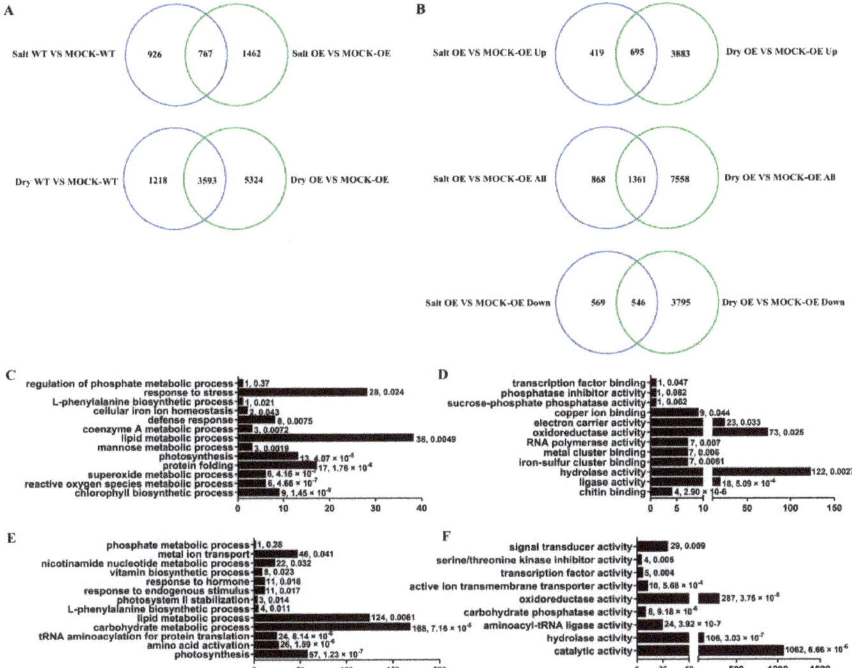

Figure 5. Transcriptomic analysis of *OX-GmbZIP15* transgenic soybean plants. (**A**) Number of specific and common salt- and drought-responsive differentially expressed genes (DEGs) in the WT and *OX-GmbZIP15-16* soybean plants. (**B**) Number of specific and common DEGs in the *OX-GmbZIP15-16* soybean plants after salt and drought-stress treatment. (**C,D**) gene annotation (GO) analysis of the DEGs downregulated in *OX-GmbZIP15-16* soybean plants after salt stress: (**C**) biological process; (**D**) molecular function. (**E,F**) GO analysis of the DEGs downregulated in *OX-GmbZIP15-16* soybean plants after drought stress; (**E**) biological process; (**F**) molecular function. The numbers next to the columns indicate the number of DEGs with corresponding annotation and the *p*-value, respectively (**C–F**).

2.7. GmbZIP15 Regulates the Expression of GmSAHH1, GmABF1, and GmWRKY12 in Soybean in Response to Abiotic Stress

On the basis of the RNA-seq data, five genes (FPKM > 100) with higher expression levels in OE plants compared with WT plants under normal conditions were selected for further analysis. However, only three of these genes (*GmSAHH1*, *GmWRKY12*, and *GmABF1*) were cloned successfully. *GmSAHH1* encodes a phosphate dehydrogenase, and expression of its *Arabidopsis* homologue (*ATSAHH1*; At4g13940) is detected in developing seeds and some anthers [34]. Moreover, the abundance of ATSAHH1 is reduced in protein extracts from salt-treated cells [35].

We performed a further qRT-PCR analysis to validate the RNA-seq data. Results consistently showed that the expression of *GmSAHH1* was higher in OE than in WT soybean plants under normal conditions, but lower in OE than in WT plants under salt- and drought-stress conditions (Figure S6A,B). This repressed expression of *GmSAHH1* under abiotic stress was similar to that of *GmbZIP15*. To further investigate the biological function of *GmSAHH1* expression changes under abiotic stress, *GmSAHH1*-overexpressing (*OX-GmSAHH1*) transgenic *Arabidopsis* plants were obtained, and two lines with higher expression levels were selected for further research (Figure 6A) and then subjected to salt- and drought-stress treatments. Before treatment, no obvious morphological differences between 5-week-old WT and *OX-GmSAHH1* plants were observed. By contrast, under salt and drought stress, *OX-GmSAHH1* transgenic plants exhibited much more pronounced wilting compared with WT plants, which was almost lethal (Figure S5A,B). Seedling growth was also significantly inhibited in *OX-GmSAHH1* plants upon salt and drought stress (Figure 6B). The similar salt- and drought-stress hypersensitivity of both *GmbZIP15*- and *GmSAHH1*-overexpressing plants suggests that GmbZIP15-regulated responses to salt and drought stress are likely mediated by *GmSAHH1* expression activation.

Previous work found that *GmWRKY12* positively regulates drought- and salt-stress responses in association with ABA and salicylic acid (SA), and *GmWRKY12* overexpression in soybean roots enhances soybean salt and drought tolerance [36]. In our RNA-seq data, *GmWRKY12* exhibited higher expression in OE plants than in WT plants under normal conditions and displayed increased expression in OE plants under stress conditions than normal conditions. We conducted a further qRT-PCR analysis and found that the expression of *GmWRKY12* was induced by salt and drought treatment in WT and OE plants (Figure S6A,B). Moreover, we cloned an ABA-responsive gene (*GmABF1*) based on RNA-seq data, and its homologue in *Arabidopsis* (*AtABF1*, AT1G49720) is an ABA-dependent TF that regulates the expression of downstream ABA-inducible genes to improve plant drought resistance [10,37]. Our result demonstrated that its expression is induced by salt and drought stress (Figure S6A,B). Furthermore, GmWRKY12-overexpressing (*OX-GmWRKY12*) and GmABF1-overexpressing (*OX-GmABF1*) seedlings and plants showed improved salt- and drought-stress tolerance compared to WT plants (Figure 6B and Figure S6A,B), which agreed with the induced expression of *GmWRKY12* and *GmABF1* under salt and drought stress. These results suggest that *GmbZIP15* regulates plant salt- and drought-stress responses partly through inhibiting the expression of *GmWRKY12* and *GmABF1* upon abiotic stress.

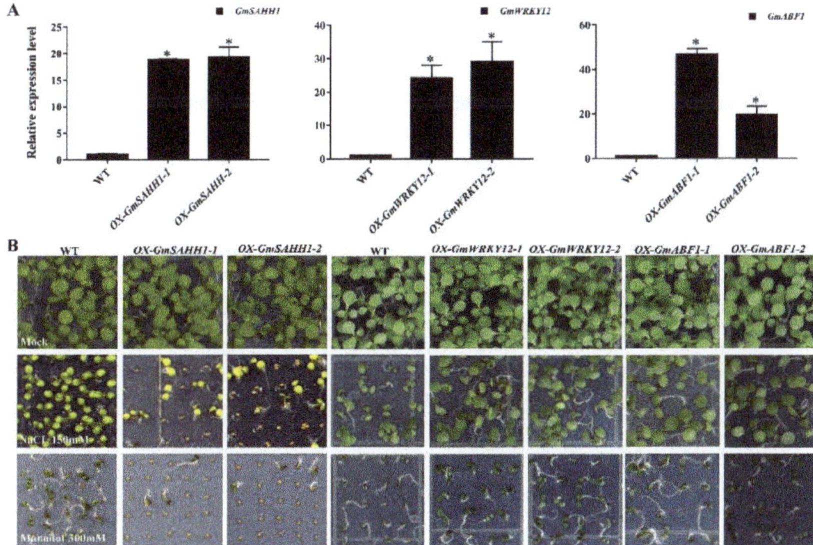

Figure 6. Phenotypic analysis of *GmSAHH1*-, *GmWRKY12*-, and *GmABF1*-overexpressed *Arabidopsis* plants in response to salt and drought stresses. (**A**) Transcript level detection of *GmSAHH1*, *GmWRKY12*, and *GmABF1* in transgenic *Arabidopsis* plants. Errors bars indicate ± SD of three biological replicates. Significant differences between samples labeled asterisks were determined by one-way ANOVA, $p < 0.05$. (**B**) Growth observation of WT and overexpression of *GmSAHH1*, *GmWRKY12*, and *GmABF1 Arabidopsis* seedlings under either normal conditions or 150-mM-NaCl- and 300-mM-mannitol-supplemented 1/2 MS medium.

3. Discussion

Abiotic stress, including salt and drought stress, has a considerable impact on the quality and yield of agricultural products. The important oilseed crop, soybean, is threatened by diverse categories of abiotic stress. Previous studies have indicated that bZIP TFs play diverse roles in response to various biotic and abiotic stress factors in different crop species, such as rice, soybean, rape, cotton, and maize [38–41]. In this study, a group-K bZIP TF, namely, GmbZIP15, was identified in soybean, and its functions in response to abiotic-stress conditions were analyzed in detail.

AtbZIP60, the *Arabidopsis GmbZIP15* homologue, positively modulates plant responses to salt, cold, and abiotic-stress conditions [33]. Here, we found that GmbZIP15 acts as a negative regulator of abiotic-stress responses. The transcription of *GmbZIP15* was suppressed by salt and drought stress (Figure S1A,B) and GmbZIP15-overexpressing soybean displayed hypersensitivity to salt and drought stress (Figures 1A and 2A). Our study suggests that *GmbZIP15* function in abiotic-stress responses differs from that of *AtbZIP60*, which is possibly due to the functional divergence of soybean and *Arabidopsis* during long-term evolution.

To adapt to abiotic stress, especially salt and drought stresses, the plants derive several strategies, including ion regulation and compartmentalization, induction of antioxidant enzymes, plant hormones and regulatory genes [42–44]. For example, the novel soybean regulatory gene GmTIP2;3 could effectively improve the tolerance of yeast to drought stress [43,45]. In addition, when under abiotic-stress conditions, plant endogenous ABA accumulates rapidly and activates the expression of stress-responsive genes, causing many physiological responses [37,46]. It has been demonstrated that ABA plays key roles in maintaining seed dormancy, inhibiting germination, and preventing seedling growth [47], and that abiotic stress is able to induce ABA biosynthesis and trigger ABA-dependent signaling pathways [48]. GmbZIP15 negatively regulates the expression of

GmABF1, and the overexpression of *GmABF1* in *Arabidopsis* confers increased resistance to salt- and drought-stress conditions (Figure 6B). In addition, *GmWRKY12* in association with ABA positively regulates drought- and salt-stress responses, and the overexpression of *GmWRKY12* in soybean roots enhances plant salt and drought tolerance [36]. These results indicate a negative regulation of GmbZIP15 for ABA signaling which might via *GmABF1* and *GmWRKY12* in response to abiotic stress. In addition, the stress-responsive genes *AtDREB2A* and *AtRD29A* in *OX-GmbZIP15 Arabidopsis* plants (Figure 4C,D) and *GmDREBb* (Figure S2A,B) in *OX-GmbZIP15* soybean plants exhibited lower transcript levels than those in WT plants under salt or drought conditions. AREBs and DREB are two groups of TFs that independently regulate the expression of genes involved in ABA-dependent and ABA-independent pathways [11]. The promoter regions of RD29 genes (RD29A and RD29B in *Arabidopsis*) are targeted by AREBs and DREBs; these genes encode hydrophilic proteins that endow plants with enhanced resistance to abiotic and cold stress [49]. Therefore, our study suggests that GmbZIP15 might act a negative regulator of plant drought- and salt-stress responses through ABA-dependent and ABA-independent pathways.

Abiotic stress potentially impairs plant cellular physiology and biochemistry via the excess generation of ROS [7,50,51]. For example, SlWRKY81 improved drought tolerance in tomato plants via the repression of SlP5CS1 transcription and thus reducing proline biosynthesis [52]. In this study, H_2O_2 contents sharply increased in *OX-GmbZIP15* plants compared to WT plants under drought- and salt-stress conditions (Figures 1B and 2B). To control the level of ROS accumulation under stress conditions, plants have evolved a number of antioxidants, such as SOD, POD, and CAT, to scavenge ROS and to restore cellular redox homeostasis [53–57]. With the development of molecular biology, our understanding of molecular and physiology mechanisms is becoming clearer. Our results showed that the activities of POD and CAT were suppressed in *OX-GmbZIP15* transgenic soybean plants under salt and drought stress (Figure S2C,D), indicating compromised ROS scavenging capability in *OX-GmbZIP15* plants in comparison with WT plants. Therefore, we hypothesize that GmbZIP15 plays a negative role in regulating these ROS-scavenging enzyme systems under abiotic stress.

Previous studies showed that bZIP TFs function in many biotic and abiotic-stress responses in plants through regulating diverse biochemical and physiological pathways [12,23,39,58]. RNA- sequencing has been widely used to investigate the molecular processes related to adaptive responses to abiotic stresses and to identify stress-resistance candidate genes by analyzing differences in transcript abundance [44]. In our research, we found that the functional annotation of DEGs that were enriched in the set of downregulated genes in OE plants after salt and drought treatment compared to that in control conditions indicated that, under salt or drought-stress conditions, GmbZIP15-regulated genes were mainly involved in processes such as oxidoreductase activity, phenylalanine biosynthesis, phosphotransferase activity, and some secondary metabolism (Figure 5). As an important polyphenolic secondary metabolite, isoflavones play a crucial role in plants facing diverse environmental-stress conditions [59–61]. *PtSAP13*, for example, enhances salt tolerance by upregulating the transcript level of stress-responsive genes and inducing multiple biological pathways, such as phenylalanine biosynthesis and dioxygenase activity [62], thus implying that the phenylalanine metabolism process is involved in GmbZIP15-regulated abiotic-stress responses (Figure 5C,E). In addition, photosynthesis is essential for plant growth and is important for plants to maintain a balance between growth and stress responses [63,64]. For example, when cyanobacteria grow under stress conditions, photosynthesis-related genes are usually downregulated, whereas stress response-related genes are upregulated [65]. Water deficiency significantly affects photosynthetic characteristics. The drought-tolerant soybean cultivar displayed the maximum values of chlorophyll fluorescence (Fv/Fm, qP, φPSII, and ETR) [66]. When under environmental stress, plants will close their stomata and thus restrict the entry of CO_2 into the leaf and reduce the rate of photosynthesis [67]. In this study, many downregulated genes in *OX-GmbZIP15* transgenic plants after drought and salt stress were associated with photosynthesis (Figure 5C–F), suggesting that stress adaption was prioritized over photosynthesis; however, impaired stomatal aperture regulation in *OX-GmbZIP15* affected plant survival. In addition, enzyme-catalyzed removal of ROS such as superoxide and H_2O_2 are important in plant survival under stress conditions [54,63]. As observed here, those genes downregulated in response to abiotic stress represented antioxidant-related processes. Taken together, multiple metabolic pathways seem to be involved in the GmbZIP15-mediated abiotic-stress response network.

In summary, overexpression of *GmbZIP15* in soybean resulted in hypersensitivity to salt and drought stresses compared with wild-type (WT) plants, which was associated with lower transcript levels of stress-responsive genes, defective stomatal aperture regulation, and reduced antioxidant enzyme activities. Furthermore, RNA-seq and qRT-PCR analyses revealed that *GmbZIP15* positively regulates *GmSAHH1* expression and negatively regulates *GmWRKY12* and *GmABF1* expression in response to salt and drought stresses (Figure 7). These data provided new information for understanding the function of *GmbZIP15* and might facilitate the improvement of plant abiotic-stress tolerance through genetic manipulation in the future.

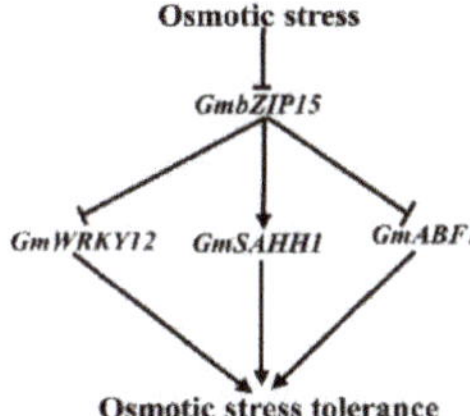

Figure 7. A schematic model of *GmbZIP15* mediated abiotic-stress tolerance in soybean. *GmbZIP15* negatively modulates the abiotic-stress tolerance: *GmbZIP15* positively regulates the expression of *GmSAHH1* and negatively regulates the expression of *GmWRKY12* and *GmABF1* in response to abiotic stresses. The arrows indicate induction or positive modulation; the blunt-end arrows represent block or suppression.

4. Materials and Methods

4.1. Vector Construction and Transformation

To generate the *OX-GmbZIP15* construct, the *GmbZIP15* (Glyma.02G161100) coding DNA sequence (CDS) was amplified and the PCR fragments were cloned into the pENTR/D-TOPO vector (Invitrogen, Carlsbad, CA, USA). The pENTR clones were recombined into the destination vector pGWB506 using LR Clonase II (Invitrogen). The resulting construct also contained the selectable marker BAR for glufosinate resistance [68].

35S: GmbZIP15-SRDX was generated by amplifying GmbZIP15 cDNA sequence and an SRDX motif was added to the end of the cDNA sequence (ctagatctggatctagaactccgtttgggtttcgcttaa). The PCR fragment was cloned into the pENTR/D-TOPO vector (Invitrogen), and the pENTR/D-TOPO clones were recombined into the destination vector pGWB506 using LR Clonase II (Invitrogen) [69]. The vectors *OX-GmbZIP15* and *35S: GmbZIP15-SRDX* were then transformed into soybean by agrobacterium-mediated method [70] and the soybean genotype C03-3 was used.

GmWRKY12 (Glyma.01G224800)-, *GmABF1* (GmbZIP157, Glyma.20G049200)-, and *GmSAHH1* (Glyma.08G108800)-overexpressing vectors were constructed as above [68]. WT *Arabidopsis* (Col-0) plants were then infected with the transformed bacteria by the floral dip method [71]. All the primers used in the article were listed in Table S2.

4.2. Plant Materials and Stress Treatments

Soybean plant seeds including WT (C03-3) and transgenic *GmbZIP15* plants were grown for 15 days in pots containing nutritional soil and vermiculite in green house. The seedlings were then exposed to drought and salt stresses. For drought stress, the soybean seedlings were watered with 300 mM mannitol to induce the rapid drought stress. For salt treatment, the seedlings were transferred to 200 mM NaCl solution. All seedlings leaves were harvested at 0, 6, 12, and 24 h under stress conditions for RNA extraction.

Arabidopsis ecotypes Col-0 was used in this study. The T-DNA mutant *Atbzip60* (SALK_050203C) was obtained from the *Arabidopsis* Biological Resource Centre (ABRC). All Seeds were germinated on 1/2 MS medium containing NaCl or mannitol, after vernalization at 4 °C for 3 days, the plates containing the seeds were placed in a growth chamber with temperature 22 °C, and a photoperiod of 16 h light/8 h dark.

4.3. Diaminobenzidine (DAB) Staining

Following previously described methods for hydrogen peroxide (H_2O_2) detection [72], the soybean leaves after salt and drought treatment for 4 days were immediately vacuum-infiltrated for 20 min with Tris-HCl (pH 7.4) containing 1% (*w/v*) DAB. Thereafter, all the leaves were placed in light for 10 h then boiled for 20 min in 75% ethanol.

4.4. Determination of Stomatal Aperture

The fully expanded leaves of 2-week-old soybean plants were floated in the stomatal opening buffer with 30 mM KCl and 10 mM MES-KOH, pH 6.15 for 2 h under a cool white light, and then 200 mM NaCl or 300 mM mannitol were added to the opening buffer [22]. After 1 h, the subepidermal peels were stripped and used for stomatal aperture measurements under the microscope. In addition, different phytohormones were added to the opening buffer and the stomatal apertures were observed at different timepoints.

4.5. RNA Extraction and Quantitative qRT-PCR

Total RNA was extracted using Trizol (Invitrogen, Carlsbad, CA, USA) then reverse-transcribed using the PrimeScript RT-PCR kit (TaKaRa) [71]. The relative expression levels of selected genes were detected by qRT-PCR using Bio-Rad QRT-PCR system (Foster City, CA, USA) and SYBR Premix Ex Taq II (TaKaRa Perfect Real Time). The qRT-PCR program was 95 °C for 30 s; 40 cycles of 95 °C for 5 s and 60 °C for 34 s; and 95 °C for 15 s [68]. GmActin was used for normalization.

4.6. Determination of Antioxidant Enzyme Activity

The antioxidants including peroxidase (POD) and catalase (CAT) were extracted from approximately 0.1 g of soybean leaves using 1 mL extraction solution. The 2-week-old seedlings were treated with 200 mM NaCl or 300 mM mannitol for 24 h and then the leaves were harvested. The enzyme activities were measured according to the protocol from Solarbio Biochemical Assay Division.

4.7. RNA-Seq Data Analysis

Leaves of 2-week-old soybean plants including WT and *OX-GmbZIP15-16* plants treated with 200 mM NaCl or 300 mM mannitol were harvested at 24 h for RNA-seq, and three biological replicates were analyzed. The libraries were constructed by BGI (Beijing Genomics Institute) then sequenced. GO analyses were performed using the agriGO online toolkit [http://bioinfo.cau.edu.cn/agriGO/index.php].

Supplementary Materials: The following are available online at http://www.mdpi.com/1422-0067/21/20/7778/s1.

Author Contributions: M.Z. performed vector construction and phenotype analysis. Y.L. performed RNA-seq and soybean transformation. H.C. and M.G. calculated all the data. M.C., Z.S., L.Y. and Y.C. performed qRT-PCR analysis. M.Z. and Y.Q. wrote the manuscript. Y.Q. and B.W. revised the manuscript. All authors have read and agreed to the published version of the manuscript.

Funding: This research was funded by NSFC (U1605212, 31970333) and a Guangxi Distinguished Experts Fellowship.

Conflicts of Interest: The authors declare no conflict of interest.

Abbreviations

ABA	Abscisic acid
JA	Jasmonic acid
SA	Salicylic acid
ETH	Ethephon
DREB	Dehydration responsive element binding protein
NAC	No Apical Meristem
bZIP	Basic leucine-zipper
MS	Murashige and Skoog Medium
GO	Gene annotation
ETR	Electron transport rate

References

1. Huang, G.T.; Ma, S.L.; Bai, L.P.; Zhang, L.; Ma, H.; Jia, P.; Liu, J.; Zhong, M.; Guo, Z.F. Signal transduction during cold, salt, and drought stresses in plants. *Mol. Biol. Rep.* **2012**, *39*, 969–987. [CrossRef] [PubMed]
2. Zhao, M.J.; Yin, L.J.; Ma, J.; Zheng, J.C.; Wang, Y.X.; Lan, J.H.; Fu, J.D.; Chen, M.; Xu, Z.S.; Ma, Y.Z. The Roles of GmERF135 in Improving Salt Tolerance and Decreasing ABA Sensitivity in Soybean. *Front. Plant. Sci.* **2019**, *10*, 940. [CrossRef] [PubMed]
3. Huang, Y.; Jiao, Y.; Xie, N.; Guo, Y.; Zhang, F.; Xiang, Z.; Wang, R.; Wang, F.; Gao, Q.; Tian, L.; et al. OsNCED5, a 9-cis-epoxycarotenoid dioxygenase gene, regulates salt and water stress tolerance and leaf senescence in rice. *Plant Sci.* **2019**, *287*, 110188. [CrossRef] [PubMed]
4. Ku, Y.S.; Sintaha, M.; Cheung, M.Y.; Lam, H.M. Plant Hormone Signaling Crosstalks between Biotic and Abiotic Stress Responses. *Int. J. Mol. Sci.* **2018**, *19*, 3206. [CrossRef] [PubMed]
5. de Carvalho, M.H.C. Drought stress and reactive oxygen species. *Plant Signal. Behav.* **2014**, *3*, 156–165. [CrossRef] [PubMed]
6. Miller, G.A.D.; Suzuki, N.; Ciftci-Yilmaz, S.; Mittler, R.O.N. Reactive oxygen species homeostasis and signalling during drought and salinity stresses. *Plant Cell Environ.* **2010**, *33*, 453–467. [CrossRef]
7. Li, S.; Wang, N.; Ji, D.; Zhang, W.; Wang, Y.; Yu, Y.; Zhao, S.; Lyu, M.; You, J.; Zhang, Y.; et al. A GmSIN1/GmNCED3s/GmRbohBs Feed-Forward Loop Acts as a Signal Amplifier That Regulates Root Growth in Soybean Exposed to Salt Stress. *Plant Cell* **2019**, *31*, 2107–2130. [CrossRef]
8. Asano, T.; Hayashi, N.; Kobayashi, M.; Aoki, N.; Miyao, A.; Mitsuhara, I.; Ichikawa, H.; Komatsu, S.; Hirochika, H.; Kikuchi, S.; et al. A rice calcium-dependent protein kinase OsCPK12 oppositely modulates salt-stress tolerance and blast disease resistance. *Plant J.* **2012**, *69*, 26–36. [CrossRef]
9. Gao, Y.F.; Liu, J.K.; Yang, F.M.; Zhang, G.Y.; Wang, D.; Zhang, L.; Ou, Y.B.; Yao, Y.A. The WRKY transcription factor WRKY8 promotes resistance to pathogen infection and mediates drought and salt stress tolerance in Solanum lycopersicum. *Physiol. Plant.* **2019**. [CrossRef]
10. Yoshida, T.; Fujita, Y.; Maruyama, K.; Mogami, J.; Todaka, D.; Shinozaki, K.; Yamaguchi-Shinozaki, K. Four Arabidopsis AREB/ABF transcription factors function predominantly in gene expression downstream of SnRK2 kinases in abscisic acid signalling in response to osmotic stress. *Plant Cell Environ.* **2015**, *38*, 35–49. [CrossRef]
11. Yoshida, T.; Mogami, J.; Yamaguchi-Shinozaki, K. ABA-dependent and ABA-independent signaling in response to osmotic stress in plants. *Curr. Opin. Plant Biol.* **2014**, *21*, 133–139. [CrossRef] [PubMed]
12. Zhang, C.; Li, C.; Liu, J.; Lv, Y.; Yu, C.; Li, H.; Zhao, T.; Liu, B. The OsABF1 transcription factor improves drought tolerance by activating the transcription of COR413-TM1 in rice. *J. Exp. Bot.* **2017**, *68*, 4695–4707. [CrossRef] [PubMed]
13. Kim, S.; Kang, J.Y.; Cho, D.I.; Park, J.H.; Kim, S.Y. ABF2, an ABRE-binding bZIP factor, is an essential component of glucose signaling and its overexpression affects multiple stress tolerance. *Plant J.* **2004**, *40*, 75–87. [CrossRef] [PubMed]
14. Chang, H.C.; Tsai, M.C.; Wu, S.S.; Chang, I.F. Regulation of ABI5 expression by ABF3 during salt stress responses in Arabidopsis thaliana. *Bot. Stud.* **2019**, *60*, 16. [CrossRef]

15. Fujita, Y.; Fujita, M.; Satoh, R.; Maruyama, K.; Parvez, M.M.; Seki, M.; Hiratsu, K.; Ohme-Takagi, M.; Shinozaki, K.; Yamaguchi-Shinozaki, K. AREB1 Is a Transcription Activator of Novel ABRE-Dependent ABA Signaling that Enhances Drought Stress Tolerance in Arabidopsis. *Plant Cell* **2005**, *17*, 3470–3488. [CrossRef]

16. Kang, J.-Y.; Choi, H.-I.; Im, M.-Y.; Kim, S.Y. Arabidopsis Basic Leucine Zipper Proteins That Mediate Stress-Responsive Abscisic Acid Signaling. *Plant Cell* **2002**, *14*, 343–357. [CrossRef]

17. Sakuma, Y.; Liu, Q.; Dubouzet, J.G.; Abe, H.; Shinozaki, K.; Yamaguchi-Shinozaki, K. DNA-binding specificity of the ERF/AP2 domain of Arabidopsis DREBs, transcription factors involved in dehydration- and cold-inducible gene expression. *Biochem. Biophys. Res. Commun.* **2002**, *290*, 998–1009. [CrossRef]

18. Li, X.-P.; Tian, A.-G.; Luo, G.-Z.; Gong, Z.-Z.; Zhang, J.-S.; Chen, S.-Y. Soybean DRE-binding transcription factors that are responsive to abiotic stresses. *Theor. Appl. Genet.* **2005**, *110*, 1355–1362. [CrossRef]

19. Hao, Y.J.; Wei, W.; Song, Q.X.; Chen, H.W.; Zhang, Y.Q.; Wang, F.; Zou, H.F.; Lei, G.; Tian, A.G.; Zhang, W.K.; et al. Soybean NAC transcription factors promote abiotic stress tolerance and lateral root formation in transgenic plants. *Plant J.* **2011**, *68*, 302–313. [CrossRef]

20. Kidokoro, S.; Watanabe, K.; Ohori, T.; Moriwaki, T.; Maruyama, K.; Mizoi, J.; Myint Phyu Sin Htwe, N.; Fujita, Y.; Sekita, S.; Shinozaki, K.; et al. Soybean DREB1/CBF-type transcription factors function in heat and drought as well as cold stress-responsive gene expression. *Plant J.* **2015**, *81*, 505–518. [CrossRef]

21. Li, X.W.; Wang, Y.; Yan, F.; Li, J.W.; Zhao, Y.; Zhao, X.; Zhai, Y.; Wang, Q.Y. Overexpression of soybean R2R3-MYB transcription factor, GmMYB12B2, and tolerance to UV radiation and salt stress in transgenic Arabidopsis. *Genet. Mol. Res.* **2016**, *15*. [CrossRef]

22. Li, J.; Besseau, S.; Toronen, P.; Sipari, N.; Kollist, H.; Holm, L.; Palva, E.T. Defense-related transcription factors WRKY70 and WRKY54 modulate osmotic stress tolerance by regulating stomatal aperture in Arabidopsis. *New Phytol.* **2013**, *200*, 457–472. [CrossRef] [PubMed]

23. Yang, S.; Xu, K.; Chen, S.; Li, T.; Xia, H.; Chen, L.; Liu, H.; Luo, L. A stress-responsive bZIP transcription factor OsbZIP62 improves drought and oxidative tolerance in rice. *BMC Plant Biol.* **2019**, *19*, 260. [CrossRef] [PubMed]

24. Zhang, M.; Liu, Y.; Shi, H.; Guo, M.; Chai, M.; He, Q.; Yan, M.; Cao, D.; Zhao, L.; Cai, H.; et al. Evolutionary and expression analyses of soybean basic Leucine zipper transcription factor family. *BMC Genomics* **2018**, *19*, 159. [CrossRef] [PubMed]

25. Gaguancela, O.A.; Zuniga, L.P.; Arias, A.V.; Halterman, D.; Flores, F.J.; Johansen, I.E.; Wang, A.; Yamaji, Y.; Verchot, J. The IRE1/bZIP60 Pathway and Bax Inhibitor 1 Suppress Systemic Accumulation of Potyviruses and Potexviruses in Arabidopsis and Nicotiana benthamiana Plants. *Mol. Plant Microbe Interact.* **2016**, *29*, 750–766. [CrossRef]

26. Hobo, T.; Kowyama, Y.; Hattori, T. A bZIP factor, TRAB1, interacts with VP1 and mediates abscisic acid-induced transcription. *Proc. Natl. Acad. Sci. USA* **1999**, *96*, 15348–15353. [CrossRef]

27. Wang, W.; Qiu, X.; Yang, Y.; Kim, H.S.; Jia, X.; Yu, H.; Kwak, S.S. Sweetpotato bZIP Transcription Factor IbABF4 Confers Tolerance to Multiple Abiotic Stresses. *Front. Plant Sci.* **2019**, *10*, 630. [CrossRef] [PubMed]

28. Liao, Y.; Zhang, J.S.; Chen, S.Y.; Zhang, W.K. Role of soybean GmbZIP132 under abscisic acid and salt stresses. *J. Int. Plant Biol.* **2008**, *50*, 221–230. [CrossRef] [PubMed]

29. Xu, Z.; Ali, Z.; Xu, L.; He, X.; Huang, Y.; Yi, J.; Shao, H.; Ma, H.; Zhang, D. The nuclear protein GmbZIP110 has transcription activation activity and plays important roles in the response to salinity stress in soybean. *Sci. Rep.* **2016**, *6*, 20366. [CrossRef] [PubMed]

30. Liao, Y.; Zou, H.F.; Wei, W.; Hao, Y.J.; Tian, A.G.; Huang, J.; Liu, Y.F.; Zhang, J.S.; Chen, S.Y. Soybean GmbZIP44, GmbZIP62 and GmbZIP78 genes function as negative regulator of ABA signaling and confer salt and freezing tolerance in transgenic Arabidopsis. *Planta* **2008**, *228*, 225–240. [CrossRef]

31. Xiong, L.; Schumaker, K.S.; Zhu, J.K. Cell signaling during cold, drought, and salt stress. *Plant Cell* **2002**, *14*, S165–S183. [CrossRef] [PubMed]

32. Verslues, P.E.; Agarwal, M.; Katiyar-Agarwal, S.; Zhu, J.; Zhu, J.K. Methods and concepts in quantifying resistance to drought, salt and freezing, abiotic stresses that affect plant water status. *Plant J.* **2006**, *45*, 523–539. [CrossRef] [PubMed]

33. Tang, W.; Page, M. Transcription factor AtbZIP60 regulates expression of Ca^{2+}-dependent protein kinase genes in transgenic cells. *Mol. Biol. Rep.* **2013**, *40*, 2723–2732. [CrossRef] [PubMed]

34. Sujatha, T.; Sivanandan, C.; Bhat, S.; Srinivasan, R. In silico and deletion analysis of upstream promoter fragment of S-Adenosyl Homocysteine Hydrolase (SAHH1) gene of Arabidopsis leads to the identification of a fragment capable of driving gene expression in developing seeds and anthers. *J. Plant Biochem. Biotechnol.* **2009**, *18*, 13–20. [CrossRef]

35. Ndimba, B.K.; Chivasa, S.; Simon, W.J.; Slabas, A.R. Identification of Arabidopsis salt and osmotic stress responsive proteins using two-dimensional difference gel electrophoresis and mass spectrometry. *Proteomics* **2005**, *5*, 4185–4196. [CrossRef] [PubMed]

36. Shi, W.Y.; Du, Y.T.; Ma, J.; Min, D.H.; Jin, L.G.; Chen, J.; Chen, M.; Zhou, Y.B.; Ma, Y.Z.; Xu, Z.S.; et al. The WRKY Transcription Factor GmWRKY12 Confers Drought and Salt Tolerance in Soybean. *Int. J. Mol. Sci.* **2018**, *19*, 4087. [CrossRef] [PubMed]

37. Wang, X.; Wang, Y.; Wang, L.; Liu, H.; Zhang, B.; Cao, Q.; Liu, X.; Lv, Y.; Bi, S.; Zhang, S.; et al. Arabidopsis PCaP2 Functions as a Linker Between ABA and SA Signals in Plant Water Deficit Tolerance. *Front. Plant Sci.* **2018**, *9*, 578. [CrossRef]

38. Gao, S.Q.; Chen, M.; Xu, Z.S.; Zhao, C.P.; Li, L.; Xu, H.J.; Tang, Y.M.; Zhao, X.; Ma, Y.Z. The soybean GmbZIP1 transcription factor enhances multiple abiotic stress tolerances in transgenic plants. *Plant Mol. Biol.* **2011**, *75*, 537–553. [CrossRef]

39. Miyamoto, K.; Nishizawa, Y.; Minami, E.; Nojiri, H.; Yamane, H.; Okada, K. Overexpression of the bZIP transcription factor OsbZIP79 suppresses the production of diterpenoid phytoalexin in rice cells. *J. Plant Physiol.* **2015**, *173*, 19–27. [CrossRef]

40. Liu, J.; Chen, N.; Chen, F.; Cai, B.; Dal Santo, S.; Tornielli, G.B.; Pezzotti, M.; Cheng, Z.M. Genome-wide analysis and expression profile of the bZIP transcription factor gene family in grapevine (*Vitis vinifera*). *BMC Genom.* **2014**, *15*, 281. [CrossRef]

41. Dong, Q.; Xu, Q.; Kong, J.; Peng, X.; Zhou, W.; Chen, L.; Wu, J.; Xiang, Y.; Jiang, H.; Cheng, B. Overexpression of ZmbZIP22 gene alters endosperm starch content and composition in maize and rice. *Plant Sci.* **2019**, *283*, 407–415. [CrossRef] [PubMed]

42. Tang, X.; Mu, X.; Shao, H.; Wang, H.; Brestic, M. Global plant-responding mechanisms to salt stress: Physiological and molecular levels and implications in biotechnology. *Crit. Rev. Biotechnol.* **2015**, *35*, 425–437. [CrossRef] [PubMed]

43. Agarwal, P.K.; Shukla, P.S.; Gupta, K.; Jha, B. Bioengineering for Salinity Tolerance in Plants: State of the Art. *Mol. Biotechnol.* **2013**, *54*, 102–123. [CrossRef] [PubMed]

44. Guo, H.; Zhang, L.; Cui, Y.N.; Wang, S.M.; Bao, A.K. Identification of candidate genes related to salt tolerance of the secretohalophyte Atriplex canescens by transcriptomic analysis. *BMC Plant Biol.* **2019**, *19*, 213. [CrossRef]

45. Dayong, Z.; Jinfeng, T.; Xiaolan, H.; Zhaolong, X.; Ling, X.; Peipei, W.; Yihong, H.; Marian, B.; Hongxiang, M.; Hongbo, S. A Novel Soybean Intrinsic Protein Gene, GmTIP2;3, Involved in Responding to Osmotic Stress. *Front. Plant Sci.* **2016**, *6*, 1237.

46. Zhang, Y.; Kang, E.; Yuan, M.; Fu, Y.; Zhu, L. PCaP2 regulates nuclear positioning in growing Arabidopsis thaliana root hairs by modulating filamentous actin organization. *Plant Cell Rep.* **2015**, *34*, 1317–1330. [CrossRef]

47. Finkelstein, R.R.; Gampala, S.S.; Rock, C.D. Abscisic acid signaling in seeds and seedlings. *Plant Cell* **2002**, *14*, S15–S45. [CrossRef]

48. Zhu, J.K. Salt and drought stress signal transduction in plants. *Annu. Rev. Plant Biol.* **2002**, *53*, 247–273. [CrossRef]

49. Jia, H.; Zhang, S.; Ruan, M.; Wang, Y.; Wang, C. Analysis and application of RD29 genes in abiotic stress response. *Acta Physiol. Plant.* **2012**, *34*, 1239–1250. [CrossRef]

50. Luan, Q.; Chen, C.; Liu, M.; Li, Q.; Wang, L.; Ren, Z. CsWRKY50 mediates defense responses to Pseudoperonospora cubensis infection in Cucumis sativus. *Plant Sci.* **2019**, *279*, 59–69. [CrossRef]

51. Xu, Z.; Raza, Q.; Xu, L.; He, X.; Huang, Y.; Yi, J.; Zhang, D.; Shao, H.B.; Ma, H.; Ali, Z. GmWRKY49, a Salt-Responsive Nuclear Protein, Improved Root Length and Governed Better Salinity Tolerance in Transgenic Arabidopsis. *Front. Plant Sci.* **2018**, *9*, 809. [CrossRef] [PubMed]

52. Li, X.; Wan, H.; Zhou, G.; Cheng, Y.; Ahammed, G.J. SlWRKY81 reduces drought tolerance by attenuating proline biosynthesis in tomato. *Sci. Hortic.* **2020**, *270*, 109444.

53. Alscher, R.G.; Erturk, N.; Heath, L.S. Role of superoxide dismutases (SODs) in controlling oxidative stress in plants. *J. Exp. Bot.* **2002**, *53*, 1331–1341. [CrossRef] [PubMed]

54. Cheng, Q.; Dong, L.; Gao, T.; Liu, T.; Li, N.; Wang, L.; Chang, X.; Wu, J.; Xu, P.; Zhang, S. The bHLH transcription factor GmPIB1 facilitates resistance to Phytophthora sojae in Glycine max. *J. Exp. Bot.* **2018**, *69*, 2527–2541. [CrossRef]

55. Bechtold, U.; Albihlal, W.S.; Lawson, T.; Fryer, M.J.; Sparrow, P.A.; Richard, F.; Persad, R.; Bowden, L.; Hickman, R.; Martin, C.; et al. Arabidopsis HEAT SHOCK TRANSCRIPTION FACTORA1b overexpression enhances water productivity, resistance to drought, and infection. *J. Exp. Bot.* **2013**, *64*, 3467–3481. [CrossRef]

56. Yu, H.; Gao, Q.; Dong, S.; Zhou, J.; Ye, Z.; Lan, Y. Effects of dietary n-3 highly unsaturated fatty acids (HUFAs) on growth, fatty acid profiles, antioxidant capacity and immunity of sea cucumber Apostichopus japonicus (Selenka). *Fish Shellfish Immunol.* **2016**, *54*, 211–219. [CrossRef]

57. Ranjan, A.; Jayaraman, D.; Grau, C.; Hill, J.H.; Whitham, S.A.; Ane, J.M.; Smith, D.L.; Kabbage, M. The pathogenic development of Sclerotinia sclerotiorum in soybean requires specific host NADPH oxidases. *Mol. Plant Pathol.* **2018**, *19*, 700–714. [CrossRef]

58. Li, Q.; Wu, Q.; Wang, A.; Lv, B.; Dong, Q.; Yao, Y.; Wu, Q.; Zhao, H.; Li, C.; Chen, H.; et al. Tartary buckwheat transcription factor FtbZIP83 improves the drought/salt tolerance of Arabidopsis via an ABA-mediated pathway. *Plant Physiol. Biochem.* **2019**, *144*, 312–323. [CrossRef]

59. Wang, Z.; Wang, S.; Xiao, Y.; Li, Z.; Wu, M.; Xie, X.; Li, H.; Mu, W.; Li, F.; Liu, P.; et al. Functional characterization of a HD-ZIP IV transcription factor NtHDG2 in regulating flavonols biosynthesis in Nicotiana tabacum. *Plant Physiol. Biochem.* **2020**, *146*, 259–268. [CrossRef]

60. Yan, J.; Wang, B.; Jiang, Y.; Cheng, L.; Wu, T. GmFNSII-controlled soybean flavone metabolism responds to abiotic stresses and regulates plant salt tolerance. *Plant Cell Physiol.* **2014**, *55*, 74–86. [CrossRef]

61. Jeong, Y.J.; An, C.H.; Park, S.C.; Pyun, J.W.; Lee, J.; Kim, S.W.; Kim, H.S.; Kim, H.; Jeong, J.C.; Kim, C.Y. Methyl Jasmonate Increases Isoflavone Production in Soybean Cell Cultures by Activating Structural Genes Involved in Isoflavonoid Biosynthesis. *J. Agric. Food Chem.* **2018**, *66*, 4099–4105. [CrossRef]

62. Li, J.; Sun, P.; Xia, Y.; Zheng, G.; Sun, J.; Jia, H. A Stress-Associated Protein, PtSAP13, From Populus trichocarpa Provides Tolerance to Salt Stress. *Int. J. Mol. Sci.* **2019**, *20*, 5782. [CrossRef]

63. Liu, W.; Zhao, B.G.; Chao, Q.; Wang, B.; Zhang, Q.; Zhang, C.; Li, S.; Jin, F.; Yang, D.; Li, X. Function analysis of ZmNAC33, a positive regulator in drought stress response in Arabidopsis. *Plant Physiol. Biochem.* **2019**, *145*, 174–183. [CrossRef] [PubMed]

64. Mutava, R.N.; Prince, S.J.K.; Syed, N.H.; Song, L.; Valliyodan, B.; Chen, W.; Nguyen, H.T. Understanding abiotic stress tolerance mechanisms in soybean: A comparative evaluation of soybean response to drought and flooding stress. *Plant Physiol. Biochem.* **2015**, *86*, 109–120. [CrossRef]

65. Ge, H.; Fang, L.; Huang, X.; Wang, J.; Chen, W.; Liu, Y.; Zhang, Y.; Wang, X.; Xu, W.; He, Q.; et al. Translating Divergent Environmental Stresses into a Common Proteome Response through the Histidine Kinase 33 (Hik33) in a Model Cyanobacterium. *Mol. Cell. Proteomics* **2017**, *16*, 1258–1274. [CrossRef] [PubMed]

66. Iqbal, N.; Hussain, S.; Raza, M.A.; Yang, C.Q.; Safdar, M.E.; Brestic, M.; Aziz, A.; Hayyat, M.S.; Asghar, M.A.; Wang, X.C. Drought Tolerance of Soybean (*Glycine max* L. Merr.) by Improved Photosynthetic Characteristics and an Efficient Antioxidant Enzyme Activities Under a Split-Root System. *Front. Physiol.* **2019**, *10*, 786. [CrossRef]

67. Zhang, F.; Zhu, G.; Du, L.; Shang, X.; Cheng, C.; Yang, B.; Hu, Y.; Cai, C.; Guo, W. Genetic regulation of salt stress tolerance revealed by RNA-Seq in cotton diploid wild species, Gossypium davidsonii. *Sci. Rep.* **2016**, *6*, 20582. [CrossRef] [PubMed]

68. Cai, H.; Zhao, L.; Wang, L.; Zhang, M.; Su, Z.; Cheng, Y.; Zhao, H.; Qin, Y. ERECTA signaling controls Arabidopsis inflorescence architecture through chromatin-mediated activation of PRE1 expression. *New Phytol.* **2017**, *214*, 1579–1596. [CrossRef] [PubMed]

69. Zhao, L.; Cai, H.; Su, Z.; Wang, L.; Huang, X.; Zhang, M.; Chen, P.; Dai, X.; Zhao, H.; Palanivelu, R.; et al. KLU suppresses megasporocyte cell fate through SWR1-mediated activation of WRKY28 expression in Arabidopsis. *Proc. Natl. Acad. Sci. USA* **2018**, *115*, E526–E535. [CrossRef]

70. Yang, X.; Yang, J.; Wang, Y.; He, H.; Niu, L.; Guo, D.; Xing, G.; Zhao, Q.; Zhong, X.; Sui, L.; et al. Enhanced resistance to sclerotinia stem rot in transgenic soybean that overexpresses a wheat oxalate oxidase. *Transgenic Res.* **2019**, *28*, 103–114. [CrossRef]

71. Cai, H.; Zhang, M.; Chai, M.; He, Q.; Huang, X.; Zhao, L.; Qin, Y. Epigenetic regulation of anthocyanin biosynthesis by an antagonistic interaction between H2A.Z and H3K4me3. *New Phytol.* **2019**, *221*, 295–308. [CrossRef]
72. Zhang, H.; Wu, Q.; Cao, S.; Zhao, T.; Chen, L.; Zhuang, P.; Zhou, X.; Gao, Z. A novel protein elicitor (SsCut) from Sclerotinia sclerotiorum induces multiple defense responses in plants. *Plant Mol. Biol.* **2014**, *86*, 495–511. [CrossRef] [PubMed]

Publisher's Note: MDPI stays neutral with regard to jurisdictional claims in published maps and institutional affiliations.

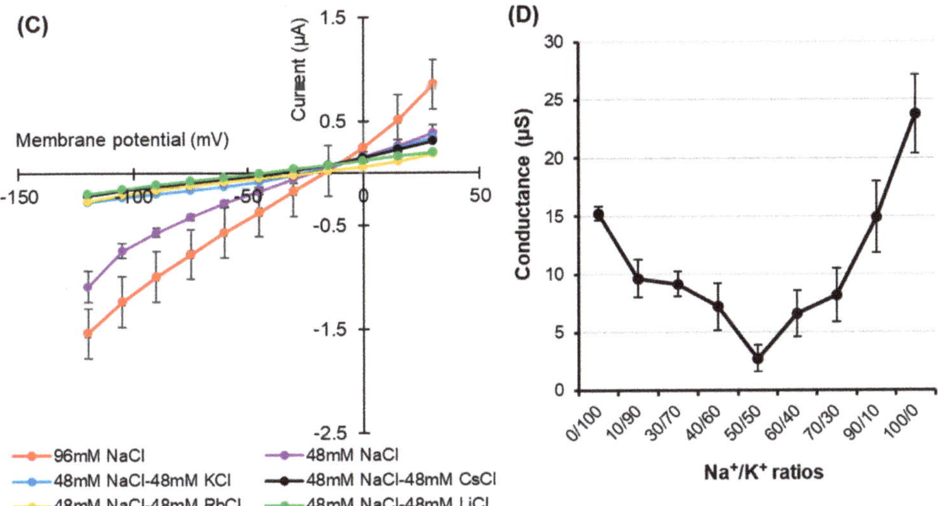

Figure 2. Monovalent alkaline cation selectivity of *HvPIP2;8* and the effect of the interaction of K^+ and Na^+ on *HvPIP2;8*-mediated ion conductance activity. Current–voltage relationships obtained from oocytes either expressing *HvPIP2;8* (**A**) or injected with water (**B**) *HvPIP2;8* displays a different monovalent alkaline cation selectivity. Oocytes were successively immersed in bath solutions with a high calcium condition, supplemented with Na^+, K^+, Cs^+, Rb^+, and Li^+ (as chloride salts) at the concentration of 96 mM. (**C**) Inhibition of *HvPIP2;8*-mediated Na^+ transport by monovalent alkaline cations in the presence of 48 mM NaCl with 48 mM of each alkaline cation. (**D**) The effect of external Na^+/K^+ concentration ratios on the conductance of *HvPIP2;8*-expressing oocytes from V (membrane potential) = −75 mV to −120 mV. The total concentration of (Na + K) was constantly 96 mM. *X. laevis* oocytes were injected with 10 ng of *HvPIP2;8* cRNA for the recording of the conductance in every experiment. Data are the means ± SE (n = 7 to 8 for **A**, n = 5 for **B**, n = 4 to 5 for **C**, and n = 5 to 6 for **D**).

HvPIP2;8-associated currents were then measured in the presence of solutions with combinations of different monovalent cations. Current–voltage relationships were obtained from *HvPIP2;8* expressed in oocytes bathing in 48 mM Na^+ solutions in the co-presence of either 48 mM K^+, Cs^+, Rb^+, or Li^+ (as chloride salt). Smaller *HvPIP2;8*-associated currents were observed in Na^+ solutions when other monovalent cations were added to the solution (Figure 2C). A positive shift in the reversal potential was observed in 96 mM NaCl solutions relative to 48 mM NaCl solutions, consistent with *HvPIP2;8* mediating Na^+ transport (Figure 2C, Supplementary Figure S2). However, in the presence of solutions containing 48 mM KCl and 48 mM NaCl, there were smaller *HvPIP2;8*-mediated Na^+ currents than in solutions with only 48 mM NaCl (Figure 2C). The use of solutions that included different combinations of Na^+ and K^+ concentrations revealed that an external Na^+:K^+ ratio of 50:50 limited the ionic conductance of the *HvPIP2;8*-expressing oocytes; the magnitude of the currents in the 50:50 ratio solutions was 88.4% and 81.9% of the magnitude of the currents in a 100:0 or 0:100 Na^+:K^+ ratio solution, respectively (Figure 2D, Supplementary Figure S3). These observations indicated that the Na^+ permeability of *HvPIP2;8* appears to be highly dependent on the external K^+ concentration.

2.3. HvPIP2;8 Was not Permeable to Cl⁻

The effect of the presence of the external anion Cl^- on *HvPIP2;8* ion transport was tested using Na-gluconate and Choline-Cl solutions (96 mM each). Similar current–voltage relationships for *HvPIP2;8*-expressing oocytes were observed regardless of whether there was Cl^- or gluconate solutions used in the bath (Figure 3A), and there was no shift in the reversal potential (−9 mV) for the different solutions, indicating that the *HvPIP2;8*-induced currents were not affected by Cl^-. In the presence of 96 mM Choline-Cl, the *HvPIP2;8*-expressing oocytes elicited minor currents (-0.68 ± 0.42 µA at -120 mV), comparable to those of the water-injected control oocytes (-0.39 ± 0.03 µA at -120 mV), which was significantly different to the currents observed in the presence of 96 mM NaCl (Figure 3B). These results indicated that the *HvPIP2;8*-associated Na^+-induced currents across the plasma membrane of the oocytes were not affected by the external Cl^- concentration.

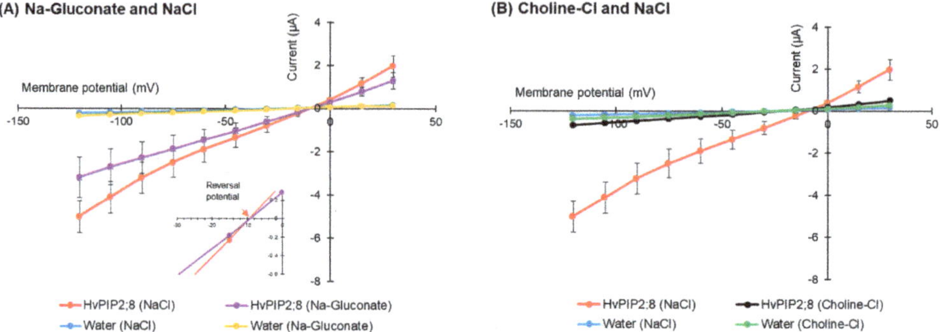

Figure 3. *HvPIP2;8*-mediated Na^+ transport is Cl^- independent. (**A**) Current–voltage relationships obtained from oocytes either expressing *HvPIP2;8* or injected with water in the presence of either 96 mM NaCl or 96 mM Na-gluconate. Inset: Expanded current voltage curves around the reversal potential. (**B**) Current–voltage relationships obtained from oocytes either expressing *HvPIP2;8* or injected with water in the presence of either 96 mM NaCl or 96 mM Choline-Cl. All solutions contained 30 µM Ca^{2+}. *X. laevis* oocytes were injected with 10 ng of *HvPIP2;8* cRNA. Data are the means $\pm$ SE (n = 7 to 8).

2.4. Effects of Divalent Cations on HvPIP2;8-Mediated Ion Transport Activity

The effect of different divalent cations on the ion conductance activity of *HvPIP2;8* was tested. Maximal ionic conductance associated with *HvPIP2;8* was observed when the oocytes were bathed in a divalent cation-free saline (86.4 mM NaCl, 9.6 mM KCl, 10 mM HEPES, pH 7.5 with Tris, and osmolality was adjusted to 200 mosmol Kg^{-1} with supplemental mannitol). However, the *HvPIP2;8* channel was inhibited by the extracellular application of 1.8 mM Ba^{2+}, Cd^{2+}, or Ca^{2+} (Figure 4A). In contrast, the application of 1.8 mM $MgCl_2$ gave rise to a weaker inhibitory effect on the *HvPIP2;8*-mediated ion currents (Figure 4A,B).

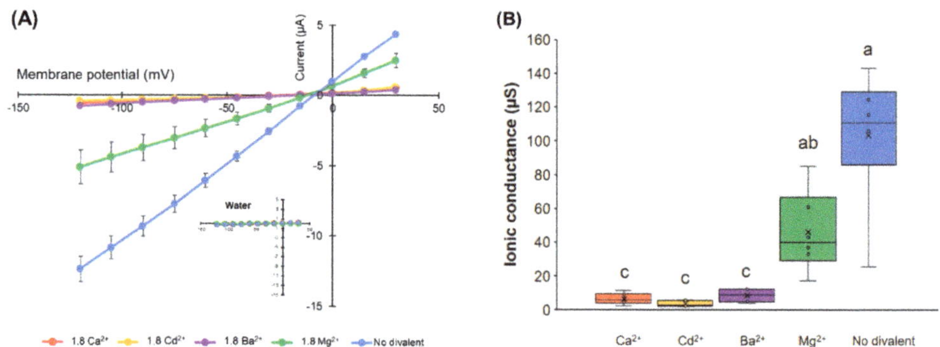

Figure 4. *Cont.*

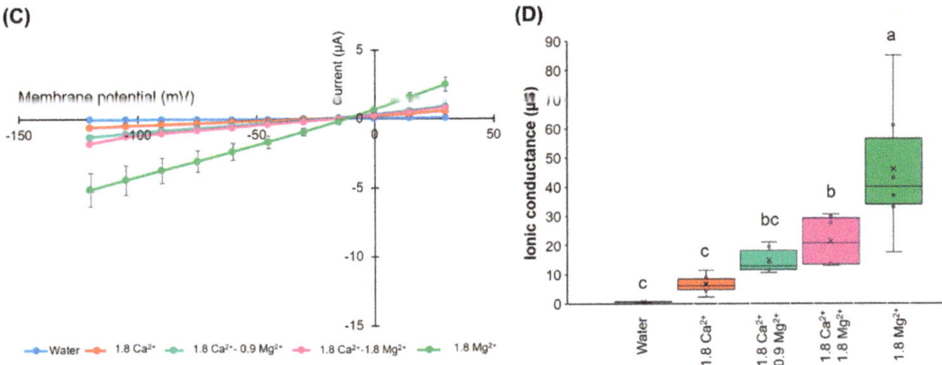

Figure 4. Effects of divalent cations on the ion current responses in oocytes expressing *HvPIP2;8*.
(**A**) Effect of divalent cations on the ion currents of the *HvPIP2;8*-transporter; bath solutions with a
high 1.8 mM Ca^{2+} background calcium conditions were successively replaced with either 1.8 mM
Ca^{2+}, Ba^{2+}, Cd^{2+}, and Mg^{2+} (as chloride salts), at concentrations of 86.4 mM NaCl and 9.6 mM KCl.
(**B**) Box plot summary of the ionic conductance presented in (**A**); the ionic conductances were calculated
from V = −75 mV to −120 mV. (**C**) Relief of Ca^{2+} inhibition by the addition of Mg^{2+} on the ion current
responses in oocytes expressing *HvPIP2;8*; note the different range of the Y-axis from the plot in (**A**).
(**D**) Box plot summary of the ionic conductance presented in (**C**). Steady-state current–voltage curves of
the *X. laevis* oocytes injected with 10 ng of cRNA per oocyte were recorded. Currents from the oocytes
injected with water were the negative controls from the same batch. Significant differences ($p < 0.05$)
are indicated by different letters using one-way ANOVA with Duncan's multiple comparisons test.
Data are the means ± SE of three independent experiments, (n = 6 for **A,B**).

An increase in the $MgCl_2$ concentration in the presence of 1.8 mM $CaCl_2$ tended to partially
cancel out the inhibitory effect of Ca^{2+} in relation to the ion channel activity (Figure 4C,D). These
results indicate that there may be a competitive interaction between Ca^{2+} and Mg^{2+}, which influences
HvPIP2;8-mediated ion channel activity and the presence of more external Mg^{2+} can partially relieve
the inhibitory effect of high Ca^{2+} on *HvPIP2;8* ionic conductance.

2.5. Co-Expression of HvPIP2;8 with HvPIP1s Limited HvPIP2;8 Ion Transport Activity

Co-expression of HvPIP1;2 and HvPIP2;1-2;5 resulted in increases in water transport across
the plasma membrane of the oocytes, relative to the water transport of oocytes expressing HvPIP2s
alone [24], but *HvPIP2;8* was not included in that work [31]. Here, we examined the effect of the
co-expression of HvPIP1s (HvPIP1;1 to HvPIP1;5) with *HvPIP2;8* in relation to ionic conductance. When
expressed alone, each HvPIP1 displayed similar currents to the water-injected controls (Figure 5A,C).
However, when each HvPIP1 was co-expressed with *HvPIP2;8*, the large *HvPIP2;8*-associated currents
were not observed in any of the co-expression combinations examined, whereas when *HvPIP2;8*
was expressed alone, the *HvPIP2;8*-associated currents were significant as expected (Figure 5B,D).
These results indicate that HvPIP1s might be interacting with *HvPIP2;8* and that this either prevents,
or significantly reduces (for HvPIP1;3 and HvPIP1;4), the *HvPIP2;8* ion channel activity.

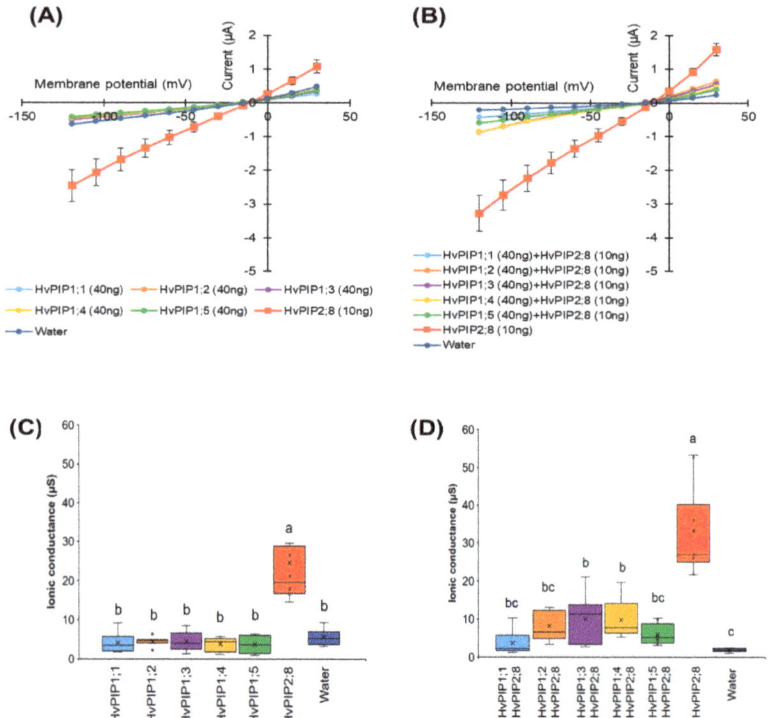

Figure 5. Co-expression of HvPIP1s and *HvPIP2;8* reduces *HvPIP2;8*-dependent currents in X. *laevis* oocytes. (**A**) Current–voltage relationships obtained from oocytes either expressing *HvPIP2;8* alone, each HvPIP1 alone, or injected with water; each HvPIP1 of the five HvPIP1s did not show ion channel activity when expressed alone. (**B**) Co-expression of *HvPIP2;8* with each HvPIP1 largely inhibited the ion channel activity of *HvPIP2;8*. (**C,D**) Box plot summary of the ionic conductance for data shown in (**A,B**), respectively. Oocytes were injected with 10 ng cRNA of *HvPIP2;8*, 40 ng cRNA of each HvPIP1 in both the solo- and co-expression analyses. The bath solution included 86.4 mM NaCl, 9.6 mM KCl, 1.8 mM $MgCl_2$, 1.8 mM EGTA, 1.8 mM $CaCl_2$, 10 mM HEPES, and a pH 7.5 with Tris, and therefore the free Ca^{2+} concentration was 30 µM. Significant differences ($p < 0.05$) are indicated by different letters using one-way ANOVA with Duncan's multiple comparisons test. Data are the means ± SE (n = 8).

2.6. An HvPIP2;8 S285D Phosphomimic Mutant Had Greater Ionic Conductance Than Wild Type HvPIP2;8

To investigate the possibility of a regulatory role for a C-terminal tail serine at residue 285, a phosphomimic mutant of *HvPIP2;8* was generated where the site coding for this residue was mutated such that it coded for aspartic acid (D) instead of serine (S). The ionic conductance and osmotic water permeability (P_{os}) of this mutant was compared to the wild type *HvPIP2;8* and water-injected controls in oocytes from three independent frogs. The ionic conductance observed for *HvPIP2;8* S285D was greater than the ionic conductance for *HvPIP2;8* and nearly seven-fold higher than the water-injected controls (Figure 6A; Supplementary Figure S4A). There was no significant difference in the mean P_{os} for *HvPIP2;8* S285D relative to *HvPIP2;8* WT, but there was a 3.5-fold variability in the magnitude of P_{os} as well as variability between batches (Figure 6B,C; Supplementary Figure S4B). To explore whether phosphorylation at *HvPIP2;8* S285 could influence the P_{os} and ionic conductance relationship, simple linear regressions were fitted for the *HvPIP2;8* wild type and *HvPIP2;8* S285D between the P_{os} and ionic conductance from oocytes harvested from three independent frogs (Figure 6C). To remove the frog batch variability in the P_{os} and ionic conductance, the mean of P_{os} and ionic conductance of the H_2O controls were subtracted from both the *HvPIP2;8* wild type and *HvPIP2;8* S285D (Figure 6C). No significant relationship between the P_{os} and ionic conductance was observed for the wild type *HvPIP2;8*, whereas for the *HvPIP2;8* S285D-expressing oocytes, a negative reciprocal relationship was observed (Figure 6C,D).

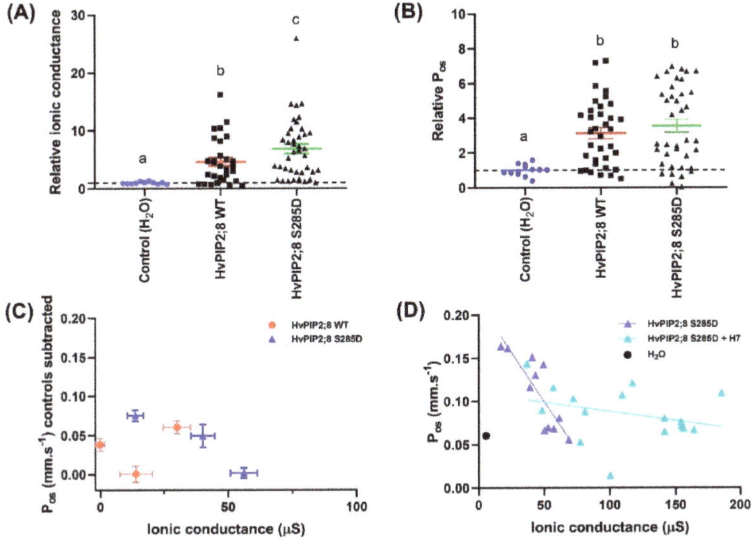

Figure 6. Phosphorylation mimic *HvPIP2;8* S285D influences *HvPIP2;8*-facilitated cation transport. Oocytes were injected with 46 nL water (Control) or with 46 nL water (n = 11) containing 23 ng *HvPIP2;8* WT (n = 30) or *HvPIP2;8* S285D (n = 40) cRNA. Ionic conductance and osmotic water permeability (P_{os}) of the cRNA-injected oocytes were determined via the TEVC and the swelling assay, respectively. (**A**) Na^+ conductance relative to H_2O-injected control (dotted line). Currents were recorded in "Na100" (100 mM NaCl; 2 mM KCl, 1 mM $MgCl_2$, 5 mM HEPES, 50 µM $CaCl_2$, 100 mM NaCl or 100 mM KCl, osmolality 220 mosmol Kg^{-1} (adjusted with D-mannitol), and pH 8.5). (**B**) P_{os} relative to H_2O-injected control (dotted line). Data in (**A**,**B**) was collected from three different frogs and is shown as the mean ± SE, where each data point represents an individual oocyte; for each oocyte, both the ionic conductance and P_{os} were measured; data from (**C**,**D**) is from one batch and again for each oocyte both the ionic conductance and P_{os} were measured. Significant differences ($p < 0.05$) are indicated by different letters (one-way ANOVA, Fisher's post-test). (**C**) Relationships between the mean P_{os} and mean ionic conductance for *HvPIP2;8* WT and *HvPIP2;8* S285D with the mean of the controls subtracted. This data is from oocytes from the same three independent frogs as shown in (**A**,**B**). (**D**) Kinase inhibitor H7 influenced the relationship between the ionic conductance and water permeability in *HvPIP2;8* S285D-expressing oocytes. Oocytes were either untreated or were pre-treated in a low Na^+ Ringer solution that contained 10 µM H7 dihydrochloride (H7) for 2 h before TEVC and the swelling assay. An individual conductance was plotted against the corresponding P_{os} for each oocyte, and the mean for the water-injected controls is shown (black circle, dotted line). Linear regression of P_{os} versus ionic conductance was only significant for *HvPIP2;8* S285D without H7 treatment ($p < 0.005$).

We investigated whether the activity of the endogenous kinases in the *X. laevis* oocytes might influence the phosphorylation state of *HvPIP2;8* or *HvPIP2;8* S285D by applying a kinase inhibitor H7. The H7 treatment significantly increased the ionic conductance and reduced the P_{os} of *HvPIP2;8* S285D relative to the *HvPIP2;8* S285D-expressing oocytes that were not subjected to the H7 treatment (Figure 6D, Supplementary Figure S4C). Analysis of the predicted phosphorylation sites in *HvPIP2;8* (http://www.cbs.dtu.dk/services/NetPhos/) indicated that there are nine candidate sites targeted by protein kinase A or C, which are the kinases involved in *Xenopus* oocyte signal transduction pathways.

S285 was not included in the nine candidate sites predicted by NetPhos, indicating that the endogenous oocyte kinases are likely to target an alternative site or sites (Supplementary Figure S4D,E; [39]).

2.7. Expression of HvPIP2;8 in Barley

Previously, *HvPIP2;8* was observed to be stably expressed in shoots, roots, pistils, and leaves [31]. In this study, to further explore the transcript regulation of *HvPIP2;8*, qPCR was used to assess expression in salt-treated and control shoot and root samples from the barley cultivar Haruna-Nijo (Figure 7). In roots, the transcript levels remained stable in both the salt-treated and control samples (Figure 7). However, in shoots from Haruna-Nijo plants, *HvPIP2;8* transcripts were more abundant in the salt-treated samples than the control samples after 1 day of either the 100 mM or 200 mM NaCl treatment. After five days of NaCl treatment, the *HvPIP2;8* transcripts were less abundant in the shoots than they were at 1 day after NaCl treatment (Figure 7). RT-PCR showed that the transcript levels of *HvPIP2;8* were increased in shoots of the salt-tolerant cultivar, K305 subjected to a 200 mM NaCl treatment relative to the controls (Supplementary Figure S5). Whereas, in the salt-sensitive cultivar, I743, there was no difference in the abundance of *HvPIP2;8* in the salt- and control-treated samples (Supplementary Figure S5). This indicates that the *HvPIP2;8* gene expression in shoots of barley cultivars might vary depending on cultivar and environmental conditions.

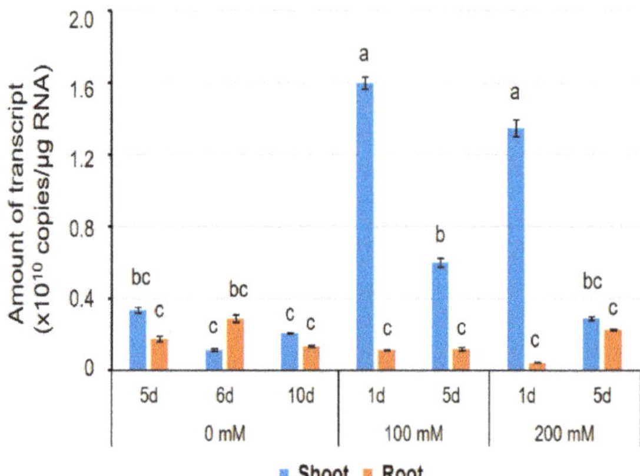

Figure 7. The expression level of the *HvPIP2;8* transcripts in a barley cultivar, Haruna-Nijo, detected by qPCR. Five-day-old barley seedlings were prepared by hydroponic culture and further grown on the culture solution with or without NaCl (100 mM or 200 mM) for 1 day or 5 days. Transcript levels of *HvPIP2;8* in shoots and roots were investigated by absolute quantification. Absolute amounts of transcripts (copies/µg RNA) were displayed. Significant differences ($p < 0.05$) are indicated by different letters using one-way ANOVA with Duncan's multiple comparisons test. Data are the means ± SE, and n = 3.

3. Discussion

HvPIP2;8 has the potential to function as both a water and a cation channel, where the channel characteristics can be influenced by K^+ and divalent cation activity, protein phosphorylation, and protein interactions, and where the *HvPIP2;8* transcript levels can be influenced by salt treatments. We observed that *Xenopus* oocytes expressing *HvPIP2;8* displayed significant ionic conductance relative to the controls and relative to the oocytes expressing any of the six other HvPIP2s and five HvPIP1 proteins. Previous studies have demonstrated water channel activity for HvPIP2;1, HvPIP2;2, HvPIP2;3, HvPIP2;4, HvPIP2;5, and HvPIP2;7 when expressed in oocytes, indicating that the low or absent ionic conductance associated with expression of these proteins is unlikely to relate to miss-folding or miss-targeting [24,35].

HvPIP2;8-associated ionic conductance was inhibited by external Ca^{2+}, Cd^{2+}, and Ba^{2+}, but less so by Mg^{2+} (Figure 4B). *HvPIP2;8* was permeable to both Na^+ and K^+, and the Na^+ permeability of *HvPIP2;8* was inhibited in the presence of external K^+, but not external Cl^-. Co-expression of *HvPIP2;8* and HvPIP1 proteins reduced the *HvPIP2;8*-induced ionic conductance with differences between the PIP1 isoforms; HvPIP1;3 and HvPIP1;4 co-expressed with *HvPIP2;8* still maintained a higher ionic conductance than the water-injected controls (Figure 5D). Greater ionic conductance was observed for an S285D mutant version of *HvPIP2;8* relative to the wild type, but no difference in the P_{os}. Treatments to manipulate oocyte kinase activity differentially influenced wild type *HvPIP2;8* relative to the S285D mutant. The salt-tolerant and salt-sensitive barley lines differed in their response to salt treatments, such that the salt tolerant cultivars tested displayed an increase in the abundance of *HvPIP2;8* transcripts within the first day after the salt treatments.

In saline conditions, excessive salt accumulation is detrimental to plant growth and limits crop productivity. This problem is often referred to as ionic toxicity, and for many cereals it is brought about by excessive Na^+ influx into roots followed by excess Na^+ accumulation, particularly in the aerial parts of the plants [40]. Uptake of Na^+ at the root–soil boundary is conferred by multiple pathways involving a range of different types of membrane transporters and channels. For example, OsHKT2;1, one of the high-affinity K^+ transporter family proteins in rice, mediates direct Na^+ absorption from the outer environment of roots when the rice plant faces K^+ starvation conditions [41]. The roles of some important Na^+ transporters, such as the SOS1, NHX, and HKT families, which contribute to salt-stress resistance, have been well characterized [28]. However, some pathways for Na^+ influx into plant roots remain unresolved at the molecular level, although we assume owing to electrophysiological studies that non-selective cation channels (NSCC) mediate significant Na^+ influx into roots following salinity stress [15–17,37]. Candidates for NSCCs include cyclic nucleotide-gated channels (CNGCs) and glutamate receptors (GLRs), but confirmation of the molecular identity of the NSCC will require further research [42,43]. In the future, determining the structure and the role of unidentified Na^+ permeable transporters/channels in plants that contribute to NSCC activity, including Na^+ permeable aquaporins, will help us to understand the complete picture of Na^+ transport and homeostasis during salinity stress.

Aquaporins are well known for their function as water channels [34–36,44]. Previous research has revealed that heterologous expression of AtPIP2;1, categorized as a plasma membrane-localized aquaporin in *Arabidopsis thaliana*, is associated with non-selective cation conductance, and this ion channel function is sensitive to Ca^{2+} [12]. More recently, the Ca^{2+} sensitivity of another water and ion channel aquaporin, AtPIP2;2, was revealed [13]. In the present study, we used TEVC experiments to screen *X. laevis* oocytes expressing barley HvPIPs and the controls, and this revealed significant *HvPIP2;8*-associated ion channel activity under low Ca^{2+} conditions. Among the set of 12 HvPIPs tested, *HvPIP2;8* stood out in relation to conferring ion channel activity in oocytes (Figures 1A and 5A). The screen revealed that HvPIP2;1 might be capable of facilitating ionic conductance, but the HvPIP2;1-associated currents were significantly smaller than the *HvPIP2;8*-associated currents (Figure 1A, Supplementary Table S2). It is also noteworthy that all the plant aquaporins so far shown to conduct cations in the *Xenopus* oocytes show different characteristics in terms of cation selectivity and divalent cation inhibition. The human AQP1, which has an ion channel function, also shows differences in channel characteristics relative to PIP2;1 and PIP2;2 [13]. This would strongly suggest that the aquaporins are not triggering a native *Xenopus* channel to be activated or recruited to the membrane. The effect of the phospho-mimic mutations in *HvPIP2;8* and AtPIP2;1 discussed below is also difficult to explain in terms of recruitment of a native *Xenopus* channel.

As *HvPIP2;8* has been demonstrated to conduct water and ions when expressed in oocytes ([31]; Figure 1), we hypothesize that this aquaporin can function as a channel that mediates both water and ion transport in plants. The Na^+ channel activity associated with *HvPIP2;8* was sensitive to external Ca^{2+} concentrations (Figure 1B,C), and the IC_{50} of the channel was calculated to be approximately 401 μM (https://www.aatbio.com/tools/ic50-calculator), which was similar to the Ca^{2+} sensitivity of the AtPIP2;1 ionic conductance (IC_{50} = 321 μM; [12]). The inhibition by Ca^{2+} is not total since even at 1.8 mM there is still a significant ion conductance (Figure 4D). The analysis of alkali monovalent cations selectivity revealed that *HvPIP2;8* mediated not only Na^+ but also K^+ transport, although *HvPIP2;8* did not mediate Rb^+, Cs^+, or Li^+ transport (Figure 2A,B). However, these monovalent cations, and K^+, blocked the Na^+ channel activity of *HvPIP2;8* when the same amount of each cation was included in the bath solutions (Figure 2C,D). Inhibition or activation of Na^+ transport activity by the presence of similar or greater concentrations of external K^+ in the TEVC bath solution has been observed for different types of high-affinity K^+ transport (HKT)-type sodium transporters. For example, *Triticum aestivum* TaHKT1;5-D and *Triticum monococcum* TmHKT1;5-A encode dual affinity Na^+-transporters and their dual affinity Na^+ transport was inhibited by raising the external K^+ concentration [45,46]; whereas, for OsHKT2;2, extracellular K^+ stimulated the OsHKT2;2-mediated Na^+ transport [47]. Additional research is needed to model how different monovalent ions interact with the pore lining residues of the ion channel aquaporins towards understanding the K^+ inhibition effect on the *HvPIP2;8* Na^+ channel activity.

The influence of divalent cations on the *HvPIP2;8*-mediated Na^+ currents was complicated. The application of 1.8 mM Ba^{2+}, Ca^{2+}, or Cd^{2+} inhibited the *HvPIP2;8* ionic conductance (Figure 4A,B). The application of the same amount of Mg^{2+} in the bath solution did not have an equivalent inhibitory influence on the ionic conductance (just an approximately 63% reduction compared to no divalent control: Figure 4A,B). However, an increase in the Mg^{2+} concentrations in the presence of Ca^{2+} seemed to ameliorate the inhibitory effect of the Ca^{2+} (Figure 4C,D). The competitive interaction between Ca^{2+} and Mg^{2+} might be due to a higher affinity for Mg^{2+} than for Ca^{2+}. Alternatively, Mg^{2+} might interact with the same binding site as Ca^{2+}. Previous research revealed that the AtPIP2;1 and AtPIP2;2 ionic conductance was significantly inhibited by 100 μM and 10 μM extracellular free Ca^{2+}, respectively, and the ionic conductance was also significantly inhibited by the addition of Ba^{2+} and Cd^{2+} [13]. There was also an interaction between Ca^{2+} inhibition and Ba^{2+} relief of block for AtPIP2;1 [13] that is similar to the interaction seen here for Ca^{2+} and Mg^{2+}. The determination of specific interaction sites for divalent cations in the structure of *HvPIP2;8* will be essential to understanding the characteristics observed.

A previous study showed that co-expression of *HvPIP2;8* with HvPIP1;2 in *X. laevis* oocytes did not enhance the water transport activity compared to that of the expression of *HvPIP2;8* alone [31]. In contrast, we observed that co-expression of HvPIP1;2 with other HvPIP2s (2;1 to 2;5) increased the water permeability coefficient [24]. This indicates that heteromerization of each HvPIP2 and HvPIP1;2 could modulate water channel activity differently. We observed here that co-expression of *HvPIP2;8* with the HvPIP1s, including HvPIP1;2, significantly decreased the ionic conductance relative to expression of *HvPIP2;8* alone (Figure 5), indicating that the *HvPIP2;8*-mediated ion channel activity might be negatively regulated through heteromerization. Some isoforms retained some ion conductance when co-expressed (HvPIP1;3 and HvPIP1;4). A previous study revealed that when AtPIP2;1 was co-expressed with AtPIP1;2 the water permeability was greater than when AtPIP2;1 or AtPIP1;2 was expressed alone [12]. However, the ionic conductance of AtPIP2;1 could be suppressed to the level of the water-injected controls when AtPIP2;1 was co-expressed with AtPIP1;2, indicating that the ionic conductance was not associated with higher water permeability [12]. However, this was done at high external Ca^{2+} concentration and it remains to be seen if the same result would be obtained at lower external Ca^{2+}. Together, these results reveal that the activity for water and ion channel PIPs could be differently regulated by independent mechanisms. At present, the mechanism of HvPIP1-dependent decreases in the ion transport activity of *HvPIP2;8* is still unknown. It might be related to changes in the central tetrameric pore dimensions comparing the ion conducting homotetramer with the heterotetramer since the central tetrameric pore is the favored pathway for ion conductance through AQP1 [48]. Elucidating the underlying molecular mechanisms will be important to understand the functions of *HvPIP2;8* as a Na^+-permeable ion channel.

The water and ion channel functions, and the membrane localization of the aquaporins, have previously been reported to be influenced by the phosphorylation status of the C-terminal domain (CTD). For example, versions of AtPIP2;1 mimicking either a phosphorylated state or unphosphorylated state of residues 280 and 283 by mutating these serine residues to either an aspartic acid (D) or an alanine (A) were used to test whether changes in phosphorylation in the CTD might influence AtPIP2;1 water and ion channel function [14]. This revealed that the phosphorylation mimic mutants S280D, S283D, and S280D/S283D had a significantly greater ion conductance for Na^+ and K^+, whereas the phosphonull mutants S280A single and S280A/S283A double had greater water permeability. Interestingly, among the HvPIP2s, *HvPIP2;8* is the only HvPIP2 lacking an AtPIP2;1 S280 equivalent site at CTD; and *HvPIP2;8* also lacks a serine on loop D that is present in AtPIP2;1 (Supplementary Figure S4F). As a first step towards exploring whether changes in phosphorylation of the *HvPIP2;8* CTD may influence water ion channel function, a phosphomimic mutant S285D was generated. The ionic conductance of *HvPIP2;8* S285D was greater than that of *HvPIP2;8* WT, but the P_{os} was similar (Figure 6A,B). Treatments with a kinase inhibitor, H7, resulted in an increase in S285D ionic conductance and a decrease in P_{os} relative to the untreated S285D (Figure 6D). H7 influences the activity of endogenous kinases, which can alter the phosphorylation state of the aquaporins that have been heterologously expressed in the oocytes [14,39,49]. The observation that treatment with H7 can increase the ionic conductance and decrease the P_{os} of *HvPIP2;8* S285D indicates that there is likely to be residues, other than S285, where differential phosphorylation of these additional residues in *HvPIP2;8* can influence the water and ion channel activity. A significant negative relationship between P_{os} and ionic conductance was observed for *HvPIP2;8* S285D (Figure 6D), suggesting a mutually exclusive gating of ion and water flow for this mutant, which was also observed for the AtPIP2;1 phospho-mimic mutants [14]. There are nine *HvPIP2;8* residues that are candidate sites for potentially being phosphorylated by the endogenous oocyte kinases PKA and PKC (Supplementary Figure S4D,E), including a serine site on loop D that could be influential in gating based on previous structural analysis of *Spinacia oleracea* SoPIP2;1 [50]. The next step towards determining which additional sites are targeted by endogenous oocyte kinases will be require testing of additional mutant versions of *HvPIP2;8*; this will assist in determining which phosphorylation sites may be part of the post-translational mechanisms for regulating *HvPIP2;8* water and ion channel function.

In barley, the *HvPIP2;8* gene expresses in both roots and shoots (Figure 7). Interestingly, *HvPIP2;8* expression in shoots was upregulated in response to salt stress (Figure 7). An RT-PCR analysis revealed that the upregulation trend for *HvPIP2;8* transcript abundance was observed in salt-tolerant barley, but not detected in a salt-sensitive barley cultivar (Supplementary Figure S5). These observations, and the characteristics of the *HvPIP2;8* observed by TEVC experiments, led us to wonder whether *HvPIP2;8* could play a positive role in shoot tissues to help cope with salt stress. We observed that the external free Ca^{2+} concentrations have a significant impact on the ion channel activity of *HvPIP2;8* (Figures 1 and 5, and Supplementary Figure S1). It is well known that Ca^{2+} plays key roles in ameliorating Na^+ toxicity under salt stress [51]. In addition, changes in free Ca^{2+} have important signaling roles, particularly in response to stress conditions [52,53]. The implications of Ca^{2+} sensitivity could be different depending on what kind of physiological role *HvPIP2;8* has in planta: for example, if the *HvPIP2;8* mediated the Na^+ influx into the cytosol of mesophyll cells in leaves, then a Ca^{2+}-dependent inhibitory effect might be a positive feature as it could prevent excess Na^+ influx; or if *HvPIP2;8* played a role analogous to the role of the HKT1s, some of which are known to function in unloading of Na^+ from the xylem to protect the leaf blades [54], then a Ca^{2+}-dependent inhibitory effect might be a negative feature as it could prevent Na^+ transport into stelar cells in shoots, such as in leaf sheaths. It is also possible that *HvPIP2;8* might be an entry point for Na^+ influx in root surface cells under salt stress when there is a low external Ca^{2+}. When equivalent external K^+ and Na^+ were available, we observed that the presence of the K^+ inhibited the Na^+ transport (Figure 2). This could have physiological relevance in relation to regulating monovalent ion transport in conditions where K^+ is abundant relative to when Na^+ is in excess, such as in saline conditions. Water and ion transport are involved in the regulation of cell expansion, and aquaporins that can transport both ions and water could also potentially have key functions in cell expansion processes [55,56]. To understand the physiological roles of *HvPIP2;8*, it will be necessary to phenotype the barley control and mutant or transgenic lines that significantly vary in the abundance of *HvPIP2;8*. It will also be essential to determine the specific location and abundance of the *HvPIP2;8* protein in both roots and shoots in control and salt-stressed conditions.

4. Materials and Methods

4.1. Plant Materials and Growth Conditions

For sterilization, the seeds of barley (*Hordeum vulgare* L., cv. Haruna-Nijo, cv. K305, cv. I743) were treated with 10% H_2O_2 for 10 min. After 1 day, seeds were immersed in distilled water with aeration, and the germinated seeds were transplanted and hydroponically cultured with aeration in 3.5 L pots with 0.25 mM $CaSO_4$ for 2 days, and then for more days after replacing the medium with the hydroponic solution as described previously [57]. With aeration, all pots were in the dark for first 1 day, and then for a 12 h dark/12 h light cycle with fluorescent lamps of 150 photons μmol m^{-2} s^{-1} in an airconditioned room (23 ± 0.5 °C). Salt stress (100 mM or 200 mM NaCl) was treated with 5-day-old seedlings by adding 20.5 g and 40.9 g NaCl to the 3.5 L of hydroponic solution, respectively.

4.2. Extraction of RNA and Gene Expression Analysis by RT-PCR and Real-Time Quantitative PCR (qPCR)

Shoots and roots were sampled from hydroponically grown barley plants at 5, 6, or 10 days old (control), and 1 or 5 days after the treatment of 100 mM or 200 mM NaCl. Samples were rinsed and immediately frozen in liquid nitrogen. Total RNA was extracted using a mortar and pestle and the RNeasy Plant Mini Kit (Qiagen, Hilden, Germany). cDNA was synthesized using the Rever Tra Ace kit (Toyobo, Osaka Japan). cDNA fragments of *HvPIP2;8* (GenBank accession number AK356299) and Elongation factor 1α (*EF1α*, GenBank accession number Z50789) as the internal control were amplified with a set of specific primers (Supplementary Table S1). Absolute quantification was performed in the qPCR analysis using the 7300 real-time PCR machine (Applied Biosystem, Foster City, CA, USA) with PCR conditions of 50 °C for 2 min, 95 °C for 10 min, 33 cycles of 95 °C for 15 s, and 58 °C for 1 min to analyze the expression level of *HvPIP2;8*. Transcript copy numbers were quantified from three technical replications, and two biological independent experiments were conducted.

4.3. Preparation of HvPIP cRNAs

The coding region of each *HvPIP* (from *Hordeum vulgare* cv. Haruna-Nijo) was cloned into the vector pXβG-ev1 [24,31]. Each construct was linearized and cRNAs were synthesized using the mMESSAGE mMACHINE T3 kit (Ambion, Austin, TX, USA), with a final concentration of 1 μg/μL.

4.4. Expression of HvPIPs in X. laevis Oocytes

Oocytes were obtained from adult female *X. laevis* frogs and placed in a modified Barth's solution (MBS: 88 mM NaCl, 1 mM KCl, 2.4 mM $NaHCO_3$, 1.5 mM Tris-HCl (pH 7.6), 0.3 mM $Ca(NO_3)_2 \cdot 4H_2O$, 0.41 mM $CaCl_2 \cdot 4H_2O$, 0.82 mM $MgSO_4 \cdot 7H_2O$, 10 μg mL^{-1} penicillin sodium salt, and 10 μg mL^{-1} streptomycin sulfate) or in a low Na$^+$ Ringer solution (62 mM NaCl, 36 mM KCl, 5 mM $MgCl_2$, 0.6 mM $CaCl_2$, 5 mM HEPES (pH 7.6), 5% (v/v) horse serum, and antibiotics (0.05 mg mL^{-1} tetracycline, 100 units mL^{-1} penicillin, and 0.1 mg mL^{-1} streptomycin)) in the experiments comparing *HvPIP2;8* S285D relative to *HvPIP2;8* WT. The lobes were torn apart and treated with 1 mg mL^{-1} collagenase B (type B, Boehringer Mannheim, Germany) in Ca-free MBS for 1.5 h. Isolated oocytes were washed several times and incubated in MBS for 1 day at 20 °C before the microinjection.

Oocytes were injected with 10 ng of *HvPIP2* cRNA. As for *HvPIP1s*, 40 ng of each cRNA was injected. Oocytes were injected with nuclear-free water as a negative control in all experiments. Injected oocytes were incubated for 24 h to 48 h at 20 °C in MBS or a low Na$^+$ Ringer solution until the electrophysiological experiments were performed. The experiments using frog oocytes were approved by the Animal Care and Use Committee, Okayama University (approval number OKU-2017271), which follows the related international and domestic regulations. Experiments testing *HvPIP2;8* wild type (WT) relative to the *HvPIP2;8* S285D mutant, and additional replication experiments confirming the HvPIP2 relative ionic conductance, were performed at the University of Adelaide, Waite Research Institute. For the experiments comparing *HvPIP2;8* WT an *HvPIP2;8* S285D, the currents were recorded in "Na100": 100 mM NaCl; 2 mM KCl, 1 mM $MgCl_2$, 5 mM HEPES, 50 μM $CaCl_2$, 100 mM NaCl or 100 mM KCl, osmolality 220 mosmol Kg^{-1} (adjusted with D-mannitol), and a pH of 8.5. Ionic conductance was calculated by taking the slope of a regression of the linear region across the reversal potential (-40 mV to $+20$ mV). Oocytes were either untreated or were pre-treated in a low Na^+ Ringer solution that contained 10 μM H7 dihydrochloride (H7) for 2 h before TEVC and the swelling assay.

4.5. Electrophysiology

Two-electrode voltage clamp (TEVC) was performed using *X. laevis* oocytes injected with water or cRNA. Borosilicate glass pipettes (Harvard Apparatus, GC150TF-10, 1.5 mm O.D. × 1.17 mm I.D.) for voltage and current injecting electrodes were pulled and filled with 3 M KCl. A bath clamp system was used to minimize the effect of series resistance in the bath solution. The bath current and voltage sensing electrodes consisted of a silver–silver chloride electrode connected to the bath by 3% agar with 3 M KCl bridges. All bath solutions contained a background of high external calcium concentration (1.8 mM $MgCl_2$, 1.8 mM Mannitol, 1.8 mM $CaCl_2$, 10 mM HEPES, pH 7.5 with Tris) or low external calcium concentration (1.8 mM $MgCl_2$, 1.8 mM EGTA (ethylene glycol-bis (β-aminoethyl ether)-*N*,*N*,*N'*,*N'*-tetraacetic acid), 1.8 mM $CaCl_2$, 10 mM HEPES, pH 7.5 with Tris), except where otherwise mentioned. Osmolality of the bath solutions was adjusted to 200 mosmol Kg^{-1} with supplemental mannitol. Free Ca^{2+} was calculated using https://somapp.ucdmc.ucdavis. edu/pharmacology/bers/maxchelator/CaMgATPEGTA-NIST-Plot.htm. Divalent cations (Ca^{2+}, Mg^{2+}, Ba^{2+} and Cd^{2+}) and monovalent cations (Na^+, K^+, Li^+, Cs^+ and Rb^+) were added as chloride salts or gluconate salts. Each oocyte was carefully pierced with the voltage and current electrodes and the membrane voltage was allowed to stabilize. Conductance responses were monitored through the experiments by the repeat of steps from -120 mV to $+30$ mV with 2 s steady states and 5 s intervals. The recording was performed and analyzed with an Axoclamp 900A amplifier and Clampex 9.0 software (Molecular Devices, CA, USA) at room temperature (20–22 °C).

For the analysis of the *HvPIP2;8* 285S mutants relative to the *HvPIP2;8* wild type (WT), the oocyte preparation, oocyte water permeability, and electrophysiology have been described [14]. The water- or cRNA-injected oocytes were incubated in a low Na^+ Ringer solution (62 mM NaCl, 36 mM KCl, 5 mM $MgCl_2$, 0.6 mM $CaCl_2$, 5 mM HEPES, 5% (*v/v*) horse serum, and antibiotics (0.05mg mL^{-1} tetracycline, 100 units mL^{-1} penicillin/0.1 mg mL^{-1} streptomycin), pH 7.6, for 24–36 h. The water- or cRNA-injected oocytes were pre-incubated in a 3 mL iso-osmotic solution (5 mM NaCl, 2 mM KCl, 1 mM $MgCl_2$, 50 μM $CaCl_2$, pH 8.5) with an osmolality of 240 mosmol Kg^{-1} (adjusted with D-mannitol) for 1 h prior to being transferred to a solution with the same ionic composition (5 mM NaCl, 2 mM KCl, 1 mM $MgCl_2$, 50 μM $CaCl_2$, pH 8.5), with an osmolality of 45 mosmol Kg^{-1} for the photometric swelling assay. Two-electrode voltage clamp (TEVC) recordings were performed on *X. laevis* oocytes 24–36 h post injection. Preparation of glass pipettes was as described [14]. TEVC experiments were performed using an Oocyte Clamp OC-725C (Warner Instruments, Hamden, CT, USA) with a Digidata 1440A data acquisition system interface (Axon Instruments, Foster City, CA, USA). Injected oocytes were continuously perfused with solution after being pierced with the voltage and current electrodes and allowed to stabilize. TEVC was performed in solutions consisting of 100 mM NaCl ("Na100") or 100 mM KCl ("K100") in a basal solution (2 mM KCl, 1 mM $MgCl_2$, and 5 mM HEPES, osmolality was adjusted to 220–230 mosmol Kg^{-1} with D-mannitol) with 50 μM $CaCl_2$ and a pH of 8.5. For experiments involving kinase inhibitor H7, the injected oocytes were incubated prior to TEVC in a low Na^+ Ringer solution (described previously) supplemented with 10 μM H7 dihydrochloride (Sigma, #17016) from concentrated stocks dissolved in water. Steady-state currents were recorded starting from a -40 mV holding potential for 0.5 s and ranging from 40 mV to -120 mV with 20 mV decrements for

0.5 s before following a -40 mV pulse for another 0.5 s. Ionic conductance was calculated by taking the slope of a regression of the linear region across the reversal potential (-40 mV to $+20$ mV). TEVC recordings were analyzed with CLAMPEX 9.0 software (pClamp 9.0 Molecular Devices, CA, USA).

Biological replication included testing of different oocytes from different batches harvested from different frogs, and the oocyte and batch replication was three or more; the representative result from one or more oocyte batch from each experiment is included in the figures.

4.6. Statistical Analysis

Statistical analysis was conducted using SPSS statistics software (version 20). Analysis of variance was identified by one-way ANOVA followed by the least significant difference (LSD) test at the 0.05 level; or one-way ANOVA followed by Fisher's post-hoc test.

5. Conclusions

Our electrophysiological analyses of barley HvPIP aquaporins expressed in *X. laevis* oocytes have shown that *HvPIP2;8* facilitates an ionic conductance at the plasma membrane in the presence of Na^+ and/or K^+ in an external Ca^{2+}-sensitive manner. Co-expression of HvPIP1s and *HvPIP2;8* significantly reduced the *HvPIP2;8*-dependent ionic conductance, and our manipulation of protein phosphorylation revealed that this channel is likely to be subject to complex regulation involving heteromerization and post-translational modification. These findings progress our insight into the potential roles of plant aquaporins under salt stress and they are likely to inspire future research to uncover the molecular and structural mechanisms that control the dual permeability of aquaporins for ions and water, and testing of the physiological role of *HvPIP2;8* in planta.

Supplementary Materials: Supplementary materials can be found at http://www.mdpi.com/1422-0067/21/19/7135/s1. Supplementary Figure S1. The inhibition of Na^+ transport by external free Ca^{2+} concentration in *Xenopus laevis* oocytes expressing *HvPIP2;8* in the presence of 86.4 mM NaCl and 9.6 mM KCl. The background solution contained (1.8 mM $MgCl_2$, 1.8 mM EGTA, 1.8 mM $CaCl_2$, 10 mM HEPES pH 7.5 with Tris) for external free 30 µM Ca^{2+} to 1 mM Ca^{2+} concentration; and (1.8 mM $MgCl_2$, 1.8 mM Mannitol, 1.8 mM $CaCl_2$, 10 mM HEPES pH 7.5 with Tris) for external free 1.8 mM Ca^{2+} concentration as control. Steady-state current-voltage curves of *X. laevis* oocytes injected with 10 ng of cRNA per oocyte from the same batch (n = 5–6 for *HvPIP2;8* cRNA). Supplementary Figure S2. Reversal potentials of *HvPIP2;8* mediated ionic currents in the presence of 48 mM NaCl with 48 mM each alkaline cation. Data are means $\pm$ SE (n = 4–5). Supplementary Figure S3. Interaction between K^+ and Na^+ on *HvPIP2;8* mediated currents. (A) Effect of external Na^+ cation on K^+ permeability. (B) Effect of external K^+ cation on Na^+ permeability through *HvPIP2;8*-transporter. The total concentration of (Na + K) was constantly 96 mM. Na^+ and K^+ external concentration (chloride salt) were 9 different ratios bath solutions with high calcium condition contained a background (1.8 mM $MgCl_2$, 1.8 mM $CaCl_2$, 1.8 mM Mannitol, 10 mM HEPES pH 7.5 with Tris). Steady-state current-voltage curves of *X. laevis* oocytes injected with 10 ng of cRNA per oocyte from the same batch. Data are means $\pm$ SE, n = 5–6. Supplementary Figure S4. (A,B) *HvPIP2;8* WT and *HvPIP2;8* S285D cRNA injected oocytes exhibited batch to batch variation in three independent frogs. Oocytes were injected with 46 nL water (Control) or with 46 nL water (n = 11) containing 23 ng *HvPIP2;8* WT (n = 30) or *HvPIP2;8* S285D (n = 40) cRNA. Ionic conductance and osmotic water permeability (P_{os}) of cRNA injected oocytes was determined via TEVC and swelling assay, respectively. Dots in black indicates 1st batch, dots in red indicates 2nd batch and dots in blue indicates 3rd batch. Data is shown as mean $\pm$ SEM where each data point represents an individual oocyte. Significant differences ($p < 0.05$) are indicated by different letters (one-way ANOVA, Fisher's post-test). (C) Kinase inhibitor H7 treatment increase Na^+ and K^+ conductance in *HvPIP2;8* S285D expressing oocytes. Oocytes were injected with 46 nL water (Control) or with 46 nL water containing 23 ng *HvPIP2;8* WT or *HvPIP2;8* S285D cRNA. Ionic conductance of cRNA injected oocytes was determined via the TEVC. For TEVC, currents were tested in solution containing 100 mM NaCl and 2 mM KCl or 100 mM KCl; and each solution contained: 1 mM $MgCl_2$, 5 mM HEPES, 50 µM $CaCl_2$, 100 mM NaCl or 100 mM KCl, osmolality of 220 mosmol Kg^{-1}, pH 8.5. There was 2 mM KCl in the Na solution, but the K^+ solution did not contain Na^+. Oocytes injected with water (with H7, n = 5; without H7, n = 4) or *HvPIP2;8* (with H7, n = 14; without H7, n = 10) or *HvPIP2;8* S285D (with H7, n = 17; without H7, n = 11) cRNA were either untreated or were pre-treated in low Na^+ Ringers solution that contained with 10 µM dihydrochloride (H7) for 2 h before TEVC. Data is shown as mean $\pm$ SE where each data point represents an individual oocyte. Significant differences are indicated by one asterisk ($p < 0.05$) or two asterisks ($p < 0.01$) (one-way ANOVA, Fisher's post-test). (D) Predicted protein kinase A (PKA) and protein kinase C (PKC), phosphorylation sites in *HvPIP2;8*. Blue, amino acids predicted to be phosphorylated by PKA. Red, amino acids predicted to be phosphorylated by PKC. NetPhos 3.1 server (http://www.cbs.dtu.dk/services/NetPhos/, access time 07/2020) was used to determine the sites. (E) Predicted amino acid sequence alignment for *HvPIP2;8* and AtPIP2;1 indicating predicted phosphorylation sites based on NetPhos (http://www.cbs.dtu.dk/services/NetPhos/)

analysis. Sites predicted to be phosphorylated by either PKA or PKC are listed, along with key sites of interest related to gating *Spinacia oleracea* SoPIP2;1 (Leu200) [58] and protein interactions AtPIP2;1 G103 [6]. (F) Predicted amino acid sequence alignment for eight HvPIP2s relative to AtPIP2;1; it is important to note that the sequence of HvPIPs in different barley varieties may differ. Supplementary Figure S5. Expression analysis of *HvPIP2;8* using RT PCR. Salt tolerant K005 (A,D), salt sensitive I743 (C,D), and moderate tolerant Haruna-Nijo where *HvPIP2;8* was isolated originally (E,F) were grown 5 days without salt stress then grow more 1 day and 5 days with or without supplemented 100 or 200 mM NaCl. Total RNA was isolated from shoots and roots and *HvPIP2;8* fragments (A,C,E) or internal standard EF1α (B,D,F) fragments were amplified. Black and white arrow heads indicate PCR products of *HvPIP2;8* and EF1α, respectively. Representative result was shown in 3 replications. M, DNA size markers. Table S1. Gene-specific primer pairs used in PCR experiments. Table S2. Ionic conductance of oocytes injected HvPIP2s or water in the presence of 86.4 mM NaCl and 9.6 mM KCl. Ionic conductance was calculated based on the data obtained from $V = -75$ mV to -120 mV of the membrane potential in Figure 1. Data are means $\pm$ SE (n = 5–7), ns (not significant), * ($p < 0.05$) using one-way ANOVA with Duncan's multiple comparisons test. Table S3. Reversal potential of ion currents in oocytes expressing *HvPIP2;8* in the presence of NaCl or KCl. Free external Ca^{2+} was calculated as about 30 μM in low Ca^{2+} and 1.8 mM in high Ca^{2+} solutions. Data are means $\pm$ SE.

Author Contributions: Conceptualization, T.H. and M.K.; methodology, T.H., C.S.B., S.D.T. and M.K.; validation, T.H., C.S.B., S.D.T. and M.K.; formal analysis, M.K.; investigation, S.T.H.T., S.I., J.Q. and S.M.; resources, M.K.; data curation, T.H., C.S.B., S.D.T. and M.K.; writing—original draft preparation, S.T.H.T.; writing—review and editing, T.H., C.S.B., S.D.T. and M.K.; visualization, S.T.H.T., C.S.B. and M.K.; supervision, S.D.T. and M.K.; project administration, M.K.; funding acquisition, C.S.B. and M.K. All authors have read and agreed to the published version of the manuscript.

Funding: This research was founded by Ohara Foundation for Agricultural Science, (to M.K.), and by the Ministry of Education, Culture, Sports, Science and Technology (MEXT) as part of the Joint Research Program implemented at the Institute of Plant Science and Resources, Okayama University in Japan (2918, 3019, 3118, R220 to T.H.), and the MEXT scholarship (Research Student, to STH.T. and S.I.). We thank the Australian Research Council for support from DP190102725 (C.B., J.Q. and S.D.T.), FT180100476 (C.B.), and CE140100008 (S.D.T.), and partially by Japanese KAKENHI Grant-in-Aid for Scientific Research (C) (20K06708 to T.H. and M.K.). We thank the Australia-Japan Foundation for granting for "Boosting barley and rice stress tolerance in Australia and Japan" (C.B., S.D.T., M.K. and T.H.).

Acknowledgments: We thank Yoshiyuki Tsuchiya for experimental assistance.

Conflicts of Interest: The authors declare no conflict of interest. The funders had no role in the design of the study; in the collection, analyses, or interpretation of data; in the writing of the manuscript, and in the decision to publish the results.

References

1. Maurel, C.; Boursiac, Y.; Luu, D.T.; Santoni, V.; Shahzad, Z.; Verdoucq, L. Aquaporins in plants. *Physiol. Rev.* **2015**, *95*, 1321–1358. [PubMed]

2. Laloux, T.; Junqueira, B.; Maistriaux, L.C.; Ahmed, J.; Jurkiewicz, A.; Chaumont, F. Plant and Mammal Aquaporins: Same but Different. *Int. J. Mol. Sci.* **2018**, *19*, 27.

3. Katsuhara, M.; Shibasaka, M. Barley root hydraulic conductivity and aquaporins expression in relation to salt tolerance. *Soil Sci. Plant Nutr.* **2007**, *53*, 466–470. [CrossRef]

4. Grondin, A.; Rodrigues, O.; Verdoucq, L.; Merlot, S.; Leonhardt, N.; Maurel, C. Aquaporins contribute to ABA-triggered stomatal closure through OST1-mediated phosphorylation. *Plant Cell* **2015**, *27*, 1945–1954. [CrossRef] [PubMed]

5. Rodrigues, O.; Reshetnyak, G.; Grondin, A.; Saijo, Y.; Leonhardt, N.; Maurel, C.; Verdoucq, L. Aquaporins facilitate hydrogen peroxide entry into guard cells to mediate ABA- and pathogen-triggered stomatal closure. *Proc. Natl. Acad. Sci. USA* **2017**, *114*, 9200–9205. [CrossRef] [PubMed]

6. Wang, C.; Hu, H.; Qin, X.; Zeise, B.; Xu, D.; Rappel, W.J.; Schroeder, J.I. Reconstitution of CO_2 regulation of SLAC1 anion channel and function of CO_2-permeable PIP2;1 aquaporin as CARBONIC ANHYDRASE4 interactor. *Plant Cell* **2016**, *28*, 568–582. [CrossRef]

7. Katsuhara, M.; Koshio, K.; Shibasaka, M.; Hayashi, Y.; Hayakawa, T.; Kasamo, K. Over-expression of a barely aquaporin increased the shoot/root ratio and raised salt sensitivity in transgenic rice plants. *Plant Cell Physiol.* **2003**, *44*, 1378–1383.

8. McGaughey, S.A.; Qiu, J.; Tyerman, S.D.; Byrt, C.S. Regulating root aquaporin function in response to changes in salinity. *Ann. Plant Rev. Online* **2018**, *1*, 381–416.

9. Prak, S.; Hem, S.; Boudet, J.; Viennois, G.; Sommerer, N.; Rossignol, M.; Santoni, V. Multiple phosphorylations in the C-terminal tail of plant plasma membrane aquaporins: Role in subcellular trafficking of AtPIP2;1 in response to salt stress. *Mol. Cell. Proteom.* **2008**, *7*, 1019–1030. [CrossRef] [PubMed]

10. Liu, S.Y.; Fukumoto, T.; Gena, P.; Feng, P.; Sun, Q.; Li, Q.; Ding, X.D. Ectopic expression of a rice plasma membrane intrinsic protein (OsPIP1;3) promotes plant growth and water uptake. *Plant J.* **2020**, *102*, 779–796. [CrossRef] [PubMed]

11. Ikeda, M.; Beitz, E.; Kozono, D.; Guggino, W.B.; Agre, P.; Yasui, M. Characterization of aquaporin-6 as a nitrate channel in mammalian cells—Requirement of pore-lining residue threonine 63. *J. Biol. Chem.* **2002**, *277*, 39873–39879.

12. Byrt, C.S.; Zhao, M.; Kourghi, M.; Bose, J.; Henderson, S.W.; Qiu, J.; Tyerman, S. Non-selective cation channel activity of aquaporin AtPIP2;1 regulated by Ca^{2+} and pH. *Plant Cell Environ.* **2017**, *40*, 802–815. [CrossRef]

13. Kourghi, M.; Nourmohammadi, S.; Pei, J.V.; Qiu, J.; McGaughey, S.; Tyerman, S.D.; Yool, A.J. Divalent cations regulate the ion conductance properties of diverse classes of aquaporins. *Int. J. Mol. Sci.* **2017**, *18*, 2323. [CrossRef]

14. Qiu, J.; McGaughey, S.A.; Groszmann, M.; Tyerman, S.D.; Byrt, C.S. Phosphorylation influences water and ion channel function of AtPIP2;1. *Plant Cell Environ.* **2020**, in press. [CrossRef] [PubMed]

15. Demidchik, V.; Tester, M. Sodium fluxes through nonselective cation channels in the plasma membrane of protoplasts from Arabidopsis roots. *Plant Physiol.* **2002**, *128*, 379–387. [CrossRef] [PubMed]

16. Tyerman, S.D.; Skerrett, M.; Garrill, A.; Findlay, G.P.; Leigh, R.A. Pathways for the permeation of Na^+ and Cl^- into protoplasts derived from the cortex of wheat roots. *J. Exp. Bot.* **1997**, *48*, 459–480. [CrossRef]

17. Essah, P.A.; Davenport, R.; Tester, M. Sodium influx and accumulation in Arabidopsis. *Plant Physiol.* **2003**, *133*, 307–318. [CrossRef] [PubMed]

18. Isayenkov, S.V.; Maathuis, F.J.M. Plant salinity stress: Many unanswered questions remain. *Front. Plant Sci.* **2019**, *10*, 80. [CrossRef] [PubMed]

19. Bienert, M.D.; Diehn, T.A.; Richet, N.; Chaumont, F.; Bienert, G.P. Heterotetramerization of Plant PIP1 and PIP2 Aquaporins Is an Evolutionary Ancient Feature to Guide PIP1 Plasma Membrane Localization and Function. *Front. Plant Sci.* **2018**, *9*, 15. [CrossRef]

20. Zelazny, E.; Borst, J.W.; Muylaert, M.; Batoko, H.; Hemminga, M.A.; Chaumont, F. FRET imaging in living maize cells reveals that plasma membrane aquaporins interact to regulate their subcellular localization. *Proc. Natl. Acad. Sci. USA* **2007**, *104*, 12359–12364. [CrossRef]

21. Otto, B.; Uehlein, N.; Sdorra, S.; Fischer, M.; Ayaz, M.; Belastegui-Macadam, X.; Kaldenhoff, R. Aquaporin Tetramer Composition Modifies the Function of Tobacco Aquaporins. *J. Biol. Chem.* **2010**, *285*, 31253–31260. [CrossRef] [PubMed]

22. Vitali, V.; Jozefkowicz, C.; Fortuna, A.C.; Soto, G.; Flecha, F.L.G.; Alleva, K. Cooperativity in proton sensing by PIP aquaporins. *FEBS J.* **2019**, *286*, 991–1002. [CrossRef] [PubMed]

23. Vandeleur, R.K.; Mayo, G.; Shelden, M.C.; Gilliham, M.; Kaiser, B.N.; Tyerman, S.D. The Role of Plasma Membrane Intrinsic Protein Aquaporins in Water Transport through Roots: Diurnal and Drought Stress Responses Reveal Different Strategies between Isohydric and Anisohydric Cultivars of Grapevine. *Plant Physiol.* **2009**, *149*, 445–460. [CrossRef] [PubMed]

24. Horie, T.; Kaneko, T.; Sugimoto, G.; Sasano, S.; Panda, S.K.; Shibasaka, M.; Katsuhara, M. Mechanisms of water transport mediated by PIP aquaporins and their regulation via phosphorylation events under salinity stress in Barley roots. *Plant Cell Physiol.* **2011**, *52*, 663–675. [CrossRef]

25. Boursiac, Y.; Chen, S.; Luu, D.T.; Sorieul, M.; van den Dries, N.; Maurel, C. Early effects of salinity on water transport in Arabidopsis roots. Molecular and cellular features of aquaporin expression. *Plant Physiol.* **2005**, *139*, 790–805. [CrossRef] [PubMed]

26. Boursiac, Y.; Boudet, J.; Postaire, O.; Luu, D.T.; Tournaire-Roux, C.; Maurel, C. Stimulus-induced downregulation of root water transport involves reactive oxygen species-activated cell signalling and plasma membrane intrinsic protein internalization. *Plant J.* **2008**, *56*, 207–218. [CrossRef] [PubMed]

27. Li, X.; Wang, X.; Yang, Y.; Li, R.; He, Q.; Fang, X.; Luu, D.T.; Maurel, C.; Lin, J. Single-molecule analysis of PIP2;1 dynamics and partitioning reveals multiple modes of Arabidopsis plasma membrane aquaporin regulation. *Plant Cell* **2011**, *23*, 3780–3797. [CrossRef]

28. Ismail, A.M.; Horie, T. Genomics, Physiology, and Molecular Breeding Approaches for Improving Salt Tolerance. *Annu. Rev. Plant Biol.* **2017**, *68*, 405–434. [CrossRef]

29. Munns, R.; Tester, M. Mechanisms of salinity tolerance. *Ann. Rev. Plant Biol.* **2008**, *59*, 651–681. [CrossRef]

30. Katsuhara, M.; Hanba, Y.T.; Shiratake, K.; Maeshima, M. Expanding roles of plant aquaporins in plasma membranes and cell organelles. *Funct. Plant Biol.* **2008**, *35*, 1–14. [CrossRef]

31. Shibasaka, M.; Sasano, S.; Utsugi, S.; Katsuhara, M. Functional characterization of a novel plasma membrane intrinsic protein2 in barley. *Plant Sigal. Behav.* **2012**, *7*, 1648–1652. [CrossRef] [PubMed]

32. Hove, R.M.; Ziemann, M.; Bhave, M. Identification and Expression Analysis of the Barley (*Hordeum vulgare* L.) Aquaporin Gene Family. *PLoS ONE* **2015**, *10*, e0128025. [CrossRef] [PubMed]

33. Kaneko, T.; Horie, T.; Nakahara, Y.; Tsuji, N.; Shibasaka, M.; Katsuhara, M. Dynamic Regulation of the Root Hydraulic Conductivity of Barley Plants in Response to Salinity/Osmotic Stress. *Plant Cell Physiol.* **2015**, *56*, 875–882. [CrossRef] [PubMed]

34. Knipfer, T.; Besse, M.; Verdeil, J.L.; Fricke, W. Aquaporin-facilitated water uptake in barley (*Hordeum vulgare* L.) roots. *J. Exp. Bot.* **2011**, *62*, 4115–4126. [CrossRef]

35. Besse, M.; Knipfer, T.; Miller, A.J.; Verdeil, J.L.; Jahn, T.P.; Fricke, W. Developmental pattern of aquaporin expression in barley (*Hordeum vulgare* L.) leaves. *J. Exp. Bot.* **2011**, *62*, 4127–4142. [CrossRef]

36. Coffey, O.; Bonfield, R.; Corre, F.; Althea Sirigiri, J.; Meng, D.; Fricke, W. Root and cell hydraulic conductivity, apoplastic barriers and aquaporin gene expression in barley (*Hordeum vulgare* L.) grown with low supply of potassium. *Ann. Bot.* **2018**, *122*, 1131–1141. [CrossRef]

37. Amtmann, A.; Laurie, S.; Leigh, R.; Sanders, D. Multiple inward channels provide flexibility in Na^+/K^+ discrimination at the plasma membrane of barley suspension culture cells. *J. Exp. Bot.* **1997**, *48*, 481–497. [CrossRef]

38. Chen, Z.; Pottosin, I.I.; Cuin, T.A.; Fuglsang, A.T.; Tester, M.; Jha, D.; Zepeda-Jazo, I.; Zhou, M.; Palmgren, M.G.; Newman, I.A.; et al. Root plasma membrane transporters controlling K^+/Na^+ homeostasis in salt-stressed barley. *Plant Physiol.* **2007**, *145*, 1714–1725. [CrossRef]

39. Kim, M.J.; Lee, Y.S.; Han, J.K. Modulation of lysophosphatidic acid-induced Cl^- currents by protein kinases A and C in the Xenopus oocyte. *Biochem. Pharmacol.* **2000**, *59*, 241–247. [CrossRef]

40. Horie, T.; Karahara, I.; Katsuhara, M. Salinity tolerance mechanisms in glycophytes: An overview with the central focus on rice plants. *Rice* **2012**, *5*, 1–18. [CrossRef]

41. Horie, T.; Costa, A.; Kim, T.H.; Han, M.J.; Horie, R.; Leung, H.Y.; Miyao, A.; Hirochika, H.; An, G.; Schroeder, I.J. Rice OsHKT2;1 transporter mediates large Na^+ influx component into K^+-starved roots for growth. *EMBO J.* **2007**, *26*, 3003–3014. [CrossRef]

42. Davenport, R.J.; Tester, M. A weakly voltage-dependent, nonselective cation channel mediates toxic sodium influx in wheat. *Plant Physiol.* **2000**, *122*, 823–834. [CrossRef] [PubMed]

43. Mori, I.C.; Nobukiyo, Y.; Nakahara, Y.; Shibasaka, M.; Furuichi, T.; Katsuhara, M. A Cyclic Nucleotide-Gated Channel, HvCNGC2-3, Is Activated by the Co-Presence of Na^+ and K^+ and Permeable to Na^+ and K^+ Non-Selectively. *Plants* **2018**, *7*, 61. [CrossRef] [PubMed]

44. Chaumont, F.; Tyerman, S.D. Aquaporins: Highly Regulated Channels Controlling Plant Water Relations. *Plant Physiol.* **2014**, *164*, 1600–1618. [CrossRef] [PubMed]

45. Byrt, C.S. Genes for Sodium Exclusion in Wheat. Ph.D. Thesis, University of Adelaide, Adelaide, Australia, 2008. Available online: https://digital.library.adelaide.edu.au/dspace/handle/2440/56208 (accessed on 26 September 2020).

46. Xu, B.; Hrmova, M.; Gilliham, M. High affinity Na^+ transport by wheat HKT1;5 is blocked by K^+. *BioRxiv* **2018**. Available online: https://www.biorxiv.org/content/10.1101/280453v2.abstract (accessed on 26 September 2020).

47. Yao, X.; Horie, T.; Xue, S.; Leung, H.Y.; Katsuhara, M.; Brodsky, D.E.; Wu, Y.; Schroeder, J.I. Differential sodium and potassium transport selectivities of the rice OsHKT2;1 and OsHKT2;2 transporters in plant cells. *Plant Physiol.* **2010**, *152*, 341–355. [CrossRef]

48. Yool, A.J.; Weinstein, A.M. New roles for old holes: Ion channel function in aquaporin-1. *News Physiol. Sci.* **2002**, *17*, 68–72. [CrossRef]

49. Engh, R.A.; Girod, A.; Kinzel, V.; Huber, R.; Bossemeyer, D. Crystal structures of catalytic subunit of cAMP-dependent protein kinase in complex with isoquinolinesulfonyl protein kinase inhibitors H7, H8, and H89 structural implications for selectivity. *J. Biol. Chem.* **1996**, *271*, 26157–26164. [CrossRef]

50. Nyblom, M.; Frick, A.; Wang, Y.; Ekvall, M.; Hallgren, K.; Hedfalk, K.; Neutze, R.; Tajkhorshid, E.; Törnroth-Horsefield, S. Structural and functional analysis of SoPIP2;1 mutants adds insight into plant aquaporin gating. *J. Mol. Biol.* **2009**, *387*, 653–668. [CrossRef]

51. Hyder, S.Z.; Greenway, H. Effects of Ca^{2+} on plant sensitivity to high NaCl concentrations. *Plant Soil* **1965**, *23*, 258–260. [CrossRef]

52. Sanders, D.; Pelloux, J.; Brownlee, C.; Harper, J.F. Calcium at the crossroads of signaling. *Plant Cell* **2002**, *14*, S401–S417. [CrossRef]

53. Choi, W.G.; Toyota, M.; Kim, S.H.; Hilleary, R.; Gilroy, S. Salt stress induced Ca^{2+} waves are associated with rapid, long-distance root-to-shoot signaling in plants. *Proc. Natl. Acad. Sci. USA* **2014**, *111*, 6497–6502. [CrossRef] [PubMed]

54. Horie, T.; Hauser, F.; Schroeder, J.I. HKT transporter-mediated salinity resistance mechanisms in Arabidopsis and monocot crop plants. *Trend. Plant Sci.* **2009**, *14*, 660–668. [CrossRef] [PubMed]

55. Liam, D.; Davies, J. Cell expansion in roots. *Curr. Opin. Plant Biol.* **2004**, *7*, 33–39.

56. Martinez-Ballesta, M.C.; Garcia-Ibañez, P.; Yepes-Molina, L.; Rios, J.J.; Carvajal, M. The expanding role of vesicles containing aquaporins. *Cells* **2018**, *7*, 179. [CrossRef]

57. Katsuhara, M.; Akiyama, Y.; Koshio, K.; Shibasaka, M.; Kasamo, K. Functional analysis of water channels in barley roots. *Plant Cell Physiol.* **2002**, *43*, 885–893. [CrossRef]

58. Törnroth-Horsefield, S.; Wang, Y.; Hedfalk, K.; Johanson, U.; Karlsson, M.; Tajkhorshid, E.; Neutze, R.; Kjellbom, P. Structural mechanism of plant aquaporin gating. *Nature* **2006**, *439*, 688–694. [CrossRef]

International Journal of
Molecular Sciences

Article

Na⁺ Transporter SvHKT1;1 from a Halophytic Turf Grass Is Specifically Upregulated by High Na⁺ Concentration and Regulates Shoot Na⁺ Concentration

Yuki Kawakami [1], Shahin Imran [2], Maki Katsuhara [2] and Yuichi Tada [3,*]

[1]   Graduate School of Bionics, Computer and Media Sciences, Tokyo University of Technology,
     1404-1 Katakura, Hachioji, Tokyo 192-0982, Japan; g11170090f@edu.teu.ac.jp
[2]   Institute of Plant Science and Resources, Okayama University, Chuo 2-20-1, Kurashiki,
     Okayama 710-0046, Japan; ptj87a5q@s.okayama-u.ac.jp (S.I.); kmaki@okayama-u.ac.jp (M.K.)
[3]   School of Biosciences and Biotechnology, Tokyo University of Technology, 1404-1 Katakura, Hachioji,
     Tokyo 192-0982, Japan
*   Correspondence: tadayui@stf.teu.ac.jp

Received: 20 July 2020; Accepted: 21 August 2020; Published: 24 August 2020

Abstract: We characterized an Na⁺ transporter SvHKT1;1 from a halophytic turf grass, *Sporobolus virginicus*. SvHKT1;1 mediated inward and outward Na⁺ transport in *Xenopus laevis* oocytes and did not complement K⁺ transporter-defective mutant yeast. SvHKT1;1 did not complement *athkt1;1* mutant *Arabidopsis*, suggesting its distinguishable function from other typical HKT1 transporters. The transcript was abundant in the shoots compared with the roots in *S. virginicus* and was upregulated by severe salt stress (500 mM NaCl), but not by lower stress. *SvHKT1;1*-expressing *Arabidopsis* lines showed higher shoot Na⁺ concentrations and lower salt tolerance than wild type (WT) plants under nonstress and salt stress conditions and showed higher Na⁺ uptake rate in roots at the early stage of salt treatment. These results suggested that constitutive expression of *SvHKT1;1* enhanced Na⁺ uptake in root epidermal cells, followed by increased Na⁺ transport to shoots, which led to reduced salt tolerance. However, Na⁺ concentrations in phloem sap of the SvHKT1;1 lines were higher than those in WT plants under salt stress. Based on this result, together with the induction of the *SvHKT1;1* transcription under high salinity stress, it was suggested that SvHKT1;1 plays a role in preventing excess shoot Na⁺ accumulation in *S. virginicus*.

Keywords: *Arabidopsis thaliana*; halophyte; high-affinity potassium transporter (HKT); Na⁺ transporter; salt tolerance; *Sporobolus virginicus*

---

1. Introduction

Soil salinity is one of the major environmental stress factors, causing significant losses in global agricultural productivity [1]. To fight this problem, it is necessary to develop salt-tolerant crops, which require a better understanding of the physiological mechanisms controlling salinity tolerance in plants. Salt stress imposes both osmotic and ionic stresses, and then oxidative stress caused by these stresses. Osmotic imbalance causes water deficit, reduced leaf area expansion, and stomatal closure, which ultimately lessen photosynthesis and growth [2]. Toxic Na⁺ can accumulate in the cytoplasm and cause imbalances in the absorption of other essential ions such as K⁺, leading to malfunction of essential biochemical and physiological processes [3]. Na⁺ has a strong inhibitory effect on K⁺ uptake by cells [4]. The increase in cytoplasmic Na⁺ and reduction of K⁺ result in changes in membrane potential, osmotic pressure, turgor pressure, calcium signaling, reactive oxygen species signaling, and transcriptional regulation, as well as alteration of gene expression and modification of protein expression pattern and spectra of small interfering RNAs (siRNAs), signaling molecules, phytohormones, and metabolites [5]. K^+ deficiency can disrupt various enzymatic processes and impose an energetic burden on the cell owing to the requirement of organic solute synthesis to compensate for the export of Na^+ for osmotic adjustment [1]. More than 50 enzymes are activated by K^+, which cannot be substituted with Na^+ [6]. Therefore, it is important to understand how Na^+ is taken up and transported in plants under saline conditions. In order to maintain a high cytosolic K^+/Na^+ ratio, plants have different K^+ and Na^+ transporters to protect the plant against damage due to toxic Na^+ accumulation [7,8]. To maintain low cytoplasmic Na^+ concentrations and cell turgor pressure, Na^+ can be sequestered into vacuoles by the tonoplast-localized Na^+/H^+ exchanger 1 (NHX1) [9]. In the root epidermis cells, a plasma membrane-localized Na^+/H^+ antiporter (SOS1) extrudes Na^+ back to the soil in a mechanism coupled to H^+ transport [10]. Other Na^+ transporters also play crucial roles in salinity tolerance by controlling the Na^+ movement throughout the plant. Na^+ enters roots passively, via non-selective cation channels [11] and possibly other Na^+ transporters such as high-affinity potassium transporters (HKTs) [6]. The HKTs permeable either to Na^+ only (HKT1) or to K^+ and Na^+ (HKT2) are thought to play major roles in controlling Na^+ accumulation in plants. For a detailed overview of the physiological roles of HKTs, see the review by Almeida et al. [7]. The HKT2s were shown to have a role in Na^+ uptake from the external medium. In rice, OsHKT2;1 catalyzes Na^+ uptake in low K^+, low Na^+ (<2 mM) conditions [12]. OsHKT2;1 functions as a relatively Na^+-specific transporter that mediates Na^+ influx in K^+-starved roots and, thus, promotes their growth [12,13]. Overexpression of *HvHKT2;1* in barley causes increased Na^+ uptake in salt stress conditions [14]. Similarly, altered expression of *TaHKT2;1* in wheat affected Na^+ accumulation in the low-affinity range [15]. We also reported that SvHKT2;1 and SvHKT2;2 from a halophyte, *Sporobolus virginicus*, mediate both K^+ and Na^+ transport in transgenic *Arabidopsis*, *Xenopus laevis* oocytes, and yeast [16]. *S. virginicus* is a halophytic C_4 grass and shows a salinity tolerance up to 1.5 M NaCl [17]. The HKT1s play major roles in Na^+ transport. In *Arabidopsis*, the HKT family comprises a single member, AtHKT1;1, which is permeable to Na^+ only [18] and contributes to Na^+ removal from the ascending xylem sap and Na^+ recirculation from the leaves to the roots via the phloem vasculature [19–22]. Similarly, Na^+ removal from the root xylem sap and/or shoot phloem was reported for other HKT1s including OsHKT1;1, OsHKT1;4, and OsHKT1;5 in rice [23–27], TaHKT1;5-D in bread wheat [28], HvHKT1;1 and HvHKT1;5 in barley [29,30], and TmHKT1;5-A in wheat [31]. On the other hand, a halophytic relative of *Arabidopsis*, *Eutrema salsuginea* (previously *Thellungiella halophila* or *Thulengiella salsuginea*), possessed three copies of *HKT1* genes [32]. Of the three, *EsHKT1;1* and *EsHKT1;2* showed high affinity for Na^+ and K^+, respectively, in yeast [33]. Another *Arabidopsis* halophytic relative, *Eutrema parvula* (*Schrenkiella parvula*), possesses two *HKT1* genes *EpHKT1;1* and *EpHKT1;2* [34]. Although EpHKT1;2 and EsHKT1;2 belong to HKT1, they show K^+ uptake ability, which makes them functionally different from other members of HKT1 [35]. Thus, the ion permeability of HKT1s differs depending on the plant species. However, there is no report on the functions of HKT1s from halophytic monocotyledonous plants.

In this study, we isolated a gene for sodium transporter *SvHKT1;1* from a halophytic turf grass, *Sporobolus virginicus*, and revealed its unique expression profile, ion permeability, and possible functions in salt tolerance.

2. Results

2.1. Comparison of Amino-Acid Sequences of SvHKT1;1 and Other HKTs

Among the unigenes of *S. virginicus* [36], we searched for genes that are homologous to known *HKT* and identified *SvHKT1;1*, which belongs to class I *HKT* genes because its deduced amino-acid (AA) sequences contain a serine in the first P-loop (Supplementary Figure S1) [37,38]. Phylogenetic analysis following alignment of the AA sequences of SvHKT1;1 with other class I HKTs indicated an evolutionarily close relationship to *Setaria italica* (foxtail millet) SiHKT1;1 and rice OsHKT1;1 (Figure 1).

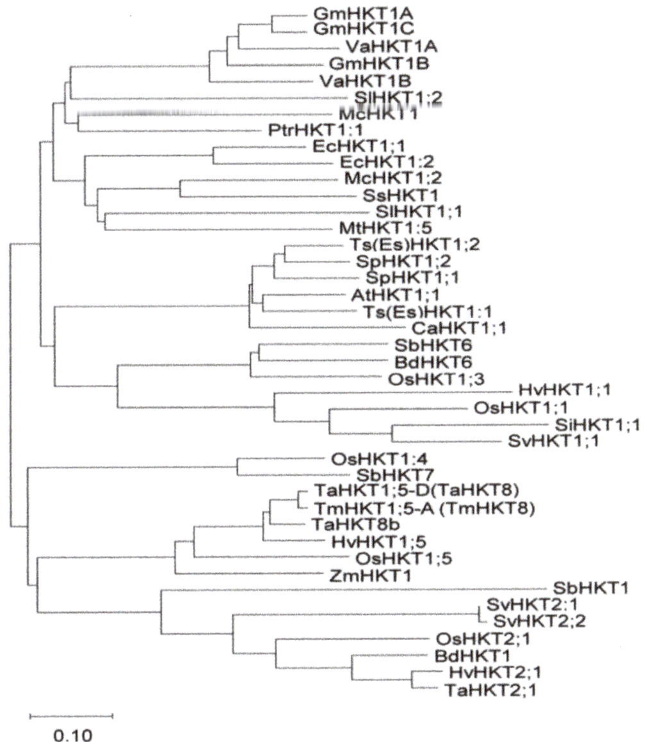

Figure 1. Phylogenetic analysis of high-affinity potassium transporters (HKTs). A phylogenetic analysis of the selected HKT amino-acid sequences was performed using the neighbor-joining method in the MEGA-X [39] software package. Accession numbers of amino-acid sequences used are listed in Supplementary Figure S1. The branch length is proportional to the evolutionary distance between the HKTs, indicating the number of amino-acid changes per site. The scale bar shows a length corresponding to 0.10 of the value.

2.2. Expression of SvHKT1;1 Gene and Na⁺ Concentration in S. virginicus under Salt Stress

The expression profile of *SvHKT1;1* gene in roots and shoots of *S. virginicus* was determined by qRT-PCR using *eIF3* as a reference gene under different salt stress conditions. *SvHKT1;1* transcript was found to preferentially accumulate in shoots compared with roots under any conditions (Figure 2A–D). In experiments at different salt concentrations, the expression stayed at similar levels under 0–300 mM NaCl conditions but significantly increased under 500 mM NaCl in both roots and shoots (Figure 2A,B). In a time-course experiment, the expression in both shoots and roots were upregulated 6 h after treatment with 500 mM NaCl, then decreased to levels several times higher than that at 0 h until 48 h after the treatment (Figure 2C,D). Similar expression profiles were observed by qRT-PCR analysis using *actin* as a reference gene (Supplementary Figure S2). Thus, *SvHKT1;1* expression is specifically upregulated by high Na⁺ concentration and preferentially detected in shoots.

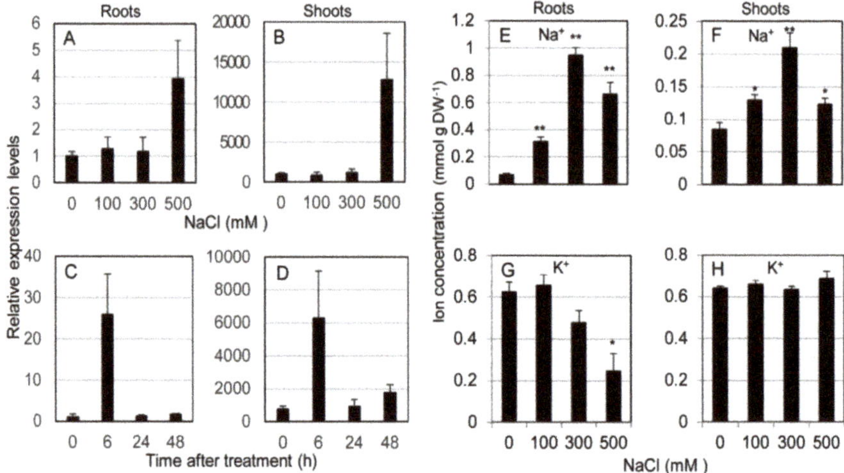

Figure 2. Expression profile of *SvHKT1;1* gene and Na⁺ concentrations in *Sporobolus virginicus* under salt stress. The expression profile of *SvHKT1;1* gene and Na⁺ concentrations in roots and shoots of hydroponically grown *S. virginicus* were determined. (**A–D**) Expression levels of *SvHKT1;1* gene determined by qRT-PCR under different NaCl concentrations (**A,B**) or at different time points after salt treatment (**C,D**). Plants grown in 1/2 Murashige and Skoog (MS) medium were transferred to 1/2 MS medium supplemented with 0, 100, 300, or 500 mM NaCl, and the roots (**A**) and shoots (**B**) were harvested at 48 h after the treatment. Plants grown in 1/2 MS medium were transferred to 1/2 MS medium supplemented with 500 mM NaCl, and the roots (**C**) and shoots (**D**) were harvested at indicated time points. Expression levels relative to that in roots at 0 h after treatment (1.0) are shown. *eIF3* was used as a reference gene. (**E–H**) Na⁺ (**E,F**) and K⁺ (**G,H**) concentrations in roots (**E,G**) and shoots (**F,H**) of hydroponically grown *S. virginicus* under different NaCl concentrations. The roots and shoots were harvested at 48 h after the treatment. Data are presented as means ± SE (n = 3 biological replicates). Single and double asterisks denote significant differences compared with the values of WT plants of the same conditions at $p < 0.05$ and $p < 0.01$, respectively, determined using the Student's t-test.

Na⁺ and K⁺ concentrations in roots and shoots of *S. virginicus* were determined under different salt stress conditions. Na⁺ was found to preferentially accumulate in roots compared with shoots under salt stress (Figure 2E,F). Shoot and root Na⁺ concentration linearly increased in accordance with the NaCl concentration up to 300 mM NaCl but dropped at 500 mM, when *SvHKT1;1* transcription was dramatically upregulated (Figure 2E,F). On the other hand, root K⁺ concentration decreased with increasing external NaCl concentration but shoot K⁺ concentration remained constant (Figure 2G,H).

2.3. Localization of SvHKT1;1 in Nicotiana benthamiana Cells

To investigate the intracellular localization of EGFP–SvHKT1;1 fusion protein, *Agrobacterium* expressing *EGFP–SvHKT1;1* or a control *EGFP*-only construct were infiltrated into *N. benthamiana* leaf cells, and the fluorescence signals were observed using confocal laser scanning microscopy (Figure 3). EGFP–SvHKT1;1 fusion protein specifically localized to the plasma membrane (Figure 3B,E,H). In contrast, when EGFP alone was expressed, it localized to the nucleus and the cytoplasm (Figure 3A,D,G). This pattern was verified by treatment with a hypertonic solution, 0.5 M mannitol, which induced plasmolysis (Figure 3C,F,I), indicating SvHKT1;1 localized to the plasma membrane.

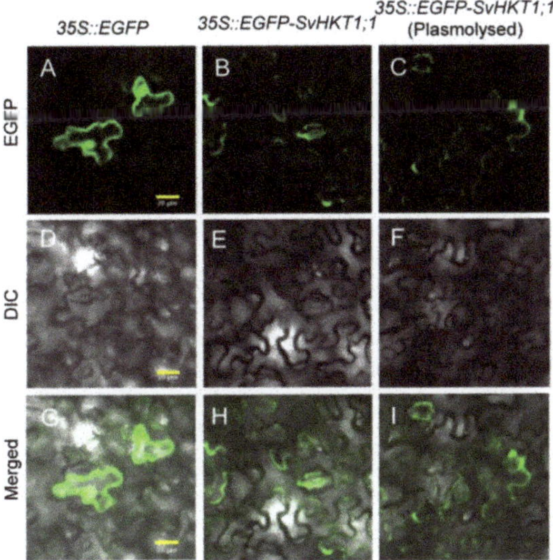

Figure 3. Subcellular localization of EGFP-fused SvHKT1;1 protein in *Nicotiana benthamiana* leaves. Confocal fluorescence images of EGFP (**A–C**), differential interference contrast images (**D–F**), and merged images (**G–I**) of *N. benthamiana* leaf cells expressing EGFP control (**A,D,G**) and EGFP–SvHKT1;1 (**B,C,E,F,H,I**). Images of non-plasmolyzed (**A–F**) and plasmolyzed (**G–I**) cells. Scale bar represents 20 μm and is applicable to all panels in this figure.

2.4. Functional Analysis of SvHKT1;1 in X. laevis Oocytes

To examine K^+ and/or Na^+ transporter activities in *X. laevis* oocytes, *SvHKT1;1* complementary RNAs (cRNAs) were injected into oocytes, and the electrophysiological profile was analyzed. In two-electrode voltage clamp (TEVC) experiments, SvHKT1;1 produced inward and outward currents (Figure 4A) when oocytes were bathed in both NaCl and Na-gluconate solutions, but ion currents were hardly detected in KCl solution (Figure 4A). These results indicate that the SvHKT1;1 transporter mediates Na^+, but not K^+ or Cl^-, transport. On the other hand, water-injected control oocytes showed small background currents in the same conditions (Figure 4B).

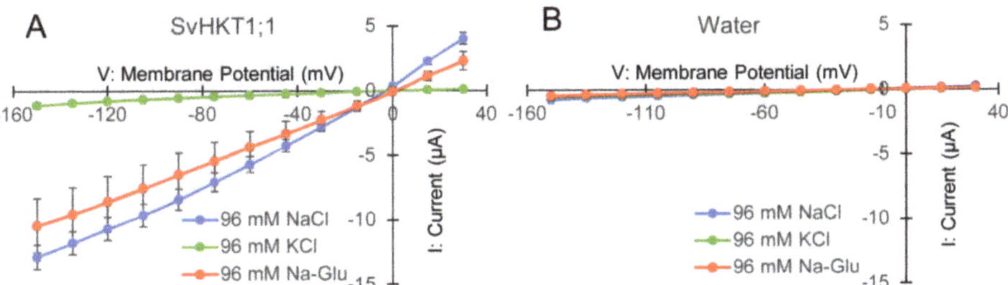

Figure 4. Analyses of SvHKT1;1-mediated ion transport by two electrode voltage clamp experiments using *Xenopus laevis* oocytes. Current–voltage relationship of oocytes injected with 12.5 ng of *SvHKT1;1* complementary RNA (cRNA) (**A**) or water (**B**) bathed in solutions containing an indicated amount of NaCl, KCl, or Na-gluconate. Voltage steps ranged from −150 to +30 mV with 15-mV increments. Data are presented as means ± SD (*n* = 3–7).

2.5. SvHKT1;1 Does Not Complement Yeast with Defective K⁺ Transporters

To examine K$^+$ channel/transporter activities in yeast, *SvHKT1;1* was expressed in yeast strain 9.3 with defective K$^+$ transporters (Supplementary Figure S3). Yeast lines harboring *SvHKT1;1* and a negative control expressing an empty vector showed similar poor growth on agar containing 0.2 mM K$^+$, whereas yeast lines harboring positive control *SvHKT2;1* grew better than *SvHKT1;1* and negative control lines (Supplementary Figure S3). These results indicate that *SvHKT1;1* does not complement the K$^+$-uptake deficiency in the mutant yeast.

2.6. SvHKT1;1 Does Not Complement athkt1;1 Mutant Arabidopsis Plants

We transformed *athkt1;1 Arabidopsis* plants in which the gene for *AtHKT1;1* sodium transporter was tagged with T-DNA. The *AtHKT1;1* transcript was not detected in the mutant line, and the *SvHKT1;1* transcript was detected in two independent transgenic mutant lines (*athkt1;1/SvHKT1;1-B* and *-C*) transformed with the *SvHKT1;1* gene driven by the *AtHKT1;1* promoter by semi qRT-PCR (Figure 5A). When transferred onto 100 mM NaCl medium, the transgenic lines, as well as *athkt1;1* mutant line, showed diminished growth (Figure 5B,C). This result indicated that *SvHKT1;1* does not complement the defected AtHKT1;1 function. Root Na$^+$ concentrations in both the mutant and the transgenic lines were significantly increased compared to that in WT plants, despite the loss of functionality of *AtHKT1;1*, whereas their shoot Na$^+$ concentrations increased, but not significantly, under salt stress (Figure 5D,F). Interestingly, root K$^+$ concentrations in the transgenic lines decreased significantly or tended to decrease compared to that in WT plants, although no difference was observed among their shoot K$^+$ concentrations (Figure 5E,G).

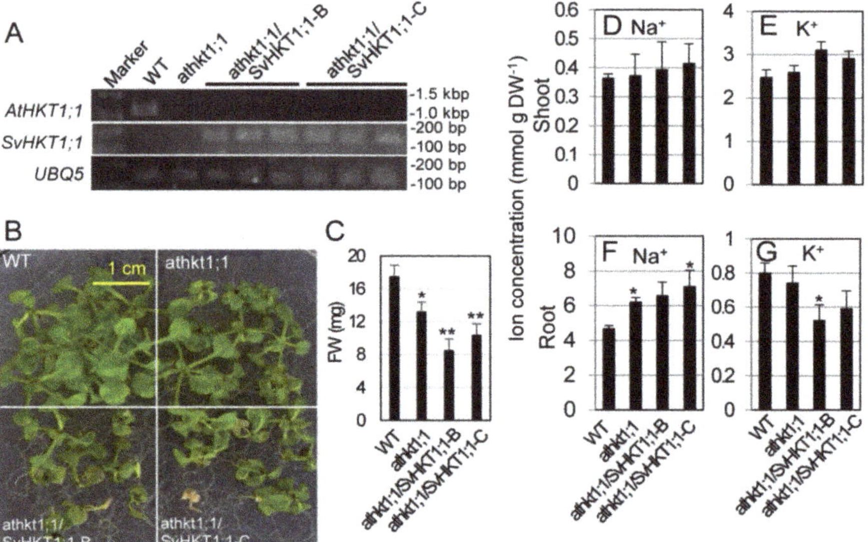

Figure 5. Complementation test of *athkt1;1* mutant *Arabidopsis* with *AtHKT1;1pro::SvHKT1;1* construct. Transcripts of *AtHKT1;1* or *SvHKT1;1* were detected in WT, mutant, and two independent lines of transformed *Arabidopsis* plants (three biological replicates) by RT-PCR (**A**). The appearance (**B**) and fresh weight (**C**) of the plants grown for two weeks on 100 mM NaCl medium. Na$^+$ (**D,F**) and K$^+$ (**E,G**) concentrations in the shoots (**D,E**) and roots (**F,G**) of the plants. Data are presented as means ± SD (*n* = 9 (**B,C**) and *n* = 3 (**D–G**)). Please note that each panel has a different Y-axis scale. Single and double asterisks denote significant differences compared with the values of WT plants of the same conditions at *p* < 0.05 and *p* < 0.01, respectively, determined using the Student's *t*-test.

2.7. Constitutive Expression of SvHKT1;1 in Wild-Type Arabidopsis Plants

We introduced *35S::SvHKT1;1* into WT *Arabidopsis* and examined the expression levels of the transgenes in four T_2 lines with putative single transgenes (#4, 7, 8, and 17), judging from the segregation ratio of the hygromycin tolerant T_3. There was a large variation in the expression level (Figure 6A). We examined root growth of the transgenic lines on 0.1 mM K^+ medium, because expression of K^+/Na^+ symporters, *SvHKT2;1* and *SvHKT2;2*, in *Arabidopsis* resulted in enhanced root growth under K^+-starved conditions in our previous study [16]. The SvHKT1;1 transgenic lines showed elongated root growth compared with WT plants, while the root growth of WT plants was severely inhibited (Figure 6B,C). A weak correlation was observed between the values of root elongation and the transcription levels.

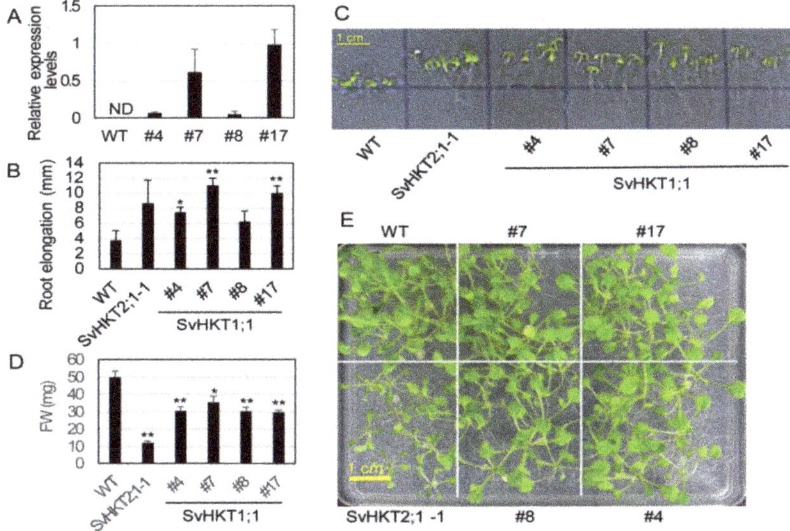

Figure 6. Expression level of the transgene, root growth, and salt tolerance of transgenic *Arabidopsis* plants expressing *SvHKT1;1*. (**A**) Expression levels of transgenes in the SvHKT1;1 transgenic lines. *Actin* was used as an internal standard. ND; not detected. Data are presented as means ± SE ($n = 3$ biological replicates). (**B**) Root elongation of transgenic and WT seedlings grown on 0.1 mM K^+ medium. Data are presented as means ± SE ($n = 3$ biological replicates). (**C**) The appearance of transgenic lines and WT seedlings on 0.1 mM K^+ medium examined in panel B. (**D**) Fresh weight (FW) of WT and the transgenic lines. One-week-old seedlings germinated on 1/2 MS agar medium were transplanted onto 1/2 MS agar medium supplemented with 50 mM NaCl, and their FW was determined after another two weeks of incubation. Data are presented as means ± SE ($n = 10$ biological replicates). (**E**) The appearance of plants examined in panel D. Single and double asterisks denote significant differences compared with the values of WT plants of the same conditions at $p < 0.05$ and $p < 0.01$, respectively, determined using the Student's *t*-test. Single and double asterisks denote significant differences compared with the values of WT plants of the same conditions at $p < 0.05$ and $p < 0.01$, respectively, determined using the Student's *t*-test.

Then, we assessed salt tolerance of the SvHKT1;1 lines at seedling stage in comparison with WT plants and the SvHKT2;1-1 line, which showed reduced salt tolerance in our previous study [40]. *Arabidopsis* seedlings were transferred onto 1/2 MS agar medium supplemented with 50 mM NaCl, incubated for a further 14 days, and their shoot fresh weight (FW) was determined. SvHKT1;1 lines showed diminished growth on 50 mM NaCl medium, although not as severely as that of the SvHKT2;1-1 line (Figure 6D,E). These results indicated that constitutive expression of *SvHKT1;1* increased salt sensitivity of the transgenic lines.

We measured shoot FW and ion concentrations in shoots and roots of the transgenic lines hydroponically cultured in 1/2 MS liquid medium supplemented with 0 or 100 mM NaCl at the bolting stage (Figure 7). Under nonstress conditions (0 mM NaCl), there was no difference between shoot growth of the transgenic lines and WT plants (Figure 7A). Under salt stress, the transgenic lines showed significantly or relatively smaller growth than WT plants (Figure 7B). Under nonstress condition, shoot Na^+ concentrations in two transgenic lines were significantly higher and the other two were higher than those in WT plants, while no significant differences were observed in shoot K^+ between SvHKT1;1 lines and WT plants (Figure 7C,D). No significant differences were observed in their root Na^+ and K^+ concentrations (Figure 7E,F). Under salt stress, shoot Na^+ and root K^+ concentrations in the transgenic lines were significantly or relatively higher than those of WT plants (Figure 7G,J). No significant differences were observed in shoot K^+ and root Na^+ concentrations between SvHKT1;1 lines and WT plants (Figure 7H,I).

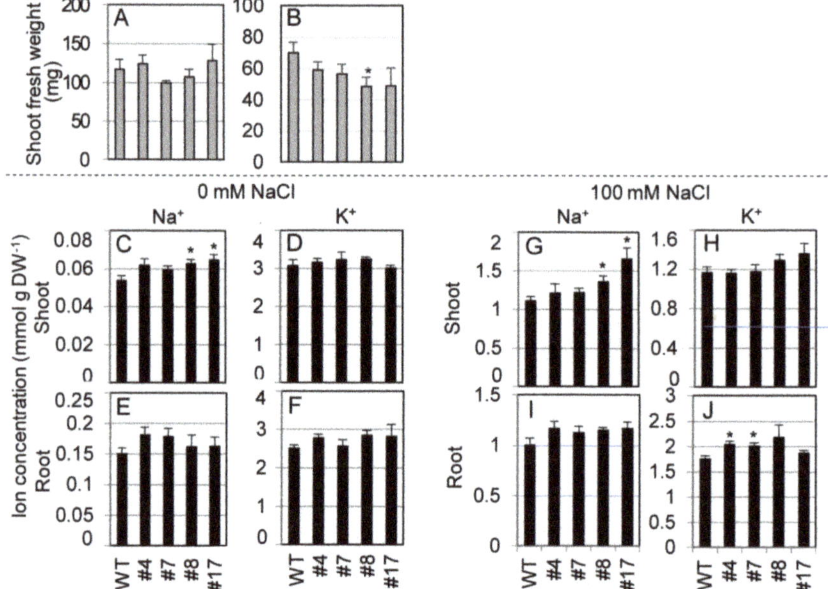

Figure 7. Shoot fresh weight and ion concentrations of *SvHKT1;1* transgenic lines and WT plants. (**A,B**) Shoot fresh weight of WT plants and the transgenic lines. Two-week-old seedlings germinated on 1/2 MS agar medium were hydroponically cultured in 1/2 MS liquid medium for another one week, and then cultured in 1/2 MS medium supplemented with 0 (**A**) or 100 mM (**B**) NaCl, and their FW was determined after one week. (**C–J**) Ion concentrations in WT plants and the transgenic lines. Na^+ (**C,E,G,I**) and K^+ (**D,F,H,J**) concentrations in their shoots (**C,D,G,H**) and roots (**E,F,I,J**) were determined. Data are presented as means ± SE (n = 3–4 biological replicates). Please note that each panel has a different Y-axis scale. Single and double asterisks denote significant differences compared with the values of WT plants of the same conditions at $p < 0.05$ and $p < 0.01$, respectively, determined using the Student's t-test.

2.8. K^+ and Na^+ Concentrations in the Xylem and Phloem Saps of Arabidopsis Expressing SvHKT1;1

To examine the mode of SvHKT1;1-mediated K^+ and Na^+ transport in *Arabidopsis* plants, *SvHKT1;1* lines and WT plants were hydroponically cultured in 1/2 Hoagland solution, and K^+ and Na^+ concentrations in the xylem and phloem saps were determined at the bolting stage (Figure 8). We used Hoagland solution in this experiment because *Arabidopsis* plants showed better growth performance in 1/2 Hoagland solution than in 1/2 MS medium. Under nonstress condition, Na^+ concentrations in the xylem saps of SvHKT1;1 lines were higher, but not significantly, than those in WT plants (Figure 8A), and Na$^+$ concentrations in the phloem saps were lower, but not significantly, in SvHKT1;1 lines than those in WT plants (Figure 8C). K$^+$ concentrations in the xylem and phloem saps were similar among the transgenic lines and WT plants, except for xylem sap of the #7 line (Figure 8B,D). In contrast, the phloem sap Na$^+$ concentrations were significantly higher in SvHKT1;1 lines than in WT plants under 100 mM NaCl conditions (Figure 8E). The phloem sap K$^+$ concentrations were similar among the transgenic lines and WT plants (Figure 8F). We could not obtain xylem sap from the transgenic lines and WT plants under salt stress.

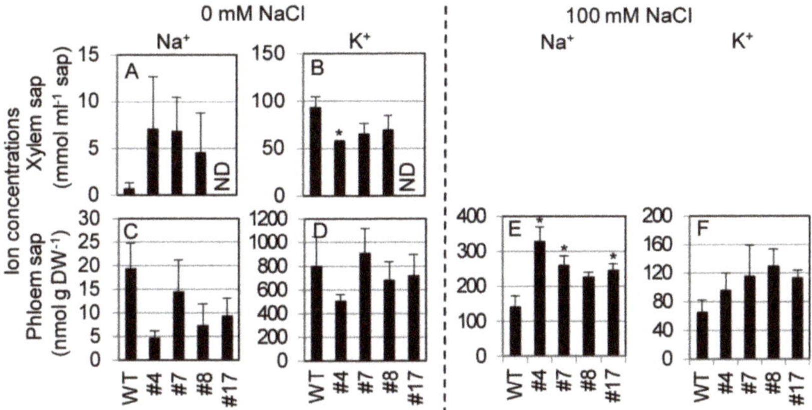

Figure 8. Na$^+$ and K$^+$ concentrations in the xylem and phloem saps from *SvHKT1;1* transgenic lines and WT plants. (**A–D**) Ion concentrations in the xylem and phloem saps of WT plants and the transgenic lines under nonstress condition. Plants were hydroponically cultured in 1/2 Hoagland liquid solution until the bolting stage, and Na$^+$ (**A,C**) and K$^+$ (**B,D**) concentrations in their xylem (**A,B**) and phloem (**C,D**) saps were determined. (**E,F**) Ion concentrations in the phloem saps of WT plants and the transgenic lines under 100 mM NaCl. ND; not determined. Three-week-old plants were subjected to 1/2 Hoagland liquid solution supplemented with 100 mM NaCl for seven days, and Na$^+$ (**E**) and K$^+$ (**F**) concentrations in their phloem saps were determined. Xylem saps were not obtained from salt-treated plants. Data are presented as means ± SE (n = 3–4 biological replicates). Please note that each panel has a different Y-axis scale. Single and double asterisks denote significant differences compared with the values of WT plants of the same conditions at $p < 0.05$ and $p < 0.01$, respectively, determined using the Student's t-test.

2.9. Ion Uptake and Translocation Rates in Arabidopsis Seedlings after Salt Treatment

To examine the mode of Na$^+$ and K$^+$ uptake, release, or translocation by SvHKT1;1 in the roots of transgenic lines, we subjected the transgenic and WT seedlings to liquid 1/2 MS medium supplemented with 100 mM NaCl for 1 h and measured the changes of their Na$^+$ and K$^+$ concentrations at the early stage of saline stress (Figure 9). While Na$^+$ uptake rate in roots of WT plants and SvHKT2;1-1 line took negative values, indicating their Na$^+$ release, SvHKT1;1 lines showed significantly increased or almost unchanged uptake rate after they were transferred to 100 mM NaCl medium (Figure 9A). Na$^+$ translocation rates in both shoots and whole plants of all transgenic lines were significantly higher than those in WT plants (Figure 9A). These results indicated enhanced Na$^+$ uptake and translocation rates in the transgenic lines compared with WT plants. There were no differences in K$^+$ uptake and translocation rates among the transgenic lines and WT plants, except for whole plants of SvHKT1;1 transgenic line #7 (Figure 9B).

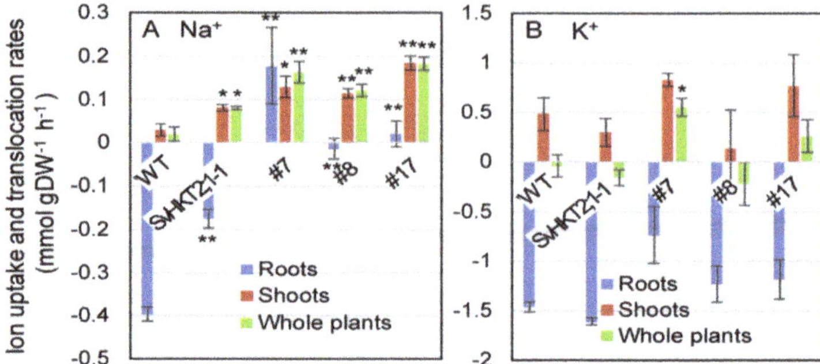

Figure 9. Na$^+$ and K$^+$ uptake and translocation rates in *SvHKT1;1* transgenic lines and WT plants under salt stress. Twelve-day-old *Arabidopsis* seedlings, which were pre-incubated in 1/2 MS liquid medium for 24 h to adapt to liquid medium, were transferred into micro cuvettes filled with 1/2 MS liquid medium supplemented with 100 mM NaCl and further incubated for one hour. Their roots and shoots were separately harvested before and after the treatment, dried overnight, and weighted. Na$^+$ (**A**) and K$^+$ (**B**) uptake and translocation rates in the roots, shoots, and whole plants under salt stress were calculated and expressed as mmol per g of dry weight per hour of salt treatment (mmol·g DW^{-1}·h^{-1}). Ten seedlings were pooled and used as one sample. Data are presented as means ± SE (n = 3 biological replicates). Single and double asterisks denote significant differences compared with the values of WT plants at $p < 0.05$ and $p < 0.01$, respectively, determined using the Student's *t*-test.

3. Discussion

We isolated a gene for class I *HKT*, *SvHKT1;1*, from a halophyte *S. virginicus*. The deduced AA sequence contains a serine in the first P-loop, which is common to HKT1 sodium transporters [37,38]. Electrophysiological analysis showed that SvHKT1;1 mediates inward and outward Na$^+$, but not K$^+$, transport in *X. laevis* oocytes (Figure 4), which was commonly observed in typical glycophytic HKT1s. SvHKT1;1 did not complement K$^+$ transport activity in K$^+$ transporter-defective mutant yeast. Thus, SvHKT1;1 was proven to be a typical Na$^+$ monoporter.

Although HKT1s were reported to have diverse expression patterns in both dicotyledonous and monocotyledonous plants, the expression profile of *SvHKT1;1* is unique compared with that of other *HKT1s*. The transcript was abundant in the shoots compared with the roots and was upregulated by severe salt stress (500 mM NaCl), but not by mild or moderate salt stress (less than 300 mM) (Figure 2). *AtHKT1;1* expression was reported to be slightly induced by mild salt stress [20,41,42]. *OsHKT1;1* expression in rice was associated with the phloem and xylem of leaves and roots, and its transcripts were induced in shoot but not in roots [23,25], while the induction of *OsHKT1;5* expression by salt stress was found in the roots but not in shoots [24,26]. *OsHKT1;4* transcripts were prominent in leaf sheaths and stem [27]. In wheat, an *OsHKT1;5*-like gene, *TmHKT1;5-A*, showed root-specific constitutive expression and was not induced by NaCl [31]. On the other hand, expression of *EsHKT1;2* and *EpHKT1;2* in halophytic *Arabidopsis* relatives was dramatically induced by salt stress (150 mM) [33,43]; however, EsHKT1;2 and EpHKT1;2 showed K$^+$ uptake ability and were, therefore, functionally distinguished from SvHKT1;1. Thus, each *HKT1* has a diverse expression profile, and their expression patterns may reflect the unique Na$^+$ management strategy of each plant.

Downregulation of *EsHKT1;2* in *E. salsuginea* leads to a hyper-salt-sensitive phenotype under K$^+$-deficient conditions [33,35], and overexpression of *EpHKT1;2* enhanced its salt stress tolerance [43]. Based on these findings, these genes could be considered major contributors to the halophytic nature of *E. salsuginea* and *E. parvula* [43]. It was pointed out that HKT1s in non-halophytes are also associated with salt tolerance. Na$^+$ removal from root xylem sap and/or shoot phloem was reported for HKT1s including AtHKT1;1 in *Arabidopsis* [19–22], OsHKT1;1 and OsHKT1;5 in rice [23–25], TaHKT1;5-D in bread wheat [28], HvHKT1;1 and HvHKT1;5 in barley [29,30], and TmHKT1;5-A in wheat [31]. Phylogenetic analysis showed that SvHKT1;1 is evolutionarily close to some of these HKT1s, such as

OsHKT1;1 and HvHKT1;1 (Figure 1); however, *SvHKT1;1* driven by the *AtHKT1;1* promoter did not complement the *athkt1;1 Arabidopsis* mutant (Figure 6), indicating the distinguished function of SvHKT1;1 from these HKT1s.

The shoot and root Na^+ concentrations in *S. virginicus* increased linearly in accordance with the NaCl concentration in culture solution up to 300 mM NaCl, where *SvHKT1;1* was not upregulated, but the Na^+ concentration decreased at 500 mM, where *SvHKT1;1* was dramatically upregulated (Figure 2). This well-synchronized pattern between the gene expression and the decrease in shoot Na^+ concentration suggested that SvHKT1;1 could be involved in Na^+ excretion from shoots in *S. virginicus* and is upregulated only when the plants need to cope with extremely severe salinity. This mechanism may make it possible for *S. virginicus* to accumulate shoot Na^+ under salinity stress but not to exceed levels required for osmotic adjustment. It was suggested that a plant's ability to exclude Na^+ is positively correlated with the overall salinity tolerance in glycophytes, including wheat, sorghum, maize, and tomato [31,44–48]. However, the essentiality of Na^+ exclusion may differ depending on severity of the stress [49] and the capacity of Na^+ sequestration in the shoot (tissue tolerance). Comparison of barley cultivars of different salt tolerance suggested that plants need to rapidly adjust their shoot osmotic potential by sending an appropriate amount of Na^+ to the shoot within the first few days and shutting down any further Na^+ delivery to the shoot [49]. We reported that *S. virginicus* also gradually accumulates a certain amount of Na^+ in shoots under 500 mM NaCl conditions over five days, but regulates so that the Na concentration does not excess a certain level [17]. A halophytic relative of *A. thaliana*, *T. halophila* (*E. salsuginea*), accumulated less Na^+ and more K^+ than *A. thaliana* during short-term (25 h) exposure to salt stress; however, after long-term exposure (5 weeks), *T. halophila* accumulated more Na^+ than *A. thaliana* [50]. Thus, halophytes have the property that they do not accumulate high concentrations of Na^+ in the short term. The expression of *SvHKT1;1* in *S. virginicus* was dramatically upregulated at 6 h after 500 mM NaCl treatment but not under 300 mM NaCl (Figure 2). This early response of *SvHKT1;1* in high Na^+ conditions may be responsible for the Na^+ accumulation in the short term. Therefore, we hypothesize that SvHKT1;1 could play a major role in preventing excess Na^+ accumulation in *S. virginicus* shoots under high-saline conditions. To test this hypothesis, more detailed spatial–temporal expression profiling of SvHKT1;1 and loss-of-function experiments in *S. virginicus* are needed.

In this study, we investigated the function of SvHKT1;1 in transgenic *Arabidopsis* because our attempt to transform *S. virginicus* was unsuccessful and, thus, the knockout or knockdown line is not available. The transgenic lines showed elongated root growth on low-K^+ medium (containing 0.1 mM K^+ and 0.725 mM /Na^+), where root growth of WT plants is severely inhibited (Figure 6B,C). A similar phenotype was observed in *Arabidopsis* expressing K^+/Na^+ symporters, SvHKT2s [16]. These results indicated that enhanced root growth of the transformants under low K^+ conditions was due to an increase in the ability to absorb Na^+ but not K^+.

Overexpression of *AtHKT1;1* specifically in the root xylem parenchyma cells of *Arabidopsis* improved Na^+ exclusion and salinity tolerance [51], and root cortical and epidermal cell-specific expression of *AtHKT1;1* in rice enhanced salinity tolerance [52]. These results indicate that the major role of AtHKT1;1 in salt tolerance is Na^+ exclusion from root xylem. However, surprisingly, constitutive overexpression of *AtHKT1* in potato also reduced Na^+ accumulation in leaves and enhanced salt tolerance [53]. In this study, shoot Na^+ concentrations in transgenic *Arabidopsis* lines were significantly higher than those of WT plants under 100 mM NaCl conditions (Figure 7G). Na^+ concentrations in xylem saps were relatively higher than those in WT plants under nonstress conditions (Figure 8A). Measurement of Na^+ uptake and translocation rates also indicated enhanced Na^+ uptake in roots of *SvHKT1;1* lines under salinity conditions (Figure 9A). Considering these results together, it was suggested that constitutive expression of *SvHKT1;1* enhanced Na^+ uptake in root epidermal cells and then increased Na^+ transport to shoots, which led to reduced salt tolerance (Figures 5 and 7). On the other hand, Na^+ concentrations in phloem sap of the SvHKT1;1 lines were not significantly different from those in WT plants under nonstress conditions (Figure 8C); however, the transgenic lines showed higher phloem sap Na^+ concentrations than WT plants under 100 mM NaCl conditions (Figure 8E). These results may suggest that SvHKT1;1 mediated Na^+ uploading into phloem when an excess amount of Na^+ was accumulated in shoots to translocate Na^+ to roots. Since shoot Na^+ concentration in the transgenic lines under 100 mM NaCl (Figure 7H) was six times higher than that in *S. virginicus* under 300 mM NaCl (Figure 2F), the condition may be adequate for SvHKT1;1 to mediate Na^+ uploading to phloem. These data further support our hypothesis that SvHKT1;1 could play a major role in preventing excess Na^+ accumulation in *S. virginicus* shoots under high-saline conditions, although tissue specificity of *SvHKT1;1* expression was not revealed.

4. Materials and Methods

4.1. Isolation of SvHKT1;1 Gene

We searched for HKT gene homologs in previously constructed unigenes assembled from *S. virginicus* RNA-Seq data [36], and we found HKT-like unigenes. Among them, one unigene sequence, which is similar to sodium transporter *AtHKT1;1* gene, *SvHKT1;1* (DDBJ (DNA Data Bank of Japan). accession number LC545616), was PCR-amplified using specific primers, SvHKT1B1F 5′–CACCATGCATCCAGCCAGTTCAGTTCTA–3′ and SvHKT1B2R 5′–TCCTTGAGGTCATGGAGTTGG–3′. Amplified sequences were cloned into pENTER vectors (Thermo Fisher Scientific, Tokyo, Japan) to form the entry vector, pENTER-SvHKT1;1.

4.2. Phylogenetic Analysis

A phylogenetic analysis of the HKT amino-acid (AA) sequences using the neighbor-joining method, following their alignment using ClustalW, was performed using the MEGA-X software package [39]. Accession numbers for amino-acid sequences of HKTs used for phylogenetic analysis are listed in Supplementary Figure S1.

4.3. Real-Time qRT-PCR

S. virginicus plants were hydroponically cultivated in 1/2 MS salt solution, and then transplanted to 1/2 MS salt solution supplemented with 0, 100, 300, or 500 mM NaCl treatments. Shoots and roots (n = 3 biological replicates) were separately harvested for RNA isolation and ion measurement 48 h after the treatments. The RNAiso plus (TakaraBio, Ohotsu, Japan) was used to extract the total RNA, and real-time qRT-PCR was performed as previously reported [36]. A pair of primer sets, qSvHKT1BF 5′–CTTGGCCCACATAGTATCAGG–3′ and qSvHKT1BR 5′–GGTGAAGATGGAGAAGGTGCATAC–3′, was used. The relative expression levels of the target to reference genes, *eukaryotic translation initiation factor 3 subunit-like protein* (*eIF3*) and *actin* from *S. virginicus*, was detected using primer sets, qSveIF1F 5′–ACATGTGAGTCTGACCTCGTCGAC–3′ and qSveIF2R 5′–TGAGCAAGCCAATGGCCTTCTCAG–3′ and SvActinF 5′–CAGATCATGTTCGAGACCTTC–3′ and SvActinR 5′–GACGGTGTGGCTGACACCAT–3′, respectively, and they were calculated using the delta-delta Ct method.

Similarly, RNA was extracted from 14-day-old *Arabidopsis* plants grown on 1/2 MS medium. Real-time qRT-PCR analysis was performed as previously reported [16] using primer sets for *SvHKT1;1* and *ubiquitin extension protein 5* (*UBQ5*), UBQ5F 5′–TGTGAAGGCGAAGATCCAAG–3′, and UBQ5R 5′–GAGACGGAGGACGAGATGAAG–3′ as a reference.

For semi-quantitative RT-PCR analysis, first-strand complementary DNA (cDNA) was synthesized from 250 ng of total RNA using a QuantiTect Reverse Transcription Kit (Qiagen, Tokyo, Japan) according to the manufacturer's instructions and diluted 20-fold, and 1 μL was used as a template. Semi qRT-PCR was carried out for 30 cycles of 98 °C for 10 s, followed by 60 °C for 10 s, and 68 °C for 60 s, using Tks Gflex DNA Polymerase (TakaraBio). In addition to primer pairs for *SvHKT1;1* and *UBQ5*, primer sets for *AtHKT1;1*, AtHKT101F 5′–GAGAACTAAAATGGACAGAGTGGTG–3′ and AtHKT102R 5′–GTACCAAGATAGCTGGGGAAAGTG–3′, were used.

4.4. Subcellular Localization of SvHKT1;1 in Nicotiana benthamiana Leaves

To examine the subcellular localization of SvHKT1;1, the entry vector pENTER-SvHKT1;1 was reacted with a destination vector, pH7WGF2.0, encoding an N-terminal EGFP fusion [54] using LR clonase reactions (Thermo Fisher Scientific). As a control, a nonfused EGFP construct was used. The recombinant plasmids were transformed into *Agrobacterium tumefaciens* strain GV3101, and then infiltrated into *N. benthamiana* leaves. Two days post infiltration, GFP fluorescence and differential interference contrast images were observed using an FX3000 confocal fluorescence microscope (Olympus, Tokyo, Japan). To observe plasmolyzed cells, leaf cells were treated with 500 mM mannitol for 30 min.

4.5. Functional Analysis of SvHKT1;1 in X. laevis Oocytes

The *SvHKT1;1* cDNA was PCR-amplified using primers to produce the entry vector pENTER-SvHKT1;1, excised from the entry vectors using the restriction enzymes *Not*I and *Asc*I, and then inserted into the *Not*I and *Asc*I sites of pXBG-NA [16]. A mMESSAGE mMACHINE in vitro transcription kit (Thermo Fisher Scientific) was used to synthesize the capped-analogue RN. Oocytes and <u>TEVC</u> experiments were prepared and performed as described previously [16], with a minor modification. In brief, 12.5 ng of cRNA of *SvHKT1;1* was injected into *X. laevis* oocytes and incubated at 18 °C for two days. Water-injected oocytes were also prepared as negative controls. The data recordings and analysis were performed using an Axoclamp 900 A amplifier and an Axon Instruments Digidata 1440 A with Clampex 10.3 and Clampfit 10.3 software (Molecular Devices, Sunnyvale, CA, USA). The analyses of ion selectivity using alkali cation salts used oocytes bathed in a background solution containing 96 mM NaCl, KCl, or Na-glutamate salts, adjusted to pH 7.5. The background solution also contained 1.8 mM $CaCl_2$, 1 mM $MgCl_2$, 1.8 mM mannitol, and 10 mM 4-(2-hydroxyethyl)-1-piperazineethanesulfonic acid (HEPES) for NaCl or KCl salt, and 1.8 mM Ca-Glu, 1 mM Mg-Glu, 1.8 mM mannitol, and 10 mM HEPES for Na-Glu salt. The experiments using frog oocytes were approved by the Animal Care and Use Committee, Okayama University (approval number OKU-2017271 on 26 June 2017) that follows the related international and domestic regulations.

4.6. Production of Transgenic Arabidopsis

The entry vector pENTR-SvHKT1;1 was reacted using LR enzyme (Thermo Fisher Scientific) with a destination vector pGH1 [16] and pAtHKT1 [42], to form pGH1-SvHKT1;1 and pAtHKT1-SvHKT1;1, in which the transgenes are driven by *CaMV35S* and *Arabidopsis AtHKT1;1* promoters, respectively. *Arabidopsis* wild-type (WT) plants (ecotype Columbia) and a T-DNA-tagged *athkt1;1* mutant line (ABRC Stock Number: CS372002 [55]) were transformed with expression vectors pGH1-SvHKT1;1 and pAtHKT1-SvHKT1;1, respectively, by floral dipping [56]. *Agrobacterium* strain GV3101 was used for transformation.

4.7. Cultivation of Arabidopsis Plants

Seeds of *Arabidopsis* were sown on 1/2 MS agar medium (1/2 MS salts, 1% sucrose and 0.8% agar at pH 5.7). The plants were grown at 23 °C under a 16-h/8-h light/dark cycle with approximately 60 $\mu mol \cdot m^{-2} \cdot s^{-1}$ light intensity. For salt stress treatment at seedling stage, seven-day-old seedlings were transplanted onto 1/2 MS agar medium supplemented with 50 mM NaCl and incubated for a further 14 days.

Hydroponic culture of transgenic *Arabidopsis* was performed using the Home Hyponica Karen (Kyowa Co., LTD, Osaka, Japan) system with 1/2 MS medium or 1/2 Hoagland salt solution [57] supplemented with 0.2% 2-(N-morpholino)ethanesulfonic acid (MES) as the hydroponic culture solution. Fourteen-day-old plants grown on 1/2 MS agar medium were transplanted to the hydroponic system. For salt stress treatment at bolting stage, the hydroponic culture solution was replaced with 1/2 MS medium supplemented with 0 or 100 mM NaCl and 0.2% MES at the age of 24 days, when plants had almost started bolting. For RNA extraction, 14-day-old *Arabidopsis* plants cultivated on 1/2 MS agar medium were used.

4.8. Measurement of Ion Concentrations in Plants

Measurement of ion concentrations in plants was performed as described previously using an Ion Analyzer IA-300 (Toa DKK, Tokyo, Japan) [16].

4.9. Collection of Xylem and Phloem Saps

Transgenic and WT *Arabidopsis* plants were grown on 1/2 MS plate medium supplemented with 1% sucrose (pH 5.7) for two weeks, and then were transplanted to liquid 1/2 Hoagland solution supplemented with 0.2% MES (pH 5.7) until they reached the bolting stage. Collection of xylem and phloem sap was carried out according to the methods of Sunarpi et al. [20]. The collected samples were used for ion measurement using an Ion Analyzer IA-300.

4.10. Measurement of Ion Uptake and Translocation Rates in Arabidopsis Seedlings after Salt Treatment

Rates of Na^+ and K^+ uptake or release in WT plants and the transgenic lines were determined based on the changes in their Na^+ and K^+ concentrations after exposure to salinity stress. Seedlings at 12 days old grown on 1/2 MS agar medium supplemented with 1% sucrose were used. Ten seedlings were pooled and used as one sample. Nine pooled samples were prepared for each line. The pooled samples were transferred into microcuvettes filled with 3.0 mL of 1/2 MS liquid medium, taking care to fully immerse the roots into the medium; finally they were incubated at 23 °C under approximately 60 μmol·m^{-2}·s^{-1} light intensity. After 24 h of incubation, three pooled samples were harvested from each line as samples before salt treatment (zero-time samples). They were briefly washed with pure water, harvested by dividing into shoots and roots, and dried at 60 °C overnight to determine the dry weight (DW). The remaining pooled samples were briefly washed in liquid 1/2 MS medium supplemented with 100 or 200 mM NaCl (three pooled samples for each condition), transferred to another microcuvette filled with 3.0 mL of 1/2 MS medium supplemented with 100 or 200 mM NaCl. After 60 min of incubation, seedlings were briefly washed in pure water, harvested as salt-treated samples (samples after 60 min) by dividing into roots and shoots, and dried at 60 °C overnight to determine their DW. The change in Na^+ and K^+ concentrations in the roots, shoots, and whole plants by 60 min of 100 or 200 mM NaCl treatment was calculated and expressed as millimoles per gram of dry weight per hour (mmol·g DW^{-1}·h^{-1}).

5. Conclusions

SvHKT1;1 from a halophytic turf grass, *S. virginicus*. SvHKT1;1 is an Na^+ transporter and its expression is abundant in the shoots compared with the roots in *S. virginicus* and interestingly upregulated only by severe salt stress (500 mM NaCl). Arabidopsis constitutively expressing *SvHKT1;1* showed higher shoot Na^+ concentrations and lower salt tolerance than WT plants by its enhanced Na^+ uptake in roots. Na^+ concentrations in phloem sap of the transgenic Arabidopsis were higher than those in WT plants under salt stress These results suggested possibility that SvHKT1;1 plays a role in preventing excess shoot Na^+ accumulation in *S. virginicus*.

Supplementary Materials: Supplementary materials can be found at http://www.mdpi.com/1422-0067/21/17/6100/s1: Figure S1: Alignment of partial amino-acid sequences of selected class I HKTs, Figure S2: Expression profiles of SvHKT1;1 gene in *Sporobolus virginicus*, Figure S3: Growth of yeast strain 9.3 transformed with the empty vector or the plasmid containing SvHKT2;1 or SvHKT1;1 gene, Table S1: Accession numbers for amino-acid sequences of HKTs used for phylogenetic analysis.

Author Contributions: Gene cloning and yeast complementation test, Y.K.; salt tolerance test and measurement of ion concentrations in plants, Y.K. and Y.T.; TEVC experiments, S.I. and M.K.; gene expression analysis and protein localization analysis, Y.T.; writing, M.K. and Y.T. All authors have read and agreed to the published version of the manuscript.

Funding: This work was partially supported by the Joint Research Program implemented at the Institute of Plant Science and Resources, Okayama University in Japan (2724 to Y.T.).

Conflicts of Interest: The authors declare no conflict of interest. The funders had no role in the design of the study; in the collection, analyses, or interpretation of data; in the writing of the manuscript, or in the decision to publish the results.

References

1. Munns, R.; Tester, M. Mechanisms of Salinity Tolerance. *Annu. Rev. Plant Biol.* **2008**, *59*, 651–681. [CrossRef]
2. Roy, S.J.; Negrão, S.; Tester, M. Salt resistant crop plants. *Curr. Opin. Biotechnol.* **2014**, *26*, 115–124. [CrossRef] [PubMed]
3. Muchate, N.S.; Nikalje, G.C.; Rajurkar, N.S.; Suprasanna, P.; Nikam, T.D. Plant Salt Stress: Adaptive Responses, Tolerance Mechanism and Bioengineering for Salt Tolerance. *Bot. Rev.* **2016**, *82*, 371–406. [CrossRef]
4. Almeida, D.M.; Oliveira, M.M.; Saibo, N.J.M. Regulation of Na^+ and K^+ homeostasis in plants: Towards improved salt stress tolerance in crop plants. *Genet. Mol. Biol.* **2017**, *40*, 326–345. [CrossRef] [PubMed]
5. Volkov, V. Salinity tolerance in plants. Quantitative approach to ion transport starting from halophytes and stepping to genetic and protein engineering for manipulating ion fluxes. *Front. Plant Sci.* **2015**, *6*, 873. [PubMed]
6. Tester, M.; Davenport, R. Na^+ Tolerance and Na^+ transport in higher plants. *Ann. Bot.* **2003**, *91*, 503–527. [CrossRef]
7. Almeida, P.; Katschnig, D.; De Boer, A.H. HKT transporters—State of the art. *Int. J. Mol. Sci.* **2013**, *14*, 20359–20385. [CrossRef]
8. Maathuis, F.J.M.; Amtmann, A. K^+ nutrition and Na^+ toxicity: The basis of cellular K^+/Na^+ ratios. *Ann. Bot.* **1999**, *84*, 123–133. [CrossRef]
9. Apse, M.P.; Aharon, G.S.; Snedden, W.A.; Blumwald, E. Salt tolerance conferred by overexpression of a vacuolar Na+/H+ antiport in *Arabidopsis*. *Science* **1999**, *285*, 1256–1258. [CrossRef]
10. Shi, H.; Lee, B.H.; Wu, S.J.; Zhu, J.K. Overexpression of a plasma membrane Na^+/H^+ antiporter gene improves salt tolerance in *Arabidopsis thaliana*. *Nat. Biotechnol.* **2003**, *21*, 81–85. [CrossRef]
11. Demidchik, V.; Tester, M. Sodium fluxes through nonselective cation channels in the plasma membrane of protoplasts from Arabidopsis roots. *Plant Physiol.* **2002**, *128*, 379–387. [CrossRef] [PubMed]
12. Horie, T.; Costa, A.; Kim, T.H.; Han, M.J.; Horie, R.; Leung, H.Y.; Miyao, A.; Hirochika, H.; An, G.; Schroeder, J.I. Rice OsHKT2;1 transporter mediates large Na^+ influx component into K^+-starved roots for growth. *EMBO J.* **2007**, *26*, 3003–3014. [CrossRef] [PubMed]
13. Horie, T.; Yoshida, K.; Nakayama, H.; Yamada, K.; Oiki, S.; Shinmyo, A. Two types of HKT transporters with different properties of Na^+ and K^+ transport in *Oryza sativa*. *Plant J.* **2001**, *27*, 129–138. [CrossRef] [PubMed]
14. Mian, A.; Oomen, R.J.F.J.; Isayenkov, S.; Sentenac, H.; Maathuis, F.J.M.; Véry, A.A. Over-expression of an Na^+- and K^+-permeable HKT transporter in barley improves salt tolerance. *Plant J.* **2011**, *68*, 468–479. [CrossRef]
15. Laurie, S.; Feeney, K.A.; Maathuis, F.J.M.; Heard, P.J.; Brown, S.J.; Leigh, R.A. A role for HKT1 in sodium uptake by wheat roots. *Plant J.* **2002**, *32*, 139–149. [CrossRef]
16. Tada, Y.; Endo, C.; Katsuhara, M.; Horie, T.; Shibasaka, M.; Nakahara, Y.; Kurusu, T. High-affinity K^+ transporters from a halophyte, *Sporobolus virginicus*, mediate both K^+ and Na^+ transport in transgenic Arabidopsis, *X. laevis* oocytes and yeast. *Plant Cell Physiol.* **2019**, *60*, 176–187. [CrossRef]
17. Tada, Y.; Komatsubara, S.; Kurusu, T. Growth and physiological adaptation of whole plants and cultured cells from a halophyte turf grass under salt stress. *AoB Plants* **2014**, *6*. [CrossRef]
18. Uozumi, N.; Kim, E.J.; Rubio, F.; Yamaguchi, T.; Muto, S.; Tsuboi, A. The *Arabidopsis* HKT1 gene homolog mediates inward Na^+ currents in *Xenopus laevis* oocytes and Na^+ uptake in *Saccharomyces cerevisiae*. *Plant Physiol.* **2000**, *122*, 1249–1260. [CrossRef]
19. Berthomieu, P.; Conéjéro, G.; Nublat, A.; Brackenbury, W.J.; Lambert, C.; Savio, C.; Uozumi, N.; Oiki, S.; Yamada, K.; Cellier, F.; et al. Functional analysis of AtHKT1 in *Arabidopsis* shows that Na^+ recirculation by the phloem is crucial for salt tolerance. *EMBO J.* **2003**, *22*, 2004–2014. [CrossRef]
20. Sunarpi; Horie, T.; Motoda, J.; Kubo, M.; Yang, H.; Yoda, K.; Horie, R.; Chan, W.Y.; Leung, H.Y.; Hattori, K.; et al. Enhanced salt tolerance mediated by AtHKT1 transporter-induced Na unloading from xylem vessels to xylem parenchyma cells. *Plant J.* **2005**, *44*, 928–938.

21. Davenport, R.J.; MuÑOz-Mayor, A.; Jha, D.; Essah, P.A.; Rus, A.N.A.; Tester, M. The Na$^+$ transporter AtHKT1;1 controls retrieval of Na$^+$ from the xylem in *Arabidopsis*. *Plant Cell Environ.* **2007**, *30*, 497–507. [CrossRef] [PubMed]

22. Rodríguez-Navarro, A.; Rubio, F. High-affinity potassium and sodium transport systems in plants. *J. Exp. Bot.* **2006**, *57*, 1149–1160. [CrossRef] [PubMed]

23. Wang, R.; Jing, W.; Xiao, L.; Jin, Y.; Shen, L.; Zhang, W. The Rice High-Affinity Potassium Transporter1;1 Is Involved in Salt Tolerance and Regulated by an MYB-Type Transcription Factor. *Plant Physiol.* **2015**, *168*, 1076–1090. [CrossRef] [PubMed]

24. Ren, Z.H.; Gao, J.P.; Li, L.G.; Cai, X.L.; Huang, W.; Chao, D.Y.; Zhu, M.Z.; Wang, Z.Y.; Luan, S.; Lin, H.X. A rice quantitative trait locus for salt tolerance encodes a sodium transporter. *Nat. Genet.* **2005**, *37*, 1141–1146. [CrossRef]

25. Jabnoune, M.; Espeout, S.; Mieulet, D.; Fizames, C.; Verdeil, J.L.; Conéjéro, G.; Rodríguez-Navarro, A.; Sentenac, H.; Guiderdoni, E.; Abdelly, C.; et al. Diversity in expression patterns and functional properties in the rice HKT transporter family. *Plant Physiol.* **2009**, *150*, 1955–1971. [CrossRef]

26. Kobayashi, N.I.; Yamaji, N.; Yamamoto, H.; Okubo, K.; Ueno, H.; Costa, A.; Tanoi, K.; Matsumura, H.; Fujii-Kashino, M.; Horiuchi, T.; et al. OsHKT1;5 mediates Na(+) exclusion in the vasculature to protect leaf blades and reproductive tissues from salt toxicity in rice. *Plant J.* **2017**, *91*, 657–670. [CrossRef]

27. Suzuki, K.; Yamaji, N.; Costa, A.; Okuma, E.; Kobayashi, N.I.; Kashiwagi, T.; Katsuhara, M.; Wang, C.; Tanoi, K.; Murata, Y.; et al. OsHKT1;4-mediated Na+ transport in stems contributes to Na+ exclusion from leaf blades of rice at the reproductive growth stage upon salt stress. *BMC Plant Biol.* **2016**, *16*, 22. [CrossRef]

28. Byrt, C.S.; Xu, B.; Krishnan, M.; Lightfoot, D.J.; Athman, A.; Jacobs, A.K.; Watson-Haigh, N.S.; Plett, D.; Munns, R.; Tester, M.; et al. The Na$^+$ transporter, TaHKT1;5-D, limits shoot Na$^+$ accumulation in bread wheat. *Plant J.* **2014**, *80*, 516–526. [CrossRef]

29. Van Bezouw, R.F.H.M.; Janssen, E.M.; Ashrafuzzaman, M.; Ghahramanzadeh, R.; Kilian, B.; Graner, A.; Visser, R.G.F.; van der Linden, C.G. Shoot sodium exclusion in salt stressed barley (*Hordeum vulgare* L.) is determined by allele specific increased expression of HKT1;5. *J. Plant Physiol.* **2019**, *241*, 153029.

30. Han, Y.; Yin, S.; Huang, L.; Wu, X.; Zeng, J.; Liu, X.; Qiu, L.; Munns, R.; Chen, Z.H.; Zhang, G. A Sodium Transporter HvHKT1;1 Confers Salt Tolerance in Barley via Regulating Tissue and Cell Ion Homeostasis. *Plant Cell Physiol.* **2018**, *59*, 1976–1989. [CrossRef]

31. Munns, R.; James, R.A.; Xu, B.; Athman, A.; Conn, S.J.; Jordans, C.; Byrt, C.S.; Hare, R.A.; Tyerman, S.D.; Tester, M.; et al. Wheat grain yield on saline soils is improved by an ancestral Na+ transporter gene. *Nat. Biotechnol.* **2012**, *30*, 360. [CrossRef] [PubMed]

32. Wu, H.J.; Zhang, Z.; Wang, J.Y.; Oh, D.H.; Dassanayake, M.; Liu, B.; Huang, Q.; Sun, H.X.; Xia, R.; Wu, Y.; et al. Insights into salt tolerance from the genome of *Thellungiella salsuginea*. *Proc. Natl. Acad. Sci. USA* **2012**, *109*, 12219–12224. [CrossRef] [PubMed]

33. Ali, Z.; Park, H.C.; Ali, A.; Oh, D.H.; Aman, R.; Kropornicka, A.; Hong, H.; Choi, W.; Chung, W.S.; Kim, W.-Y.; et al. TsHKT1;2, a HKT1 homolog from the extremophile arabidopsis relative *Thellungiella salsuginea*, shows K$^+$ specificity in the presence of NaCl. *Plant Physiol.* **2012**, *158*, 1463–1474. [CrossRef] [PubMed]

34. Dassanayake, M.; Oh, D.H.; Haas, J.S.; Hernandez, A.; Hong, H.; Ali, S.; Yun, D.J.; Bressan, R.A.; Zhu, J.K.; Bohnert, H.J.; et al. The genome of the extremophile crucifer *Thellungiella parvula*. *Nat. Genet.* **2011**, *43*, 913–918. [CrossRef] [PubMed]

35. Ali, A.; Maggio, A.; Bressan, R.A.; Yun, D.J. Role and Functional Differences of HKT1-Type Transporters in Plants under Salt Stress. *Int. J. Mol. Sci.* **2019**, *20*, 1059. [CrossRef] [PubMed]

36. Yamamoto, N.; Takano, T.; Tanaka, K.; Ishige, T.; Terashima, S.; Endo, C.; Kurusu, T.; Yajima, S.; Yano, K.; Tada, Y. Comprehensive analysis of transcriptome response to salinity stress in the halophytic turf grass *Sporobolus virginicus*. *Front. Plant Sci.* **2015**, *6*, 241. [CrossRef] [PubMed]

37. Mäser, P.; Hosoo, Y.; Goshima, S.; Horie, T.; Eckelman, B.; Yamada, K.; Yoshida, K.; Bakker, E.P.; Shinmyo, A.; Oiki, S.; et al. Glycine residues in potassium channel-like selectivity filters determine potassium selectivity in four-loop-per-subunit HKT transporters from plants. *Proc. Natl. Acad. Sci. USA* **2002**, *99*, 6428–6433. [CrossRef]

38. Platten, J.D.; Cotsaftis, O.; Berthomieu, P.; Bohnert, H.; Davenport, R.J.; Fairbairn, D.J. Nomenclature for HKT transporters, key determinants of plant salinity tolerance. *Trends Plant Sci.* **2006**, *11*, 372–374. [CrossRef]

39. Kumar, S.; Stecher, G.; Li, M.; Knyaz, C.; Tamura, K. MEGA X: Molecular Evolutionary Genetics Analysis across Computing Platforms. *Mol. Biol. Evol.* **2018**, *35*, 1547–1549. [CrossRef]

40. Tada, Y.; Ohnuma, A. Comparative Functional Analysis of Class II Potassium Transporters, SvHKT2;1, SvHKT2;2, and HvHKT2;1, on Ionic Transport and Salt Tolerance in Transgenic *Arabidopsis*. *Plants* **2020**, *9*, 786. [CrossRef]

41. Wang, Q.; Guan, C.; Wang, P.; Ma, Q.; Bao, A.K.; Zhang, J.L.; Wang, S.M. The Effect of *AtHKT1;1* or *AtSOS1* mutation on the expressions of Na$^+$ or K$^+$ transporter genes and ion homeostasis in *Arabidopsis thaliana* under salt stress. *Int. J. Mol. Sci.* **2019**, *20*, 1085. [CrossRef] [PubMed]

42. Tada, Y. The HKT Transporter Gene from Arabidopsis, AtHKT1;1, Is Dominantly Expressed in Shoot Vascular Tissue and Root Tips and Is Mild Salt Stress-Responsive. *Plants* **2019**, *8*, 204. [CrossRef] [PubMed]

43. Ali, A.; Khan, I.U.; Jan, M.; Khan, H.A.; Hussain, S.; Nisar, M.; Chung, W.S.; Yun, D.J. The High-Affinity Potassium Transporter EpHKT1;2 From the Extremophile Eutrema parvula Mediates Salt Tolerance. *Front. Plant Sci.* **2018**, *9*, 1108. [CrossRef] [PubMed]

44. Davenport, R.J.; Reid, R.J.; Smith, F.A. Sodium-calcium interactions in two wheat species differing in salinity tolerance. *Physiol. Plant.* **1997**, *99*, 323–327. [CrossRef]

45. Cuin, T.A.; Bose, J.; Stefano, G.; Jha, D.; Tester, M.; Mancuso, S.; Shabala, S. Assessing the role of root plasma membrane and tonoplast Na+/H+ exchangers in salinity tolerance in wheat: In planta quantification methods. *Plant Cell Environ.* **2011**, *34*, 947–961. [CrossRef]

46. Yang, Y.W.; Newton, R.J.; Miller, F.R. Salinity Tolerance in Sorghum. I. Whole Plant Response to Sodium Chloride in S. bicolor and S. halepense. *Crop Sci.* **1990**, *30*, 775–781. [CrossRef]

47. Fortimeier, R.; Schubert, S. Salt tolerance of maize (*Zea mays* L.): The role of sodium exclusion. *Plant Cell Environ.* **1995**, *18*, 1041–1047.

48. Al-Karaki, G.N. Growth, water use efficiency, and sodium and potassium acquisition by tomato cultivars grown under salt stress. *J. Plant Nutr.* **2000**, *23*, 1–8. [CrossRef]

49. Zhu, M.; Zhou, M.; Shabala, L.; Shabala, S. Physiological and molecular mechanisms mediating xylem Na+ loading in barley in the context of salinity stress tolerance. *Plant Cell Environ.* **2017**, *40*, 1009–1020. [CrossRef]

50. Volkov, V.; Wang, B.; Dominy, P.J.; Fricke, W.; Amtmann, A. *Thellungiella halophila*, a salt-tolerant relative of *Arabidopsis thaliana*, possesses effective mechanisms to discriminate between potassium and sodium. *Plant Cell Environ.* **2004**, *27*, 1–14. [CrossRef]

51. Møller, I.S.; Gilliham, M.; Jha, D.; Mayo, G.M.; Roy, S.J.; Coates, J.C.; Haseloff, J.; Tester, M. Shoot Na$^+$ Exclusion and Increased Salinity Tolerance Engineered by Cell Type–Specific Alteration of Na$^+$ Transport in *Arabidopsis*. *Plant Cell* **2009**, *21*, 2163–2178. [CrossRef]

52. Plett, D.; Safwat, G.; Gilliham, M.; Skrumsager Møller, I.; Roy, S.; Shirley, N.; Jacobs, A.; Johnson, A.; Tester, M. Improved salinity tolerance of rice through cell type-specific expression of AtHKT1;1. *PLoS ONE* **2010**, *5*, e12571. [CrossRef] [PubMed]

53. Wang, L.; Liu, Y.; Li, D.; Feng, S.; Yang, J.; Zhang, J.; Zhang, J.; Wang, D.; Gan, Y. Improving salt tolerance in potato through overexpression of AtHKT1 gene. *BMC Plant Biol.* **2019**, *19*, 357. [CrossRef] [PubMed]

54. Karimi, M.; Inzé, D.; Depicker, A. Gateway vectors for *Agrobacterium*-mediated plant transformation. *Trends Plant Sci.* **2002**, *7*, 193–195. [CrossRef]

55. Rosso, M.G.; Li, Y.; Strizhov, N.; Reiss, B.; Dekker, K.; Weisshaar, B. An *Arabidopsis thaliana* T-DNA mutagenized population (GABI-Kat) for flanking sequence tag-based reverse genetics. *Plant Mol. Biol.* **2003**, *53*, 247–259. [CrossRef] [PubMed]

56. Clough, S.J.; Bent, A.F. Floral dip: A simplified method for *Agrobacterium*-mediated transformation of *Arabidopsis thaliana*. *Plant J.* **1998**, *16*, 735–743. [CrossRef] [PubMed]

57. Hoagland, D.R.; Arnon, D.I. *The Water Culture Method for Growing Plants Without Soil*; University of California, College of Agriculture, Agricultural Experiment Station: Berkeley, CA, USA, 1938; Volume 347, pp. 1–39.

International Journal of
Molecular Sciences

Article

Transcriptomic Analysis of Short-Term Salt Stress Response in Watermelon Seedlings

Qiushuo Song [1,2], Madhumita Joshi [2] and Vijay Joshi [2,*] 

1   Department of Horticultural Sciences, Texas A&M University, College Station, TX 77843, USA; qiushuo1995@gmail.com
2   Texas A&M AgriLife Research and Extension Center, Uvalde, TX 78801, USA; Madhumita.Joshi@ag.tamu.edu
*   Correspondence: Vijay.Joshi@tamu.edu; Tel.: +1-830-278-9151 (ext. 236)

Received: 6 August 2020; Accepted: 19 August 2020; Published: 21 August 2020

Abstract: Watermelon (*Citrullus lanatus* L.) is a widely popular vegetable fruit crop for human consumption. Soil salinity is among the most critical problems for agricultural production, food security, and sustainability. The transcriptomic and the primary molecular mechanisms that underlie the salt-induced responses in watermelon plants remain uncertain. In this study, the photosynthetic efficiency of photosystem II, free amino acids, and transcriptome profiles of watermelon seedlings exposed to short-term salt stress (300 mM NaCl) were analyzed to identify the genes and pathways associated with response to salt stress. We observed that the maximal photochemical efficiency of photosystem II decreased in salt-stressed plants. Most free amino acids in the leaves of salt-stressed plants increased many folds, while the percent distribution of glutamate and glutamine relative to the amino acid pool decreased. Transcriptome analysis revealed 7622 differentially expressed genes (DEGs) under salt stress, of which 4055 were up-regulated. The GO analysis showed that the molecular function term "transcription factor (TF) activity" was enriched. The assembled transcriptome demonstrated up-regulation of 240 and down-regulation of 194 differentially expressed TFs, of which the members of ERF, WRKY, NAC bHLH, and MYB-related families were over-represented. The functional significance of DEGs associated with endocytosis, amino acid metabolism, nitrogen metabolism, photosynthesis, and hormonal pathways in response to salt stress are discussed. The findings from this study provide novel insights into the salt tolerance mechanism in watermelon.

Keywords: watermelon; salt stress; RNA-seq; amino acids; endocytosis

1. Introduction

Soil salinization is recognized as a major problem for agricultural production and sustainability at a global level. The area of degraded saline soils has rapidly increased due to climate change and limited rainfall, posing a great challenge to global food security [1]. It is estimated that around 20% to 50% of irrigated land is salt-affected in arid and semi-arid regions [2–4]. Salt adversely impacts plant growth and development as it holds water and nutrients in the soil at high tension making these components unavailable for plants at the root zone.

Watermelon (*Citrullus lanatus* L.) is a widely popular vegetable fruit crop for human consumption worldwide. Although China tops the watermelon production, it is grown widely across arid and semi-arid environments in the world. Watermelons can tolerate some degree of soil acidity [5] but grow best in non-saline sandy loam or silt loam soils. The research efforts to improve salt tolerance using conventional or transgenic breeding have had limited success due to the genetically and physiologically complex nature of salt-induced responses [3,6]. In watermelon, approaches such as salt-tolerant rootstocks [5,7,8] and agronomical practices [9,10] have been explored. Understanding the molecular mechanisms underscoring the salt stress response in model xerophyte plant species would contribute to finding conserved response cues in plants and identifying salt-tolerant traits [11–13]. Watermelon, being relatively tolerant of drought and salt stress, makes an excellent model crop to study salt stress-induced responses. There has been little research on transcriptome analysis to understand molecular regulation of salt stress-induced responses in watermelon. In the present study, we examined gene expression changes in watermelon seedlings due to short-term salt stress using RNA sequencing (RNA-seq) and discuss putative candidate genes and pathways associated with salt stress induced responses. The results from this study will provide a foundation to understand salt tolerance mechanisms and its exploitation to allow development of salt-tolerant watermelon cultivars.

2. Results

2.1. Validation of Salt Stress Treatment

We studied how short-term exposure of watermelon seedlings to salt stress changed chlorophyll fluorescence parameters (as determined on dark-adapted and illuminated leaves) and amino acid metabolism. Six-week-old seedlings of the cultivar Crimson Sweet subjected to salt stress treatment were monitored for photosystem II (PSII) performance to measure the maximal quantum yield of PSII photochemistry (Fv/Fm) and the efficiency of excitation capture of the open PSII center (Fv′/Fm′) (Figure 1). Measurements of photosystem II efficiency showed a consistent reduction in the PSII efficiency (Qy) in both dark-adapted and illuminated leaf samples. The decrease in Qy in treated plants was much more rapid than the control ones over time. Early decrease in Qy in our experiment is consistent with reports showing inhibition of PSII activity due to salt stress [8,14,15]. In the light-adapted leaves, Qy of salt-stressed tissues was dramatically lower than that of control plants, confirming a negative effect on the quantum yield of PSII electron transport by salinity stress after seven hours of exposure.

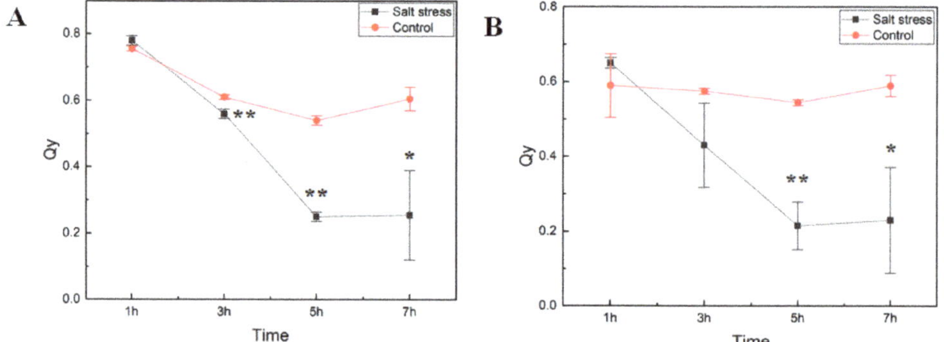

Figure 1. Photosynthetic efficiency of photosystem II. The maximal quantum yield of photosystem II (PSII) photochemistry (Fv/Fm) (**A**) and the efficiency of excitation capture of the open PSII center (Fv′/Fm′) (**B**) in *cv.* Crimson Sweet leaves under salt stress were measured using FluorPen (PAR-FluorPen FP 110/D). Asterisks ** and * represent significant difference at $p < 0.05$ and $p < 0.1$, respectively. Qy, equal to Fv/Fm in the dark (**A**) or light-adapted (**B**) samples, photosystem II efficiency. The error bars represent the standard deviation.

During salt stress, plants accumulate high concentrations of compatible osmolytes, such as nitrogen-containing compounds, mainly amino acids. In this study, most free amino acids showed several-fold increases in response to salt stress implying that the accumulation of free amino acids is crucial to salt stress in watermelon (Figure 2). The changes in the percent of amino acids relative to the pool size are shown in Figure S1 (Supplementary Materials) and the absolute amounts of each amino acid in Table S1 (Supplementary Materials). The branched-chain amino acids (valine, Val; isoleucine, Ile; leucine, Leu), serine (Ser), and Asparagine (Asn) showed much higher fold change increases than glutamate (Glu) and glutamine (Gln). The relative proportion of most amino acids also increased in the leaves after salt stress, except Glu, Gln, tryptophan (Trp), and glycine (Gly). No significant changes were observed in the content of most amino acids in roots due to short-term salt stress (Table S1), implying their limited localized synthesis or transport. It has been suggested that high abundant amino acids (proline, Pro; Arginine, Arg; asparagine, Asn, Glu) are synthesized during abiotic stress, while the low abundant amino acids (BCAAs) accumulate due to increased protein turnover or degradation under conditions such as salt stress [16,17]. However, such an increase in amino acid accumulation through proteolysis happens to a lesser extent in salt stress than drought stress [18]. Consistent with the meta-analysis [1], we also observed a significant decrease in the proportion of Glu and Gln, which serve as precursors for the synthesis of Pro, citrulline (Cit), and Arg, as well as polyamines. There was a limited increase in terms of fold change or change in percent accumulation in Pro or Cit, which are a major drought stress-induced amino acids in watermelon leaves [19]. Our results are consistent with studies that showed non-overlapping patterns of amino acid accumulation in salt stress and drought stress as well as a smaller increase in Cit due to salt stress [20] than drought stress.

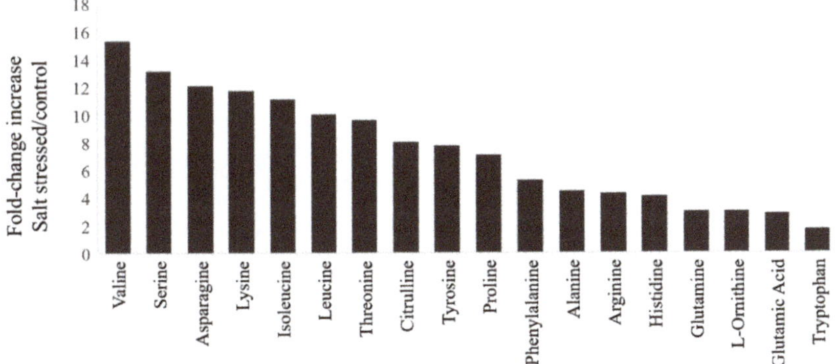

Figure 2. Fold-change increases in amino acids in watermelon leaves due to salt stress: The fold-change in plants treated with NaCl relative to the control group for amino acids showing significant changes (*t*-test, $p < 0.05$). The absolute amino acid quantities were normalized using internal standards and expressed as fold-change relative to control.

2.2. Transcriptome Profiling of Salt-Stressed Seedlings

To understand the transcriptomic changes due to salt stress, we performed RNA-seq analysis of seedlings exposed to salt stress. A total of six libraries were sequenced using the Illumina HiSeq platform comprising three replicates each of the control and salt-treated plants. On average, 42.20 to 47.67 million paired-end raw reads were generated from leaf tissues in both treatments, of which more than 96% mapped to the reference watermelon genome (Supplementary Materials, Table S2). The RNA-seq dataset is accessible through GEO Series accession number GSE146087 (https://www.ncbi.nlm.nih.gov/geo/).

2.3. Identification of Differentially Expressed Genes (DEGs)

The relative expression levels of genes were evaluated as the fragment per kilobase of transcript sequence per millions base pairs sequenced (FPKM) values, calculated based on the uniquely mapped reads for under control or salt stress condition. The RNA-seq data identified a total of 7622 differentially expressed genes in response to salt stress using comparative analysis when a cutoff of adjusted *p*-value (padj) < 0.05 and |log2fold change [L2fc]| > 1 were used. Out of the total DEGs, 4055 (53.2%) of those

DEGs were up-regulated, while 3,567 (46.8%) of them were down-regulated. A volcano scatter plot showing the number of DEGs (Figure 3) and a list of DEGs is presented in the Supplementary Materials (Table S3). A wider dispersion indicates the presence of a higher level of difference regarding gene expression in response to salt stress. A higher number of up-regulated than down-regulated DEGs is consistent with a meta-analysis that included 25 independent salt stress transcriptomic studies [12], suggesting activation of a set of conserved genes regulating intrinsic salt-stress induced responses.

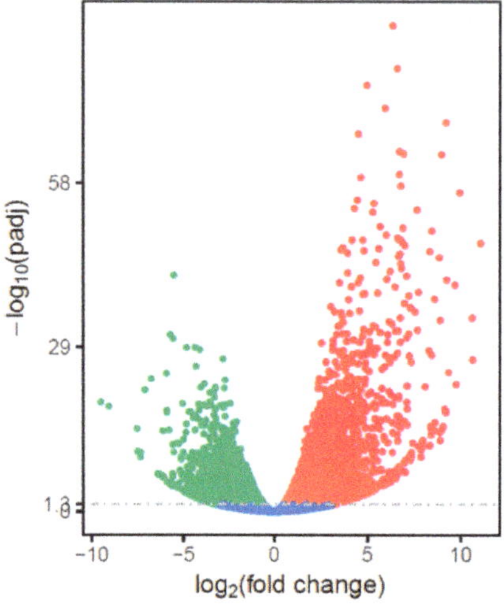

Figure 3. Summary of differentially expressed genes (DEGs) in the watermelon leaves during salt stress. Each point represents a gene; blue dots indicate no significant difference; red dots indicate up-regulated DEGs; green dots indicate down-regulated DEGs. The horizontal axis shows the fold change of genes between different samples (padj < 0.05), and the vertical coordinate indicates the statistically significant degree of changes in gene expression levels at −log10 (padj *p*-value).

2.4. GO, and KEGG Enrichment Results of DEGs

To uncover the molecular mechanisms underlying the salt tolerance in watermelon leaves, the DEGs were characterized using the Gene Ontology (GO) knowledgebase (http://geneontology.org/). GO enrichment scatterplots show the top 20 enriched functions for up- or down-regulated DEGs due to salt stress (Figure 4). Among the biological process (BP) terms the function "protein folding" (GO:0006457) and among the molecular function (MF) the function "transcription factor activity", sequence-specific DNA binding (GO:0003700) were enriched in the up-regulated unigenes with the cutoff of adjusted *q*-value < 0.05. On the contrary, in the cellular component (CC) category, "thylakoid" (GO:0009579), "photosystem" (GO:0009521), "photosynthetic membrane" (GO:0034357), "chromosomal part" (GO:0044427), "photosystem I" (GO:0009522), "chromosome" (GO:0005694), and "photosystem II" (GO:0009523) were the most enriched GO terms among the down-regulated DEGs.

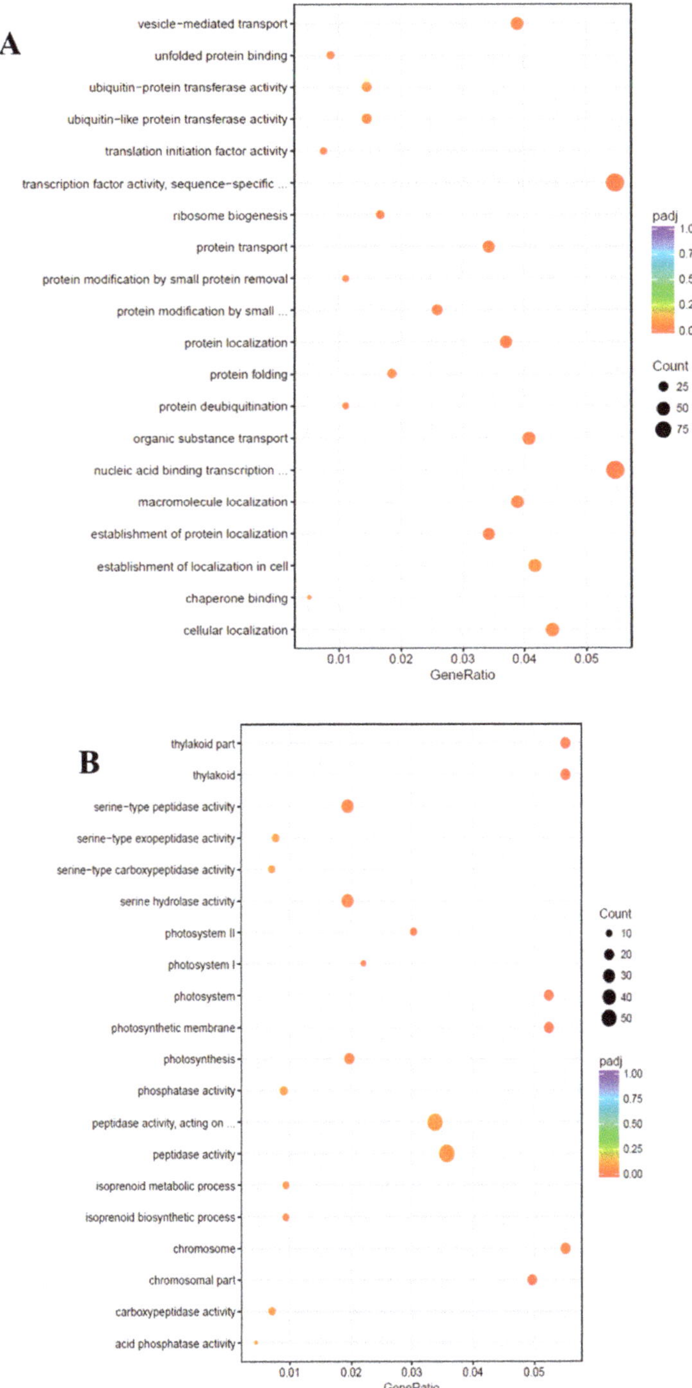

Figure 4. Gene Ontology (GO) enrichment scatter plot. The GO enrichment analysis showing the top 20 enriched functions for up-regulated (**A**) and down-regulated (**B**) DEGs. The horizontal axis is GeneRatio (the ratio between the number of differentially expressed genes in each GO term, and all

differentially expressed genes that can be found in the GO database). The vertical axis is the description of GO terms. The significance showing padj *q*-values are shown as a color scale, where the color and size of the dots represent the range of *q*-value, and the number of DEGs mapped to the indicated functions, respectively.

Pathway analysis of DEGs was performed using the Kyoto Encyclopedia of Genes and Genomes (KEGG) pathway database with KOBAS [21]. The up-regulated and down-regulated DEGs were assigned to 106 and 114 pathways, respectively. The KEGG enrichment analysis showing the top 20 enriched functions is shown in Figure 5. The KEGG pathway annotations like "Ribosome biogenesis in eukaryotes", "Endocytosis", "Protein processing in endoplasmic reticulum" and "Spliceosome" were enriched in the up-regulated unigenes due to salinity stress. However, among the down-regulated unigenes, several KEGG pathways were significantly enriched such as "glycan degradation", "beta-Alanine metabolism", "Terpenoid backbone biosynthesis", "Photosynthesis", "Photosynthesis-antenna proteins", "Histidine metabolism", "Glycosaminoglycan degradation", "Valine, leucine, and isoleucine degradation", "Folate biosynthesis" and "Homologous recombination".

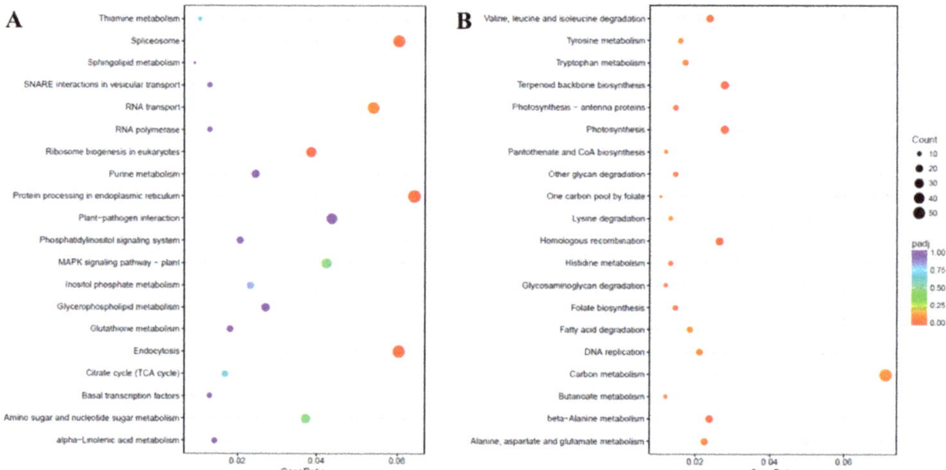

Figure 5. Kyoto Encyclopedia of Genes and Genomes (KEGG) enrichment scatter plot. The KEGG enrichment analysis showing the top 20 enriched pathways for up-regulated (**A**) and down-regulated (**B**) DEGs. The horizontal axis is GeneRatio (the ratio between the number of differentially expressed genes in each pathway, and all differentially expressed genes that can be found in the KEGG database). The vertical axis is the description of the KEGG term. The significance showing padj *q*-values are shown as a color scale, where the color and size of the dots represent the range of *q*-value and the number of DEGs mapped to the individual pathways, respectively.

2.5. Differentially Expressed Transcription Factors (TF) in Response to Salinity Stress

TFs play a critical role in salt stress-induced responses via transcriptional regulation of several genes in plants [22]. The assembled transcriptome demonstrated a total of 240 differentially expressed TFs were up-regulated in NaCl-treated leaf samples, while 194 TFs showed decreased expression due to salinity stress. The distribution of transcription factor families identified among DEGs in watermelon leaves in response to salt stress is shown in Figure 6. Of the up-regulated TFs, the largest number was found in ERF (36 unigenes), followed by WRKY (23 unigenes) and NAC (19 unigenes) families. In contrast, the largest number of down-regulated TFs were found in bHLH (24 unigenes), MYB-related (16 unigenes), and C2H2 (15 unigenes) families. The up-regulation of members of ERF TFs in this study indicates the significant involvement of the ethylene signaling pathway in response to salt stress in watermelon. Ethylene signaling modulates salt response via membrane receptors, components in the cytoplasm, and transcription factors [23]. The salinity stress promotes ethylene biosynthesis activating the downstream network and expression of ERFs [24]. The role of several tomato ERF.E2, ERF.F5, ERF.E3, ERF.B3, and ERF84 genes in enhancing salt tolerance has been demonstrated [25–29]. The ectopic expression of the barley [30], wheat [31,32], and rice [33] ERF genes also enhanced tolerance to salt stress. Consistent with our results, recent transcriptomic studies in cotton [34] and potato [35] plants revealed the induction of a high proportion of ERFs in response to salt stress, suggesting the essential role of ERFs in salt response mechanisms in plants. Despite its significance, with few exceptions [36,37], there is not much information available about the role of ERF TFs in cucurbits. The WRKY TF family is one of the largest families in higher plants and plays a crucial role in plant development and stress responses, including salt stress. Our results are consistent with a study showing the up-regulation of most WRKY genes using NaCl treatment in watermelon [38] and *Cucurbita pepo* [39]. Transcriptomic studies in other plants have demonstrated the differential expression of several members of the WRKY family in response to salt stress [40,41]. The functional role of WRKY TFs in enhancing salt tolerance has been validated using transgenic approaches by overexpressing WRKY genes from maize [42], cotton [43], soybean [44] and grapevine [45]. Similarly, NAC TFs have been implicated in a wide range of stresses, including salinity. Our data is in agreement with studies validating the induction of several NAC TFs during salt stress in watermelon [46] and melon [47].

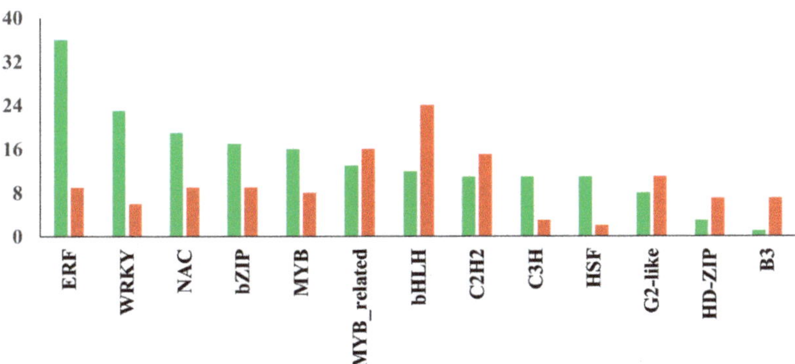

Figure 6. Distribution of transcription factor families. The red and green bars show the total number of down-regulated and up-regulated TFs in the respective family, respectively.

Reliability of Transcriptome Sequencing Data

To validate the reliability of transcriptome sequencing data, relative gene expression analysis of selected genes associated with Cit metabolism (*AAT*, N-acetylornithine; *AOD2*, N-acetylornithine deacetylase; *ArgD*, arginine decarboxylase; ASL1, arginosuccinate lyase; ASS1, arginosuccinate synthase; *CPS1* and *CPS2*, carbamoyl phosphate synthase; *OTC*, ornithine carbamoyltransferase) was performed using real-time quantitative PCR (qRT-PCR). Cit is a major non-protein amino acid in watermelon and accounts for almost 50% of the leaf amino acid pool in response to abiotic stress [19]. Cit, being an intermediate of the master metabolic pathway that synthesizes several salt stress-associated metabolites (spermine and spermidine, Pro, GABA, Arg), selected genes associated with its metabolism were expected to perturb in response to salt stress. The qRT-PCR data (Figure 7) were very consistent with the transcriptome sequencing data. Additionally, the linear regression equation y = 0.9229x − 0.395 with a high correlation (R^2 = 0.97) showed a positive correlation and significant similarity between the two analysis techniques (Figure S2). The expression of *AOD2* and *ASS1* was significantly induced in salt-stressed samples, while the expression of *CPS2* was down-regulated. The induction in the expression of *AOD2* in salt-treated samples explains the enhanced accumulation of Cit, while the upregulation of *ASS1* supports the enhanced Arg accumulation.

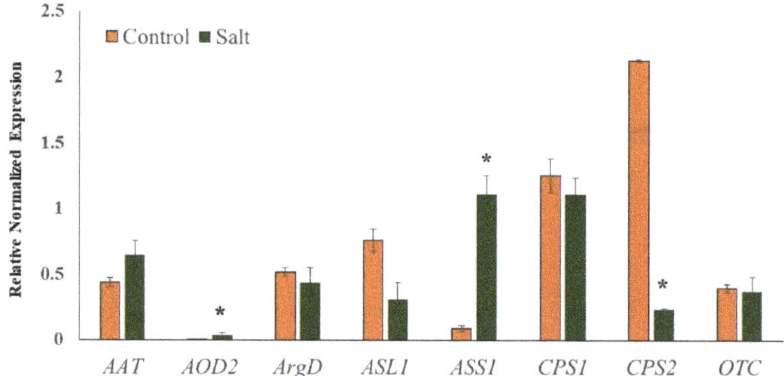

Figure 7. Quantitative real-time PCR (RT-qPCR) analysis.). Relative expression profiles of genes involved in the citrulline metabolic pathways in seedling leaf tissues of Crimson Sweet due to salt stress. The error bars are the means ± SE ($n = 3$), and asterisks (*) represent significant differences between treated and control tissues ($p < 0.05$).

The reliability of RNA-seq data was further validated by comparing the correlations among biological replicates using the Pearson correlation coefficient (Figure S3). Unlike the high correlation among the libraries for the same treatment (i.e., biological replicates), the weak correlation across treatments (control vs. salt treatments) suggests a larger effect of salt stress on the gene expression profiles of watermelon leaf tissues. Additionally, to demonstrate the source of variance in the RNA-seq data, principal component analysis (PCA) with three principal components (PC1, 2, and 3) was performed (Figure S4). The PC score plots showed that the contribution of PC1 alone was 74.94%, followed by PC2, (9.44%), and PC3 (6.60%). The three biological replicates were collected after salt-stress and control samples were clustered together, validating the minimal variance in the analysis and suitability of data for the subsequent analysis.

3. Discussion

3.1. DEGs Associated with Endocytosis

High concentrations of salt lower the water potential and lead to ionic disequilibria across the plasma membrane, which subsequently inhibits cellular activities by entering the cytoplasm [48]. Endocytosis involves the internalization of plasma membrane proteins into the cell via a series of vesicle compartments and plays an essential role in cellular responses to environmental stimuli [49]. KEGG enrichment analysis revealed that the "Endocytosis" pathway was significantly activated in response to salt stress (Figure S5, Supplementary Materials, Table S4). Besides its role in plant growth and development, endocytosis is involved in inducing abiotic stress responses by regulating vacuolar transport [50]. Endocytosis controls cell polarity and signaling by regulating plasma membrane-associated receptors and transporters proteins [51] and the production of ROS needed for salinity tolerance [52]. The role of vesicle trafficking in adaptation against salinity stress has been validated [53–58]. Our data confirmed the induction of several *Rab* genes (ClCG02G019840, ClCG10G007150, ClCG10G012520, ClCG09G002000) in response to salt stress. The expression of native Rab7 from *Pennisetum glaucum* and its overexpression in tobacco [57] and rice [58] were greatly induced by salt stress. Exposure of plants to the salinity stress also activates phospholipase D (PLD), a phosphatidyl choline-hydrolyzing enzyme that triggers the activation of the downstream adaptive responses to relieve the damage caused by stress, including salinity [59]. Consistent with studies reporting activation of PLDs due to salt stress [60–63], expression of watermelon *PLDs* (ClCG00G000210, ClCG08G014000, ClCG06G004910) were also induced due to salt stress. Further, vacuolar protein sorting (VPS) components play an important role in maintaining osmo-homeostasis of vacuoles by sequestering toxic ions, like sodium and chloride, or other compounds involved in osmoregulation. Several VPS genes were up-regulated due to salt stress in this study. Additionally, expression of genes involved in the formation of clathrin-coated vesicles such as *clathrin* proteins (ClCG11G006070), *ADP ribosylation factors* (ClCG05G025400, ClCG07G011900, ClCG10G001580, ClCG10G021000), *ARF-guanine nucleotide exchange factors* (ClCG01G025140, ClCG11G018450, ClCG01G014890), adaptor protein (ClCG02G021820), and products in phosphatidylinositol signaling (ClCG03G009330) were up-regulated due to salt stress. The molecular chaperones *HSP70* mediate uncoating of vesicles before merging with early endosomes. In our study, four DEGs encoding *HSP70* (ClCG04G008300, ClCG09G019940, ClCG09G020000, ClCG11G011300) assigned to the endocytosis pathway were strongly up-regulated. Studies have confirmed the induction and up-regulation of HSP70 in saline stress situations rice [64], wheat [65], and potato [66].

Additionally, we found several genes in the SNARE interaction in the vascular proteins pathway were up-regulated due to salt stress (Figure S6, Supplementary Materials, Table S5). The members of the superfamily of N-ethylmaleimide-sensitive factor adaptor protein receptor (SNARE)-domain-containing proteins are involved in transport processes between individual compartments, including endocytosis [67]. Although a high number of SNARE proteins are present in the plant kingdom, their role in plant biotic and abiotic stress has only been recently understood [68]. We identified several *Syntaxin* family proteins (ClCG02G015160, ClCG07G004070, ClCG08G012870, ClCG09G005420, ClCG10G004910, ClCG10G018250) and *Golgi SNAP receptor complex member 1, target SNARE coiled-coil domain, vesicle-trafficking SEC22b,* and *vesicle transport v-SNARE 11-like* genes that were up-regulated. Several studies [69–72] have reported the functional role of SNARE interactions in salt stress, justifying the significance of SNARE proteins in salt stress-induced responses in watermelon.

3.2. DEGs Related to Amino Acid Metabolism

To counter the detrimental effects of salt-induced stress, plants produce compatible solutes like free amino acids to minimize high salinity-caused osmotic stress. Positive correlations between increased salt tolerance and accumulations of total free amino acids have been reported in several crops [73–76].

3.2.1. Branched-Chain Amino Acids (BCAAs)

Proline accumulation is known as an important mechanism in osmotic regulation in plants under a wide range of abiotic stresses [77]. However, it has been recognized that the levels of other amino acids, like BCAAs, are often greater or comparable to proline [19,78,79]. A partial deficiency of BCAAs resulted in increasing the sensitivity to salt stress [80]. Induction in the expression of *threonine dehydratase* (ClCG04G009590) along with the repression of both Thr catabolic *threonine aldolases* (ClCG02G017030 and ClCG06G009580) are in agreement with increased accumulation of BCAAs. The up-regulation of *branched-chain amino acid aminotransferase* (BCAT, ClCG08G016800), involved in both the synthesis and degradation of BCAAs, suggests its role in maintaining the non-toxic levels of free BCAAs and alleviating the injury caused by salt stress. The down-regulation of several genes involved in the degradation of BCAAs such as *2-oxoisovalerate dehydrogenase* (ClCG03G014630), *3-hydroxyisobutyrate dehydrogenase* (ClCG03G006140, ClCG05G009680), *3-hydroxyisobutyryl-CoA hydrolase* (ClCG05G002680, ClCG05G016290, ClCG11G016500, ClCG03G012360), *3-ketoacyl-CoA thiolase* (ClCG01G011180, ClCG02G002930), *methylcrotonoyl-CoA carboxylase* (ClCG02G006000), and several *aldehyde dehydrogenases* substantiates the role of BCAA accumulation during salt stress. A partial deficiency of BCAAs resulted in increasing the sensitivity to salt stress in Arabidopsis [80]. Although several studies have reported stress-induced accumulation [75], the metanalysis of transcriptome and metabolome datasets revealed that the low abundant BCAAs could also accumulate due to increased protein degradation [16]. Nevertheless, it has been suggested that BCAAs can serve either as substrates for stress-induced protein biosynthesis or as signaling molecules for regulating stress-responsive gene expression [80]. Intriguingly, *acetolactate synthase* (ALS) small subunit (ClCG03G010140), involved in BCAAs synthesis, was also down-regulated. However, though the feedback inhibition of ALS by BCAAs lacks recent experimental evidence, the non-overlapping sub-cellular localization of ALS subunits and their functional roles in Na^+ homeostasis suggest the need for additional studies to understand the significance of ALS in salt stress-induced responses [80].

3.2.2. Arginine-Polyamine-β-Alanine Pathway

Increased Arg accumulation, which is a precursor for polyamine synthesis, is supported by up-regulation in the expression of *argininosuccinate synthase* (ClCG06G017780), along with down-regulation of *arginine decarboxylase* (ClCG06G014050) and *arginine biosynthesis bifunctional* protein (ArgJ; ClCG10G020940), suppressing a futile cyclic version of arginine biosynthesis. Besides their role in plant growth and development, polyamines like putrescine (Put), spermine (Spm), and spermidine (Spd) play an important role in response to abiotic stress [81]. Put is synthesized directly from arginine by catabolic enzymes *arginine decarboxylase* (ADC) or *agmatine deiminase* (ADI) or from ornithine, catalyzed by *ornithine decarboxylase* (ODC). Put then converts to Spm and Spd via *spermidine synthase* and *spermine synthase* in the presence of decarboxylated S-adenosylmethionine (dcSAM) [82], which is synthesized by *SAM decarboxylase* (SAMDC) via decarboxylation of S-adenosylmethionine (SAM) [83]. In our study, RNA-seq analysis showed up-regulation of *SAMDC* (ClCG05G011880) and relatively abundant *spermidine synthases* (ClCG05G025220 and ClCG05G008800) but down-regulation of less abundant *spermidine synthases* (ClCG06G016890 and ClCG05G005220) genes. Unlike *ADC* (ClCG06G014050), expression of *ADI* (ClCG06G015290) and *ODC* (ClCG08G013990) involved in Put synthesis were also up-regulated. Although the expression of *ADC* expression was reduced, an induction of *ODC*, which catalyzes an alternative pathway of Put synthesis, seems to partly compensate the need for production of Put during salt stress. The polyamines Spm and Spd also serve as precursors of β-alanine synthesis in plants. β-alanine is converted to a quaternary ammonium osmoprotective compound called β-alanine betaine participating in tolerance to high salt concentration [84,85]. Up-regulation of *polyamine oxidase 1* (ClCG09G003930) and *polyamine oxidase 2* (ClCG07G010820, ClCG11G016630) further supports the possible involvement of β-alanine in salt-induced responses.

3.2.3. Amino Acid Transporters

Altered amino acid compositions in response to salt stress subsequently result in alterations in the expression of amino acid transporters. Several amino acid transporters have been identified in plants [86–88] and are grouped into two subfamilies based on sequence similarities and biochemical properties. Under salt stress, we identified a total of 45 up-regulated DEGs and 17 down-regulated DEGs associated with amino acid transport function (Supplementary Materials, Table S6). Salt stress induces changes in amino acid compositions, and the enhanced expression of amino acid transporters has been validated in Arabidopsis [89], rice [87], and wheat [90]. The DEGs involved in amino acid transport may play important roles in regulating the partitioning of different amino acids and maintain osmotic potential in response to salt stress.

3.3. DEGs Associated with Nitrogen Metabolism

Nitrogen (N) metabolism, a central process for plant growth and development, is strongly influenced by salinity. Excess salt disturbs different steps of N metabolism, namely nitrate (NO_3^-) or ammonium (NH_4^+) uptake, N transport and assimilation into amino acids, and protein synthesis [91]. The absorbed N is reduced to nitrite by *nitrate reductase* (NR) and then to NH_4^+ by *nitrite reductase* (NiR). N is further assimilated into Gln and Glu via *glutamine synthetase* (GS) and *glutamate synthase* (GOGAT) and used for further biosynthesis of other nitrogenous compounds. NH_4^+ can be incorporated into Glu by *glutamate dehydrogenase* (GDH). Although the activities of GOGAT, GS, and GDH exhibit salt-dependent regulation, their regulation (induction or repression) varies among species, cultivars, tissues, and developmental changes [92,93]. A salt-induced reduction [94] and stimulation [95] of NR activity have been reported in plants. Our results showed that salt stress selectively inhibited *high affinity nitrate transporters* (ClCG05G025540, ClCG03G003060) and *NRT1* like genes (ClCG02G009090, ClCG10G002910) but induced expression of *low-affinity nitrate transporters* (ClCG06G016390, ClCG11G002980), suggesting differential impacts of salinity on the transporters.

Further, the down-regulation of two *glutamate dehydrogenases* (ClCG01G004910, ClCG04G005320, ClCG07G013590), *glutamine synthetase* (ClCG09G004580), and *carbonic anhydrases* (ClCG05G025330, ClCG10G018930) validated the negative impacts of salt stress on nitrogen assimilation. The expression of the gene encoding *alanine transaminase* (ClCG09G001390) was up-regulated, causing 2-oxoglutarate to generate Glu. *Glutamate decarboxylase* (*GDC*) promotes the synthesis of Pro and GABA from Glu. Our data showed that the expression of *GDCs* (ClCG00G006020, ClCG01G006890, ClCG01G006910) was highly induced in response to salt stress, implying increased accumulation of Pro and downstream metabolites such as citrulline or polyamines contribute towards salt tolerance. Differential responses of various members of the same gene family due to salt stress are consistent with a study in rice [96]. Accumulation of excess Ser can be attributed to the activation of phosphorylated pathways of Ser synthesis as the expression of *D-glycerate 3-kinase* (ClCG09G002370), *D-3-phosphoglycerate dehydrogenase* (ClCG05G010250) and *phosphoserine aminotransferase* (ClCG10G000330) were strongly up-regulated. Ser is considered as a critical player in biochemical responses for the regulation of intracellular redox, energy levels, and cellular pH, particularly in stress conditions [97].

3.4. Disruption of the Energy Metabolisms by the Salt Stresses

Effective photosynthesis results through coordinated activities of four protein complexes—PSI, PSII, the cytochrome b6/f complex, and ATP synthase. As confirmed in the GO and KEGG pathway enrichment analysis, several DEGs identified in this study associated with these complexes were down-regulated due to salt stress (Figure S7, Supplementary Materials, Table S7). The down-regulation of PsbO (ClCG01G016370), PsbP (ClCG07G010800), PsbQ (ClCG05G000900, ClCG03G005130), PsbS (ClCG08G005640), and PsbW (ClCG02G016710, ClCG09G007590) in PSII complex were consistent with decreasing magnitude of Fv/Fm, suggesting impaired chlorophyll fluorescence of PSII during salt stress progression. Although both the PSI and PSII reaction centers are affected by salt stress, studies in cucumber found that PSI is more vulnerable to injury than PSII [98]. Several genes encoding PSI protein complex viz PsaE (ClCG11G010230), PsaF (ClCG01G009000), PsaG (ClCG10G004510), PsaH (ClCG07G013150), PsaK(ClCG01G025030), PsaL(ClCG11G010740), PsaN (ClCG01G015380), and PsaO (ClCG01G011760), along with proteins involved in photosynthetic electron transfer (PetE, PetF, PetH), were also down-regulated confirming inhibition of photosynthetic activities under salt stress. A decrease in electron transfer rate accumulates excess electrons leading to electron leakage, which results in the outbreak of reactive oxygen species (ROS) and damage to the PSII reaction center [99,100]. Further, nearly all the proteins (Lhca1 to Lhca5, Lhcb1 to 4, Lhcb7) involved in light-harvesting chlorophyll (LHC) were down-regulated (Figure S8, Supplementary Materials, Table S7). The inactivation of photosynthesis and LHC complex due to salt stress observed in this study is consistent with several studies [56,101,102] and validates the role of PSI and PSII complexes in balancing energy supply and ROS generation under salt stress in watermelon.

3.5. DEGs Associated with Hormonal Regulation

In the present study, the functional analysis identified many DEGs associated with hormone signaling transduction pathways emphasizing the involvement of plant hormones in regulating the response to salt stress in watermelon leaves. We grouped the DEGs into various phytohormone signaling pathways, such as auxin (AUX), cytokinin (CTK), gibberellin (GA), abscisic acid (ABA), ethylene (ETH), brassinosteroid (BR), jasmonic acid (JA), and salicylic acid (SA) (Figure 8, Supplementary Materials, Table S8). Studies have confirmed the reduced auxin levels and decreased auxin transporter expression in plants under saline conditions [103,104]. The expression of most of the genes involved in auxin signal transduction pathways such as auxin transporter protein 1 (AUX1), transport inhibitor response (TIR1), auxin response factor gene (ARF), auxin early response gene (Aux/IAA), and small auxin-up RNAs (SAUR) were significantly down-regulated in response to salt stress. Up-regulation of GH3 genes due to salt stress is consistent with previous studies [105] and is possibly responsible for triggering cellular mechanisms to protect cell auxin homeostasis during changes in extracellular auxin levels. Although cytokinins play an important role in plant growth and development, numerous pieces of evidence indicate both positive and negative effects on stress tolerance. Salt-treated plants showed increased or decreased accumulation of active cytokinins in plants [106–108]. In our study, expression of genes CRE1, B-ARR, and A-ARR were up-regulated, while B-ARR was down-regulated due to salt stress. The changes in the expression of genes associated with cytokinins are consistent with salt-induced changes in tomato [109,110] and Arabidopsis [108] plants.

The ABA signaling pathway is associated with salt stress-induced responses and helps plants by reducing the buildup of Na+ and improving osmotic adjustment. In the present study, three of the ABA receptors (PYR/PYL) and serine/threonine-protein kinase 2 (SNRK2) were significantly down-regulated in salt treatment, suggesting inhibition of the ABA signaling pathway by saline treatments. The only salt-induced changes in the expression of two F-box gibberellin-insensitive dwarf2 (GID2) and transcription factor genes associated with the GA signaling pathway suggest a sub-optimal role of the GA pathway in salt stress-induced responses in watermelon.

Although ethylene is a stress hormone regulating numerous stress responses, its specific roles in salt stress tolerance in plants remain unclear [111]. In this study, we observed up-regulation of ethylene receptors (ETR) and serine/threonine-protein kinase CTR1 (CTR1), which serve as negative regulators of the ethylene signaling transduction pathway. On the contrary, a positive regulator of ethylene signal transduction, ethylene-insensitive protein 3 (EIN3), was also up-regulated. No significant differences in expression levels of EIN2 and ERF1/2 genes were observed. Taken together, salt stress seems to have either a negative or trivial impact on the ethylene signaling pathway.

In the present study, genes involved in the BR pathway viz. BRI1-associated receptor kinase (BAK1), BR-signaling kinase (BSK), brassinosteroid insensitive 2 (BIN2), and brassinosteroid resistant 1/2 (BZR1/2) were significantly up-regulated, suggesting activation of the BR signaling pathway in response to salt stress. Similar activation of genes involved in the BR signaling pathway during salt stress has been reported in Arabidopsis [112]. Up-regulation of the cyclin gene CycD3 in our study is consistent with a study showing similar responses due to the exogenous application of 24-EBR [113]. Although the results confirm the activation of the BR signaling pathway, the exact functional relevance of the BR pathway in salt-induced responses needs further investigation.

In the JA signal pathway, the expression of jasmonate ZIM domain-containing protein (JAZ) genes and JASMONATE INSENSITIVE 1/MYC2 (JIN1/MYC2) were significantly up-regulated due to salinity, indicating activation of JA signaling transduction by saline stress. The up-regulation of COI1-dependent JA-responsive JAZ genes due to salt stress has been reported in Arabidopsis [114]. It is plausible to assume that salt stress-induced JA-Ile promotes the interaction of JAZ proteins with COI1, followed by their degradation via the 26S proteasome and de-repression of MYC2 to induce transcription of JA-responsive genes in watermelon.

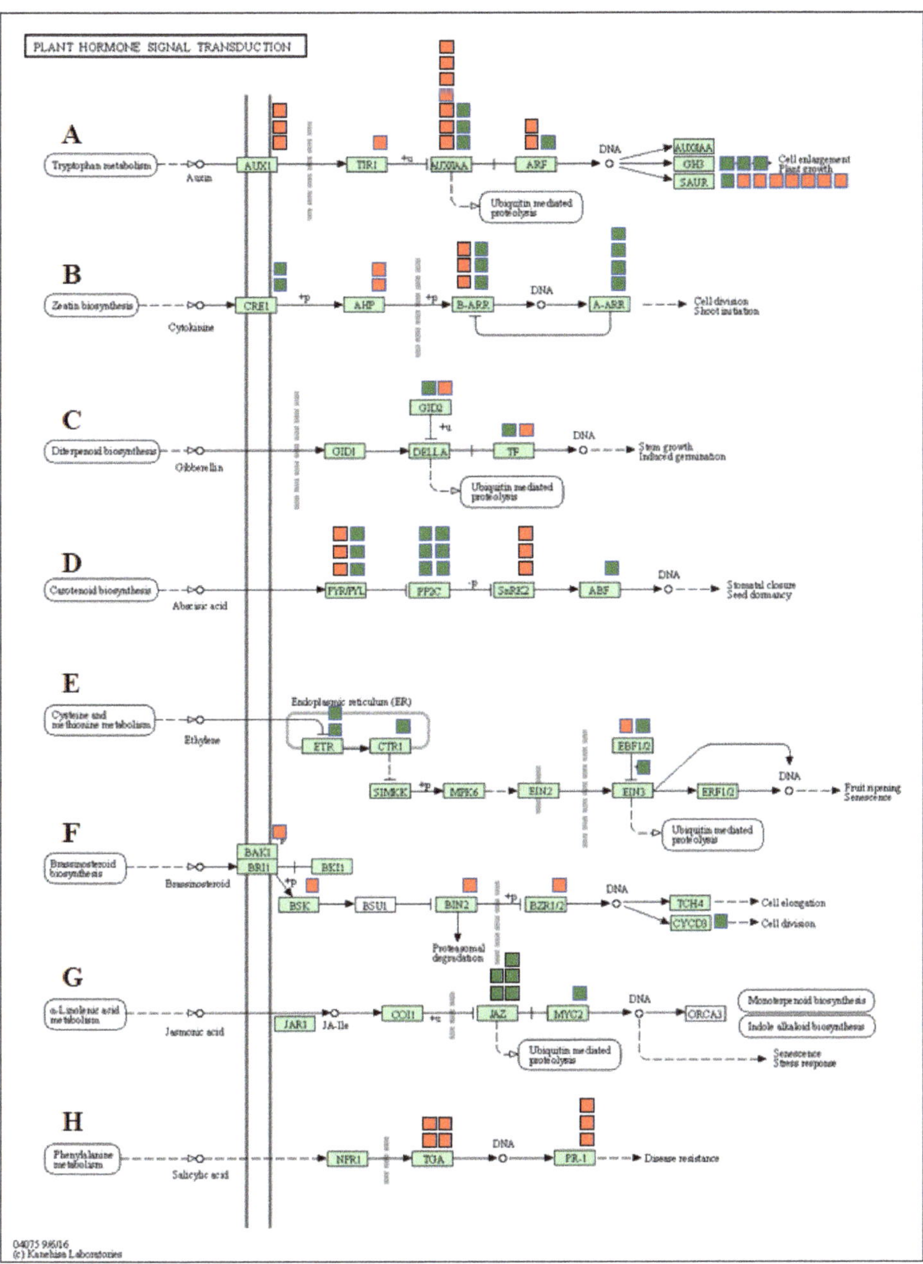

Figure 8. DEGs mapped to the plant hormone signaling transduction pathways in watermelon leaves. (**A**) Auxin signaling pathway, (**B**) cytokinin pathway, (**C**) gibberellin (GA) pathway, (**D**) abscisic acid pathway, (**E**) ethylene signaling pathway, (**F**) brassinosteroid pathway, (**G**) jasmonic acid pathway, (**H**) salicylic acid pathway. The red and green boxes show the number of down- or up-regulated DEGs, respectively.

4. Materials and Methods

4.1. Salt Stress Experiment and Photochemical Efficiency Measurement

Watermelon (*C. lanatus* L. cv. Crimson Sweet) seeds were sown in trays filled with a soilless media (Quick Dry Infield Conditioner, Turface Atheletics™, Buffalo Grove, IL, USA) and placed in the greenhouse at the Texas A&M AgriLife Research and Extension Center, Uvalde, TX, USA. Six-week-old seedlings with three fully expanded leaves were carefully lifted out from Turface media, washed under running water, and incubated in 50 mL tubes (VWR®, Radnor Corporate Center, Radnor Township, PA, USA) containing 300 mM NaCl and deionized water as a control (Barnstead™ Smart2Pure™ Water Purification System, Thermo Scientific, Waltham, MA, USA). The concentration of NaCl used in this study was previously validated for its sensitivity in watermelon [115,116]. Chlorophyll fluorescence was measured at the end of the experiment using a portable fluorometer (PAR-FluorPen FP 110/D; PSI (Photon Systems Instruments), Drasov, Czech Republic) after dark adaption for 30 min. This measurement was initiated from 1 h after initiation of the treatment and was continued every 2 h. The maximum photochemical efficiency of PSII (Fv/Fm) was calculated according to the manufacturer's protocol. After 7 h of exposure to salt stress, leaf and root tissue samples were collected from four independent seedlings for each treatment and flash-frozen using liquid nitrogen before storing at −80 °C for further processing.

4.2. Extraction Method and Quantification of Free Amino Acids with Ultra-Performance Liquid Chromatography-Electron Spray Ionization Tandem Mass Spectrometry (UPLC-ESI-MS/MS)

Approximately 20 mg frozen tissue samples collected into 2 mL microcentrifuge tubes were homogenized into fine powder in a paint shaker (Harbil model 5G-HD paint shaker) using 3 mm stainless steel beads (Demag stainless steel balls, Abbott Ball Company, Inc., West Hartford, CT, USA) to quantify free amino acids. Amino acids were extracted using an established protocol [34] by suspending the homogenized samples in 100 mM cold HCl extraction buffer, followed by incubation on ice (~20 min) and then centrifuging at a speed of 14,609× *g* for 20 min at 4 °C. The supernatants were collected and filtered through a 96-well 0.45-μm-pore filter plate (Pall® Life Sciences Filter, Pall Corporation, Port Washington, NY, USA). The eluents collected in 96-well trap plates were stored at −20 °C for further amino acid quantification.

The derivatization of filtrates was carried out with an AccQ•Tag 3X Ultra-Fluor™ derivatization kit (Waters Corporation, Milford, MA, USA) following the standard protocol. L-Norvaline (Sigma, St. Louis, MO, USA) was used as an internal standard at a fixed concentration of at 25 pmol/μl. Amino acid calibrators were purchased from Kairos™ Amino Acid Kit (Waters Corporation, Milford, MA, USA). Lyophilized amino acid powder of six amino acid calibrators representing different concentrations (5.0 pmol/μL to 1000 pmol/μL) was reconstituted with 2 mL of 0.1 M HCl to establish detection limits ranging from 0.45 pmol/μL to 90 pmol/μL. Calibration curves were built in TargetLynxTM Application Manager (Waters Corporation, Milford, MA, USA).

UPLC-ESI-MS/MS analysis was performed using a Waters Acquity H-class UPLC system equipped with a Waters Xevo TQ mass spectrometer by using electrospray ionization (ESI) probe. The Waters Acquity H-class UPLC system was composed of an autosampler, a binary solvent manager, a Waters® ACQUITY UPLC® Fluorescence (FLR) detector, a column heater and a Water's AccQ•Tag Ultra column (2.1 mm i.d. × 140 mm, 1.7 μm particles). The mobile phase consists of water phase (A) (0.1% formic acid *v/v*) and acetonitrile (B) (0.1% formic acid *v/v*) with a stable flow rate at 0.5 mL/min and column temperature setting at 55 °C. The gradient of non-linear separation was set as follows: 0–1 min (99% A), 3.2 min (87.0% A), 8 min (86.5% A), and 9 min (5% A). Finally, 2 μL of the derivatized sample was injected onto the column for analysis. IntelliStart software (Waters Corp, Milford, MA, USA) was used to optimize amino acid multiple reaction monitoring (MRM) transitions, collision energy values, and cone voltage. The ESI source was set to 150 °C with the gas desolvation flow rate at 1000 L/h, gas flow cone at 20 L/h, desolvation temperature at 500 °C, the capillary voltage at 2.0 kV, gas collision energy varied from 15 to 30 V, and cone voltage at 30 V for detecting all amino acids. MRM was operated in positive mode. Water's MassLynx™ 4.1 software was used for instrument monitoring and data acquisition. The data integration, calibration curves, and quantitation (0.45–90 pmol/µL) were carried out with TargetLynx™ Application Manager (Waters Corporation, Milford, MA, USA).

4.3. cDNA Library Preparation and RNA-Seq Analysis of Salt-Stressed Seedling Leaves

Six independent libraries were created by using a total of 6 RNA samples from 3 replicate leaf tissues of cv. Crimson Sweet under the control and salt-treated condition. The samples were flash-frozen in liquid nitrogen and ground to a fine powder using 3-mm-diameter steel balls (Abbott Ball, West Hartford, CT, USA) in a paint shaker (Harbil, Wheeling, IL, USA). Total RNA was extracted using an RNeasy® Plant Mini Kit (QIAGEN Sciences, Germantown, MD, USA) as per the manufacturer's protocol. The purity of the RNA was confirmed using a NanoPhotometer® spectrophotometer (IMPLEN GmbH, Inc., München, Germany). The RNA Nano 6000 Assay Kit of the Bioanalyzer 2100 system (Agilent Technologies, Santa Clara, CA, USA) was used to assess RNA integrity and quantitation. Sequencing libraries were generated using a NEBNext® Ultra™ RNA Library Prep Kit for Illumina® (New England Biolabs, Ipswich, MA USA) following the manufacturer's protocol. The clustering of the index-coded samples was performed on a cBot Cluster Generation System using PE Cluster Kit cBot-HS (Illumina) according to the manufacturer's instructions. After cluster generation, the libraries were sequenced on an Illumina Hiseq platform, and 150 bp paired-end reads were generated. Raw reads of fastq format were processed to obtain clean reads by removing the adapter, reads containing poly N (reads when uncertain nucleotides constitute more than 10 percent of either read; N > 10%), and low-quality reads (reads when low-quality nucleotides (base quality less than 20) constitute more than 50% of the read). The Qscore (quality value) of over 50% bases of these reads is ≤5) from raw data. At the same time, Q20, Q30, and GC content of the clean data were calculated. Watermelon reference genome version 2 (cv. Charleston Gray) and gene model annotation files were downloaded from CuGenDB (http://cucurbitgenomics.org/). Index of the reference genome was built using Bowtie v2.2.3, and paired-end clean reads were aligned to the reference genome using TopHat v2.0.12. HTSeq v0.6.1 was used to count the reads mapped to each gene. FPKM [117] of each gene was calculated based on the length of the gene and reads count mapped to this gene. Differential expression analysis of genes was performed using the DESeq R package (1.18.0) [118]. Genes with *p*-value < 0.05 found by DESeq were considered as differentially expressed. Gene ontology (GO) [119] enrichment analysis of differentially expressed genes was implemented using the GOseq R package, in which gene length bias was corrected. GO terms with a *p*-value less than 0.05 were considered significantly enriched by DEGs. To test the statistical enrichment of differential expression genes, KOBAS software in the Kyoto Encyclopedia of Genes and Genomes (KEGG) pathways database [120] was used. To identify the source of variance in the expressed transcripts between control and salt treatment, and the repeatability of samples within a group, principal component analysis (PCA) was used. PCA was performed using the scikit-learn package [121] and plotted using Matplotlib [122]. The RNA-seq dataset is accessible through GEO Series accession number GSE146087 (https://www.ncbi.nlm.nih.gov/geo/).

4.4. Validation by Quantitative Real-Time PCR

To validate the RNA-seq data, total RNA was extracted from three replicate leaf tissues of salt-stressed seedlings (Crimson Sweet). The expression pattern of selected DEGs was examined using quantitative real-time PCR (RT-qPCR). The gene-specific primers based on the selected unigene sequences (Table S9) were designed using Primer Premier 3.0 software. Total RNA was extracted with the Quick-RNA™ Miniprep Kit (Zymo Research Corporation, Irvine, CA, USA) followed by DNase1 (Zymo Research Corporation, Irvine, CA, USA) treatment, and subjected to reverse transcription using iScript RT Supermix (Bio-Rad Laboratories, Inc., Hercules, CA, USA). The quality and quantity of the RNA were examined using a Denovix DS-11+ spectrophotometer (DeNovix Inc. Wilmington, DE, USA). Gene expression analysis via reverse transcription-qPCR was performed using a BioRad CFX96 qPCR instrument and by using a SsoAdv Univer SYBR GRN Master Kit (Bio-Rad Laboratories, Inc., Hercules, CA, USA). Watermelon β-actin and α-tubulin5 genes [123] were used as the internal controls, and the relative expression levels (Cq values) for each gene were normalized by taking an average of three biological replicates. The relative expression levels of each gene were calculated using the $2^{-\Delta\Delta Ct}$ method. The primers for qPCR used in this chapter are listed in Supplementary Materials, Table S9.

5. Conclusions

In conclusion, this study represents comprehensive information regarding the transcriptome of watermelon seedlings in response to salt stress. The transcriptome profiling generated over 43 million reads from the control and salinity-treated libraries. The transcriptome assembly detected 7622 genes that were expressed differentially in response to salinity. These differentially expressed genes included transcription factors, genes related to primary metabolism, endocytosis, hormonal pathways, and transporters involved in responses to salinity. The gene expression patterns of the TFs identified in this study help in improving our understanding of the significance of transcriptional regulation in watermelon during salt stress. These results provide a basis for future studies aimed at discovering novel genes, their functional validation in model species, and finding molecular mechanisms associated with salt tolerance in watermelon.

Supplementary Materials: The following are available online at http://www.mdpi.com/1422-0067/21/17/6036/s1.

Author Contributions: Conceptualization, Q.S., M.J., and V.J.; methodology, Q.S. and M.J.; formal analysis, Q.S. and M.J.; investigation, V.J.; resources, M.J. and V.J.; data curation, V.J.; writing—original draft preparation, Q.S., M.J., and V.J.; writing—review and editing, V.J.; visualization, V.J.; supervision, V.J.; project administration, V.J.; funding acquisition, V.J. All authors have read and agreed to the published version of the manuscript.

Funding: This research is supported by the USDA National Institute of Food and Agriculture [Hatch Project No TEX09647/project accession no. 1011513].

Acknowledgments: We acknowledge Novogene Corporation Inc., Sacramento, CA, USA for its assistance in RNA sequencing and processing of samples.

Conflicts of Interest: The authors declare no conflict of interest.

Abbreviations

GO	Gene Ontology
KEGG	Kyoto Encyclopedia of Genes and Genomes

References

1. Godfray, H.C.J.; Beddington, J.R.; Crute, I.R.; Haddad, L.; Lawrence, D.; Muir, J.F.; Pretty, J.; Robinson, S.; Thomas, S.M.; Toulmin, C. Food security: The challenge of feeding 9 billion people. *Science* **2010**, *327*, 812–818. [CrossRef] [PubMed]
2. Pitman, M.; Lauchli, A. Global impact of salinity and agricultural ecosystems. In *Salinity: Environment-plants-Molecules*; Springer: Dordrecht, The Netherlands, 2002; pp. 3–20.
3. Flowers, T. Improving crop salt tolerance. *J. Exp. Bot.* **2004**, *55*, 307–319. [CrossRef] [PubMed]
4. Shahid, S.A.; Zaman, M.; Heng, L. Soil Salinity: Historical Perspectives and a World Overview of the Problem. In *Guideline for Salinity Assessment, Mitigation and Adaptation Using Nuclear and Related Techniques*; Springer International Publishing: Cham, Switzerland, 2018; pp. 43–53.
5. Yetisir, H.; Uygur, V. Responses of grafted watermelon onto different gourd species to salinity stress. *J. Plant Nutr.* **2010**, *33*, 315–327. [CrossRef]
6. Yamaguchi, T.; Blumwald, E. Developing salt-tolerant crop plants: Challenges and opportunities. *Trends Plant Sci.* **2005**, *10*, 615–620. [CrossRef] [PubMed]
7. Simpson, C.R.; King, S.; Nelson, S.D.; Jifon, J.; Schuster, G.; Volder, A. Salinity Evaluation for watermelon (*Citrullus lanatus*) grafted with different rootstocks. *Subtrop. Agric. Environ.* **2015**, *66*, 1–6.
8. Yan, Y.; Wang, S.; Wei, M.; Gong, B.; Shi, Q. Effect of Different Rootstocks on the Salt Stress Tolerance in Watermelon Seedlings. *Hortic. Plant J.* **2018**, *4*, 239–249. [CrossRef]

9. Simpson, C.R.; Franco, J.G.; King, S.R.; Volder, A. Intercropping Halophytes to Mitigate Salinity Stress in Watermelon. *Sustainability* **2018**, *10*, 681. [CrossRef]

10. Romic, D.; Ondrasek, G.; Romic, M.; Josip, B.; Vranjes, M.; Petosic, D. Salinity and irrigation method affect crop yield and soil quality in watermelon (*Citrullus lanatus* L.) growing. *Irrig. Drain.* **2008**, *57*, 463–469. [CrossRef]

11. Zhu, J.-K. Plant salt tolerance. *Trends Plant Sci.* **2001**, *6*, 66–71. [CrossRef]

12. Zhang, L.; Zhang, X.; Fan, S. Meta-analysis of salt-related gene expression profiles identifies common signatures of salt stress responses in Arabidopsis. *Plant Syst. Evol.* **2017**, *303*, 757–774. [CrossRef]

13. Yuan, F.; Leng, B.; Wang, B. Progress in studying salt secretion from the salt glands in recretohalophytes: How do plants secrete salt? *Front Plant Sci.* **2016**, *7*, 977. [CrossRef] [PubMed]

14. Belkhodja, R.; Morales, F.; Abadia, A.; Gomez-Aparisi, J.; Abadia, J. Chlorophyll Fluorescence as a Possible Tool for Salinity Tolerance Screening in Barley (*Hordeum vulgare* L.). *Plant Physiol.* **1994**, *104*, 667–673. [CrossRef] [PubMed]

15. Tsai, Y.-C.; Chen, K.-C.; Cheng, T.-S.; Lee, C.; Lin, S.-H.; Tung, C.-W. Chlorophyll fluorescence analysis in diverse rice varieties reveals the positive correlation between the seedlings salt tolerance and photosynthetic efficiency. *BMC Plant Biol.* **2019**, *19*, 403. [CrossRef]

16. Hildebrandt, T.M. Synthesis versus degradation: Directions of amino acid metabolism during Arabidopsis abiotic stress response. *Plant Mol. Biol.* **2018**, *98*, 121–135. [CrossRef] [PubMed]

17. Huang, T.; Jander, G. Abscisic acid-regulated protein degradation causes osmotic stress-induced accumulation of branched-chain amino acids in Arabidopsis thaliana. *Planta* **2017**, *246*, 737–747. [CrossRef]

18. Araújo, W.L.; Tohge, T.; Ishizaki, K.; Leaver, C.J.; Fernie, A.R. Protein degradation—An alternative respiratory substrate for stressed plants. *Trends Plant Sci.* **2011**, *16*, 489–498. [CrossRef] [PubMed]

19. Song, Q.; Joshi, M.; DiPiazza, J.; Joshi, V. Functional Relevance of Citrulline in the Vegetative Tissues of Watermelon During Abiotic Stresses. *Front Plant Sci.* **2020**, *11*, 512. [CrossRef] [PubMed]

20. Yokota, A.; Kawasaki, S.; Iwano, M.; Nakamura, C.; Miyake, C.; Akashi, K. Citrulline and DRIP-1 protein (ArgE homologue) in drought tolerance of wild watermelon. *Ann. Bot.* **2002**, *89*, 825–832. [CrossRef]

21. Xie, C.; Mao, X.; Huang, J.; Ding, Y.; Wu, J.; Dong, S.; Kong, L.; Gao, G.; Li, C.-Y.; Wei, L. KOBAS 2.0: A web server for annotation and identification of enriched pathways and diseases. *Nucleic Acids Res.* **2011**, *39*, W316–W322. [CrossRef]

22. Zhu, J.K. Salt and drought stress signal transduction in plants. *Annu. Rev. Plant Biol.* **2002**, *53*, 247–273. [CrossRef]

23. Cao, Y.-R.; Chen, S.-Y.; Zhang, J.-S. Ethylene signaling regulates salt stress response: An overview. *Plant Signal Behav.* **2008**, *3*, 761–763. [CrossRef] [PubMed]

24. Zhang, M.; Smith, J.A.C.; Harberd, N.P.; Jiang, C. The regulatory roles of ethylene and reactive oxygen species (ROS) in plant salt stress responses. *Plant Mol. Biol.* **2016**, *91*, 651–659. [CrossRef] [PubMed]

25. Zhang, H.; Huang, Z.; Xie, B.; Chen, Q.; Tian, X.; Zhang, X.; Zhang, H.; Lu, X.; Huang, D.; Huang, R. The ethylene-, jasmonate-, abscisic acid-and NaCl-responsive tomato transcription factor JERF1 modulates expression of GCC box-containing genes and salt tolerance in tobacco. *Planta* **2004**, *220*, 262–270. [CrossRef] [PubMed]

26. Pan, Y.; Seymour, G.B.; Lu, C.; Hu, Z.; Chen, X.; Chen, G. An ethylene response factor (ERF5) promoting adaptation to drought and salt tolerance in tomato. *Plant Cell Rep.* **2012**, *31*, 349–360. [CrossRef]

27. Wang, H.; Huang, Z.; Chen, Q.; Zhang, Z.; Zhang, H.; Wu, Y.; Huang, D.; Huang, R. Ectopic overexpression of tomato JERF3 in tobacco activates downstream gene expression and enhances salt tolerance. *Plant Mol. Biol.* **2004**, *55*, 183–192. [CrossRef]

28. Klay, I.; Pirrello, J.; Riahi, L.; Bernadac, A.; Cherif, A.; Bouzayen, M.; Bouzid, S. Ethylene response factor Sl-ERF. B. 3 is responsive to abiotic stresses and mediates salt and cold stress response regulation in tomato. *Sci. World J.* **2014**, *2014*. [CrossRef]

29. Li, Z.; Tian, Y.; Xu, J.; Fu, X.; Gao, J.; Wang, B.; Han, H.; Wang, L.; Peng, R.; Yao, Q. A tomato ERF transcription factor, SlERF84, confers enhanced tolerance to drought and salt stress but negatively regulates immunity against Pseudomonas syringae pv. tomato DC3000. *Plant Physiol. Biochem.* **2018**, *132*, 683–695. [CrossRef]

30. Jung, K.-H.; Seo, Y.-S.; Walia, H.; Cao, P.; Fukao, T.; Canlas, P.E.; Amonpant, F.; Bailey-Serres, J.; Ronald, P.C. The submergence tolerance regulator Sub1A mediates stress-responsive expression of AP2/ERF transcription factors. *Plant Physiol.* **2010**, *152*, 1674–1692. [CrossRef]

31. Xu, Z.-S.; Xia, L.-Q.; Chen, M.; Cheng, X.-G.; Zhang, R.-Y.; Li, L.-C.; Zhao, Y.-X.; Lu, Y.; Ni, Z.-Y.; Liu, L. Isolation and molecular characterization of the Triticum aestivum L. ethylene-responsive factor 1 (TaERF1) that increases multiple stress tolerance. *Plant Mol. Biol.* **2007**, *65*, 719–732. [CrossRef]

32. Rong, W.; Qi, L.; Wang, A.; Ye, X.; Du, L.; Liang, H.; Xin, Z.; Zhang, Z. The ERF transcription factor Ta ERF 3 promotes tolerance to salt and drought stresses in wheat. *Plant Biotechnol. J.* **2014**, *12*, 468–479. [CrossRef]

33. Schmidt, R.; Mieulet, D.; Hubberten, H.-M.; Obata, T.; Hoefgen, R.; Fernie, A.R.; Fisahn, J.; San Segundo, B.; Guiderdoni, E.; Schippers, J.H. SALT-RESPONSIVE ERF1 regulates reactive oxygen species–dependent signaling during the initial response to salt stress in rice. *Plant Cell* **2013**, *25*, 2115–2131. [CrossRef] [PubMed]

34. Long, L.; Yang, W.-W.; Liao, P.; Guo, Y.-W.; Kumar, A.; Gao, W. Transcriptome analysis reveals differentially expressed ERF transcription factors associated with salt response in cotton. *Plant Sci.* **2019**, *281*, 72–81. [CrossRef] [PubMed]

35. Li, Q.; Qin, Y.; Hu, X.; Li, G.; Ding, H.; Xiong, X.; Wang, W. Transcriptome analysis uncovers the gene expression profile of salt-stressed potato (*Solanum tuberosum* L.). *Sci. Rep.* **2020**, *10*, 5411. [CrossRef] [PubMed]

36. Ma, Y.; Zhang, F.; Bade, R.; Daxibater, A.; Men, Z.; Hasi, A. Genome-Wide Identification and Phylogenetic Analysis of the ERF Gene Family in Melon. *J. Plant Growth Regul.* **2015**, *34*, 66–77. [CrossRef]

37. Hu, L.; Liu, S. Genome-wide identification and phylogenetic analysis of the ERF gene family in cucumbers. *Genet Mol. Biol.* **2011**, *34*, 624–633. [CrossRef] [PubMed]

38. Yang, X.; Li, H.; Yang, Y.; Wang, Y.; Mo, Y.; Zhang, R.; Zhang, Y.; Ma, J.; Wei, C.; Zhang, X. Identification and expression analyses of WRKY genes reveal their involvement in growth and abiotic stress response in watermelon (*Citrullus lanatus*). *PLoS ONE* **2018**, *13*, e0191308. [CrossRef]

39. Bankaji, I.; Sleimi, N.; Vives-Peris, V.; Gómez-Cadenas, A.; Pérez-Clemente, R.M. Identification and expression of the Cucurbita WRKY transcription factors in response to water deficit and salt stress. *Sci. Hortic.* **2019**, *256*, 108562. [CrossRef]

40. Song, H.; Wang, P.; Hou, L.; Zhao, S.; Zhao, C.; Xia, H.; Li, P.; Zhang, Y.; Bian, X.; Wang, X. Global Analysis of WRKY Genes and Their Response to Dehydration and Salt Stress in Soybean. *Front Plant Sci.* **2016**, *7*, 9. [CrossRef]

41. Gu, L.; Wang, H.; Wei, H.; Sun, H.; Li, L.; Chen, P.; Elasad, M.; Su, Z.; Zhang, C.; Ma, L.; et al. Identification, Expression, and Functional Analysis of the Group IId WRKY Subfamily in Upland Cotton (*Gossypium hirsutum* L.). *Front Plant Sci.* **2018**, *9*, 1684. [CrossRef]

42. Li, H.; Gao, Y.; Xu, H.; Dai, Y.; Deng, D.; Chen, J. ZmWRKY33, a WRKY maize transcription factor conferring enhanced salt stress tolerances in Arabidopsis. *Plant Growth Regul.* **2013**, *70*, 207–216. [CrossRef]

43. Fan, X.; Guo, Q.; Xu, P.; Gong, Y.; Shu, H.; Yang, Y.; Ni, W.; Zhang, X.; Shen, X. Transcriptome-wide identification of salt-responsive members of the WRKY gene family in *Gossypium aridum*. *PLoS ONE* **2015**, *10*, e0126148. [CrossRef] [PubMed]

44. Zhou, Q.Y.; Tian, A.G.; Zou, H.F.; Xie, Z.M.; Lei, G.; Huang, J.; Wang, C.M.; Wang, H.W.; Zhang, J.S.; Chen, S.Y. Soybean WRKY-type transcription factor genes, GmWRKY13, GmWRKY21, and GmWRKY54, confer differential tolerance to abiotic stresses in transgenic Arabidopsis plants. *Plant Biotechnol. J.* **2008**, *6*, 486–503. [CrossRef] [PubMed]

45. Mzid, R.; Zorrig, W.; Ayed, R.B.; Hamed, K.B.; Ayadi, M.; Damak, Y.; Lauvergeat, V.; Hanana, M. The grapevine VvWRKY2 gene enhances salt and osmotic stress tolerance in transgenic *Nicotiana tabacum*. *3 Biotech* **2018**, *8*, 277. [CrossRef] [PubMed]

46. Lv, X.; Lan, S.; Guy, K.M.; Yang, J.; Zhang, M.; Hu, Z. Global Expressions Landscape of NAC Transcription Factor Family and Their Responses to Abiotic Stresses in Citrullus lanatus. *Sci. Rep.* **2016**, *6*, 30574. [CrossRef] [PubMed]

47. Wei, S.; Gao, L.; Zhang, Y.; Zhang, F.; Yang, X.; Huang, D. Genome-wide investigation of the NAC transcription factor family in melon (*Cucumis melo* L.) and their expression analysis under salt stress. *Plant Cell Rep.* **2016**, *35*, 1827–1839. [CrossRef] [PubMed]

48. Maathuis, F.J. Sodium in plants: Perception, signalling, and regulation of sodium fluxes. *J. Exp. Bot.* **2014**, *65*, 849–858. [CrossRef]

49. Murphy, A.S.; Bandyopadhyay, A.; Holstein, S.E.; Peer, W.A. Endocytotic cycling of PM proteins. *Annu. Rev. Plant Biol.* **2005**, *56*, 221–251. [CrossRef]

50. Valencia, J.P.; Goodman, K.; Otegui, M.S. Endocytosis and Endosomal Trafficking in Plants. *Annu. Rev. Plant Biol.* **2016**, *67*, 309–335. [CrossRef]

51. Bar, M.; Avni, A. EHD2 inhibits ligand-induced endocytosis and signaling of the leucine-rich repeat receptor-like protein LeEix2. *Plant J.* **2009**, *59*, 600–611. [CrossRef]

52. Leshem, Y.; Seri, L.; Levine, A. Induction of phosphatidylinositol 3-kinase-mediated endocytosis by salt stress leads to intracellular production of reactive oxygen species and salt tolerance. *Plant J.* **2007**, *51*, 185–197. [CrossRef]

53. Mazel, A.; Leshem, Y.; Tiwari, B.S.; Levine, A. Induction of salt and osmotic stress tolerance by overexpression of an intracellular vesicle trafficking protein AtRab7 (AtRabG3e). *Plant Physiol.* **2004**, *134*, 118–128. [CrossRef] [PubMed]

54. George, S.; Parida, A. Over-expression of a Rab family GTPase from phreatophyte Prosopis juliflora confers tolerance to salt stress on transgenic tobacco. *Mol. Biol. Rep.* **2011**, *38*, 1669–1674. [CrossRef] [PubMed]

55. Liu, F.; Guo, J.; Bai, P.; Duan, Y.; Wang, X.; Cheng, Y.; Feng, H.; Huang, L.; Kang, Z. Wheat TaRab7 GTPase is part of the signaling pathway in responses to stripe rust and abiotic stimuli. *PLoS ONE* **2012**, *7*, e37146. [CrossRef] [PubMed]

56. Sui, J.-M.; Li, R.; Fan, Q.-C.; Song, L.; Zheng, C.-H.; Wang, J.-S.; Qiao, L.-X.; Yu, S.-L. Isolation and characterization of a stress responsive small GTP-binding protein AhRabG3b in peanut (*Arachis hypogaea* L.). *Euphytica* **2013**, *189*, 161–172. [CrossRef]

57. Agarwal, P.K.; Agarwal, P.; Jain, P.; Jha, B.; Reddy, M.K.; Sopory, S.K. Constitutive overexpression of a stress-inducible small GTP-binding protein PgRab7 from Pennisetum glaucum enhances abiotic stress tolerance in transgenic tobacco. *Plant Cell Rep.* **2008**, *27*, 105–115. [CrossRef] [PubMed]

58. Tripathy, M.K.; Tiwari, B.S.; Reddy, M.K.; Deswal, R.; Sopory, S.K. Ectopic expression of PgRab7 in rice plants (*Oryza sativa* L.) results in differential tolerance at the vegetative and seed setting stage during salinity and drought stress. *Protoplasma* **2017**, *254*, 109–124. [CrossRef] [PubMed]

59. Hong, Y.; Zhao, J.; Guo, L.; Kim, S.-C.; Deng, X.; Wang, G.; Zhang, G.; Li, M.; Wang, X. Plant phospholipases D and C and their diverse functions in stress responses. *Prog. Lipid Res.* **2016**, *62*, 55–74. [CrossRef] [PubMed]

60. Bargmann, B.O.; Laxalt, A.M.; Riet, B.T.; Van Schooten, B.; Merquiol, E.; Testerink, C.; Haring, M.A.; Bartels, D.; Munnik, T. Multiple PLDs required for high salinity and water deficit tolerance in plants. *Plant Cell Physiol.* **2009**, *50*, 78–89. [CrossRef]

61. Laxalt, A.M.; Ter Riet, B.; Verdonk, J.C.; Parigi, L.; Tameling, W.I.; Vossen, J.; Haring, M.; Musgrave, A.; Munnik, T. Characterization of five tomato phospholipase D cDNAs: Rapid and specific expression of LePLDβ1 on elicitation with xylanase. *Plant J.* **2001**, *26*, 237–247. [CrossRef]

62. Yu, H.Q.; Yong, T.M.; Li, H.J.; Liu, Y.P.; Zhou, S.F.; Fu, F.L.; Li, W.C. Overexpression of a phospholipase Dα gene from *Ammopiptanthus nanus* enhances salt tolerance of phospholipase Dα1-deficient Arabidopsis mutant. *Planta* **2015**, *242*, 1495–1509. [CrossRef]

63. Ji, T.; Li, S.; Huang, M.; Di, Q.; Wang, X.; Wei, M.; Shi, Q.; Li, Y.; Gong, B.; Yang, F. Overexpression of Cucumber Phospholipase D alpha Gene (CsPLDα) in Tobacco Enhanced Salinity Stress Tolerance by Regulating Na+–K+ Balance and Lipid Peroxidation. *Front Plant Sci.* **2017**, *8*, 499. [CrossRef] [PubMed]

64. Ngara, R.; Ndimba, B.K. Understanding the complex nature of salinity and drought-stress response in cereals using proteomics technologies. *Proteomics* **2014**, *14*, 611–621. [CrossRef] [PubMed]

65. Sobhanian, H.; Aghaei, K.; Komatsu, S. Changes in the plant proteome resulting from salt stress: Toward the creation of salt-tolerant crops? *J. Proteom.* **2011**, *74*, 1323–1337. [CrossRef] [PubMed]

66. Liu, J.; Pang, X.; Cheng, Y.; Yin, Y.; Zhang, Q.; Su, W.; Hu, B.; Guo, Q.; Ha, S.; Zhang, J.; et al. The Hsp70 Gene Family in Solanum tuberosum: Genome-Wide Identification, Phylogeny, and Expression Patterns. *Sci. Rep.* **2018**, *8*, 16628. [CrossRef]

67. Lipka, V.; Kwon, C.; Panstruga, R. SNARE-Ware: The Role of SNARE-Domain Proteins in Plant Biology. *Annu. Rev. Cell Dev. Biol.* **2007**, *23*, 147–174. [CrossRef]

68. Kwon, C.; Lee, J.-H.; Yun, H.S. SNAREs in Plant Biotic and Abiotic Stress Responses. *Mol. Cells* **2020**, *43*, 501–508.

69. Chen, L.-M.; Fang, Y.-S.; Zhang, C.-J.; Hao, Q.-N.; Cao, D.; Yuan, S.-L.; Chen, H.-F.; Yang, Z.-L.; Chen, S.-L.; Shan, Z.-H.; et al. GmSYP24, a putative syntaxin gene, confers osmotic/drought, salt stress tolerances and ABA signal pathway. *Sci. Rep.* **2019**, *9*, 5990. [CrossRef]

70. Salinas-Cornejo, J.; Madrid-Espinoza, J.; Ruiz-Lara, S. Identification and transcriptional analysis of SNARE vesicle fusion regulators in tomato (*Solanum lycopersicum*) during plant development and a comparative analysis of the response to salt stress with wild relatives. *J. Plant Physiol.* **2019**, *242*, 153018. [CrossRef]

71. Pan, L.; Yu, X.; Shao, J.; Liu, Z.; Gao, T.; Zheng, Y.; Zeng, C.; Liang, C.; Chen, C. Transcriptomic profiling and analysis of differentially expressed genes in asparagus bean (*Vigna unguiculata* ssp. sesquipedalis) under salt stress. *PLoS ONE* **2019**, *14*, e0219799. [CrossRef]

72. Singh, D.; Yadav, N.S.; Tiwari, V.; Agarwal, P.K.; Jha, B. A SNARE-like superfamily protein SbSLSP from the halophyte *Salicornia brachiata* confers salt and drought tolerance by maintaining membrane stability, K+/Na+ ratio, and antioxidant machinery. *Front Plant Sci.* **2016**, *7*, 737. [CrossRef]

73. Ahmad, P.; Abdel Latef, A.A.; Hashem, A.; Abd_Allah, E.F.; Gucel, S.; Tran, L.-S.P. Nitric oxide mitigates salt stress by regulating levels of osmolytes and antioxidant enzymes in chickpea. *Front Plant Sci.* **2016**, *7*, 347. [CrossRef] [PubMed]

74. Geng, G.; Lv, C.; Stevanato, P.; Li, R.; Liu, H.; Yu, L.; Wang, Y. Transcriptome Analysis of Salt-Sensitive and Tolerant Genotypes Reveals Salt-Tolerance Metabolic Pathways in Sugar Beet. *Int. J. Mol. Sci.* **2019**, *20*, 5910. [CrossRef] [PubMed]

75. Zhang, Y.; Li, D.; Zhou, R.; Wang, X.; Dossa, K.; Wang, L.; Zhang, Y.; Yu, J.; Gong, H.; Zhang, X.; et al. Transcriptome and metabolome analyses of two contrasting sesame genotypes reveal the crucial biological pathways involved in rapid adaptive response to salt stress. *BMC Plant Biol.* **2019**, *19*, 66. [CrossRef] [PubMed]

76. Guo, R.; Shi, L.; Yan, C.; Zhong, X.; Gu, F.; Liu, Q.; Xia, X.; Li, H. Ionomic and metabolic responses to neutral salt or alkaline salt stresses in maize (*Zea mays* L.) seedlings. *BMC Plant Biol.* **2017**, *17*, 41. [CrossRef]

77. Verbruggen, N.; Hermans, C. Proline accumulation in plants: A review. *Amino Acids* **2008**, *35*, 753–759. [CrossRef]

78. Jander, G.; Joshi, V. Aspartate-Derived Amino Acid Biosynthesis in Arabidopsis thaliana. *Arab. Book* **2009**, *7*, e0121. [CrossRef]

79. Joshi, V.; Joung, J.-G.; Fei, Z.; Jander, G. Interdependence of threonine, methionine and isoleucine metabolism in plants: Accumulation and transcriptional regulation under abiotic stress. *Amino Acids* **2010**, *39*, 933–947. [CrossRef]

80. Zhang, C.; Pang, Q.; Jiang, L.; Wang, S.; Yan, X.; Chen, S.; He, Y. Dihydroxyacid dehydratase is important for gametophyte development and disruption causes increased susceptibility to salinity stress in Arabidopsis. *J. Exp. Bot.* **2015**, *66*, 879–888. [CrossRef]

81. Liu, J.H.; Wang, W.; Wu, H.; Gong, X.; Moriguchi, T. Polyamines function in stress tolerance: From synthesis to regulation. *Front Plant Sci.* **2015**, *6*, 827. [CrossRef]

82. Mehta, R.A.; Cassol, T.; Li, N.; Ali, N.; Handa, A.K.; Mattoo, A.K. Engineered polyamine accumulation in tomato enhances phytonutrient content, juice quality, and vine life. *Nat. Biotechnol.* **2002**, *20*, 613–618. [CrossRef]

83. Roje, S. S-Adenosyl-L-methionine: Beyond the universal methyl group donor. *Phytochemistry* **2006**, *67*, 1686–1698. [CrossRef] [PubMed]

84. Hanson, A.D.; Rathinasabapathi, B.; Chamberlin, B.; Gage, D.A. Comparative physiological evidence that β-alanine betaine and choline-O-sulfate act as compatible osmolytes in halophytic Limonium species. *Plant Physiol.* **1991**, *97*, 1199–1205. [CrossRef] [PubMed]

85. Hanson, A.D.; Rathinasabapathi, B.; Rivoal, J.; Burnet, M.; Dillon, M.O.; Gage, D.A. Osmoprotective compounds in the Plumbaginaceae: A natural experiment in metabolic engineering of stress tolerance. *Proc. Natl. Acad. Sci. USA* **1994**, *91*, 306–310. [CrossRef]

86. Liu, X.; Bush, D. Expression and transcriptional regulation of amino acid transporters in plants. *Amino Acids* **2006**, *30*, 113–120. [CrossRef] [PubMed]

87. Zhao, H.; Ma, H.; Yu, L.; Wang, X.; Zhao, J. Genome-wide survey and expression analysis of amino acid transporter gene family in rice (*Oryza sativa* L.). *PLoS ONE* **2012**, *7*, e49210. [CrossRef] [PubMed]

88. Tegeder, M.; Offler, C.E.; Frommer, W.B.; Patrick, J.W. Amino acid transporters are localized to transfer cells of developing pea seeds. *Plant Physiol.* **2000**, *122*, 319–326. [CrossRef]

89. Wang, T.; Chen, Y.; Zhang, M.; Chen, J.; Liu, J.; Han, H.; Hua, X. Arabidopsis AMINO ACID PERMEASE1 contributes to salt stress-induced proline uptake from exogenous sources. *Front Plant Sci.* **2017**, *8*, 2182. [CrossRef]

90. Wan, Y.; King, R.; Mitchell, R.A.C.; Hassani-Pak, K.; Hawkesford, M.J. Spatiotemporal expression patterns of wheat amino acid transporters reveal their putative roles in nitrogen transport and responses to abiotic stress. *Sci. Rep.* **2017**, *7*, 5461. [CrossRef]

91. Meng, S.; Su, L.; Li, Y.; Wang, Y.; Zhang, C.; Zhao, Z. Nitrate and Ammonium Contribute to the Distinct Nitrogen Metabolism of Populus simonii during Moderate Salt Stress. *PLoS ONE* **2016**, *11*, e0150354. [CrossRef]

92. Gu, R.; Fonseca, S.; Puskás, L.G.; Hackler, L., Jr.; Zvara, Á.; Dudits, D.; Pais, M.S. Transcript identification and profiling during salt stress and recovery of Populus euphratica. *Tree Physiology* **2004**, *24*, 265–276. [CrossRef]

93. Wei, Z.; Qing-Jie, S.U.N.; Chu-Fu, Z.; Yong-Ze, Y.; Ji, Z.; Bin-Bin, L.U. Effect of Salt Stress on Ammonium Assimilation Enzymes of the Roots of Rice (*Oryza sativa*) Cultivars Differing in Salinity Resistance. *Acta Bot. Sin.* **2004**, *46*, 921–927.

94. Parida, A.K.; Das, A.B. Salt tolerance and salinity effects on plants: A review. *Ecotoxicol. Environ. Saf.* **2005**, *60*, 324–349. [CrossRef] [PubMed]

95. Bourgeais-Chaillou, P.; Perez-Alfocea, F.; Guerrier, G. Comparative Effects of N–Sources on Growth and Physiological Responses of Soyabean Exposed to NaCI–Stress. *J. Exp. Bot.* **1992**, *43*, 1225–1233. [CrossRef]

96. Wang, H.; Zhang, M.; Guo, R.; Shi, D.; Liu, B.; Lin, X.; Yang, C. Effects of salt stress on ion balance and nitrogen metabolism of old and young leaves in rice (*Oryza sativa* L.). *BMC Plant Biol.* **2012**, *12*, 194. [CrossRef]

97. Igamberdiev, A.U.; Kleczkowski, L.A. The Glycerate and Phosphorylated Pathways of Serine Synthesis in Plants: The Branches of Plant Glycolysis Linking Carbon and Nitrogen Metabolism. *Front. Plant Sci.* **2018**, *9*, 318. [CrossRef]

98. Sonoike, K.; Terashima, I. Mechanism of photosystem-I photoinhibition in leaves of *Cucumis sativus* L. *Planta* **1994**, *194*, 287–293. [CrossRef]

99. Białasek, M.; Górecka, M.; Mittler, R.; Karpiński, S. Evidence for the involvement of electrical, calcium and ROS signaling in the systemic regulation of non-photochemical quenching and photosynthesis. *Plant Cell Physiol.* **2017**, *58*, 207–215. [CrossRef]

100. Mitsuya, S.; Takeoka, Y.; Miyake, H. Effects of sodium chloride on foliar ultrastructure of sweet potato (*Ipomoea batatas* Lam.) plantlets grown under light and dark conditions in vitro. *J. Plant Physiol.* **2000**, *157*, 661–667. [CrossRef]

101. Xu, J.; Lan, H.; Fang, H.; Huang, X.; Zhang, H.; Huang, J. Quantitative Proteomic Analysis of the Rice (*Oryza sativa* L.) Salt Response. *PLoS ONE* **2015**, *10*, e0120978. [CrossRef]

102. Shu, S.; Yuan, Y.; Chen, J.; Sun, J.; Zhang, W.; Tang, Y.; Zhong, M.; Guo, S. The role of putrescine in the regulation of proteins and fatty acids of thylakoid membranes under salt stress. *Sci. Rep.* **2015**, *5*, 14390. [CrossRef]

103. Fahad, S.; Hussain, S.; Matloob, A.; Khan, F.A.; Khaliq, A.; Saud, S.; Hassan, S.; Shan, D.; Khan, F.; Ullah, N. Phytohormones and plant responses to salinity stress: A review. *Plant Growth Regul.* **2015**, *75*, 391–404. [CrossRef]

104. Ribba, T.; Garrido-Vargas, F.; O'Brien, J.A. Auxin-mediated responses under salt stress: From developmental regulation to biotechnological applications. *J. Exp. Bot.* **2020**, *71*, 3843–3853. [CrossRef] [PubMed]

105. Korver, R.A.; Koevoets, I.T.; Testerink, C. Out of shape during stress: A key role for auxin. *Trends Plant Sci.* **2018**, *23*, 783–793. [CrossRef] [PubMed]

106. Prerostova, S.; Dobrev, P.I.; Gaudinova, A.; Hosek, P.; Soudek, P.; Knirsch, V.; Vankova, R. Hormonal dynamics during salt stress responses of salt-sensitive *Arabidopsis thaliana* and salt-tolerant *Thellungiella salsuginea*. *Plant Sci.* **2017**, *264*, 188–198. [CrossRef]

107. Žižková, E.; Dobrev, P.I.; Muhovski, Y.; Hošek, P.; Hoyerová, K.; Haisel, D.; Procházková, D.; Lutts, S.; Motyka, V.; Hichri, I. Tomato (*Solanum lycopersicum* L.) SlIPT3 and SlIPT4 isopentenyltransferases mediate salt stress response in tomato. *BMC Plant Biol.* **2015**, *15*, 85. [CrossRef]

108. Nishiyama, R.; Watanabe, Y.; Fujita, Y.; Le, D.T.; Kojima, M.; Werner, T.; Vankova, R.; Yamaguchi-Shinozaki, K.; Shinozaki, K.; Kakimoto, T. Analysis of cytokinin mutants and regulation of cytokinin metabolic genes reveals important regulatory roles of cytokinins in drought, salt and abscisic acid responses, and abscisic acid biosynthesis. *Plant Cell* **2011**, *23*, 2169–2183. [CrossRef]

109. Keshishian, E.A.; Hallmark, H.T.; Ramaraj, T.; Plačková, L.; Sundararajan, A.; Schilkey, F.; Novák, O.; Rashotte, A.M. Salt and oxidative stresses uniquely regulate tomato cytokinin levels and transcriptomic response. *Plant Direct* **2018**, *2*, e00071. [CrossRef]

110. Wang, J.; Xia, J.; Song, Q.; Liao, X.; Gao, Y.; Zheng, F.; Yang, C. Genome-wide identification, genomic organization and expression profiles of SlARR-B gene family in tomato. *J. Appl. Genet.* **2020**, *61*, 391–404. [CrossRef]

111. Cao, W.-H.; Liu, J.; He, X.-J.; Mu, R.-L.; Zhou, H.-L.; Chen, S.-Y.; Zhang, J.-S. Modulation of ethylene responses affects plant salt-stress responses. *Plant Physiol.* **2007**, *143*, 707–719. [CrossRef]

112. Zeng, H.; Tang, Q.; Hua, X. Arabidopsis Brassinosteroid Mutants det2-1 and bin2-1 Display Altered Salt Tolerance. *J. Plant Growth Regul.* **2010**, *29*, 44–52. [CrossRef]

113. Nakaya, M.; Tsukaya, H.; Murakami, N.; Kato, M. Brassinosteroids control the proliferation of leaf cells of Arabidopsis thaliana. *Plant Cell Physiol.* **2002**, *43*, 239–244. [CrossRef] [PubMed]

114. Valenzuela, C.E.; Acevedo-Acevedo, O.; Miranda, G.S.; Vergara-Barros, P.; Holuigue, L.; Figueroa, C.R.; Figueroa, P.M. Salt stress response triggers activation of the jasmonate signaling pathway leading to inhibition of cell elongation in Arabidopsis primary root. *J. Exp. Bot.* **2016**, *67*, 4209–4220. [CrossRef] [PubMed]

115. Li, H.; Chang, J.; Chen, H.; Wang, Z.; Gu, X.; Wei, C.; Zhang, Y.; Ma, J.; Yang, J.; Zhang, X. Exogenous Melatonin Confers Salt Stress Tolerance to Watermelon by Improving Photosynthesis and Redox Homeostasis. *Front. Plant Sci.* **2017**, *8*, 295. [CrossRef] [PubMed]

116. Wei, C.; Zhang, R.; Yang, X.; Zhu, C.; Li, H.; Zhang, Y.; Ma, J.; Yang, J.; Zhang, X. Comparative Analysis of Calcium-Dependent Protein Kinase in Cucurbitaceae and Expression Studies in Watermelon. *Int. J. Mol. Sci.* **2019**, *20*, 2527. [CrossRef]

117. Trapnell, C.; Williams, B.A.; Pertea, G.; Mortazavi, A.; Kwan, G.; van Baren, M.J.; Salzberg, S.L.; Wold, B.J.; Pachter, L. Transcript assembly and quantification by RNA-Seq reveals unannotated transcripts and isoform switching during cell differentiation. *Nat. Biotechnol.* **2010**, *28*, 511–515. [CrossRef]

118. Anders, S.; Huber, W. Differential expression analysis for sequence count data. *Genome Biol.* **2010**, *11*, R106. [CrossRef]

119. Ashburner, M.; Ball, C.A.; Blake, J.A.; Botstein, D.; Butler, H.; Cherry, J.M.; Davis, A.P.; Dolinski, K.; Dwight, S.S.; Eppig, J.T.; et al. Gene Ontology: Tool for the unification of biology. *Nat. Genet.* **2000**, *25*, 25–29. [CrossRef]

120. Kanehisa, M.; Goto, S.; Kawashima, S.; Okuno, Y.; Hattori, M. The KEGG resource for deciphering the genome. *Nucleic Acids Res.* **2004**, *32*, D277–D280. [CrossRef]

121. Pedregosa, F.; Varoquaux, G.; Gramfort, A.; Michel, V.; Thirion, B.; Grisel, O.; Blondel, M.; Prettenhofer, P.; Weiss, R.; Dubourg, V. Scikit-learn: Machine learning in Python. *J. Mach. Learn. Res.* **2011**, *12*, 2825–2830.

122. Hunter, J.D. Matplotlib: A 2D graphics environment. *Comput. Sci. Eng.* **2007**, *9*, 90–95. [CrossRef]

123. Kong, Q.; Yuan, J.; Gao, L.; Zhao, S.; Jiang, W.; Huang, Y.; Bie, Z. Identification of suitable reference genes for gene expression normalization in qRT-PCR analysis in watermelon. *PLoS ONE* **2014**, *9*, e90612. [CrossRef] [PubMed]

MDPI
St. Alban-Anlage 66
4052 Basel
Switzerland
Tel. +41 61 683 77 34
Fax +41 61 302 89 18
www.mdpi.com

Int. J. Mol. Sci. Editorial Office
E-mail: ijms@mdpi.com
www.mdpi.com/journal/ijms